Fujio Yamaguchi

Curves and Surfaces in Computer Aided Geometric Design

With 244 Figures and 9 Tables

Springer-Verlag
Berlin Heidelberg NewYork
London Paris Tokyo

Professor Fujio Yamaguchi

Department of Mechanical Engineering
School of Science and Engineering
Waseda University
3-4-1, Okubo, Shinjuku-ku, Tokyo 160/Japan

Originally published in Japanese as
Keizyo Syori Kogaku (1) & Keizyo Syori Kogaku (2)
© Nikkan Kogyo Shinbun Co.

ISBN 3-540-17449-4 Springer-Verlag Berlin Heidelberg New York
ISBN 0-387-17449-4 Springer-Verlag New York Berlin Heidelberg

Library of Congress Cataloging-in-Publication Data
Yamaguchi, Fujio, 1935-
[Konpyūta disupurei ni yoru keijō shori kōgaku. English]
Curves and surfaces in computer aided geometric design / Fujio Yamaguchi. P. cm.
Translation of: Konpyūta disupurei ni yoru keijō shori kōgaku.
Bibliography: Includes index.
ISBN 0-387-17449-4 (U.S.)
1. Geometrical drawing—Data processing. 2. Engineering design—Data processing.
3. Computer graphics. I. Title.
QA464.Y2813 1988 620'.00425'0285—dc19 88-26569

Printed in Germany

Typesetting: H. Hagedorn, Berlin. Offsetprinting: Kutschbach, Berlin.
Bookbinding: Lüderitz & Bauer, Berlin.

2161/3020-543210/Printed on acid-free paper

Preface

This is an English version of the book in two volumes, entitled "Keijo Shori Kogaku (1), (2)" (Nikkan Kogyo Shinbun Co.) written in Japanese. The purpose of the book is a unified and systematic exposition of the wealth of research results in the field of mathematical representation of curves and surfaces for computer aided geometric design that have appeared in the last thirty years.

The material for the book started life as a set of notes for computer aided geometric design courses which I had at the graduate schools of both computer science, the university of Utah in U.S.A. and Kyushu Institute of Design in Japan. The book has been used extensively as a standard text book of curves and surfaces for students, practical engineers and researchers.

With the aim of systematic exposition, the author has arranged the book in 8 chapters:

Chapter 0: The significance of mathematical representations of curves and surfaces is explained and historical research developments in this field are reviewed.

Chapter 1: Basic mathematical theories of curves and surfaces are reviewed and summarized.

Chapter 2: A classical interpolation method, the Lagrange interpolation, is discussed. Although its use is uncommon in practice, this chapter is helpful in understanding Chaps. 4 and 6.

Chapter 3: This chapter discusses the Coons surface in detail, which is one of the most important contributions in this field.

Chapter 4: The fundamentals of spline functions, spline curves and surfaces are discussed in some detail.

Chapter 5: This chapter discusses the so-called Bézier curve and surface which are frequently used in practical applications.

Chapter 6: This chapter deals with *B*-spline curves and surfaces, which are more general than the Bézier ones. Special emphasis is placed on this chapter in this book because *B*-spline curves and surfaces are expected to play an important role in the future.

Chapter 7: The last chapter discusses the rational polynomial curves which are capable of representing conic sections exactly.

The present author has learnt much from the published work of others, particularly from late Prof. S.A. Coons, Prof. M. Hosaka (Tokyo Denki University), Prof. A.R. Forrest (University of East Anglia), Prof. R.F. Riesenfeld (University of Utah), Prof. I.D. Faux (Cranfield Institute of Technology),

Prof. M.J. Pratt (Cranfield Institute of Technology) and he has drawn extensively on them. He has done his best to give credit where it is due.

Finally, the author's thanks go to Mr. Harold Solomon for the translation from the original Japanese.

Hino, Tokyo, September 1988 Fujio Yamaguchi

Contents

On the Symbols Used in This Book

Among the symbols used in these books, several which require caution are discussed below.

(1) t is the standard symbol for a parameter representing a curve, and u and w for parameters representing a surface. In some places the length of a curve is used as a parameter in the discussion; it is s.

(2) Derivatives with respect to parameters other than s, such as t, are denoted by dots (·); derivatives with respect to s are denoted by primes ($'$). For example,

$$\frac{d\boldsymbol{P}}{dt} \equiv \dot{\boldsymbol{P}}, \qquad \frac{d\boldsymbol{P}}{ds} \equiv \boldsymbol{P}'$$

t, u and w can be regarded as time parameters in the motion of mass points. Since in physics it is customary for derivatives with respect to time to be denoted by dots, this convention has been followed in the present book.

(3) An ordinary coordinate vector is denoted by a slanted letter, for example $\boldsymbol{P}$. A homogeneous coordinate vector is denoted by a vertical letter, such as $\mathbf{P}$.

(4) In mathematical representations of curves and surfaces, a vector that is given for the purpose of definition is denoted by $\boldsymbol{Q}$, and a defined curve or surface vector by $\boldsymbol{P}$. For example in

$$\boldsymbol{P}(t) = H_{0,0}(t)\boldsymbol{Q}_0 + H_{0,1}(t)\boldsymbol{Q}_1 + H_{1,0}(t)\dot{\boldsymbol{Q}}_0 + H_{1,1}(t)\dot{\boldsymbol{Q}}_1,$$

the position vectors $\boldsymbol{Q}_0$, $\boldsymbol{Q}_1$ and the tangent vectors $\dot{\boldsymbol{Q}}_0$, $\dot{\boldsymbol{Q}}_1$ are given, and the curve $\boldsymbol{P}(t)$ is defined.

0. Mathematical Description of Shape Information

0.1 Description and Transmission of Shape Information

Shapes of industrial products can be roughly classified into those that consist of combinations of elementary geometrical surfaces and those that cannot be expressed in terms of elementary surfaces, but vary in a complicated manner. Many examples of the former type are found among parts of machines. Most machine parts have elementary geometrical shapes such as planes and cylinders. This is because, as long as a more complicated shape is not functionally required, simpler shapes are far simpler from the point of view of production. In this book, these shapes are called Type 1 shapes. Meanwhile, the shapes of such objects as automobile bodies, telephone receivers, ship hulls and electric vacuum cleaners contain many curved surfaces that vary freely in a complicated manner. Let us call these Type 2 shapes.

A designer draws his concept of a shape on paper and proceeds with the design while checking it against the shape that he has drawn. Sometimes during the design work it becomes necessary to build a physical model of the shape. In such a case, blueprints are prepared and given to a model builder. The final step in design is to prepare a set of blueprints on which are written all of the information needed to produce the item that has been designed. The designer must write all of the information needed to produce a 3-dimensional shape on 2-dimensional paper. In the case of a Type 1 shape, these 2-dimensional drawings are called three orthogonal views; in the case of a Type 2 shape they are frequently called curve diagrams.

It has always been difficult to express the considerable amount of information needed to describe a 3-dimensional shape on a limited number of 2-dimensional drawings. In the case of Type 1 shapes the task has been simplified by defining a number of conventions for drawing preparation which the designer can learn and, as he accumulates experience, learn to transmit information effectively to other people on 2-dimensional diagrams. Since Type 1 shapes are geometrically regular and familiar in everyday life, it can be expected that a person who reads the diagram will be able to infer the correct shape from what is on the paper, so that the transmission of shape information proceeds relatively smoothly. Consequently, as far as Type 1 shapes are concerned, if it can be assumed that people will look at the diagrams, this method is very effective for describing and transmitting a considerable amount of 3-dimensional shape information on a limited number of drawings. And since the drawings are in standard formats, they are easy to be manipulated.

What, then, is the relation between shape and drawings for Type 2 shapes? Since Type 2 shapes include curved surfaces which vary in a complicated manner, in general there are many cases in which a small number of curves drawn on paper are not in themselves sufficient to describe the shape. In the case of a Type 2 shape, it is not possible to rely on the ability of the person reading the drawings to infer the necessary shape information from it. If not enough curves are given, the shape between one curve and the next will vary depending on the subjective judgment of the person reading the drawings. Often the model builder's interpretation of the drawings differs somewhat from what the designer intended. When the designer sees the model, he tries to tell the model builder what shape it is that he really intended, but in the absence of adequate tools for this purpose there is no way to give a really accurate explanation.

0.2 Processing and Analysis of Shapes

In the design of an industrial product, various types of processing and analysis are performed with respect to the shape to make sure that it not only has a beautiful exterior appearance, but also satisfies a number of necessary technical conditions. For example, calculation of the surface area, volume, weight and moment of inertia of a shape, structural strength analysis, vibration analysis, fluid flow analysis, thermal conductivity analysis and NC tape preparation also may become necessary. All of these processings and analyses are performed by having a person read shape information from drawings and then using the resulting data. Recently, through technological advances such as the Finite Element Method (FEM) and the Finite Boundary Method (FBM), it has become possible to perform these analyses even on complicated shapes. In the case of a Type 1 shape, it is possible to assume that there will be no problem of ambiguity in the representation of a shape by drawings, so there are few problems in shape processing and analysis. Meanwhile, in the case of a Type 2 shape, there are many ambiguities in the drawing representations, so it is very difficult for a person to read the data necessary for shape processing and analysis.

In addition, viewing this problem from the point of view of amount of information, in general a Type 2 shape involves a much greater amount of information than a Type 1 shape. It is very troublesome for a person to read such a large amount of information and then perform calculations using that large amount of information, and it is easy for mistakes to occur. Let us consider the case of a container as an example. The internal volume of the container must be rigorously adjusted to a specified value. If the container that is designed has an actual volume that is smaller than the nominal volume, the customers will probably complain. If the volume is too large, the company that bottles drinks will suffer a loss. Since volume calculation for a complicated shape cannot be carried out easily, normally it is done by the end summing rule. Based on curve information on the drawings, the cross-sectional shape which the container is expected to have at a certain height is drawn on graph paper, and the number of squares enclosed by the curve is counted to give the area of the cross-section. The shape of the container is approximated as a collection of a large number of thin slices, the cross-sectional area of each of which is calculated by the above method, and then the volumes of the slices are added up to give the total volume. Calculating the volume in this manner takes a great deal of time and is very troublesome, and then the shape must be adjusted to bring the actual volume into agreement with the specified volume, which is also very difficult.

From the point of view of description and transmission of shape information and processing and analysis with respect to that shape, many problems lurk in a design and production system based on drawings, particularly in the case of Type 2 shapes.

0.3 Mathematical Description of Free Form Shapes

One method of solving this kind of problem is to describe shapes mathematically. When this is done shape description becomes objective rather than subjective, and it becomes possible to use the power of computers to do the various processings and analyses with respect to the shape.

Technology for describing shapes mathematically was first developed in connection with numerical control processing technology. Since the world's first NC 3-dimensional milling machine was perfected at M.I.T. in 1951, concentrated research, mainly at M.I.T., began on software technology for describing shapes mathematically and preparing NC tapes using a computer. The large-scale software system called "APT III" was perfected in the early 1960s. In the APT system, a program reads shape information off of drawings. When the shape and the movement of a tool that processes that shape are programmed, the computer creates a numerical model of the shape; then, based on that model, it calculates accurate coordinates of the tool path and outputs an NC tape. At the stage at which the APT system is used, design work related to the shape has been completed. To put it another way, the APT system deals with the last stage of design; it uses a computer to create an NC tape which contains what might be called "production command information". Use of the APT system still requires drawings; consequently the problem of "ambiguity" of drawings in shape design remains. In this design method, shape creation and modeling and in fact all decision-making operations in the basic design and detailed design stages are done completely by human beings, and a finished physical model is produced. Using measurement of the model a mathematical model is created on the computer to produce a N/C machining tape for the part. In other words, in this method a copy of the physical model is produced mathematically; that is, this is a design method "from a physical model to mathematical model".

In 1963, Ivan E. Sutherland of M.I.T. announced a revolutionary system called "Sketchpad". It created the possibility that basic design and detailed design could be done under computer control.[1)] Even if a shape is described mathematically, it is hard for a person to understand what kind of shape is being described just by looking at the mathematical description. The mathematical description of the shape is stored in a computer memory device. If it can be displayed on a CRT screen, then it becomes visible to people and can be understood easily by a human being. In this case, the shape description and transmission can be regarded as being carried out objectively. Since it is easy to create a drawing of the shape from its mathematical model, by creating various drawings of the shape as seen from various locations and directions, a person's understanding of the shape can be deepened. The person can draw a freehand sketch by stylus of the desired shape on the CRT screen, and can instruct the computer to perform specialized processing of the shape by pointing to a particular point on the CRT screen image with the stylus. Consequently, if a mathematical model can be created based on the picture drawn by a person and on information indicated by a stylus, it becomes possible to do even basic design and detailed design on the CRT screen through conversational interaction with the computer (Fig. 0.1). That is to say, in place of the conventional design process based on drawings, a new design process can be conceived which is based on a mathematical model of a shape stored in a computer. In this method, in contrast to the method described above in which a physical model is converted to a mathematical model, mathematical model is created in the very beginning stage of design; if necessary a physical model can then be created from the mathematical model.

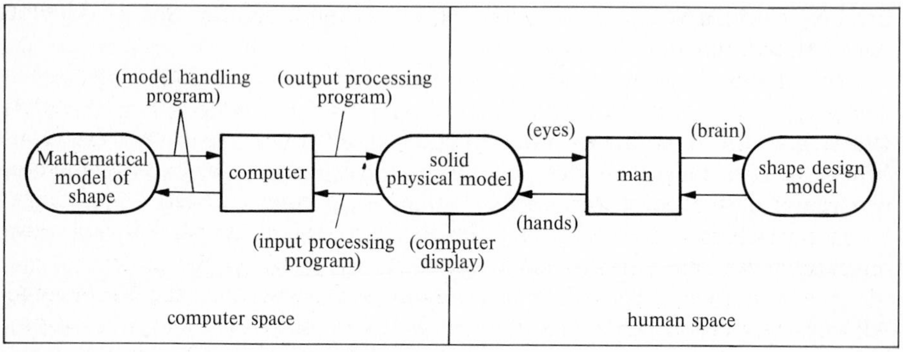

Fig. 0.1. Shape design by cooperation between man and computer

0.4 The Development of Mathematical Descriptions of Free Form Curves and Surfaces

At the start of the 1960s, J. C. Ferguson of Boeing Aircraft Company in the U.S.A. announced a method of describing curve segments as vectors, using parameters (refer to Chap. 3). A Ferguson curve segment is a cubic vector function with respect to a parameter obtained by specifying the position vectors and tangent vectors of the starting and end points of the curve segment. An actual curve is created by joining these curve segments smoothly. In addition, Ferguson, using these curve segments, proposed a method of creating a portion of a surface (called a surface patch) that satisfies the conditions imposed by specifying position vectors and tangent vectors at 4 points, and put this method to practical use in Boeing's surface creation program FMILL. FMILL is a system intended to create NC tapes. Before this work by Ferguson, mathematical representation descriptions of curves had been the form $y=f(x)$ or $F(x, y)=0$. In contrast, Ferguson curves have the following advantages.

1) Not only curves in a plane, but curves in space can be expressed by simple functions.
2) The part of a curve that is needed can easily be specified by a parameter range.
3) Since a slope parallel to the y-axis can be expressed by $dx/dt=0$, it is not necessary to use $dy/dx=\infty$.
4) Transformations of a curve such as translation and rotation can be carried out simply by multiplying by a transformation matrix.

Subsequently, parametric description of curves and surfaces became the standard method of mathematical representation.

It can be seen from NC-processed surface shapes that when Ferguson surface patches are taken to be relatively large, the surface tends to be flattened in the vicinity of the 4 corners of the patch.

Ferguson curves and surfaces are an example of the use of Hermite interpolation functions.

In 1964, S. A. Coons of M.I.T. announced a surface description method in which one considers the position vectors of the 4 corner points of a surface patch and the 4 boundary curves, and derives a mathematical description which satisfies those boundary conditions (refer to Sect. 3.3.2). A generalized version of this concept was announced in 1967 (refer to Sect. 3.3.3). Coon's surface patch is defined not only by the position vectors and higher order differential vectors with respect to the 4 corner points of the patch, but also the position vectors related to the 4 boundary curves and the higher order differential vectors with respect to the directions across the boundary curves. In practice, it is sufficient to give positions with respect to the 4 corner points of the surface patch, tangent vectors in 2 directions, mutual partial differential vectors (twist vectors) and positions with respect to the 4 boundary curves and tangent vector functions in the directions across the boundary curves. In this method, if the 4 boundary curves and tangent vector functions are expressed by Hermite interpolation formulas, a representation resembling the Ferguson patch is obtained. This is called a bi-cubic Coons surface patch; it is expressed in a simple form and is widely used. If the twist vector is set to zero, the bi-cubic Coons surface patch agrees with the Ferguson surface patch. That is, the Ferguson surface patch is a special case of the Coons surface patch. The reason that the Ferguson surface patch flattens the surface in the vicinity of the 4 corners is that the twist vector has been taken to be zero. In his paper,

Coons discussed the conditions which must be satisfied to join surface patches together to form a continuous surface, and gave a method for connecting them so as to be continuous up to the n-th order differential vector in the directions across the boundary curves.

Ferguson curve segments and surface patches and Coons surface patches share a number of problems in control and connection of segments and patches, such as the following.

1) It is difficult for a person to control the shape of a surface directly.
2) When curve segments or surface patches are connected, it is necessary to be concerned not only with local mathematical continuity at the contact points, but also with overall smooth connection of the whole curve or surface, but neither method specifies the conditions necessary for this.

Coons' intention in conceiving the surface patch was to make it possible for a person to do design work interactively on a computer while watching a CRT screen connected to the computer. However, in practice Coons' surface patch has found its greatest application in physical model surface fitting.

A technique for solving the "connection" problem is the spline (refer to Chap. 4). A spline is a flexible band made of wood, plastic or steel. In the design of products having free form surface shapes such as ships, airplanes and automobiles, splines held in shape by "weights" have long been used to obtain free form curve shapes. It is known from experience that curves produced by splines are fair. A curve produced by a spline is described by different cubic degree curves in different segments between "weight" and "weight". At the positions of "weights", that is, at the connecting points between curve segments, connections are continuous up to the curvature. Moreover, along the entire length of the spline, the integral of the square of the curvature must be a minimum among all possible curves passing through the "weight" points. This means that the total bending energy stored in the spline is a minimum.

Mathematical curves which approximate spline shapes by means of parametric vector functions are very important in CAD. In particular, the natural spline (refer to Sect. 4.4) has the property that it gives the interpolation that minimizes the integral of the square of the curvature. Ferguson's and Coons' conditions for connection of curve segments and surface patches insure local mathematical continuity. However, an infinite number of curves which are mathematically continuous at the contact points can be obtained by varying the magnitude of the tangent vector, resulting in an infinite number of possible curve shapes. The minimum interpolation property of the natural spline indicates what conditions should be satisfied in connecting curve segments and what shape the overall curve should have. In a spline curve, the connection conditions are determined simultaneously at all connection points. By repeated application of a method similar to the spline curve method, spline surfaces can be created. For example, in a bi-cubic spline surface (refer to Sect. 4.11), continuity on the patch boundary curves is obtained up to the curvature in the direction across the boundary curves. Then the spline surface is uniquely determined. The spline method can be thought of as automatically solving the "connection" problem that exists with Ferguson and Coons curves and surfaces.

The shape of a spline curve is controlled by varying the positions of the connection points between curve segments (corresponding to "weight" positions) and by increasing the number of curve segments. Since the position of one point on a spline curve is determined by all data relating to the initially given points through which the curve must pass (which are also connecting points between curve segments), the effect of changing the position of one point through which the curve must pass extends throughout the whole curve. In addition, in some cases, curves are produced which vary in ways that are hard to predict from the given series of points through which the curve must pass. For example, there can be a loop in the curve that is produced even though there is no such a shape in the given series of points. As long as one wishes to use spline curves and surfaces as they are produced, there is a problem of "control".

D. G. Schweikert proposed the use of a "spline under tension"[2]. This was an attempt to improve the controllability of the spline. By suitably adjusting the parameter that corresponds to tension, the production of loops in the spline can be prevented. The curve representation takes the form of a hyperbolic function.

P. Bézier of the Renault Company in France announced a curve representation that is defined by giving one polygon (refer to Chap. 5). This Bézier curve segment can be regarded as a curve obtained by smoothing the corners of the given polygon. Bézier curves have been put to practical use in Renault's automobile body design system UNISURF[3]. Bézier curve segments and surface patches are defined only by the position vectors of polygon vertices; unlike the Ferguson and Coons methods, this method does not require analytical data that are hard to understand intuitively such as tangent vectors and twist vectors. Bézier curve segments are expressed as a convex combination of the polygon vertex position vectors which define the curve, and possess a variation diminishing property. Consequently the curve shape can be approximately anticipated from the polygon shape. In addition, it is also possible to increase the degree of the polynomial curve; for example, a curve segment can be split into two segments without changing the shape of the curve, or the degree of the curve can be formally increased also without changing its shape. That is to say, Bézier curves and surfaces are in a form that is easy for a person to control.

In actual design of a curve or surface, it is necessary to connect a number of curve segments or surface patches smoothly. In this case, in order to connect with continuity up to the curvature, troublesome restrictive conditions must be applied to the connecting sections in the vicinity of the polygon vertices. Bézier curves and surfaces have superior "controllability", but have a problem with "connection". Bézier curves and surfaces use Bernstein Basis functions as blending functions.

Gordon and Riesenfeld proposed curves and surfaces which use Basis splines as blending functions. These are called *B*-spline curves and surfaces (refer to Chap. 6). *B*-spline curves, like Bézier curves, are defined by polygon vertices and have properties similar to those of Bézier curves. That is, *B*-spline curves are expressed as a convex combination of polygon vertex position vectors, and also have the variation diminishing property. Curve shapes are smoothed versions of the polygon shapes and can be roughly predicted from the polygon shapes. In contrast to Bézier curve, which is a convex combination of all of the vertex position vectors, *B*-spline curve differs in that it is a convex combination of a number of vertex position vectors in their immediate vicinity. It follows from this that the shape variation properties of the polygon show up even more clearly in *B*-spline curves than in Bézier curves. If n is the number of sides of the given polygon, the *B*-spline curve is formed by smoothly joining $(n-M+2)$ $(M-1)$-degree curve segments. A basis spline is determined by specifying the order M and the knot vector. The curve segments which make up a *B*-spline curve are defined by the M polygon vertices in the vicinity of each. Consequently, when one polygon vertex position is varied to control the curve shape, the effect of the change is locally confined. This property is very important in designing curves.

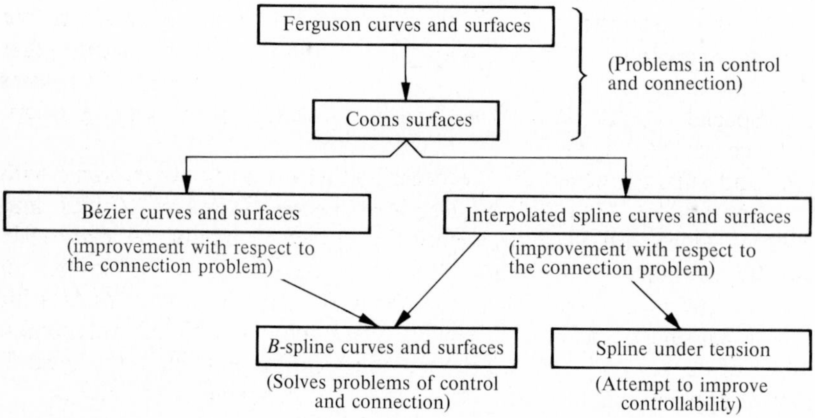

Fig. 0.2. Sequence of development of mathematical curve and surface description methods

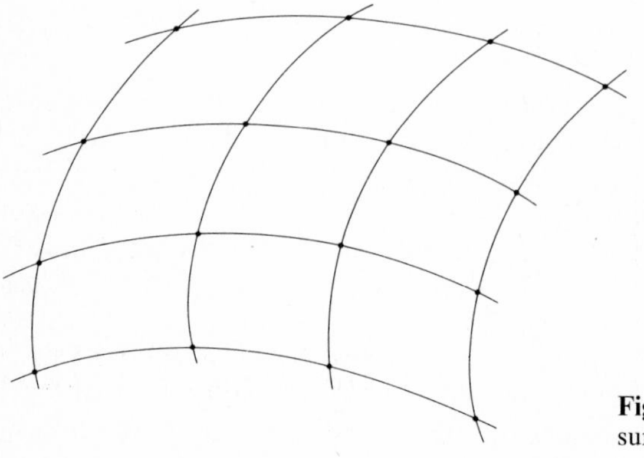

Fig. 0.3. A surface formed by joining surface patches in a matrix

B-spline curves are superior curves which combine the superior controllability of Bézier curves with the connection properties of spline curves. By specially specifying the knot vector in a *B*-spline curve segment, it can be made to agree with the Bézier curve segment.

The sequence of development of mathematical curve and surface description methods described above is shown schematically in Fig. 0.2.

B-spline curves can perhaps be thought of as possessing ideal "control" and "connection" properties. How well do *B*-spline surfaces describe surface shapes? As shown in Fig. 0.3, an ordinary, simple surface can be expressed by joining a number of surface patches together in a matrix. In the case of such a simple surface, the surface can be described and processed as a single *B*-spline surface (a surface defined by $m \times n$ position vectors). When it comes to actual surfaces, it sometimes happens that it is difficult to describe and process the surface shape as such a simple surface. Figure 0.4 shows an example of the kind of surface shape that occurs in the vicinity of a corner of a solid object formed by the convergence of three ridges. In this case, there is a problem of how to describe a triangular area of a surface. Figure 0.5 shows another example, this time of a rounded concave corner formed by the convergence of 3 ridges. In this case the surface becomes pentagonal in shape. It is not impossible to express such special shapes other than standard quadrangular shapes by mathematical functions, but special, complicated treatment becomes necessary. Special considerations also become necessary when joining neighboring surface patches together.

Curves and surfaces which can be described as ordinary polynomials with respect to parameters have been described above. However, circles and circular arcs, which are very important in industry, cannot be rigorously expressed by ordinary polynomials. Rigorous mathematical expressions of conic section curves is possible in the form of rational polynomials with respect to a parameter (refer to Chap. 7). Consequently, rational polynomial descriptions of curves and surfaces are very interesting, but full-scale research and practical applications along this line remain as problems for the future.

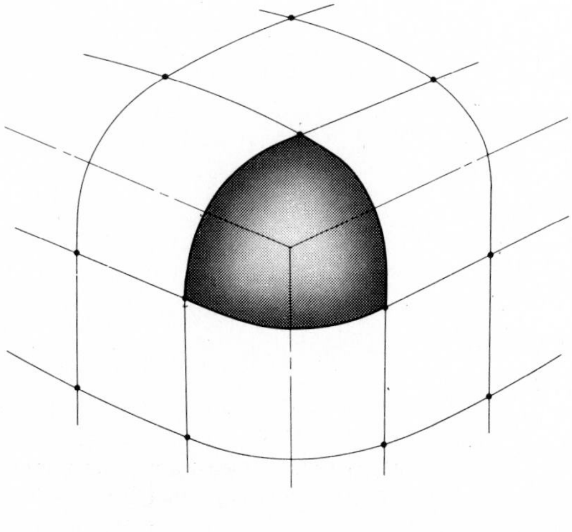

Fig. 0.4. Example of a surface in which a triangular surface patch occurs

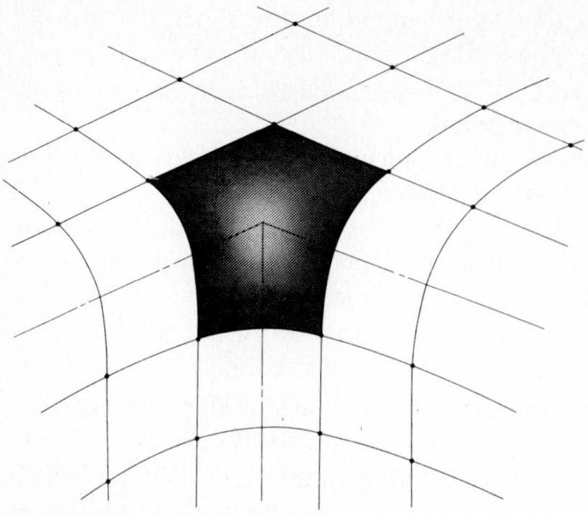

Fig. 0.5. Example of a surface in which a pentagonal surface patch occurs

References (Chap. 0)

1) Fujio Yamaguchi: Computer Graphics, Nikkan Kogyo Shimbun Sha, Introductory Chapter (in Japanese).
2) Schweikert, D. G.: "An Interpolation Curve using a Spline in Tension", *J. Math. & Phys.* **45,** 312—317, 1966.
3) Bézier, P.: "Example of an Existing System in the Motor Industry: The UNISURF System", *Proc. Roy. Soc. Lond.* **A 321,** 207—218, 1971.

1. Basic Theory of Curves and Surfaces

1.1 General

1.1.1 Properties of Object Shapes and Their Mathematical Representation

Before a computer can perform processing relating to a shape, a mathematical description of that shape must be provided on the computer's memory. Such a description should preserve as many of the properties of the actual object shape as possible. From the point of view of computer processing the following properties are particularly important.

(1) Spatial Uniqueness

Two objects cannot occupy the same space at the same time. This property is called *spatial uniqueness*. Object shapes are expressed in a computer by data structures using numerical shape data and functions, but spatial uniqueness is very difficult to express. Therefore a program is needed to test for overlap between positions of two bodies.

(2) Boundedness and Continuity

An object shape occupies a bounded space defined by a number of mutually adjacent surfaces of finite extent. In this book, this property is called the *boundedness* of the object shape. In performing computer-aided design it is extremely important for the range over which the shape being designed extends to be conveniently expressed. Let us consider the case of a curve. The curve in Fig. 1.1 is multivalued: it takes on several y-values for some values of x, and also several x-values for some values of y. If this curve is described by

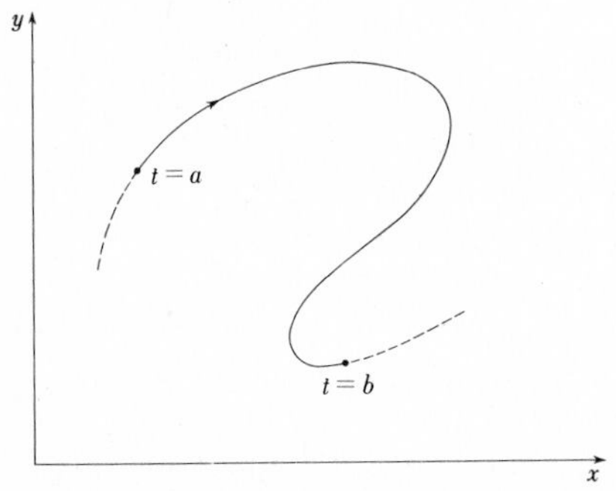

Fig. 1.1. A curve segment for which y is multi-valued with respect to x and x is multi-valued with respect to y

a function $y=F(x)$, trying to describe its range can be very difficult. On the other hand, if it is described by a parametric vector function $\boldsymbol{P}(t)=[x(t)\ y(t)]$, then the range can be described simply as $a \leqq t \leqq b$ where $t=a$ and $t=b$ at the two ends of the curve.

Curved (or plane) object surfaces are in continuous connection, without a break, along curved boundary lines or ridge lines. This property of an object is called *continuity*. Many types of processing on shapes make effective use of this continuity property. Consequently, in expressing an object shape on a computer, it is important to express how the surfaces connect to each other along boundary curves. This description is performed by a *data structure*.

(3) Independence of Coordinate Axes

An object shape is independent of the coordinate system in which it is described. This seems natural enough, but it is an important condition in formulating a mathematical model of the object. As an example, consider the problem of selecting points on a clay model, measuring their coordinates and then fitting them to a mathematical representation of a curve. This is normally done by letting the measured points define the curve and *interpolating* between them. The final curve must be the same regardless of the coordinate system in which the coordinates of the points were measured. This property is called *shape invariance under transformation*. If the method of arriving at a mathematical representation is inappropriate, the condition of shape invariance under transformation will not be satisfied.

For example, consider the following parametric curve.

$$\boldsymbol{P}(t)=A_0(t)\,\boldsymbol{P}_0+A_1(t)\,\boldsymbol{P}_1 \quad (0 \leqq t \leqq 1) \tag{1.1}$$

where $A_0(t)=(1-t)^3$ and $A_1(t)=t^3$. In this case, the relations $A_0(0)=1$, $A_0(1)=0$, $A_1(0)=0$, $A_1(1)=1$ hold, so we see that $\boldsymbol{P}(t)$ is a curve which passes through the two points $\boldsymbol{P}_0$ and $\boldsymbol{P}_1$. Taking $\boldsymbol{P}_0=[0\ 0]$ and $\boldsymbol{P}_1=[10\ 10]$, $\boldsymbol{P}(t)$ describes the line segment connecting $\boldsymbol{P}_0$ and $\boldsymbol{P}_1$ (refer to Fig. 1.2). Expressing $\boldsymbol{P}_0$ and $\boldsymbol{P}_1$ in the x^*y^* coordinate system, $\boldsymbol{P}_0^*=[-3\ -2]$ and $\boldsymbol{P}_1^*=[7\ 8]$; then if we let

$$\boldsymbol{P}^*(t)=A_0(t)\,\boldsymbol{P}_0^*+A_1(t)\,\boldsymbol{P}_1^*$$

$\boldsymbol{P}^*(t)$ passes through the same points as $\boldsymbol{P}(t)$ at $t=0$ and $t=1$, but for $0<t<1$ $\boldsymbol{P}^*(t)$ describes a curve different from the straight line of $\boldsymbol{P}(t)$. That is, the straight line representation of Eq. (1.1) is not desirable.

In Eq. (1.1) a curve is described as a vector using the parameter t. There is an alternate way to describe curves, the non-parametric method in the form $y=F(x)$. In the latter representation, at a point at which the curve becomes parallel to the y-axis the slope dy/dx becomes infinite, making it difficult to handle on a computer. In practice it becomes necessary to use another suitable coordinate system in which the slope does not become infinite. However, depending on the shape of the curve being represented, there are

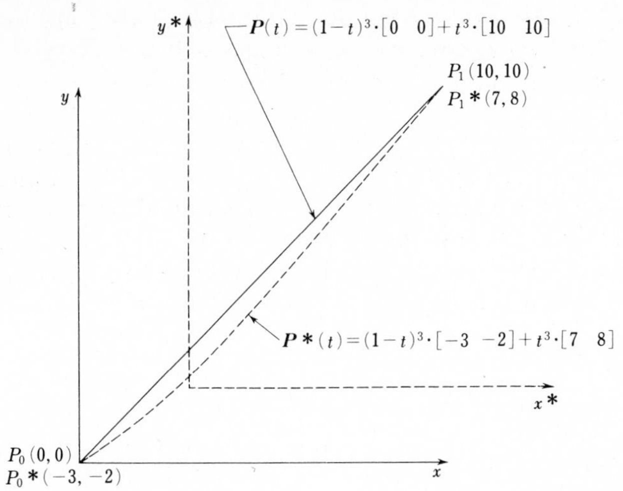

Fig. 1.2. An example in which a shape does not remain invariant under coordinate transformation

some cases in which the slope will become infinite no matter what coordinate system is chosen. This happens when the curve is closed. On the other hand, with a parametric representation the slope is:

$$\frac{dy}{dx}=\frac{dy}{dt}\bigg/\frac{dx}{dt}=\frac{\dot{y}}{\dot{x}}$$

so the direction parallel to the y-axis can be simply represented by:

$$\dot{x}=0.$$

The shape of an object is independent, unaffected by the coordinate system in which it is described. This property of an object is called *axis independence*. Axis independence must exist in order for a shape to be expressed on a computer.

In the above we have studied a relation between two types of mathematical representation of a curve and the properties of a shape. We have seen that an expression for a curve in the form $y=F(x)$ is inadequate to represent the properties possessed by shapes. A parametric representation overcomes this difficulty.

The advantages which a parametric representation of a curve has over a non-parametric representation can be summarized as follows.

1) Axis independence can be expressed (refer to Sect. 1.1.3).
2) It is easier to specify the regions of a multivalued curve.
3) Since it is an independent expression with respect to x, y and z (x and y in the case of a plane curve), affine transformations and projective

transformations can easily be performed by 4×4 matrices (3×3 matrices for plane curves).

4) Representation of space curves is easy.

In order to express a space curve by a non-parametric representation, it must be defined as the intersection of two curved surfaces: $y = F(x)$ and $z = G(x)$, or $\varphi_1(x, y, z) = 0$ and $\varphi_2(x, y, z) = 0$. In a parametric representation it can be represented in the form $\boldsymbol{P}(t) = [x(t)\ y(t)\ z(t)]$.

5) Since x, y and z are expressed as explicit functions of the parameter t, a point on the curve can be easily computed.

Since curved surfaces are basically regarded as being represented by groups of curves, the above arguments with respect to curves can be thought of as also applying to curved surfaces.

1.1.2 Design and Mathematical Representations

A mathematical representation of a curve or curved surface must not only express the properties possessed by a shape, they must also satisfy certain necessary conditions for design.

(1) Abundant Expression Power Having Generality

The designer's demands with regard to shapes are not limited; they change frequently, and a mathematical representation of a shape must be able to respond to them flexibly. For a mathematical representation of a curve, in some cases a straight line is appropriate, while in other cases a conic section such as a circle or ellipse might be appropriate. In addition, in some cases it is desirable not to be limited to fixed shapes, but rather for the representation to have the capability to respond flexibly to any arbitrary shape that the designer draws freely with a stylus.

Ordinarily, a curve or curved surface is thought of as consisting of a number of curve segments or surface patches which are connected together. When designing a smooth curve or curved surface, it is necessary to make sure that the curve segments or surface patches are connected smoothly. Mathematically, continuity to a sufficiently high order derivative (usually up to the slope or the curvature) is required. The smoothness of a curve involves not only this kind of local continuity but also the overall large-scale smoothness of the curve. Figure 1.3 shows examples of curves which are connected in such a manner that there is (a) continuity of position only (class C^0), (b) continuous to the slope (class C^1), (c) continuous to the curvature (class C^2). Figure 1.4 shows a corner R of a curve. In (a), the connection is done with straight line segments and a circular arc, so the slope is continuous at the connection point, but the curvature is discontinuous (class C^1). When the discontinuity of curvature reaches a certain level, it becomes noticeable to the naked eye. (b) shows a corner R with the connection performed with a curve in such a manner that the curvature is continuous (class C^2).

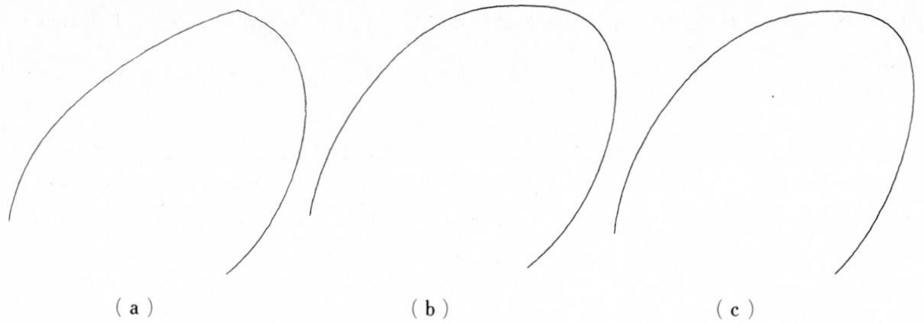

Fig. 1.3. (a) A class C^0 curve connection; (b) a class C^1 curve connection; (c) a class C^2 curve connection

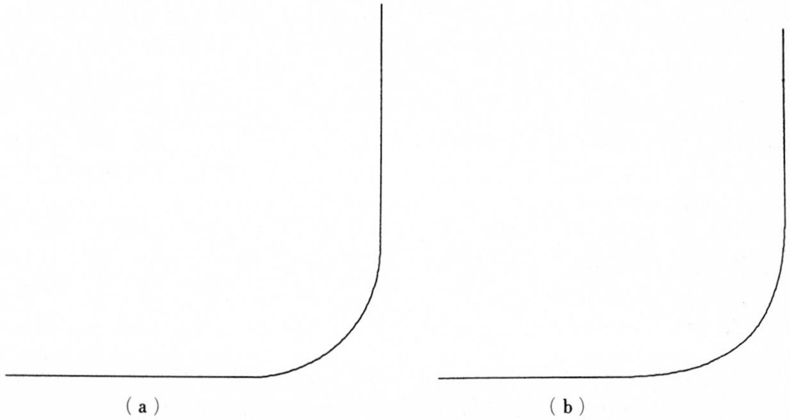

Fig. 1.4. Expression of a corner *R*. (a) Class C^1; (b) class C^2

In designing a curve or curved surface, it sometimes happens that it is desired to have the slope be discontinuous at a certain point. In fact, the discontinuity of slope is sometimes necessary. A point where the slope is discontinuous is called a *cusp*. Figure 1.3(a) shows a cusp. A line of discontinuity in a curved surface is called a *crease* (refer to Fig. 1.5).

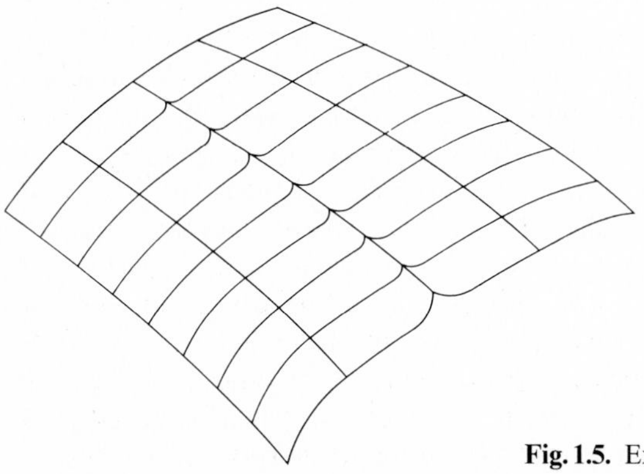

Fig. 1.5. Example of a crease in a curved surface

(2) Ease of Shape Control

Shapes of curves and curved surfaces which appear on the display could ideally be manipulated at will by the designer. At the start of the design work, the designer focuses on the overall shape that is displayed and forms his judgments. Consequently, it is desirable for the designer to be able to exercise *global* shape control (Fig. 1.6). Along the curve there are a number of *control points* which control the shape. Usually the control points are the same points that were used to define the curve. It would be very troublesome if it were necessary to change the parameters of all of the control points in order to change the shape of the curve. Ideally one would be able to change the shape of the entire curve by controlling just one suitable control point. As the design work progresses, the designer's attention shifts from the overall curve to localized sections of it. Then the designer wants to be able to change the shapes of only those sections while leaving the rest of the curve unchanged. Consequently, it is desirable to have the capability for not only global control of the shape of a curve, but also *local* control (Fig. 1.7). A local control capability is a necessary condition for being able to converge the trial-and-error design process.

There are cases in which it is impossible to control the locations of control points in such a way as to obtain the desired shape with only one

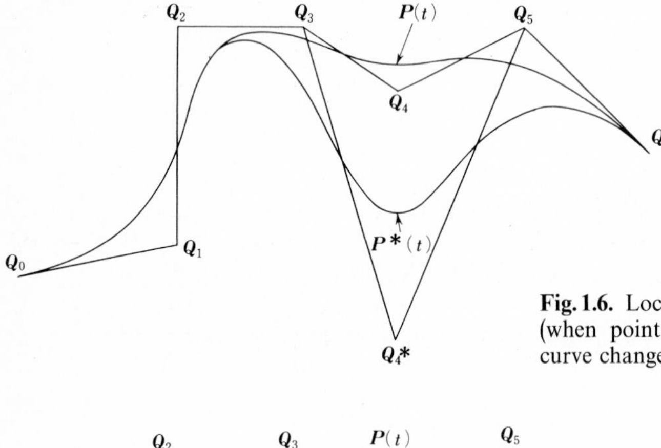

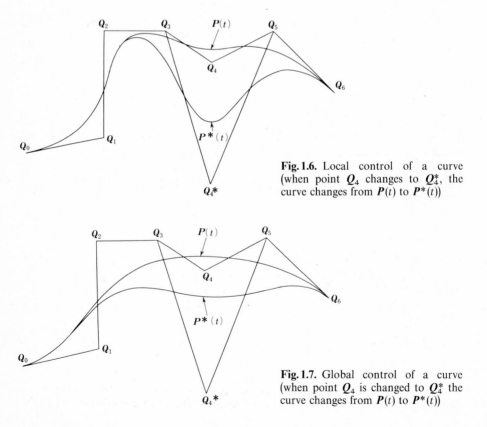

Fig. 1.6. Local control of a curve (when point Q_4 changes to Q_4^*, the curve changes from $P(t)$ to $P^*(t)$)

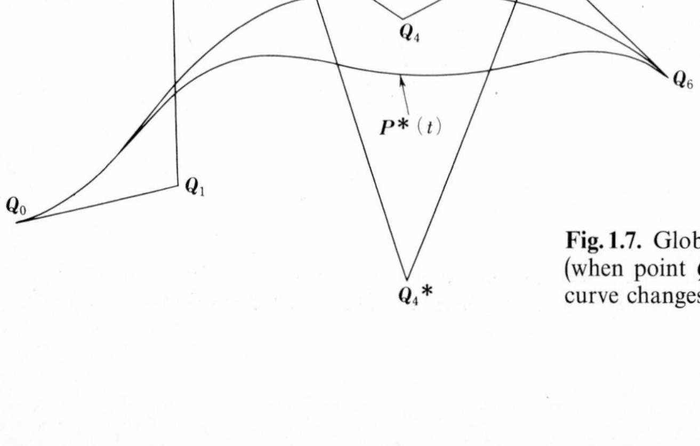

Fig. 1.7. Global control of a curve (when point Q_4 is changed to Q_4^* the curve changes from $P(t)$ to $P^*(t)$)

mathematical expression. In such cases, it is necessary to have some method to improve the capability of the mathematical expression to express a shape. Figures 1.8 and 1.9 show examples of ways in which the ability to express shapes can be increased. In Fig. 1.8 the shape is preserved while one curve segment is split into two segments. In Fig. 1.9 the shape is preserved while the degree of the curve segments is increased by one. What is important here is that the capability to express shapes is increased without changing the shapes themselves.

Some curves have special properties such as the *convex hull property* or the *variation diminishing property*. In such cases shape control is made easier because the changes that a shape will undergo when the control point parameters are varied can be predicted (refer to Chaps. 5 and 6).

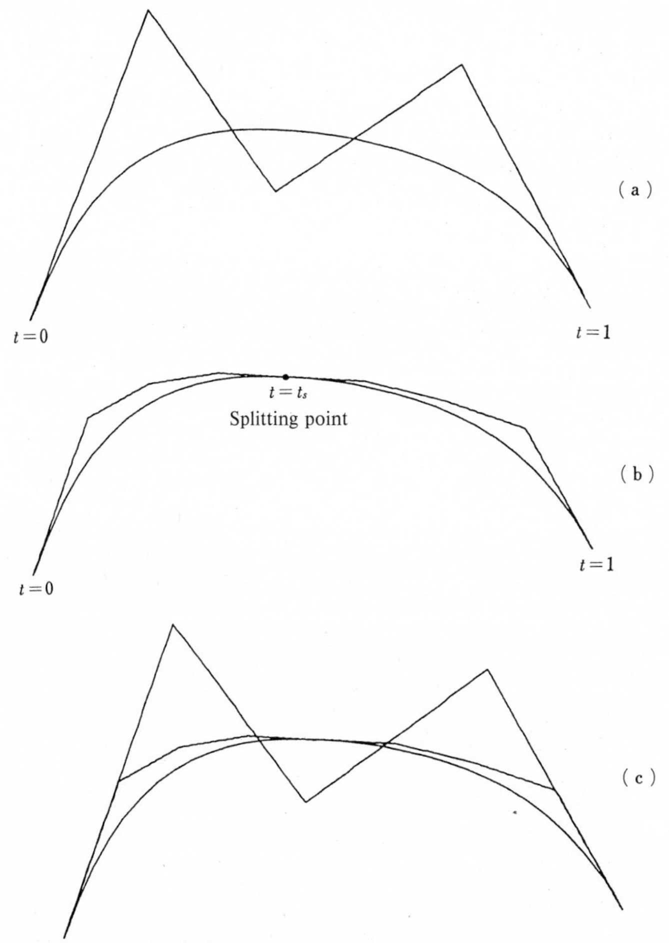

Fig. 1.8. An example of how a curve can be split into two segments while preserving its shape (a Bézier curve). (a) Before splitting; (b) after splitting; (c) after splitting

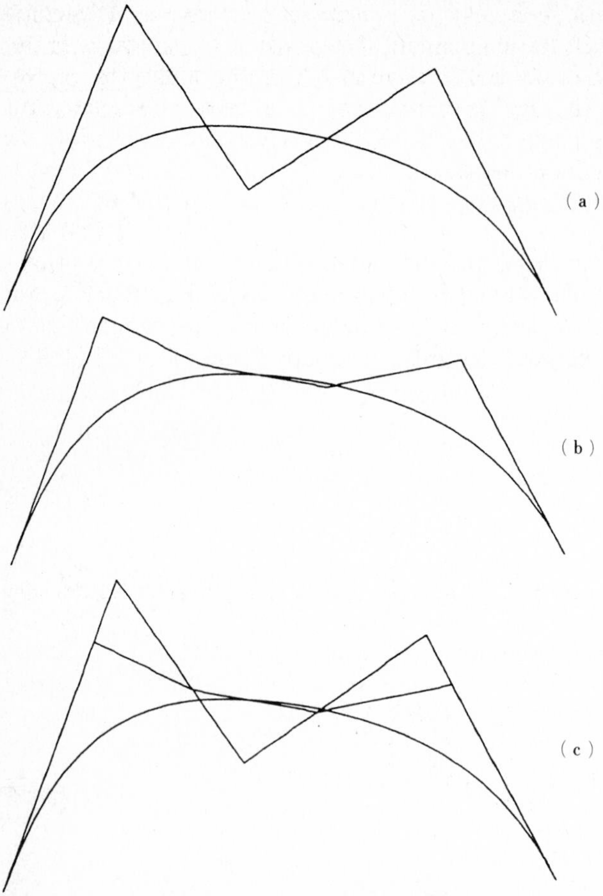

Fig. 1.9. An example in which the shape of the curve is preserved and (in form) the degree of the curve is increased (Bézier curve). (a) Before degree is increased; (b) after degree is increased; (c) before and after degree is increased

1.1.3 Invariance of a Shape Under Coordinate Transformation

In Sect. 1.1.1 we referred to the invariance of a shape under coordinate transformation. Now let us look at the conditions for such invariance to be satisfied.

Let a curve be represented by

$$\begin{aligned} \boldsymbol{P}(t) &= [x(t)\ y(t)\ z(t)] \\ &= [\phi_0(t)\ \phi_1(t) \ldots \phi_n(t)]\ [\boldsymbol{Q}_0\ \boldsymbol{Q}_1 \ldots \boldsymbol{Q}_n]^T. \end{aligned} \tag{1.2}$$

$\phi_0(t)$, $\phi_1(t)$, ..., $\phi_n(t)$ are linearly independent polynomial functions of t; $\boldsymbol{Q}_0$, $\boldsymbol{Q}_1$, ..., $\boldsymbol{Q}_n$ are the position vectors that define the curve.

Let us separately consider the 3-dimensional rotation matrix $\boldsymbol{T}$ (a 3×3 matrix) and the translation $\boldsymbol{m}=[m_x\ m_y\ m_z]$. In order for the shape to be invariant, it is necessary that the curve expressed in terms of $\boldsymbol{Q}_0^*, \boldsymbol{Q}_1^*, \ldots, \boldsymbol{Q}_n^*$ in the transformed coordinate system transform by the same amount as $\boldsymbol{Q}_0, \boldsymbol{Q}_1, \ldots, \boldsymbol{Q}_n$ do.

For the rotation matrix $\boldsymbol{T}$ we have:

$$\boldsymbol{P}^*(t)=[\phi_0(t)\ \phi_1(t)\ \ldots\ \phi_n(t)]\begin{bmatrix}\boldsymbol{Q}_0^*\\ \boldsymbol{Q}_1^*\\ \vdots\\ \boldsymbol{Q}_n^*\end{bmatrix}=[\phi_0(t)\ \phi_1(t)\ \ldots\ \phi_n(t)]\begin{bmatrix}\boldsymbol{Q}_0\boldsymbol{T}\\ \boldsymbol{Q}_1\boldsymbol{T}\\ \vdots\\ \boldsymbol{Q}_n\boldsymbol{T}\end{bmatrix}$$

$$=[\phi_0(t)\ \phi_1(t)\ \ldots\ \phi_n(t)]\begin{bmatrix}\boldsymbol{Q}_0\\ \boldsymbol{Q}_1\\ \vdots\\ \boldsymbol{Q}_n\end{bmatrix}\boldsymbol{T}=\boldsymbol{P}(t)\,\boldsymbol{T}$$

so that in the case of rotation the condition of invariance is automatically satisfied.

Next let us consider the translation.

$$\boldsymbol{P}^*(t)=[\phi_0(t)\ \phi_1(t)\ \ldots\ \phi_n(t)]\begin{bmatrix}\boldsymbol{Q}_0^*\\ \boldsymbol{Q}_1^*\\ \vdots\\ \boldsymbol{Q}_n^*\end{bmatrix}$$

$$=[\phi_0(t)\ \phi_1(t)\ \ldots\ \phi_n(t)]\begin{bmatrix}\boldsymbol{Q}_0+\boldsymbol{m}\\ \boldsymbol{Q}_1+\boldsymbol{m}\\ \vdots\\ \boldsymbol{Q}_n+\boldsymbol{m}\end{bmatrix}$$

$$=[\phi_0(t)\ \ \phi_1(t)\ \ldots\ \phi_n(t)]\begin{bmatrix}\boldsymbol{Q}_0\\ \boldsymbol{Q}_1\\ \vdots\\ \boldsymbol{Q}_n\end{bmatrix}+(\phi_0(t)+\phi_1(t)+\ \ldots\ +\phi_n(t))\,\boldsymbol{m}$$

$$=\boldsymbol{P}(t)+(\phi_0(t)+\phi_1(t)+\ \ldots\ +\phi_n(t))\,\boldsymbol{m}. \tag{1.3}$$

From Eq. (1.3), in order for the shape to be invariant the relation

$$\phi_0(t)+\phi_1(t)+\ \ldots\ +\phi_n(t)\equiv 1 \tag{1.4}$$

must be satisfied. Equation (1.4) is sometimes called Cauchy's relation.

1.2 Curve Theory

1.2.1 Parametric Representation of Curves; Tangent Lines and Osculating Planes

Let us express a curve as a vector using the parameter t:

$$\boldsymbol{P}(t)=[x(t)\ \ y(t)\ \ z(t)].$$

Next consider the derivative vector evaluated at $t=t_0$:*)

$$\dot{\boldsymbol{P}}(t_0)=[\dot{x}(t_0)\ \ \dot{y}(t_0)\ \ \dot{z}(t_0)].$$

If the condition

$$\dot{\boldsymbol{P}}(t_0)\neq\boldsymbol{0} \tag{1.5}$$

is satisfied, then the curve is said to be *regular* at $t=t_0$. A point at which a curve is regular is called an *ordinary point;* a point at which it is not regular is called a *singular point*. For condition (1.5) for regularity to be satisfied, $\dot{x}(t_0)$, $\dot{y}(t_0)$ and $\dot{z}(t_0)$ must not all be zero at once. This condition is expressed mathematically by the following inequality:

$$\dot{x}(t_0)^2+\dot{y}(t_0)^2+\dot{z}(t_0)^2>0. \tag{1.6}$$

If the derivatives of $x(t)$, $y(t)$ and $z(t)$ of a curve $\boldsymbol{P}(t)$ $(a\leqq t\leqq b)$ to order r exist and are continuous, and if the curve is regular throughout the interval, then that curve is said to be of class C^r.

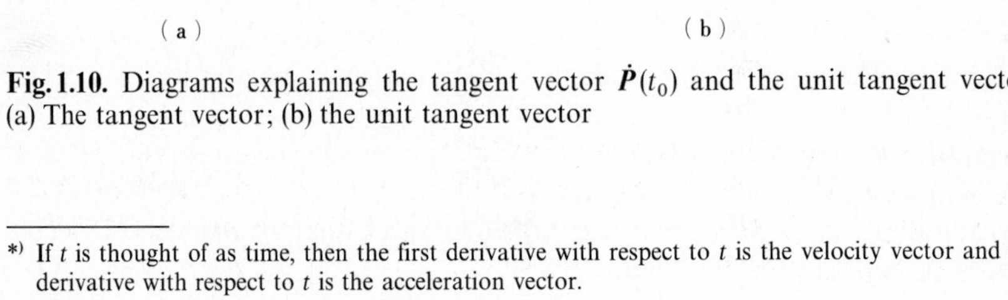

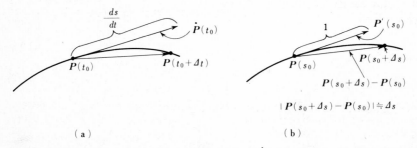

Fig. 1.10. Diagrams explaining the tangent vector $\dot{\boldsymbol{P}}(t_0)$ and the unit tangent vector $\boldsymbol{P}'(s_0)$. (a) The tangent vector; (b) the unit tangent vector

*) If t is thought of as time, then the first derivative with respect to t is the velocity vector and the second derivative with respect to t is the acceleration vector.

In Fig. 1.10(a), consider the vector $\boldsymbol{P}(t_0+\Delta t)-\boldsymbol{P}(t_0)$, from $\boldsymbol{P}(t_0)$ to $\boldsymbol{P}(t_0+\Delta t)$ on the curve $\boldsymbol{P}(t)$, divided by Δt:

$$\frac{\boldsymbol{P}(t_0+\Delta t)-\boldsymbol{P}(t_0)}{\Delta t}.$$

This vector indicates the direction from $\boldsymbol{P}(t_0)$ to $\boldsymbol{P}(t_0+\Delta t)$. If the curve is regular at $t=t_0$, then this vector converges to the finite magnitude vector $\dot{\boldsymbol{P}}(t_0)$ in the limit as $\Delta t \to 0$. $\dot{\boldsymbol{P}}(t_0)$ is called the tangent vector at the point $t=t_0$.

Let us now use the length s of the curve as a parameter. We will indicate the derivative with respect to s by the prime symbol $'$ to distinguish it from the derivative with respect to t.

Since

$$\frac{d\boldsymbol{P}}{ds} \equiv \boldsymbol{P}' = \frac{d\boldsymbol{P}}{dt}\,\frac{dt}{ds} = \frac{\dot{\boldsymbol{P}}}{\dot{s}} \tag{1.7}$$

$$\frac{ds}{dt} \equiv \dot{s} = \sqrt{\left(\frac{dx}{dt}\right)^2 + \left(\frac{dy}{dt}\right)^2 + \left(\frac{dz}{dt}\right)^2} = \sqrt{\dot{\boldsymbol{P}}^2} \tag{1.8}$$

then if $\dot{\boldsymbol{P}} \neq \boldsymbol{0}$, we have:

$$\boldsymbol{P}' = \frac{\dot{\boldsymbol{P}}}{\dot{s}} = \frac{\dot{\boldsymbol{P}}}{\sqrt{\dot{\boldsymbol{P}}^2}} \tag{1.9}$$

$$\therefore\ |\boldsymbol{P}'| = 1. \tag{1.10}$$

That is, $\boldsymbol{P}'$ is in the same direction as $\dot{\boldsymbol{P}}$, and its magnitude is 1. $\boldsymbol{P}'$ is called the *unit tangent vector* (refer to Fig. 1.10(b)). It can be seen from Eq. (1.8) that $\dot{s}$ is the magnitude of the tangent vector. In this book, the magnitude of the tangent vector will sometimes be denoted by α, and the unit tangent vector by $\boldsymbol{t}$. Hence we have:

$$\dot{s} \equiv \alpha \tag{1.11}$$

$$\boldsymbol{P}' \equiv \boldsymbol{t}. \tag{1.12}$$

The equation of a tangent line can be expressed in terms of a parameter u as:

$$\boldsymbol{R}(u) = \boldsymbol{P}(t_0) + u\boldsymbol{t}(t_0). \tag{1.13}$$

Here the parameter u is the distance from the tangent point.

At a point where a curve is regular, the tangent line is unique. However, at a singular point there occur various anomalous cases. Examples of singular points are shown in Figs. 1.11 and 1.12. At a cusp, as shown in Fig. 1.11, $\dot{\boldsymbol{P}}$

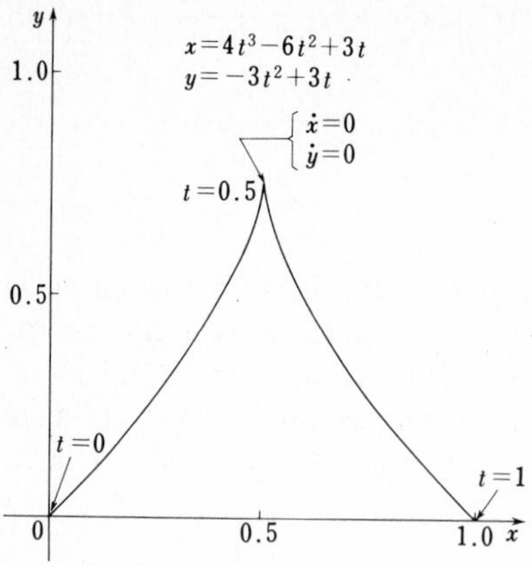

Fig. 1.11. Example of a singular point (1)

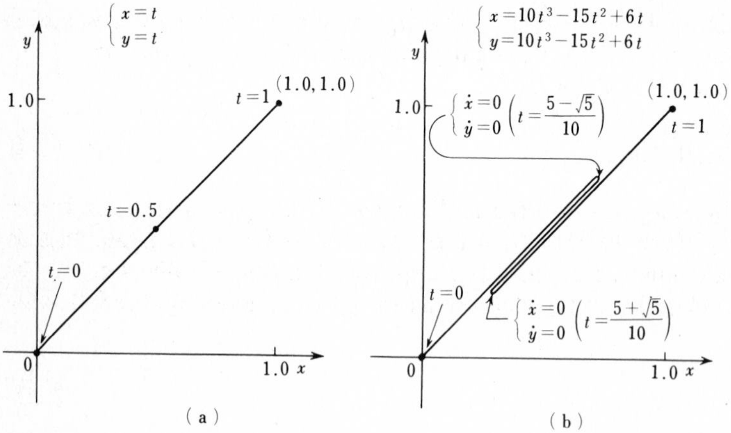

Fig. 1.12. Example of a singular point (2)

can become discontinuous or a **0** vector. Figure 1.12 shows examples in which line segments are expressed parametrically. With the expression used in Fig. 1.12(a), the line segments are regular over the entire interval. In contrast, in Fig. 1.12(b), although the line being expressed is the same two singular points occur within the interval. In this case, as t increases from the starting point $t=0$ to $t=(5-\sqrt{5})/10$, the point on the line progresses steadily toward the end point. At $t=(5-\sqrt{5})/10$ $\dot{x}=\dot{y}=0$, so this point is a singular point. From $t=(5-\sqrt{5})/10$ to $t=(5+\sqrt{5})/10$, as t increases the point on the line progresses back toward the starting point. $t=(5+\sqrt{5})/10$ is another singular point; there the direction reverses again and the point on the line progresses toward the end point $t=1$. The possibilities for such *parameterization,* the assigning of a relation between points on a line and a parameter, are numerous. It is desirable to choose a parameterization such that the curve is normal at as many points as possible.

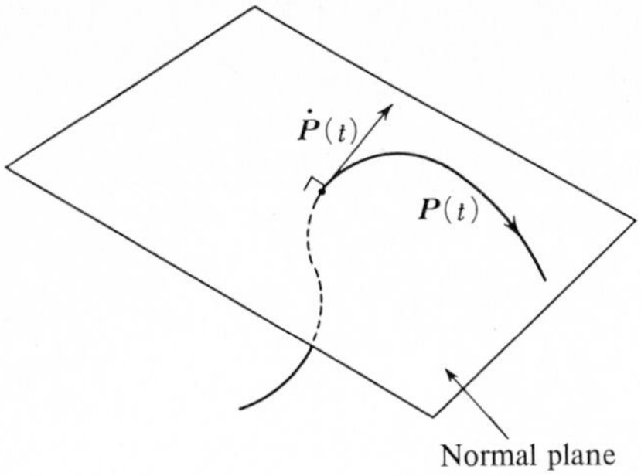

Fig. 1.13. The normal plane

Another type of singular point is one at which the curve *degenerates* to a single point.

A plane which passes through point $\boldsymbol{P}(t_0)$ on a curve and is perpendicular to the tangent line to the curve at that point is called *normal plane* (Fig. 1.13). The equation of a normal plane is:

$$(\boldsymbol{R}-\boldsymbol{P}(t_0))\cdot\dot{\boldsymbol{P}}(t_0)=0. \tag{1.14}$$

Consider a point $\boldsymbol{P}(t_0)$ on a curve and a plane which passes through it and two points very close to it, $\boldsymbol{P}(t_0+\Delta_1 t)$ and $\boldsymbol{P}(t_0+\Delta_2 t)$. The plane that is approached in the limit as $\Delta_1 t$ and $\Delta_2 t$ approach 0 independently is called the *osculating plane* at point $\boldsymbol{P}(t_0)$. The osculating plane is given by the following equation.

$$[\boldsymbol{R}-\boldsymbol{P}(t_0),\ \dot{\boldsymbol{P}}(t_0),\ \ddot{\boldsymbol{P}}(t_0)]=0. \tag{1.15}$$

The brackets indicate a triple scalar product; $\dot{\boldsymbol{P}}(t_0)\times\ddot{\boldsymbol{P}}(t_0)\neq 0$. Equation (1.15) is the condition for the three vectors $\boldsymbol{R}-\boldsymbol{P}(t_0)$, $\dot{\boldsymbol{P}}(t_0)$ and $\ddot{\boldsymbol{P}}(t_0)$ to lie in the same plane.

A general plane passing through the point $\boldsymbol{P}(t_0)$ can be written in terms of its unit normal vector $\boldsymbol{a}$ as $\boldsymbol{a}\cdot(\boldsymbol{R}-\boldsymbol{P}(t_0))=0$. The length of the perpendicular distance from the point $\boldsymbol{P}(t_0+\Delta t)$ on the curve $\boldsymbol{P}(t)$ to this plane is:

$$\boldsymbol{a}\cdot(\boldsymbol{P}(t_0+\Delta t)-\boldsymbol{P}(t_0))=\boldsymbol{a}\cdot\left(\dot{\boldsymbol{P}}(t_0)\,\Delta t+\frac{\ddot{\boldsymbol{P}}(t_0)}{2!}\,\Delta t^2+\dots\right)$$

(Fig. 1.14). In the special case when $\boldsymbol{a}\cdot\dot{\boldsymbol{P}}(t_0)=0$, $\boldsymbol{a}\cdot\ddot{\boldsymbol{P}}(t_0)=0$, that is, when the plane is the osculating plane, this length is a 3rd-order infinitesimal with respect to Δt. Consequently, the osculating plane is the plane that best fits the curve at the point $\boldsymbol{P}(t_0)$.

The line that lies in the osculating plane, passes through the point $\boldsymbol{P}(t_0)$ and is perpendicular to the tangent vector $\dot{\boldsymbol{P}}(t_0)$ at that point is called the *principal normal*. The line that passes through point $\boldsymbol{P}(t_0)$ and is perpendicular to the osculating plane is called the *binormal*, and the plane determined by the

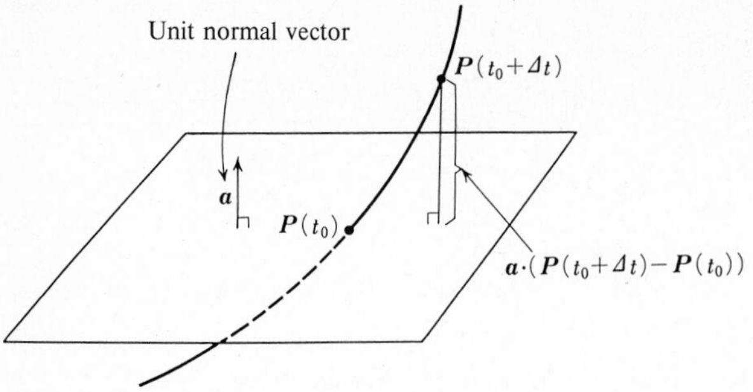

Fig. 1.14. Perpendicular distance from a point on a curve to a plane

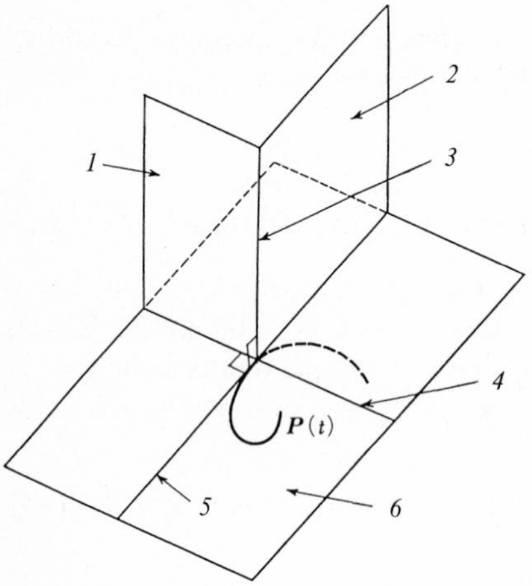

Fig. 1.15. Relation between the normal plane, osculating plane and rectifying plane. *1* normal plane; *2* rectifying plane; *3* binormal; *4* principal normal; *5* tangent line; *6* osculating plane

tangent and the binormal is called the *rectifying plane*. These relations are shown in Fig. 1.15.

1.2.2 Curvature and Torsion

Let us consider the meaning of the second derivative $\boldsymbol{P}''(s) \equiv d^2\boldsymbol{P}/ds^2$. From the definition of a derivative, we have:

$$\boldsymbol{P}''(s_0) = \lim_{\Delta s \to 0} \frac{\boldsymbol{P}'(s_0+\Delta s) - \boldsymbol{P}'(s_0)}{\Delta s}. \tag{1.16}$$

As shown in Fig. 1.16(a) and (b), in the limit as $\Delta s \to 0$ the numerator $\boldsymbol{P}'(s_0+\Delta s) - \boldsymbol{P}'(s_0)$ is perpendicular to the tangent vector at the point $\boldsymbol{P}(s_0)$ and points toward the center of curvature of the curve. It can be seen from

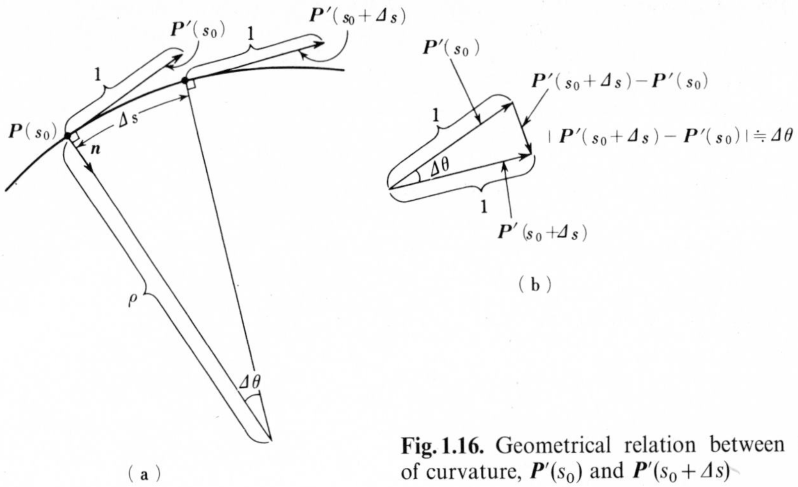

Fig. 1.16. Geometrical relation between the center of curvature, $\boldsymbol{P}'(s_0)$ and $\boldsymbol{P}'(s_0+\Delta s)$

Fig. 1.16(b) that its magnitude is $\Delta\theta$. Consequently, the magnitude of $\boldsymbol{P}''(s_0)$ is:

$$|\boldsymbol{P}''(s_0)| = \lim_{\Delta s \to 0} \frac{\Delta\theta}{\Delta s} = \lim_{\Delta s \to 0} \frac{\frac{1}{\varrho}\Delta s}{\Delta s} = \frac{1}{\varrho} \equiv \kappa .$$

Here ϱ is the radius of curvature and κ is the curvature. $\boldsymbol{P}''$ can be expressed in terms of them as:

$$\boldsymbol{P}'' = \frac{1}{\varrho}\boldsymbol{n} \equiv \kappa \boldsymbol{n} . \tag{1.17}$$

Here $\boldsymbol{n}$ is the unit vector pointing toward the center of curvature. Since $\boldsymbol{P}''(s)$ is a vector that has a magnitude equal to the curvature at point s and points toward the center of curvature, it is sometimes called the curvature vector. Curvature is the rate of turning of the unit tangent vector $\boldsymbol{t}$ with respect to the length of the curve s, in other words, a quantity that indicates how rapidly or slowly the curve is turning.

Next, let us find the relationship between the curvature vector $\boldsymbol{P}''(s)$ and the derivative vectors $\dot{\boldsymbol{P}}$ and $\ddot{\boldsymbol{P}}$ with respect to the parameter t. From Eq. (1.9):

$$\boldsymbol{P}'(s) = \frac{\dot{\boldsymbol{P}}}{\sqrt{\dot{\boldsymbol{P}}^2}} .$$

Differentiating the above equation with respect to s gives:

$$P''(s)=\frac{d}{dt}\left(\frac{\dot{P}}{\sqrt{\dot{P}^2}}\right)\frac{dt}{ds}=\frac{\ddot{P}\sqrt{\dot{P}^2}-\dot{P}\left(\frac{2\dot{P}}{2\sqrt{\dot{P}^2}}\cdot\ddot{P}\right)}{\dot{P}^2}\frac{1}{\sqrt{\dot{P}^2}}$$

$$=\frac{1}{\dot{P}^2}\left[\ddot{P}-\frac{\dot{P}}{\sqrt{\dot{P}^2}}\left(\frac{\dot{P}}{\sqrt{\dot{P}^2}}\cdot\ddot{P}\right)\right]^{4)} \tag{1.18}$$

$$\equiv\frac{B}{\dot{P}^2} \tag{1.19}$$

where:

$$B\equiv\ddot{P}-\frac{\dot{P}}{\sqrt{\dot{P}^2}}\left(\frac{\dot{P}}{\sqrt{\dot{P}^2}}\cdot\ddot{P}\right)\equiv|B|\,n. \tag{1.20}$$

Equation (1.18) gives the relation between the curvature vector and the derivative vectors $\dot{P}(t)$ and $\ddot{P}(t)$ with respect to the parameter t.

Taking the absolute value of Eq. (1.19) gives:

$$|P''(s)|=\frac{1}{\varrho}=\frac{|B|}{|\dot{P}|^2}. \tag{1.21}$$

By using Eqs. (1.18) and (1.21), the center of curvature can be determined graphically as shown in Fig. 1.17.

If we use the relation:

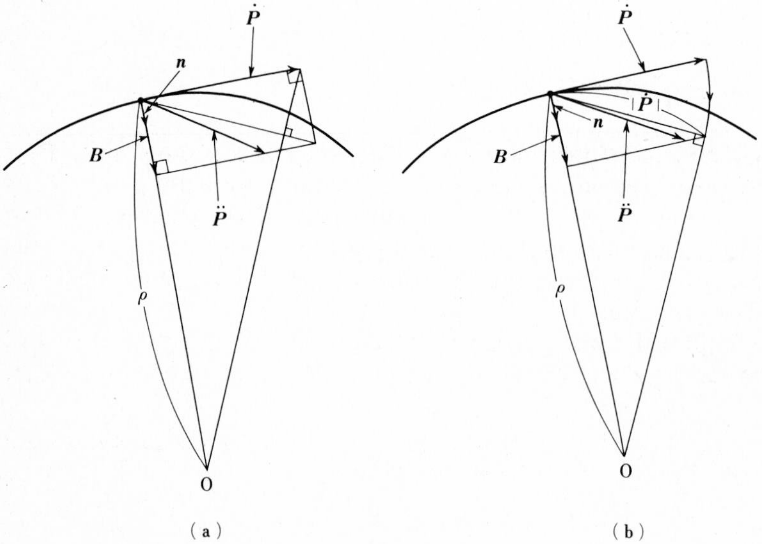

Fig. 1.17. How to find the center of curvature graphically

$$\ddot{s} = \frac{\dot{\boldsymbol{P}}}{\sqrt{\dot{\boldsymbol{P}}^2}} \cdot \ddot{\boldsymbol{P}} \tag{1.22}$$

then Eq. (1.18) becomes:

$$\boldsymbol{P}'' = \frac{1}{\dot{s}^2} (\ddot{\boldsymbol{P}} - \ddot{s}\boldsymbol{t}).$$

Combining this equation with Eq. (1.17) gives:

$$\ddot{\boldsymbol{P}} = \frac{\dot{s}^2}{\varrho} \boldsymbol{n} + \ddot{s}\boldsymbol{t}. \tag{1.23}$$

If the parameter t is regarded as time, Eq. (1.23) expresses the well-known dynamical relation that the acceleration $\ddot{\boldsymbol{P}}$ acting on a mass point equals the sum of the tangential acceleration $\ddot{s}\boldsymbol{t}$ and the normal acceleration $(\dot{s}^2/\varrho)\boldsymbol{n}$. This relation is shown in Fig. 1.18.

The form of Eq. (1.18) for the curvature vector can be changed as follows:

$$\boldsymbol{P}''(s) \equiv \boldsymbol{t}' = \frac{1}{\dot{\boldsymbol{P}}^4} [\dot{\boldsymbol{P}}^2 \ddot{\boldsymbol{P}} - \dot{\boldsymbol{P}}(\dot{\boldsymbol{P}} \cdot \ddot{\boldsymbol{P}})]$$

$$= \frac{(\dot{\boldsymbol{P}} \times \ddot{\boldsymbol{P}}) \times \dot{\boldsymbol{P}}}{\dot{\boldsymbol{P}}^4} \tag{1.24)*}$$

$$= \frac{|\dot{\boldsymbol{P}} \times \ddot{\boldsymbol{P}}|}{|\dot{\boldsymbol{P}}|^3} \boldsymbol{n} \tag{1.25)**}$$

$$\equiv \kappa \boldsymbol{n}. \tag{1.25'}$$

Consequently, the curvature κ becomes:

$$\kappa = \frac{|\dot{\boldsymbol{P}} \times \ddot{\boldsymbol{P}}|}{|\dot{\boldsymbol{P}}|^3} = \frac{|\dot{\boldsymbol{P}} \times \ddot{\boldsymbol{P}}|}{\dot{s}^3}. \tag{1.26}$$

A point on a curve at which $\kappa = 0$ is called a *point of inflection*.

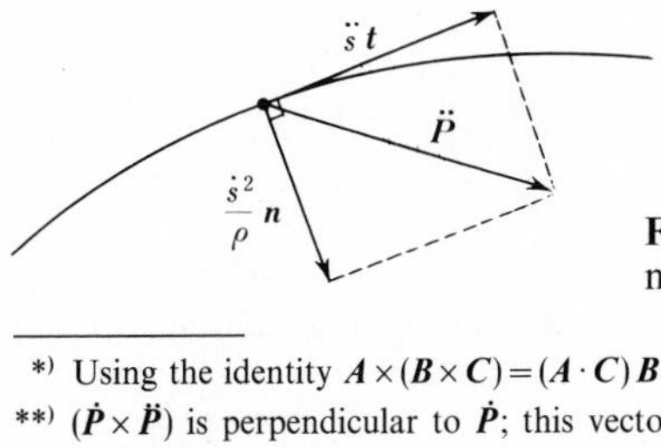

Fig. 1.18. $\ddot{\boldsymbol{P}}$ is the sum of the tangent acceleration and normal acceleration

*) Using the identity $\boldsymbol{A} \times (\boldsymbol{B} \times \boldsymbol{C}) = (\boldsymbol{A} \cdot \boldsymbol{C})\boldsymbol{B} - (\boldsymbol{A} \cdot \boldsymbol{B})\boldsymbol{C}$ for the triple vector product.

**) $(\dot{\boldsymbol{P}} \times \ddot{\boldsymbol{P}})$ is perpendicular to $\dot{\boldsymbol{P}}$; this vector product is itself a vector which points toward the center of curvature.

The vector $\boldsymbol{b}=\boldsymbol{t}\times\boldsymbol{n}$ is called the *unit binormal vector*. The unit binormal vector can be expressed as follows:

$$\begin{aligned}
\boldsymbol{b}&=\boldsymbol{t}\times\boldsymbol{n}\\
&=\frac{\dot{\boldsymbol{P}}}{|\dot{\boldsymbol{P}}|}\times\left[\frac{1}{\kappa}\,\frac{(\dot{\boldsymbol{P}}\times\ddot{\boldsymbol{P}})\times\dot{\boldsymbol{P}}}{|\dot{\boldsymbol{P}}|^4}\right]=\frac{1}{\kappa}\,\frac{1}{|\dot{\boldsymbol{P}}|^5}\,[\dot{\boldsymbol{P}}\times((\dot{\boldsymbol{P}}\times\ddot{\boldsymbol{P}})\times\dot{\boldsymbol{P}})]\\
&=\frac{1}{\kappa}\,\frac{1}{|\dot{\boldsymbol{P}}|^5}\,[|\dot{\boldsymbol{P}}|^2\,(\dot{\boldsymbol{P}}\times\ddot{\boldsymbol{P}})-(\dot{\boldsymbol{P}}\cdot(\dot{\boldsymbol{P}}\times\ddot{\boldsymbol{P}}))\,\dot{\boldsymbol{P}}]
\end{aligned}$$

(using the identity for the triple vector product)

$$=\frac{1}{\kappa}\,\frac{\dot{\boldsymbol{P}}\times\ddot{\boldsymbol{P}}}{|\dot{\boldsymbol{P}}|^3}\qquad(\because\ \dot{\boldsymbol{P}}\cdot(\dot{\boldsymbol{P}}\times\ddot{\boldsymbol{P}})=0) \tag{1.27}$$

$$=\frac{\dot{\boldsymbol{P}}\times\ddot{\boldsymbol{P}}}{|\dot{\boldsymbol{P}}\times\ddot{\boldsymbol{P}}|}. \tag{1.28}$$

Figure 1.19 shows the relation between the unit tangent vector $\boldsymbol{t}$, the unit principal normal vector $\boldsymbol{n}$ and the unit binormal vector $\boldsymbol{b}$.

Letting $\Delta\phi$ be the angle between the osculating planes at two slightly separated points $\boldsymbol{P}(s_0)$ and $\boldsymbol{P}(s_0+\Delta s)$, the *torsion* of the curve at point $\boldsymbol{P}(s_0)$ is defined as:

$$\tau=\lim_{\Delta s\to 0}\frac{\Delta\phi}{\Delta s}. \tag{1.29}$$

Torsion is a measure of the amount of rotation of the osculating plane with respect to the length of the curve s. In other words, torsion is a quantity that indicates whether the curve is twisting rapidly or slowly. Torsion can be expressed mathematically in terms of the derivatives $\dot{\boldsymbol{P}}$, $\ddot{\boldsymbol{P}}$ and $\dddot{\boldsymbol{P}}$ with respect to the parameter t as follows:

$$\tau=\frac{(\dot{\boldsymbol{P}}\times\ddot{\boldsymbol{P}})\cdot\dddot{\boldsymbol{P}}}{(\dot{\boldsymbol{P}}\times\ddot{\boldsymbol{P}})^2}=\frac{[\dot{\boldsymbol{P}},\ddot{\boldsymbol{P}},\dddot{\boldsymbol{P}}]}{(\dot{\boldsymbol{P}}\times\ddot{\boldsymbol{P}})^2}=\frac{[\dot{\boldsymbol{P}},\ddot{\boldsymbol{P}},\dddot{\boldsymbol{P}}]}{\dot{s}^6\kappa^2}. \tag{1.30}$$

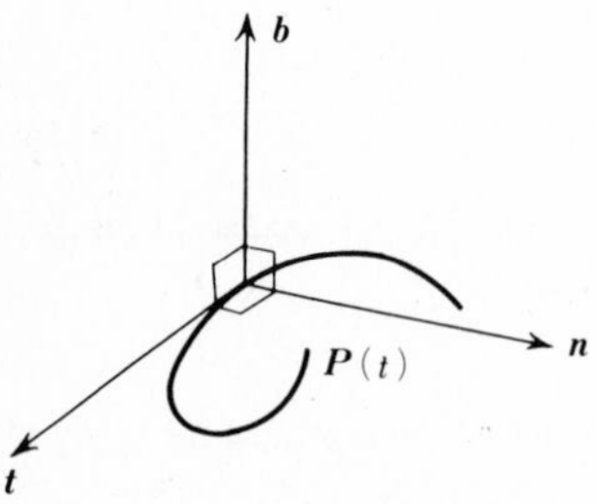

Fig. 1.19. Relation between the unit tangent vector, unit principal normal vector, and unit binormal vector $\boldsymbol{b}$, unit binormal vector; $\boldsymbol{n}$, unit principal normal vector; $\boldsymbol{t}$, unit tangent vector

The condition for a curve to be a plane curve, in other words a curve with zero torsion, is:

$$(\dot{\boldsymbol{P}} \times \ddot{\boldsymbol{P}}) \cdot \dddot{\boldsymbol{P}} = \begin{vmatrix} \dot{x} & \dot{y} & \dot{z} \\ \ddot{x} & \ddot{y} & \ddot{z} \\ \dddot{x} & \dddot{y} & \dddot{z} \end{vmatrix} = 0. \tag{1.31}$$

1.2.3 Frenet Frames and the Frenet-Serret Equations

If we take the directions of the unit tangent vector $\boldsymbol{t}$, the unit principal normal vector $\boldsymbol{n}$ and the unit binormal vector $\boldsymbol{b}$ to be the positive directions along the tangent, the principal normal and the binormal, respectively, then the coordinate system formed by the tangent, the principal normal and the binormal is a right-handed system. These coordinate axes move as the parameter t varies (refer to Fig. 1.19). The set $(\boldsymbol{P}; \boldsymbol{t}, \boldsymbol{n}, \boldsymbol{b})$ is called a *moving frame* or a *Frenet frame*. When properties of a space curve are investigated in the vicinity of various points on it, the Frenet frame of each point is used.

Let us now look at the changes in the Frenet frame as the curve parameter is varied. First from the relation for the unit tangent vector, we have:

$$\frac{d\boldsymbol{P}}{ds} \equiv \boldsymbol{P}' \equiv \boldsymbol{t}.$$

Then, from another of the relations which we derived (refer to Eq. (1.17)):

$$\frac{d\boldsymbol{t}}{ds} \equiv \boldsymbol{t}' = \kappa \boldsymbol{n}.$$

Next, let us find $d\boldsymbol{n}/ds \equiv \boldsymbol{n}'$. Since $\boldsymbol{n}$ is a unit vector,

$$\boldsymbol{n}^2 = 1.$$

Differentiating both sides of this equation with respect to s gives:

$$\boldsymbol{n} \cdot \frac{d\boldsymbol{n}}{ds} = 0.$$

We see from this that $d\boldsymbol{n}/ds$ is perpendicular to $\boldsymbol{n}$. Therefore $d\boldsymbol{n}/ds$ can be written in the following form:

$$\frac{d\boldsymbol{n}}{ds} = a_1 \boldsymbol{t} + a_2 \boldsymbol{b}. \tag{1.32}$$

Taking the scalar product of both sides of this equation with $\boldsymbol{t}$ gives:

$$\boldsymbol{t} \cdot \frac{d\boldsymbol{n}}{ds} = a_1. \tag{1.33}$$

Since we also have the relation:

$$\boldsymbol{t} \cdot \boldsymbol{n} = 0$$

differentiating by s gives:

$$\frac{d\boldsymbol{t}}{ds} \cdot \boldsymbol{n} + \boldsymbol{t} \cdot \frac{d\boldsymbol{n}}{ds} = 0.$$

Consequently we have:

$$\boldsymbol{t} \cdot \frac{d\boldsymbol{n}}{ds} = -\kappa. \tag{1.34}$$

Comparing Eqs. (1.33) and (1.34), we see that $a_1 = -\kappa$. Next, take the scalar product of $\boldsymbol{b}$ with both sides of Eq. (1.32):

$$\boldsymbol{b} \cdot \frac{d\boldsymbol{n}}{ds} = a_2. \tag{1.35}$$

Since we also have the relation:

$$\boldsymbol{b} \cdot \boldsymbol{n} = 0$$

differentiating by s gives:

$$\frac{d\boldsymbol{b}}{ds} \cdot \boldsymbol{n} + \boldsymbol{b} \cdot \frac{d\boldsymbol{n}}{ds} = 0. \tag{1.36}$$

The unit binormal vector $\boldsymbol{b}$ is perpendicular to the osculating plane. Since the torsion is the rate of rotation of the osculating plane with respect to s,

$$\left| \frac{d\boldsymbol{b}}{ds} \right| = \tau. \tag{1.37}$$

Differentiating the relation

$$\boldsymbol{b} \cdot \boldsymbol{t} = 0$$

by s gives:

$$\frac{d\boldsymbol{b}}{ds} \cdot \boldsymbol{t} + \boldsymbol{b} \cdot \frac{d\boldsymbol{t}}{ds} = 0.$$

The second term on the left-hand side of this equation is:

$$\boldsymbol{b} \cdot \frac{d\boldsymbol{t}}{ds} = \boldsymbol{b} \cdot \kappa \boldsymbol{n} = 0,$$

so that:

$$\frac{d\boldsymbol{b}}{ds} \cdot \boldsymbol{t} = 0.$$

From this relation, we see that $d\boldsymbol{b}/ds$ is a vector that is perpendicular to both $\boldsymbol{b}$ and $\boldsymbol{t}$, that is, a vector that is directed along the principal normal. Its magnitude is given by Eq. (1.37) as τ. If we adopt the convention:

$$\frac{d\boldsymbol{b}}{ds} = -\tau \boldsymbol{n}$$

then Eq. (1.36) becomes:

$$-\tau + \boldsymbol{b} \cdot \frac{d\boldsymbol{n}}{ds} = 0.$$

Comparing this equation with Eq. (1.35) we see that $a_2 = \tau$, so that Eq. (1.32) becomes:

$$\frac{d\boldsymbol{n}}{ds} \equiv \boldsymbol{n}' = -\kappa \boldsymbol{t} + \tau \boldsymbol{b}. \tag{1.38}$$

The above results can be summarized as follows:

$$\boldsymbol{P}' = \boldsymbol{t} \tag{1.39}$$

$$\boldsymbol{t}' = \kappa \boldsymbol{n} \tag{1.40}$$

$$\boldsymbol{n}' = -\kappa \boldsymbol{t} + \tau \boldsymbol{b} \tag{1.41}$$

$$\boldsymbol{b}' = -\tau \boldsymbol{n}. \tag{1.42}$$

These equations are called the *Frenet-Serret* equations. The differential geometric properties of space curves are all derived from these equations.

1.2.4 Calculation of a Point on a Curve[6)]

Consider a curve that is expressed as a polynomial in the parameter t, such as the following cubic expression:

$$\boldsymbol{P}(t) = \boldsymbol{A} t^3 + \boldsymbol{B} t^2 + \boldsymbol{C} t + \boldsymbol{D} \quad (0 \leqq t \leqq 1). \tag{1.43}$$

A point on the curve can be calculated by dividing the t-interval up into n segments as follows:

$$\left.\begin{aligned} t &= 0, \delta, 2\delta, \ldots, n\delta \\ \delta &= 1/n \end{aligned}\right\} \tag{1.44}$$

and substituting into Eq. (1.43) n sequence. In this case it is convenient to change Eq. (1.43) into the following form:

$$\boldsymbol{P}(t) = ((\boldsymbol{A}t + \boldsymbol{B})t + \boldsymbol{C})t + \boldsymbol{D}. \tag{1.45}$$

Equation (1.43) requires 6 multiplications per coordinate per point, but Eq. (1.45) requires only 3, so computations can be performed more rapidly.

Next, let us consider a method of calculating points on a curve at high speed using no multiplications at all, by using finite difference formulas.

$$\begin{aligned}
\boldsymbol{P}(t) &= \boldsymbol{A}t^3 + \boldsymbol{B}t^2 + \boldsymbol{C}t + \boldsymbol{D} \\
\Delta\boldsymbol{P}(t) &= \boldsymbol{P}(t+\delta) - \boldsymbol{P}(t) \\
&= \boldsymbol{A}(t+\delta)^3 + \boldsymbol{B}(t+\delta)^2 + \boldsymbol{C}(t+\delta) + \boldsymbol{D} - \boldsymbol{A}t^3 - \boldsymbol{B}t^2 - \boldsymbol{C}t - \boldsymbol{D} \\
&= 3\boldsymbol{A}\delta t^2 + (3\boldsymbol{A}\delta^2 + 2\boldsymbol{B}\delta)t + \boldsymbol{A}\delta^3 + \boldsymbol{B}\delta^2 + \boldsymbol{C}\delta \\
\Delta^2\boldsymbol{P}(t) &= \Delta\boldsymbol{P}(t+\delta) - \Delta\boldsymbol{P}(t) \\
&= 3\boldsymbol{A}\delta(t+\delta)^2 + (3\boldsymbol{A}\delta^2 + 2\boldsymbol{B}\delta)(t+\delta) + \boldsymbol{A}\delta^3 + \boldsymbol{B}\delta^2 + \boldsymbol{C}\delta \\
&\quad - 3\boldsymbol{A}\delta t^2 - (3\boldsymbol{A}\delta^2 + 2\boldsymbol{B}\delta)t - \boldsymbol{A}\delta^3 - \boldsymbol{B}\delta^2 - \boldsymbol{C}\delta \\
&= 6\boldsymbol{A}\delta^2 t + 6\boldsymbol{A}\delta^3 + 2\boldsymbol{B}\delta^2 \\
\Delta^3\boldsymbol{P}(t) &= \Delta^2\boldsymbol{P}(t+\delta) - \Delta^2\boldsymbol{P}(t) \\
&= 6\boldsymbol{A}\delta^2(t+\delta) + 6\boldsymbol{A}\delta^3 + 2\boldsymbol{B}\delta^2 - 6\boldsymbol{A}\delta^2 t - 6\boldsymbol{A}\delta^3 - 2\boldsymbol{B}\delta^2 \\
&= 6\boldsymbol{A}\delta^3.
\end{aligned}$$

The 3rd-order finite difference is a constant vector that is independent of the parameter. In general, for an nth-order polynomial espression, the nth-order finite difference is a constant vector.

Let us use the notation $\boldsymbol{P}_0$, $\Delta\boldsymbol{P}_0$, $\Delta^2\boldsymbol{P}_0$, $\Delta^3\boldsymbol{P}_0$ and $\boldsymbol{P}_i$, $\Delta\boldsymbol{P}_i$, $\Delta^2\boldsymbol{P}_i$, $\Delta^3\boldsymbol{P}_i$ to denote the various order finite differences at $t=0$ and $t=i\delta$ respectively. Then we have the relations:

$$\left.\begin{aligned}
\boldsymbol{P}_1 &= \boldsymbol{P}_0 + \Delta\boldsymbol{P}_0 \\
\Delta\boldsymbol{P}_1 &= \Delta\boldsymbol{P}_0 + \Delta^2\boldsymbol{P}_0 \\
\Delta^2\boldsymbol{P}_1 &= \Delta^2\boldsymbol{P}_0 + \Delta^3\boldsymbol{P}_0 \\
\Delta^3\boldsymbol{P}_1 &= \Delta^3\boldsymbol{P}_0
\end{aligned}\right\} \tag{1.46}$$

so that if the various order finite differences at $t=0$, that is, $\boldsymbol{P}_0$, $\Delta\boldsymbol{P}_0$, $\Delta^2\boldsymbol{P}_0$, $\Delta^3\boldsymbol{P}_0$ are known, then the points on the curve at $t=\delta, 2\delta, \ldots, i\delta, \ldots$ can be found in sequence solely by repeated additions. However, if, for example,

$\delta = 1/2^5$, finite differences such as $\Delta^3 \boldsymbol{P}_0$ become vectors with very small magnitudes, so that in some cases the length of a computer word is not adequate to maintain sufficient accuracy. In such cases this problem can be solved by storing $\Delta \boldsymbol{P}_0$, $\Delta^2 \boldsymbol{P}_0$, $\Delta^3 \boldsymbol{P}_0$ as $1/\delta$ precision, $1/\delta^2$ precision, $1/\delta^3$ precision numbers, respectively, and applying Eqs. (1.46) in the form:

$$\left.\begin{aligned} \boldsymbol{P}_1 &= \boldsymbol{P}_0 + \delta\left(\frac{1}{\delta}\,\Delta \boldsymbol{P}_0\right) \\ \frac{1}{\delta}\,\Delta \boldsymbol{P}_1 &= \frac{1}{\delta}\,\Delta \boldsymbol{P}_0 + \delta\left(\frac{1}{\delta^2}\,\Delta^2 \boldsymbol{P}_0\right) \\ \frac{1}{\delta^2}\,\Delta^2 \boldsymbol{P}_1 &= \frac{1}{\delta^2}\,\Delta^2 \boldsymbol{P}_0 + \delta\left(\frac{1}{\delta^3}\,\Delta^3 \boldsymbol{P}_0\right) \\ \frac{1}{\delta^3}\,\Delta^3 \boldsymbol{P}_1 &= \frac{1}{\delta^3}\,\Delta^3 \boldsymbol{P}_0 \end{aligned}\right\} \qquad (1.47)$$

If we take $\delta = 1/2^m$ (where m is a positive integer) in the above equations, then the necessary computations can be performed at high speed using exclusively shift operations and additions, without time-consuming multiplications.

In calculations with Eqs. (1.47) it is convenient to express the various order finite differences at $t=0$ as a 4×3 matrix:

$$\begin{bmatrix} \boldsymbol{P}_0 \\ \frac{1}{\delta}\,\Delta \boldsymbol{P}_0 \\ \frac{1}{\delta^2}\,\Delta^2 \boldsymbol{P}_0 \\ \frac{1}{\delta^3}\,\Delta^3 \boldsymbol{P}_0 \end{bmatrix} = \begin{bmatrix} \boldsymbol{D} \\ \boldsymbol{A}\delta^2 + \boldsymbol{B}\delta + \boldsymbol{C} \\ 6\boldsymbol{A}\delta + 2\boldsymbol{B} \\ 6\boldsymbol{A} \end{bmatrix} \qquad (1.48)$$

In this book, the matrix in Eq. (1.48) will be referred to as the finite difference matrix.

1.2.5 Connection of Curve Segments

Let us now consider the problem of connecting a curve segment $\boldsymbol{P}_{\mathrm{II}}(t)$ $(0 \leqq t \leqq 1)$ to the curve segment $\boldsymbol{P}_{\mathrm{I}}(t)$ $(0 \leqq t \leqq 1)$ (Fig. 1.20).

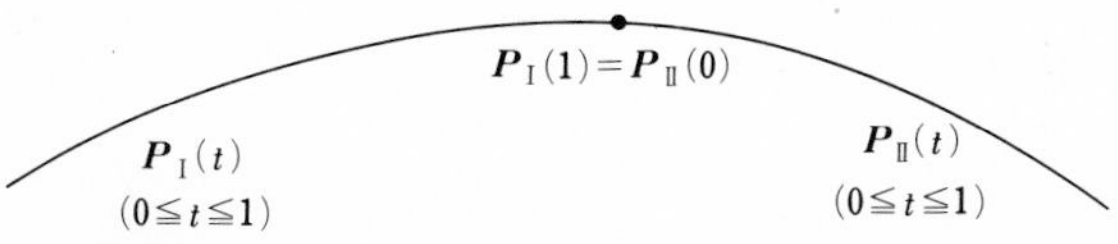

Fig. 1.20. Continuity conditions for curves. Conditions for continuity up to curvature:
① $\boldsymbol{P}_{\mathrm{II}}(0) = \boldsymbol{P}_{\mathrm{I}}(1)$; ② $\dot{\boldsymbol{P}}_{\mathrm{II}}(0) = \frac{\alpha_2}{\alpha_1}\,\dot{\boldsymbol{P}}_{\mathrm{I}}(1)$; ③ $\ddot{\boldsymbol{P}}_{\mathrm{II}}(0) = \left(\frac{\alpha_2}{\alpha_1}\right)^2 \ddot{\boldsymbol{P}}_{\mathrm{I}}(1) + \beta\,\dot{\boldsymbol{P}}_{\mathrm{I}}(1)$

First, we have the condition of continuity of position:

$$\boldsymbol{P}_{\mathrm{II}}(0) = \boldsymbol{P}_{\mathrm{I}}(1). \tag{1.49}$$

If the magnitudes of the tangent vectors of $\boldsymbol{P}_{\mathrm{I}}(t)$ and $\boldsymbol{P}_{\mathrm{II}}(t)$ at the connection point are α_1 and α_2 respectively, then the condition for the slope to be continuous is:

$$\frac{1}{\alpha_2}\dot{\boldsymbol{P}}_{\mathrm{II}}(0) = \frac{1}{\alpha_1}\dot{\boldsymbol{P}}_{\mathrm{I}}(1) = \boldsymbol{t} \tag{1.50}$$

or:

$$\dot{\boldsymbol{P}}_{\mathrm{II}}(0) = \frac{\alpha_2}{\alpha_1}\dot{\boldsymbol{P}}_{\mathrm{I}}(1) = \lambda\dot{\boldsymbol{P}}_{\mathrm{I}}(1) \quad \left(\lambda = \frac{\alpha_2}{\alpha_1}\right). \tag{1.51}$$

Next, let us find the condition for the centers of curvature of the two curve segments at the connection point to vary continuously. Equation (1.17) for the curvature is:

$$\boldsymbol{P}'' = \kappa\boldsymbol{n}.$$

In this equation κ is the magnitude of the curvature and $\boldsymbol{n}$ is the unit principal normal vector, pointing toward the center of curvature. From the above equation, for the center of curvature of the curve to vary smoothly at the connection point, it is necessary for both κ and $\boldsymbol{n}$ to be continuous.

If the binormal $\boldsymbol{b} = \boldsymbol{t} \times \boldsymbol{n}$ and $\boldsymbol{t}$ are continuous then $\boldsymbol{n}$ is also continuous. Consequently, to find the condition κ and $\boldsymbol{n}$ to be continuous, we require a condition for both κ and $\boldsymbol{b}$ to be continuous. From Eq. (1.27):

$$\kappa\boldsymbol{b} = \frac{\dot{\boldsymbol{P}} \times \ddot{\boldsymbol{P}}}{|\dot{\boldsymbol{P}}|^3}.$$

Consequently, in order for the center of curvature of a curve to vary continuously at a connection point, we must have:

$$\frac{\dot{\boldsymbol{P}}_{\mathrm{II}}(0) \times \ddot{\boldsymbol{P}}_{\mathrm{II}}(0)}{|\dot{\boldsymbol{P}}_{\mathrm{II}}(0)|^3} = \frac{\dot{\boldsymbol{P}}_{\mathrm{I}}(1) \times \ddot{\boldsymbol{P}}_{\mathrm{I}}(1)}{|\dot{\boldsymbol{P}}_{\mathrm{I}}(1)|^3}.$$

Substituting the relation in Eq. (1.51) into this equation and rearranging gives:

$$\dot{\boldsymbol{P}}_{\mathrm{I}}(1) \times \left[\ddot{\boldsymbol{P}}_{\mathrm{II}}(0) - \left(\frac{\alpha_2}{\alpha_1}\right)^2 \ddot{\boldsymbol{P}}_{\mathrm{I}}(1)\right] = \boldsymbol{0}.$$

From this equation, the following equation is obtained, with β an arbitrary scalar[7].

$$\ddot{P}_{II}(0) = \left(\frac{\alpha_2}{\alpha_1}\right)^2 \ddot{P}_I(1) + \beta \dot{P}_I(1)$$

$$= \lambda^2 \ddot{P}_I(1) + \beta \dot{P}_I(1) \quad \left(\lambda = \frac{\alpha_2}{\alpha_1}\right). \tag{1.52}$$

1.2.6 Parameter Transformation

Normally, a curve segment $\boldsymbol{P}(t)$ is normalized so that $t=0$ at one end and $t=1$ at the other end. If we are to use only one part of this curve, from $t=t_0$ to $t=t_1$, or perform some processing on this part, it is convenient to normalize the parameter over this part of the curve (refer to Fig. 1.21). In

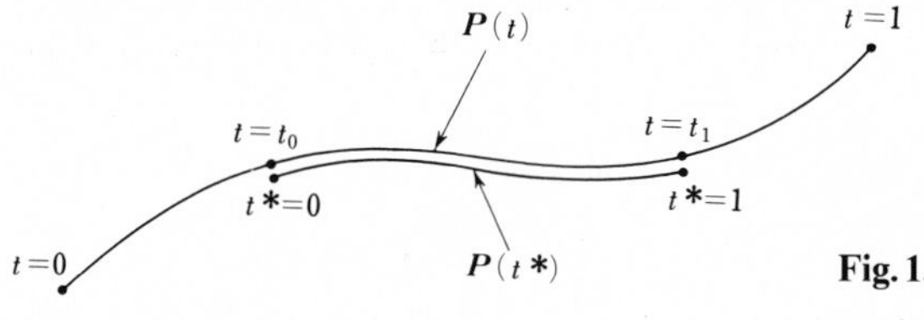

Fig. 1.21. Curve parameter transformation

order to keep the degree of the curve unchanged when performing this transformation, the following linear relation must hold between the old parameter t and the parameter after transformation t^*.

$$t = (1-t^*)\,t_0 + t^*\,t_1 = t_0 + (t_1 - t_0)\,t^*. \tag{1.53}$$

Substituting this relation for t into the original equation of the curve gives the normalized curve equation in terms of t^*, with t^* varying from 0 to 1 over the interval of interest. For example, in the case of the cubic curve:

$$\boldsymbol{P}(t) = [t^3 \;\; t^2 \;\; t \;\; 1] \; [\boldsymbol{A} \;\; \boldsymbol{B} \;\; \boldsymbol{C} \;\; \boldsymbol{D}]^T \tag{1.54}$$

the normalized curve equation over the interval $t_0 \leqq t \leqq t_1$ is:

$$\boldsymbol{P}(t^*) = [\{t_0 + (t_1 - t_0)\,t^*\}^3 \quad \{t_0 + (t_1 - t_0)\,t^*\}^2 \quad \{t_0 + (t_1 - t_0)\,t^*\} \quad 1] \begin{bmatrix} \boldsymbol{A} \\ \boldsymbol{B} \\ \boldsymbol{C} \\ \boldsymbol{D} \end{bmatrix}$$

$$= [(t_1 - t_0)^3 t^{*3} + 3(t_1 - t_0)^2 t_0 t^{*2} + 3(t_1 - t_0) t_0^2 t^* + t_0^3 \quad (t_1 - t_0)^2 t^{*2} + 2(t_1 - t_0) t_0 t^* + t_0^2 \quad (t_1 - t_0) t^* + t_0 \quad 1] \begin{bmatrix} \boldsymbol{A} \\ \boldsymbol{B} \\ \boldsymbol{C} \\ \boldsymbol{D} \end{bmatrix} =$$

$$= [t^{*3}\ t^{*2}\ t^*\ 1] \begin{bmatrix} (t_1-t_0)^3 & 0 & 0 & 0 \\ 3(t_1-t_0)^2 t_0 & (t_1-t_0)^2 & 0 & 0 \\ 3(t_1-t_0)t_0^2 & 2(t_1-t_0)t_0 & t_1-t_0 & 0 \\ t_0^3 & t_0^2 & t_0 & 1 \end{bmatrix} \begin{bmatrix} \boldsymbol{A} \\ \boldsymbol{B} \\ \boldsymbol{C} \\ \boldsymbol{D} \end{bmatrix} \tag{1.55}$$

$$= [t^{*3}\ t^{*2}\ t^*\ 1] \begin{bmatrix} (t_1-t_0)^3 \boldsymbol{A} \\ 3(t_1-t_0)^2 t_0 \boldsymbol{A} + (t_1-t_0)^2 \boldsymbol{B} \\ 3(t_1-t_0)t_0^2 \boldsymbol{A} + 2(t_1-t_0)t_0 \boldsymbol{B} + (t_1-t_0)\boldsymbol{C} \\ t_0^3 \boldsymbol{A} + t_0^2 \boldsymbol{B} + t_0 \boldsymbol{C} + \boldsymbol{D} \end{bmatrix} \tag{1.56}$$

$$\equiv [t^{*3}\ t^{*2}\ t^*\ 1] \begin{bmatrix} \boldsymbol{A}^* \\ \boldsymbol{B}^* \\ \boldsymbol{C}^* \\ \boldsymbol{D}^* \end{bmatrix}.$$

Let us now see how the tangent vector of the curve changes under the parameter transformation of Eq. (1.53). The tangent vector after transformation becomes:

$$\frac{d\boldsymbol{P}}{dt^*} = \frac{d\boldsymbol{P}}{dt}\frac{dt}{dt^*} = (t_1-t_0)\frac{d\boldsymbol{P}}{dt}. \tag{1.57}$$

That is, the magnitude of the tangent vector is reduced in proportion to the reduction of the range through which the original parameter varies ($0 \leqq t_1 - t_0 \leqq 1$). For the second derivative vector we have:

$$\frac{d^2\boldsymbol{P}}{dt^{*2}} = (t_1-t_0)^2 \frac{d^2\boldsymbol{P}}{dt^2} \tag{1.58}$$

so that it is reduced in proportion to the square of the reduction of the parameter range.

Let us now consider a more general parameter transformation, the rational transformation given by the following equation:

$$t = \frac{at^* + b}{ct^* + 1}.$$

This transformation is sometimes called *bilinear transformation* or a *homographic transformation*. With $t^* = 0$ and $t^* = 1$ at $t = t_0$ and $t = t_1$, respectively, this transformation takes the following form, with η an arbitrary scalar:

$$t = \frac{(\eta t_1 - t_0)t^* + t_0}{(\eta - 1)t^* + 1}. \tag{1.59}$$

The shape within the old parameter range $t_0 \leqq t \leqq t_1$ is now expressed without shape change in terms of the new parameter which varies in the domain $0 \leqq t^* \leqq 1$. Compared to the transformation of Eq. (1.53), the transformation of Eq. (1.59) has an increased degree of freedom because η can be chosen freely. When $\eta=1$, the transformation of Eq. (1.59) reduces to the transformation of Eq. (1.53). As η varies, what changes is the relation between a point on the curve and the parameter value; the shape does not change.

In the special case $t_0=0$, $t_1=1$ we have:

$$t=\frac{\eta t^*}{(\eta-1)t^*+1}. \tag{1.60}$$

Equation (1.59) is frequently used in rational polynomial parameter transformations, because only the parameter range changes, while the shape and the degree of the equation remain unchanged. Equation (1.60) is used when one wishes to change only the relation between points on the curve and parameter values, without changing the parameter range.

When the following curve:

$$\left.\begin{aligned} x(t)&=\frac{a_{11}t^2+a_{21}t+a_{31}}{a_{13}t^2+a_{23}t+a_{33}} \\ y(t)&=\frac{a_{12}t^2+a_{22}t+a_{32}}{a_{13}t^2+a_{23}t+a_{33}} \end{aligned}\right\} \tag{1.61}$$

expressed in terms of rational polynomials in which both the numerator and denominators are quadratic expressions is transformed using Eq. (1.60), the result is similar in which both the numerators and denominators are quadratic expressions. Let us demonstrate this.

The homogeneous vector representation $\boldsymbol{P}(t)=[x(t)\ y(t)\ 1]$ for the rational polynomial curve expressed by Eqs. (1.61) is*):

$$w(t)\boldsymbol{P}(t)=[t^2\ t\ 1]\begin{bmatrix} a_{11} & a_{12} & a_{13} \\ a_{21} & a_{22} & a_{23} \\ a_{31} & a_{32} & a_{33} \end{bmatrix}. \tag{1.62}$$

Substituting the parameter relation of Eq. (1.60) into this equation gives:

$$w(t^*)\boldsymbol{P}(t^*)=\left[\left\{\frac{\eta t^*}{(\eta-1)t^*+1}\right\}^2 \ \frac{\eta t^*}{(\eta-1)t^*+1}\ 1\right]\begin{bmatrix} a_{11} & a_{12} & a_{13} \\ a_{21} & a_{22} & a_{23} \\ a_{31} & a_{32} & a_{33} \end{bmatrix} \doteq$$

*) As for homogeneous coordinates, refer to Newman & Sproull: Principles of Interactive Computer Graphics, Second edition, McGraw-Hill.

$$\doteq [\eta^2 t^{*2} \quad \eta t^* \{(\eta-1)t^*+1\} \quad \{(\eta-1)t^*+1\}^2] \begin{bmatrix} a_{11} & a_{12} & a_{13} \\ a_{21} & a_{22} & a_{23} \\ a_{31} & a_{32} & a_{33} \end{bmatrix}^{**)}$$

$$= [t^{*2} \; t^* \; 1] \begin{bmatrix} \eta^2 & \eta(\eta-1) & (\eta-1)^2 \\ 0 & \eta & 2(\eta-1) \\ 0 & 0 & 1 \end{bmatrix} \begin{bmatrix} a_{11} & a_{12} & a_{13} \\ a_{21} & a_{22} & a_{23} \\ a_{31} & a_{32} & a_{33} \end{bmatrix}. \tag{1.63}$$

Consequently, in terms of ordinary coordinates we have:

$$x(t^*) = \frac{\{\eta^2 a_{11} + \eta(\eta-1) a_{21} + (\eta-1)^2 a_{31}\} t^{*2} + \{\eta a_{21} + 2(\eta-1) a_{31}\} t^* + a_{31}}{\{\eta^2 a_{13} + \eta(\eta-1) a_{23} + (\eta-1)^2 a_{33}\} t^{*2} + \{\eta a_{23} + 2(\eta-1) a_{33}\} t^* + a_{33}}$$

$$y(t^*) = \frac{\{\eta^2 a_{12} + \eta(\eta-1) a_{22} + (\eta-1)^2 a_{32}\} t^{*2} + \{\eta a_{22} + 2(\eta-1) a_{32}\} t^* + a_{32}}{\{\eta^2 a_{13} + \eta(\eta-1) a_{23} + (\eta-1)^2 a_{33}\} t^{*2} + \{\eta a_{23} + 2(\eta-1) a_{33}\} t^* + a_{33}}.$$

Next, let us look at the change in the tangent vector under this parameter transformation. The tangent vector after the transformation is:

$$\frac{d\boldsymbol{P}}{dt^*} = \frac{d\boldsymbol{P}}{dt} \frac{dt}{dt^*}.$$

From Eq. (1.60), we have:

$$\frac{dt}{dt^*} = \frac{\eta}{\{(\eta-1)t^*+1\}^2}$$

$$\therefore \frac{d\boldsymbol{P}}{dt^*} = \frac{\eta}{\{(\eta-1)t^*+1\}^2} \frac{d\boldsymbol{P}}{dt}. \tag{1.64}$$

Therefore, the tangent vectors at points $t^*=0$ and $t^*=1$ are, respectively:

$$\left.\frac{d\boldsymbol{P}}{dt^*}\right|_{t^*=0} = \eta \left.\frac{d\boldsymbol{P}}{dt}\right|_{t=0}, \qquad \left.\frac{d\boldsymbol{P}}{dt^*}\right|_{t^*=1} = \frac{1}{\eta} \left.\frac{d\boldsymbol{P}}{dt}\right|_{t=1}. \tag{1.65}$$

As a result of using transformation (1.60), the shape is unchanged, the range within which the parameter varies remains the same, and the variation of a point on the curve for a given change in the parameter is changed. As a result, the magnitudes of the tangent vectors at both ends are changed, but the product of the magnitudes of the two tangent vectors remains the same. That is:

**) Multiplying a homogeneous vector by a nonzero scalar constant does not alter its ordinary coordinate vector. We shall define the equality in ordinary coordinate vector with the symbol $\doteq$.

$$\left.\frac{dP}{dt^*}\right|_{t^*=0} \cdot \left.\frac{dP}{dt^*}\right|_{t^*=1} = \left.\frac{dP}{dt}\right|_{t=0} \cdot \left.\frac{dP}{dt}\right|_{t=1} . \tag{1.66}$$

1.2.7 Partitioning of a Curve Segment

Let us consider a case in which a curve $\boldsymbol{P}(t)$ is split into two curves $\boldsymbol{P}_1(u)$ $(0 \leqq u \leqq 1)$ and $\boldsymbol{P}_2(u)$ $(0 \leqq u \leqq 1)$ at a point corresponding to $t = t_s$. We will demonstrate this using the cubic curve (1.43) (Fig. 1.22).

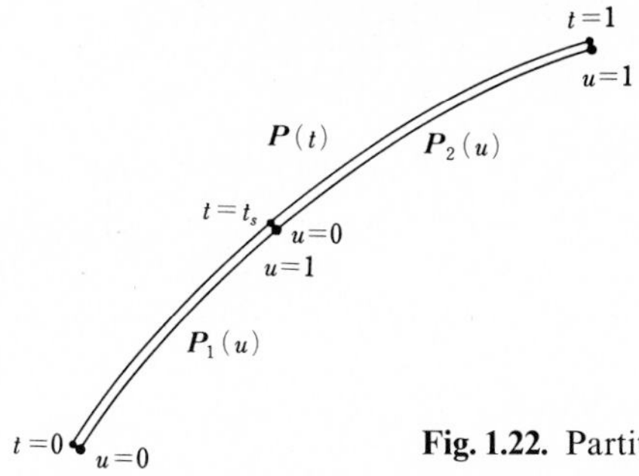

Fig. 1.22. Partitioning of a curve segment

The first curve $\boldsymbol{P}_1(u)$ is obtained by setting $t_0 = 0$, $t_1 = t_s$ and $t^* = u$ in Eq. (1.56). This gives the following expression:

$$\boldsymbol{P}_1(u) = [u^3 \; u^2 \; u \; 1] \begin{bmatrix} t_s^3 \boldsymbol{A} \\ t_s^2 \boldsymbol{B} \\ t_s \boldsymbol{C} \\ \boldsymbol{D} \end{bmatrix} . \tag{1.67}$$

The second curve $\boldsymbol{P}_2(u)$ is obtained by setting $t_0 = t_s$, $t_1 = 1$ and $t^* = u$ in Eq. (1.56) to give the expression:

$$\boldsymbol{P}_2(u) = [u^3 \; u^2 \; u \; 1] \begin{bmatrix} (1-t_s)^3 \boldsymbol{A} \\ 3(1-t_s)^2 t_s \boldsymbol{A} + (1-t_s)^2 \boldsymbol{B} \\ 3(1-t_s) t_s^2 \boldsymbol{A} + 2(1-t_s) t_s \boldsymbol{B} + (1-t_s) \boldsymbol{C} \\ t_s^3 \boldsymbol{A} + t_s^2 \boldsymbol{B} + t_s \boldsymbol{C} + \boldsymbol{D} \end{bmatrix} . \tag{1.68}$$

Partitioning the curve in this way makes it possible to apply local control to its shape.

1.2.8 Parametric Cubic Curves

The parametric cubic curve segment (1.43) is completely determined by setting the parameter equal to 0, the position vector to $\boldsymbol{Q}_0$ and the tangent vector to $\dot{\boldsymbol{Q}}_0$ at one end, and the parameter to 1, the position vector to $\boldsymbol{Q}_1$ and the tangent vector to $\dot{Q}_1$ at the other end (for further details refer to Sect. 3.2.1). This gives:

$$\begin{aligned}
A &= 2(Q_0 - Q_1) + \dot{Q}_0 + \dot{Q}_1 \\
B &= 3(Q_1 - Q_0) - 2\dot{Q}_0 - \dot{Q}_1 \\
C &= \dot{Q}_0 \\
D &= Q_0 .
\end{aligned}$$

If we have two curve segments such that the two end points of one curve and the directions of the tangent vectors are the same as those of the other,

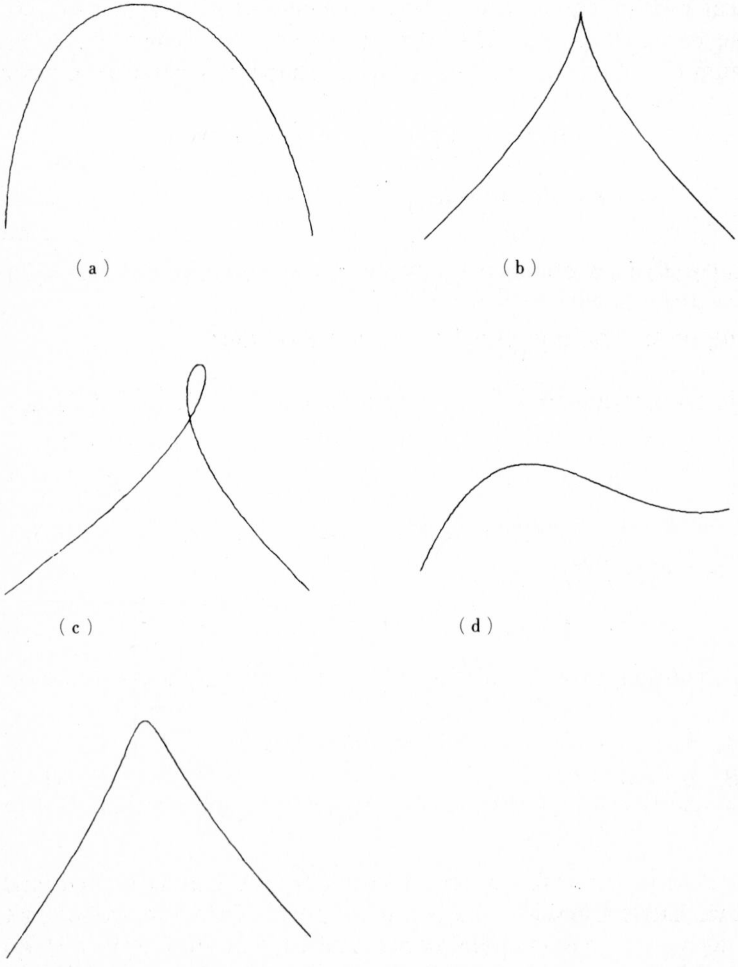

Fig. 1.23. Shapes of parametric cubic curves. (a) case with no abnormalities; (b) cusp; (c) loop; (d) one inflection point; (e) two inflection points

then, if the magnitudes of the tangent vectors at these end points are different, the shapes will be different. The magnitudes as well as directions of the tangent vectors affect the shapes of the curves.

Let us focus attention on a point of inflection of a curve segment. From Eq. (1.26), a point of inflection is a point at which:

$$\dot{\boldsymbol{P}} \times \ddot{\boldsymbol{P}} = \boldsymbol{0}.$$

If we exclude points at which $\dot{\boldsymbol{P}} = \boldsymbol{0}$, then a point of inflection is a point at which:

$$\ddot{\boldsymbol{P}} = g\dot{\boldsymbol{P}} \tag{1.69}$$

where g is a scalar. Since this is a second-degree equation with respect to t, we see that as the parameter varies from 0 to 1, a parametric cubic curve can have a maximum of 2 inflection points. A typical shape of a parametric cubic curve is shown in Fig. 1.23.

Next, let us look at the torsion of a parametric cubic curve.

$$\left.\begin{aligned} \dot{\boldsymbol{P}}(t) &= [\dot{x}\ \dot{y}\ \dot{z}] = 3\boldsymbol{A}t^2 + 2\boldsymbol{B}t + \boldsymbol{C} \\ \ddot{\boldsymbol{P}}(t) &= [\ddot{x}\ \ddot{y}\ \ddot{z}] = 6\boldsymbol{A}t + 2\boldsymbol{B} \\ \dddot{\boldsymbol{P}}(t) &= [\dddot{x}\ \dddot{y}\ \dddot{z}] = 6\boldsymbol{A} \end{aligned}\right\} . \tag{1.70}$$

Substituting these relations into Eq. (1.31), we obtain:

$$\begin{vmatrix} \dot{x} & \dot{y} & \dot{z} \\ \ddot{x} & \ddot{y} & \ddot{z} \\ \dddot{x} & \dddot{y} & \dddot{z} \end{vmatrix} = \begin{vmatrix} 3A_xt^2+2B_xt+C_x & 3A_yt^2+2B_yt+C_y & 3A_zt^2+2B_zt+C_z \\ 6A_xt+2B_x & 6A_yt+2B_y & 6A_zt+2B_z \\ 6A_x & 6A_y & 6A_z \end{vmatrix}$$

$$= -12 \begin{vmatrix} A_x & A_y & A_z \\ B_x & B_y & B_z \\ C_x & C_y & C_z \end{vmatrix}$$

Therefore, the condition for a parametric cubic curve to be a plane curve is:

$$\begin{vmatrix} A_x & A_y & A_z \\ B_x & B_y & B_z \\ C_x & C_y & C_z \end{vmatrix} = 0. \tag{1.71}$$

From this equation a curve for which $A_x = A_y = A_z = 0$, that is a quadratic parametric curve, has zero torsion and is a plane curve. Therefore, we see that the minimum degree that a polynomial curve must have in order to be a space curve is 3. In addition, even for a cubic curve, if the 2nd-degree coefficients (B_x, B_y, B_z) or the 1st-degree coefficients (C_x, C_y, C_z) are 0, the curve is a plane curve.

1.2.9 Length and Area of a Curve

A very small segment of a curve has approximately the same length as the chord which subtends that segment. The chord vector obtained by varying the parameter by Δt is approximately $\dot{\boldsymbol{P}}(t)\Delta t$, so the length l of the curve from the parameter value t_0 to t_1 can be calculated as:

$$l=\int_{t_0}^{t_1} |\dot{\boldsymbol{P}}(t)|\,dt. \tag{1.72}$$

If we assume that the curve is a plane curve that can be represented by Eq. (1.43), then:

$$\begin{aligned} l&=\int_{t_0}^{t_1} \sqrt{A_4t^4+A_3t^3+A_2t^2+A_1t+A_0}\,dt \\ &\equiv\int_{t_0}^{t_1} f(t)\,dt. \end{aligned} \tag{1.73}$$

Here:

$$\begin{aligned} &f(t)=\sqrt{A_4t^4+A_3t^3+A_2t^2+A_1t+A_0}\\ &A_4=9(A_x^2+A_y^2)\\ &A_3=12(A_xB_x+A_yB_y)\\ &A_2=6(A_xC_x+A_yC_y)+4(B_x^2+B_y^2)\\ &A_1=4(B_xC_x+B_yC_y)\\ &A_0=C_x^2+C_y^2. \end{aligned}$$

We apply the following parameter transformation:

$$t=t_0+(t_1-t_0)u$$

in order to be able to use the Gauss quadrature method. The transformation is applied to Eq. (1.73) to give:

$$\begin{aligned} l&=\int_{t_0}^{t_1} f(t)\,dt=(t_1-t_0)\int_0^1 f(t_0+(t_1-t_0)u)\,du=(t_1-t_0)\int_0^1 g(u)\,du \\ &\doteqdot(t_1-t_0)\sum_{i=1}^{m} w_i g(u_i). \end{aligned} \tag{1.74}$$

In this equation, m is the number of knots in Gauss' formula, the u_i are the knots and the w_i are weight factors.

Next, in the case of a plane curve, let us find the area of the "pie slice" formed by the origin and a curve segment (Fig. 1.24). The pie slice defined by

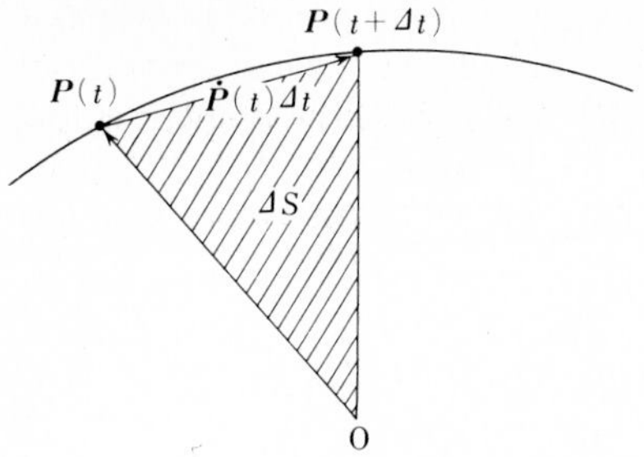

Fig. 1.24. Area enclosed by a curve

the origin and the two points on the curve $\boldsymbol{P}(t)$ and $\boldsymbol{P}(t+\Delta t)$ is approximated by the triangle $O-\boldsymbol{P}(t)-\boldsymbol{P}(t+\Delta t)$. Since this area ΔS is $1/2\,|\boldsymbol{P}(t)\times\dot{\boldsymbol{P}}(t)|\,\Delta t$, the area S swept out by the curve as the parameter varies from t_0 to t_1 is:

$$S=\frac{1}{2}\int_{t_0}^{t_1}|\boldsymbol{P}(t)\times\dot{\boldsymbol{P}}(t)|\,dt. \tag{1.75}$$

1.2.10 Intersection of a Curve with a Plane

A plane is represented by the equation:

$$ax+by+cz+d=0.$$

Since the curve $\boldsymbol{P}(t)=[x(t)\;\; y(t)\;\; z(t)]$ satisfies this equation, we have:

$$ax(t)+by(t)+cz(t)+d=0. \tag{1.76}$$

The roots of this equation can be found by the Newton-Raphson method. If the curve $\boldsymbol{P}(t)$ has been normalized in the range $0\leqq t\leqq 1$, the roots will be confined to this interval. The calculation can be made very efficient by pointing at the intersection points on the display with a stylus to obtain the first iterations.

1.2.11 Intersection of Two Curves

For simplicity let us restrict ourselves to the problem of finding intersections of two plane curves

$$\boldsymbol{P}_1(t)=[x_1(t)\; y_1(t)\; z_1(t)] \quad\text{and}\quad \boldsymbol{P}_2(u)=[x_2(u)\; y_2(u)\; z_2(u)].$$

In this case, we have:

$$\left.\begin{array}{l} x_1(t)=x_2(u) \\ y_1(t)=y_2(u) \end{array}\right\} \qquad (0\leqq t\leqq 1,\; 0\leqq u\leqq 1). \tag{1.77}$$

Let us seek solutions of the simultaneous equations for the two unknowns t and u $(0 \leqq t \leqq 1,\ 0 \leqq u \leqq 1)$ by the Newton-Raphson method. The calculation can be made very efficient by pointing at the intersection points on the display with a stylus to obtain the first iterations.

If both curves are given by cubic polynomials of the form (1.43), then Eq. (1.77) becomes the following:

$$\left.\begin{array}{l} A_{1x}t^3+B_{1x}t^2+C_{1x}t+D_{1x}-(A_{2x}u^3+B_{2x}u^2+C_{2x}u+D_{2x})=0 \\ A_{1y}t^3+B_{1y}t^2+C_{1y}t+D_{1y}-(A_{2y}u^3+B_{2y}u^2+C_{2y}u+D_{2y})=0 \end{array}\right\}. \tag{1.78}$$

1.3 Theory of Surfaces

1.3.1 Parametric Representation of Surfaces

Suppose that a curved surface is described by a polynomial in terms of the two parameters u and w which vary on its surface:

$$\boldsymbol{P}(u, w)=[x(u, w) \;\; y(u, w) \;\; z(u, w)]. \tag{1.79}$$

If a curved surface satisfies the conditions that the three Jacobians:

$$J_x \equiv \frac{\partial(y, z)}{\partial(u, w)}\text{*)}, \qquad J_y \equiv \frac{\partial(z, x)}{\partial(u, w)}, \qquad J_z \equiv \frac{\partial(x, y)}{\partial(u, w)}$$

are not all zero at the same time, in other words, that:

$$J_x^2+J_y^2+J_z^2 \neq 0 \tag{1.80}$$

that the curved surface $\boldsymbol{P}(u, w)$ has all derivatives with respect to x, y and z up to r-th order, and, moreover, that all such derivatives are continuous, then the curved surface defined by Eq. (1.79) is said to be of class C^r.

Condition (1.80) is the condition for the curved surface to not degenerate to a point or a curve, and to not contain any singular points such as spikes. This condition requires constraints on both the curve itself and the parameters. Even for the same curve, it is possible that for one parameter expression condition (1.80) will be satisfied, and for another parameter expression it will not.**)

---

*) This notation is defined by:

$$\frac{\partial(y, z)}{\partial(u, w)} = \begin{vmatrix} \dfrac{\partial y}{\partial u} & \dfrac{\partial z}{\partial u} \\[2ex] \dfrac{\partial y}{\partial w} & \dfrac{\partial z}{\partial w} \end{vmatrix}$$

**) For example, if the xy plane is expressed as $x=u$, $y=w$, $z=0$ then $J_z=1$, but if the same plane is expressed as $x=u^3$, $y=w^3$, $z=0$ then at the origin $J_x=J_y=J_z=0$.

In the discussion that follows, unless otherwise stated it is assumed that condition (1.80) holds.

If $w = w_0$ (fixed), the curved surface representation describes a curve $\boldsymbol{P}(u, w_0)$ on the curved surface. This curve is called the u-curve. Similarly, the curve $\boldsymbol{P}(u_0, w)$ on the curved surface corresponding to $u = u_0$ (fixed) is called the w-curve.

At a point $\boldsymbol{P}(u_0, w_0)$ on the curved surface, the tangent vector in the direction of the u-curve is:

$$\boldsymbol{P}_u = \left[\frac{\partial x(u, w_0)}{\partial u} \quad \frac{\partial y(u, w_0)}{\partial u} \quad \frac{\partial z(u, w_0)}{\partial u}\right]_{u=u_0}.$$

Similarly, the tangent vector in the direction of the w-curve is:

$$\boldsymbol{P}_w = \left[\frac{\partial x(u_0, w)}{\partial w} \quad \frac{\partial y(u_0, w)}{\partial w} \quad \frac{\partial z(u_0, w)}{\partial w}\right]_{w=w_0}.$$

If these two vectors are drawn starting from the point $\boldsymbol{P}(u_0, w_0)$, if condition (1.80) is satisfied the angle between them is non-zero (Fig. 1.25). That is,

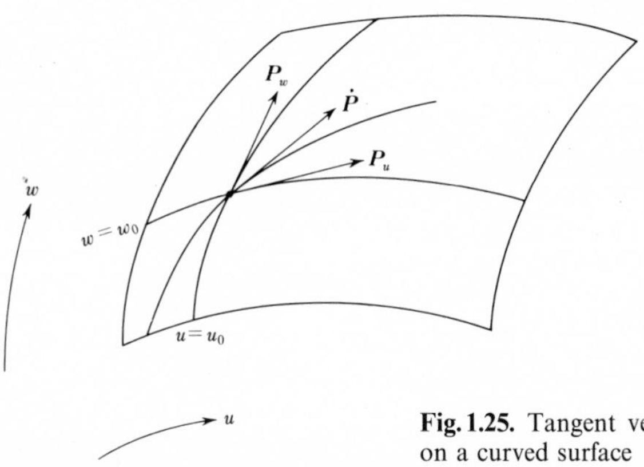

Fig. 1.25. Tangent vectors in two parameter directions on a curved surface

condition (1.80) is equivalent to saying that the vectors $\boldsymbol{P}_u$ and $\boldsymbol{P}_w$ are linearly independent at each point on the curved surface. Condition (1.80) is also equivalent to $\boldsymbol{P}_u \times \boldsymbol{P}_w \neq \boldsymbol{0}$ *).

A curve in an arbitrary direction on a curved surface can be expressed in terms of a parameter t as $\boldsymbol{P}(u(t), w(t))$ $(a \leqq t \leqq b)$. Therefore, at point $\boldsymbol{P}(u_0, w_0)$, the tangent vector in the direction of that curve becomes:

$$\left.\frac{d\boldsymbol{P}}{dt}\right|_{t=t_0} = \boldsymbol{P}_u \left.\frac{du}{dt}\right|_{t=t_0} + \boldsymbol{P}_w \left.\frac{dw}{dt}\right|_{t=t_0}. \tag{1.81}$$

*) Note that $(\boldsymbol{P}_u \times \boldsymbol{P}_w)^2 = J_x^2 + J_y^2 + J_z^2$.

Note that $u_0 = u(t_0)$ and $w_0 = w(t_0)$. The above formula expresses the fact that a tangent vector in an arbitrary direction on the curved surface lies on the plane formed by $\boldsymbol{P}_u$ and $\boldsymbol{P}_w$. This plane is called the tangent plane at the point $\boldsymbol{P}(u_0, w_0)$. Letting $\boldsymbol{R}$ be a position vector on the tangent plane, the tangent plane is given by the following equation.

$$[\boldsymbol{R} - \boldsymbol{P}(u_0, w_0), \boldsymbol{P}_u, \boldsymbol{P}_w] = 0. \tag{1.82}$$

The straight line that passes through point $\boldsymbol{P}(u_0, w_0)$ and is perpendicular to the tangent plane at that point is called the normal line. The normal vector $\boldsymbol{e}$ can be expressed in terms of $\boldsymbol{P}_u$ and $\boldsymbol{P}_w$ in a right-hand systems as follows:

$$\boldsymbol{e} = \frac{\boldsymbol{P}_u \times \boldsymbol{P}_w}{\sqrt{(\boldsymbol{P}_u \times \boldsymbol{P}_w)^2}} = \frac{\boldsymbol{P}_u \times \boldsymbol{P}_w}{\sqrt{J_x^2 + J_y^2 + J_z^2}}. \tag{1.83}$$

$\boldsymbol{e}$ is called the unit normal vector.

1.3.2 The First Fundamental Matrix of a Surface

As was stated in the preceding section, a curve on the curved surface $\boldsymbol{P}(u, w)$ can be given by $u = u(t)$, $w = w(t)$, $(a \leqq t \leqq b)$. The following notation is used for this.

$$\boldsymbol{u}(t) = [u(t)\ w(t)]. \tag{1.84}$$

From Eq. (1.81), the tangent vector to this curve is given by:

$$\begin{aligned} \dot{\boldsymbol{P}} &= \boldsymbol{P}_u \dot{u} + \boldsymbol{P}_w \dot{w} = [\dot{u}\ \dot{w}] \begin{bmatrix} \boldsymbol{P}_u \\ \boldsymbol{P}_w \end{bmatrix} \\ &\equiv \dot{\boldsymbol{u}} \boldsymbol{A}. \end{aligned} \tag{1.85}$$

Here $\boldsymbol{A}$ is defined by:

$$\boldsymbol{A} = \begin{bmatrix} \boldsymbol{P}_u \\ \boldsymbol{P}_w \end{bmatrix}. \tag{1.86}$$

The square of the magnitude $\dot{s}$ of the tangent vector is given by:

$$\begin{aligned} \dot{s}^2 &= \dot{\boldsymbol{P}}(t)^2 = \dot{\boldsymbol{P}}(t) \cdot \dot{\boldsymbol{P}}(t)^T \\ &= [\dot{u}\ \dot{w}] \begin{bmatrix} \boldsymbol{P}_u \\ \boldsymbol{P}_w \end{bmatrix} [\boldsymbol{P}_u\ \boldsymbol{P}_w] \begin{bmatrix} \dot{u} \\ \dot{w} \end{bmatrix} = [\dot{u}\ \dot{w}] \begin{bmatrix} \boldsymbol{P}_u^2 & \boldsymbol{P}_u \cdot \boldsymbol{P}_w \\ \boldsymbol{P}_u \cdot \boldsymbol{P}_w & \boldsymbol{P}_w^2 \end{bmatrix} \begin{bmatrix} \dot{u} \\ \dot{w} \end{bmatrix} \\ &\equiv \dot{\boldsymbol{u}} \boldsymbol{F} \dot{\boldsymbol{u}}^T. \end{aligned} \tag{1.87}$$

Here $\boldsymbol{F}$ is the matrix defined by:

$$\boldsymbol{F}=\begin{bmatrix} \boldsymbol{P}_u^2 & \boldsymbol{P}_u\cdot\boldsymbol{P}_w \\ \boldsymbol{P}_u\cdot\boldsymbol{P}_w & \boldsymbol{P}_w^2 \end{bmatrix}. \tag{1.88}$$

This matrix $\boldsymbol{F}$ is called the first fundamental matrix of the curved surface.

Using the first fundamental matrix, the unit tangent vector $\boldsymbol{t}$ of a curve on a curved surface is:

$$\boldsymbol{t}=\frac{\dot{\boldsymbol{P}}(t)}{|\dot{\boldsymbol{P}}(t)|}=\frac{\dot{\boldsymbol{u}}\boldsymbol{A}}{(\dot{\boldsymbol{u}}\boldsymbol{F}\dot{\boldsymbol{u}}^T)^{\frac{1}{2}}}. \tag{1.89}$$

Therefore, the cosine of the angle θ between the two curves $\boldsymbol{u}_1=\boldsymbol{u}_1(t)$ and $\boldsymbol{u}_2=\boldsymbol{u}_2(t)$ on the curved surface is, using (1.89):

$$\cos\theta=\boldsymbol{t}_1\cdot\boldsymbol{t}_2=\frac{\dot{\boldsymbol{u}}_1\boldsymbol{A}\cdot\boldsymbol{A}^T\dot{\boldsymbol{u}}_2^T}{(\dot{\boldsymbol{u}}_1\boldsymbol{F}\dot{\boldsymbol{u}}_1^T)^{\frac{1}{2}}(\dot{\boldsymbol{u}}_2\boldsymbol{F}\dot{\boldsymbol{u}}_2^T)^{\frac{1}{2}}}=\frac{\dot{\boldsymbol{u}}_1\boldsymbol{F}\dot{\boldsymbol{u}}_2^T}{(\dot{\boldsymbol{u}}_1\boldsymbol{F}\dot{\boldsymbol{u}}_1^T)^{\frac{1}{2}}(\dot{\boldsymbol{u}}_2\boldsymbol{F}\dot{\boldsymbol{u}}_2^T)^{\frac{1}{2}}}. \tag{1.90}$$

The length of a curve on the curved surface is given by:

$$l=\int_{t_0}^{t_1}\dot{s}\,dt=\int_{t_0}^{t_1}(\dot{\boldsymbol{u}}\boldsymbol{F}\dot{\boldsymbol{u}}^T)^{\frac{1}{2}}\,dt. \tag{1.91}$$

Next, let us consider a small area of a curved surface surrounded by the parametric curves $u=u_0$, $u=u_0+\Delta u$, $w=w_0$, $w=w_0+\Delta w$. Referring to Fig. 1.26, we see that the area ΔS of this small segment of surface is:

$$\Delta S\risingdotseq|\boldsymbol{P}_u\times\boldsymbol{P}_w|\,\Delta u\,\Delta w.$$

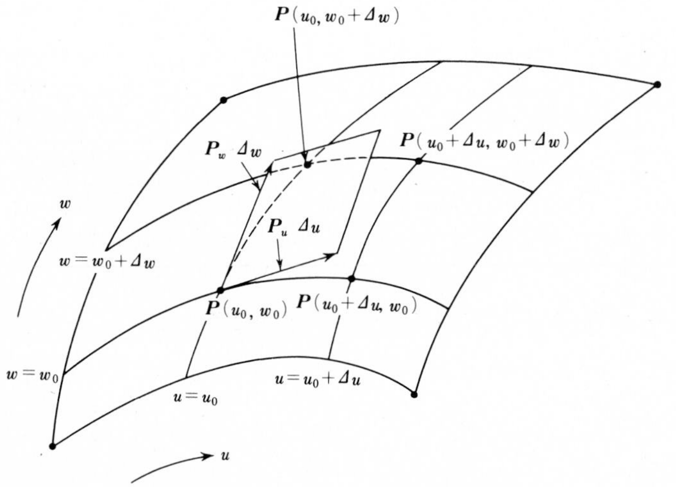

Fig. 1.26. Area of an infinitesimal curved surface patch

We also have:

$$|\boldsymbol{P}_u \times \boldsymbol{P}_w|^2 = \boldsymbol{P}_u^2 \boldsymbol{P}_w^2 - (\boldsymbol{P}_u \cdot \boldsymbol{P}_w)^{2\,*)} = |\boldsymbol{F}|. \tag{1.92}$$

Therefore, the curved surface area corresponding to the region R on the uw plane is:

$$S = \iint_R |\boldsymbol{F}|^{\frac{1}{2}}\, du\, dw. \tag{1.93}$$

1.3.3 Determining Conditions for a Tangent Vector to a Curve on a Surface

As shown in the preceding section, since the unit tangent vector of a curve on a curved surface is given by Eq. (1.89), the condition that determines the unit tangent vector $\boldsymbol{t}$ is:

$$\dot{\boldsymbol{u}} \boldsymbol{F} \dot{\boldsymbol{u}}^T > 0. \tag{1.94}$$

In the case $\boldsymbol{P}_u^2 \neq 0$, changing the form of the equation slightly gives:

$$\begin{aligned} \dot{\boldsymbol{u}} \boldsymbol{F} \dot{\boldsymbol{u}}^T &= [\dot{u}\ \dot{w}] \begin{bmatrix} \boldsymbol{P}_u^2 & \boldsymbol{P}_u \cdot \boldsymbol{P}_w \\ \boldsymbol{P}_u \cdot \boldsymbol{P}_w & \boldsymbol{P}_w^2 \end{bmatrix} \begin{bmatrix} \dot{u} \\ \dot{w} \end{bmatrix} = \boldsymbol{P}_u^2 \dot{u}^2 + 2 \boldsymbol{P}_u \cdot \boldsymbol{P}_w \dot{u} \dot{w} + \boldsymbol{P}_w^2 \dot{w}^2 \\ &= \boldsymbol{P}_u^2 \left(\dot{u} + \frac{\boldsymbol{P}_u \cdot \boldsymbol{P}_w}{\boldsymbol{P}_u^2} \dot{w} \right)^2 + \frac{\boldsymbol{P}_u^2 \boldsymbol{P}_w^2 - (\boldsymbol{P}_u \cdot \boldsymbol{P}_w)^2}{\boldsymbol{P}_u^2} \dot{w}^2 \\ &= \boldsymbol{P}_u^2 \left(\dot{u} + \frac{\boldsymbol{P}_u \cdot \boldsymbol{P}_w}{\boldsymbol{P}_u^2} \dot{w} \right)^2 + \frac{|\boldsymbol{P}_u \times \boldsymbol{P}_w|^2}{\boldsymbol{P}_u^2} \dot{w}^2 > 0 \quad (\boldsymbol{P}_u^2 \neq 0). \end{aligned} \tag{1.95}$$

This equation implies that if the conditions

$$\dot{\boldsymbol{u}} \neq \boldsymbol{0} \quad \text{and} \quad \boldsymbol{P}_u \times \boldsymbol{P}_w \neq \boldsymbol{0} \tag{1.96}$$

are satisfied, the unit tangent vectors are determined for all curves $\boldsymbol{u} = \boldsymbol{u}(t)$. If the second condition is satisfied but the first is not, then, as can be seen from Eq. (1.81) and (1.83), the tangent plane and normal line are determined, but the unit tangent vector $\boldsymbol{t}$ is not determined due to twisting of the curve $\boldsymbol{u} = \boldsymbol{u}(t)$. If the first condition is satisfied but the second is not, there can be a cusp or the u-curve and the w-curve can be parallel, with the result that the tangent plane is undetermined[7].

*) Using the vector identity $(\boldsymbol{A} \times \boldsymbol{B})(\boldsymbol{C} \times \boldsymbol{D}) = (\boldsymbol{AC})(\boldsymbol{BD}) - (\boldsymbol{AD})(\boldsymbol{BC})$.

1.3.4 Curvature of a Surface

From Eq. (1.9) we have:

$$\dot{\boldsymbol{P}} = \dot{s}\boldsymbol{P}' = \dot{s}\boldsymbol{t}.$$

Differentiating by t we obtain the equation:

$$\begin{aligned}\ddot{\boldsymbol{P}} &= \ddot{s}\boldsymbol{t} + \dot{s}\dot{\boldsymbol{t}} = \ddot{s}\boldsymbol{t} + \dot{s}^2\boldsymbol{t}' \\ &= \ddot{s}\boldsymbol{t} + \dot{s}^2\kappa\boldsymbol{n}.\end{aligned} \tag{1.97}$$

Now let us find $\ddot{\boldsymbol{P}}$ for the curve $\boldsymbol{u}(t) = [u(t)\ w(t)]$ on the curved surface $\boldsymbol{P}(u, w)$. From Eq. (1.81) we have:

$$\dot{\boldsymbol{P}} = \boldsymbol{P}_u\dot{u} + \boldsymbol{P}_w\dot{w}.$$

Therefore:

$$\begin{aligned}\ddot{\boldsymbol{P}} &= \frac{d\boldsymbol{P}_u}{dt}\dot{u} + \boldsymbol{P}_u\ddot{u} + \frac{d\boldsymbol{P}_w}{dt}\dot{w} + \boldsymbol{P}_w\ddot{w} \\ &= \boldsymbol{P}_{uu}\dot{u}^2 + \boldsymbol{P}_{uw}\dot{u}\dot{w} + \boldsymbol{P}_u\ddot{u} + \boldsymbol{P}_{ww}\dot{w}^2 + \boldsymbol{P}_{wu}\dot{w}\dot{u} + \boldsymbol{P}_w\ddot{w} \\ &= \boldsymbol{P}_{uu}\dot{u}^2 + \boldsymbol{P}_{uw}\dot{u}\dot{w} + \boldsymbol{P}_{wu}\dot{u}\dot{w} + \boldsymbol{P}_{ww}\dot{w}^2 + \boldsymbol{P}_u\ddot{u} + \boldsymbol{P}_w\ddot{w}.\end{aligned} \tag{1.98}$$

Taking the inner product of the unit normal vector $\boldsymbol{e}$ to the curved surface and $\ddot{\boldsymbol{P}}$ from Eq. (1.97), since $\boldsymbol{e}$ and $\boldsymbol{t}$ are perpendicular with each other:

$$\boldsymbol{e} \cdot \ddot{\boldsymbol{P}} = \dot{s}^2\kappa\boldsymbol{n} \cdot \boldsymbol{e}. \tag{1.99}$$

Also taking the inner product of the unit normal vector $\boldsymbol{e}$ and $\ddot{\boldsymbol{P}}$ from Eq. (1.98), since $\boldsymbol{e}$ is perpendicular to both $\boldsymbol{P}_u$ and $\boldsymbol{P}_w$:

$$\begin{aligned}\boldsymbol{e} \cdot \ddot{\boldsymbol{P}} &= \boldsymbol{e} \cdot \boldsymbol{P}_{uu}\dot{u}^2 + \boldsymbol{e} \cdot \boldsymbol{P}_{uw}\dot{u}\dot{w} + \boldsymbol{e} \cdot \boldsymbol{P}_{wu}\dot{u}\dot{w} + \boldsymbol{e} \cdot \boldsymbol{P}_{ww}\dot{w}^2 \\ &= [\dot{u}\ \dot{w}]\begin{bmatrix}\boldsymbol{e} \cdot \boldsymbol{P}_{uu} & \boldsymbol{e} \cdot \boldsymbol{P}_{uw} \\ \boldsymbol{e} \cdot \boldsymbol{P}_{wu} & \boldsymbol{e} \cdot \boldsymbol{P}_{ww}\end{bmatrix}\begin{bmatrix}\dot{u} \\ \dot{w}\end{bmatrix} \\ &\equiv \dot{\boldsymbol{u}}\boldsymbol{G}\dot{\boldsymbol{u}}^T\end{aligned} \tag{1.100}$$

where:

$$\boldsymbol{G} = \begin{bmatrix}\boldsymbol{e} \cdot \boldsymbol{P}_{uu} & \boldsymbol{e} \cdot \boldsymbol{P}_{uw} \\ \boldsymbol{e} \cdot \boldsymbol{P}_{wu} & \boldsymbol{e} \cdot \boldsymbol{P}_{ww}\end{bmatrix}. \tag{1.101}$$

From Eq. (1.99) and (1.100) we obtain:

$$\dot{s}^2\kappa\boldsymbol{n} \cdot \boldsymbol{e} = \dot{\boldsymbol{u}}\boldsymbol{G}\dot{\boldsymbol{u}}^T. \tag{1.102}$$

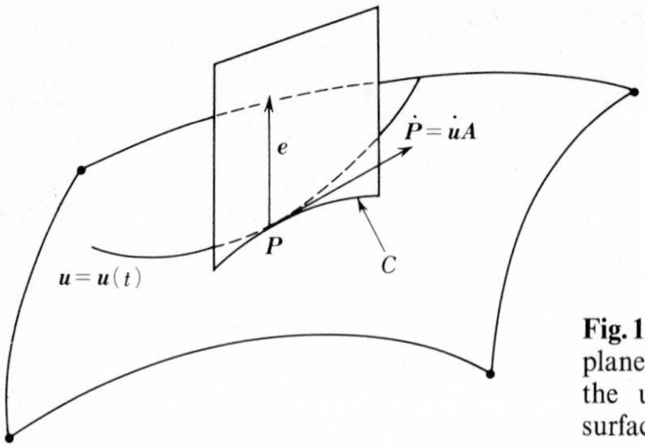

Fig. 1.27. The intersection curve C of the plane containing the tangent vector $\dot{\boldsymbol{P}}$ and the unit normal vector $\boldsymbol{e}$ with a curved surface

The matrix $\boldsymbol{G}$ is called the second fundamental matrix of the curved surface. Normally, in the curved surface representation encountered in CAD, it often happens that $\boldsymbol{P}_{uw} = \boldsymbol{P}_{wu}$. In such a case, $\boldsymbol{G}$ is a symmetrical matrix.

Let a space curve $\boldsymbol{u} = \boldsymbol{u}(t)$ be given on the curved surface $\boldsymbol{P}(u, w)$ and let C be the curve formed by the intersection of the curved surface with the plane which includes the tangent vector $\dot{\boldsymbol{P}} = \dot{\boldsymbol{u}}\boldsymbol{A}$ at a point P on the curve and the unit normal vector $\boldsymbol{e}$ (refer to Fig. 1.27). The curvature of the curve C is called the normal curvature relative to the direction $\dot{\boldsymbol{u}}\boldsymbol{A}$ at point P. The normal curvature is the projected length of the curvature vector of the curve $\boldsymbol{u}$ to $\boldsymbol{e}$. Letting κ_n be the normal curvature, from Eq. (1.102) we have:

$$\dot{s}^2 \kappa_n = \dot{\boldsymbol{u}} \boldsymbol{G} \dot{\boldsymbol{u}}^T$$

$$\therefore \quad \kappa_n = \frac{\dot{\boldsymbol{u}} \boldsymbol{G} \dot{\boldsymbol{u}}^T}{\dot{s}^2} = \frac{\dot{\boldsymbol{u}} \boldsymbol{G} \dot{\boldsymbol{u}}^T}{\dot{\boldsymbol{u}} \boldsymbol{F} \dot{\boldsymbol{u}}^T}. \tag{1.103}$$

The sign of κ_n is plus if the curve C is concave in the direction of $\boldsymbol{e}$, negative if it is convex. As shown in Fig. 1.27, $\kappa_n < 0$ at point P.

In general, if the direction of the intersection curve at point P changes, the curvature κ_n will also change. The direction in which κ_n takes an extreme value is called the principal direction of the normal curvature. In Eq. (1.103), if we take:

$$\xi = \dot{u}, \quad \eta = \dot{w}, \quad L = \boldsymbol{e} \cdot \boldsymbol{P}_{uu}, \quad M = \boldsymbol{e} \cdot \boldsymbol{P}_{uw} = \boldsymbol{e} \cdot \boldsymbol{P}_{wu},$$
$$N = \boldsymbol{e} \cdot \boldsymbol{P}_{ww}, \quad E = \boldsymbol{P}_u^2, \quad F = \boldsymbol{P}_u \cdot \boldsymbol{P}_w, \quad G = \boldsymbol{P}_w^2$$

then:

$$L\xi^2 + 2M\xi\eta + N\eta^2 - \kappa_n(E\xi^2 + 2F\xi\eta + G\eta^2) = 0. \tag{1.104}$$

To find the extreme value of κ_n, take the partial derivatives with respect to ξ and η and set $\partial\kappa_n/\partial\xi = 0$ and $\partial\kappa_n/\partial\eta = 0$ to obtain:

$$\left. \begin{aligned} (L - \kappa_n E)\xi + (M - \kappa_n F)\eta = 0 \\ (M - \kappa_n F)\xi + (N - \kappa_n G)\eta = 0 \end{aligned} \right\}. \tag{1.105}$$

Eliminating ξ and η from these equations gives:

$$(EG-F^2)\kappa_n^2-(EN+GL-2FM)\kappa_n+LN-M^2=0. \qquad (1.106)$$

This quadratic equation always has two real solutions $\kappa_{n\max}$ and $\kappa_{n\min}$. $\kappa_{n\max}$ and $\kappa_{n\min}$ are the maximum and minimum values of the normal curvature, called the principal curvatures.

From the relation between the roots and the coefficients we have:

$$K \equiv \kappa_{n\max}\kappa_{n\min}=\frac{LN-M^2}{EG-F^2}=\frac{(e\cdot P_{uu})(e\cdot P_{ww})-(e\cdot P_{uw})^2}{P_u^2P_w^2-(P_u\cdot P_w)}$$

$$=\frac{|G|}{|F|} \qquad (1.107)$$

$$2H \equiv \kappa_{n\min}+\kappa_{n\max}=\frac{EN+GL-2FM}{EG-F^2}$$

$$=\frac{P_u^2(e\cdot P_{ww})-2(P_u\cdot P_w)(e\cdot P_{uw})+P_w^2(e\cdot P_{uu})}{|F|}. \qquad (1.108)$$

K is called the total curvature or the Gauss curvature, while H is called the mean curvature. It is easy to show that the two directions in which the principal curvatures are obtained are perpendicular to each other.

1.3.5 Calculation of a Point on a Surface

Many of the curved surface patch equations which are normally used are of the bi-cubic surface type:

$$P(u,w)=\sum_{i=0}^{3}\sum_{j=0}^{3}A_{3-i,3-j}u^iw^j \qquad (1.109)$$

$$=(A_{00}w^3+A_{01}w^2+A_{02}w+A_{03})u^3+(A_{10}w^3+A_{11}w^2+A_{12}w+A_{13})u^2$$
$$+(A_{20}w^3+A_{21}w^2+A_{22}w+A_{23})u+A_{30}w^3+A_{31}w^2+A_{32}w+A_{33} \qquad (1.110)$$

$$=[u^3\ u^2\ u\ 1]\begin{bmatrix}A_{00}&A_{01}&A_{02}&A_{03}\\A_{10}&A_{11}&A_{12}&A_{13}\\A_{20}&A_{21}&A_{22}&A_{23}\\A_{30}&A_{31}&A_{32}&A_{33}\end{bmatrix}\begin{bmatrix}w^3\\w^2\\w\\1\end{bmatrix}. \qquad (1.111)$$

For a curved surface of this type, just as in the case of a curve, a point on a curved surface can be calculated rapidly by finite difference calculation.

Taking w to be fixed in Eq. (1.110), we have:

$$P(u, w) = Au^3 + Bu^2 + Cu + D$$

where:

$$\begin{aligned} A &= A_{00}w^3 + A_{01}w^2 + A_{02}w + A_{03} \\ B &= A_{10}w^3 + A_{11}w^2 + A_{12}w + A_{13} \\ C &= A_{20}w^3 + A_{21}w^2 + A_{22}w + A_{23} \\ D &= A_{30}w^3 + A_{31}w^2 + A_{32}w + A_{33} \end{aligned}$$

which is the equation of a cubic curve. Then, using the finite difference matrix in Eq. (1.48), a point on a curve on the curved surface for which w is fixed can be found.

Letting δ be the finite difference increment in the u-direction and denoting a forward finite difference by Δ, the finite difference matrix for $u=0$ is, from Eq. (1.48):

$$\begin{bmatrix} P_{0,w} \\ \frac{1}{\delta}\Delta P_{0,w} \\ \frac{1}{\delta^2}\Delta^2 P_{0,w} \\ \frac{1}{\delta^3}\Delta^3 P_{0,w} \end{bmatrix} = \begin{bmatrix} D \\ A\delta^2 + B\delta + C \\ 6A\delta + 2B \\ 6A \end{bmatrix} = \begin{bmatrix} A_{30}w^3 + A_{31}w^2 + A_{32}w + A_{33} \\ (\delta^2 A_{00} + \delta A_{10} + A_{20})w^3 + (\delta^2 A_{01} + \delta A_{11} + A_{21})w^2 \\ \quad + (\delta^2 A_{02} + \delta A_{12} + A_{22})w + \delta^2 A_{03} + \delta A_{13} + A_{23} \\ (6\delta A_{00} + 2A_{10})w^3 + (6\delta A_{01} + 2A_{11})w^2 \\ \quad + (6\delta A_{02} + 2A_{12})w + 6\delta A_{03} + 2A_{13} \\ 6A_{00}w^3 + 6A_{01}w^2 + 6A_{02}w + 6A_{03} \end{bmatrix} \tag{1.112}$$

Therefore, the finite difference matrix for $w=0$ is:

$$\begin{bmatrix} P_{0,0} \\ \frac{1}{\delta}\Delta P_{0,0} \\ \frac{1}{\delta^2}\Delta^2 P_{0,0} \\ \frac{1}{\delta^3}\Delta^3 P_{0,0} \end{bmatrix} = \begin{bmatrix} A_{33} \\ \delta^2 A_{03} + \delta A_{13} + A_{23} \\ 6\delta A_{03} + 2A_{13} \\ 6A_{03} \end{bmatrix}. \tag{1.113}$$

Next, in Eq. (1.112), let us take the finite difference in the w-direction at $w=0$. Letting δ also be the finite difference increment in this direction and denoting the forward finite difference in this direction by ∇, then the first order finite differences are:

$$\begin{bmatrix} \frac{1}{\delta}\nabla \boldsymbol{P}_{0,0} \\ \frac{1}{\delta^2}\nabla(\Delta \boldsymbol{P}_{0,0}) \\ \frac{1}{\delta^3}\nabla(\Delta^2 \boldsymbol{P}_{0,0}) \\ \frac{1}{\delta^4}\nabla(\Delta^3 \boldsymbol{P}_{0,0}) \end{bmatrix} = \begin{bmatrix} \boldsymbol{A}_{30}\delta^2+\boldsymbol{A}_{31}\delta+\boldsymbol{A}_{32} \\ (\delta^2\boldsymbol{A}_{00}+\delta\boldsymbol{A}_{10}+\boldsymbol{A}_{20})\delta^2+(\delta^2\boldsymbol{A}_{01}+\delta\boldsymbol{A}_{11}+\boldsymbol{A}_{21})\delta \\ +\delta^2\boldsymbol{A}_{02}+\delta\boldsymbol{A}_{12}+\boldsymbol{A}_{22} \\ (6\delta\boldsymbol{A}_{00}+2\boldsymbol{A}_{10})\delta^2+(6\delta\boldsymbol{A}_{01}+2\boldsymbol{A}_{11})\delta \\ +6\delta\boldsymbol{A}_{02}+2\boldsymbol{A}_{12} \\ 6\boldsymbol{A}_{00}\delta^2+6\boldsymbol{A}_{01}\delta+6\boldsymbol{A}_{02} \end{bmatrix}. \tag{1.114}$$

The second and third order finite differences are:

$$\begin{bmatrix} \frac{1}{\delta^2}\nabla^2 \boldsymbol{P}_{0,0} \\ \frac{1}{\delta^3}\nabla^2(\Delta \boldsymbol{P}_{0,0}) \\ \frac{1}{\delta^4}\nabla^2(\Delta^2 \boldsymbol{P}_{0,0}) \\ \frac{1}{\delta^5}\nabla^2(\Delta^3 \boldsymbol{P}_{0,0}) \end{bmatrix} = \begin{bmatrix} 6\boldsymbol{A}_{30}\delta+2\boldsymbol{A}_{31} \\ 6(\delta^2\boldsymbol{A}_{00}+\delta\boldsymbol{A}_{10}+\boldsymbol{A}_{20})\delta+2(\delta^2\boldsymbol{A}_{01}+\delta\boldsymbol{A}_{11}+\boldsymbol{A}_{21}) \\ 6(6\delta\boldsymbol{A}_{00}+2\boldsymbol{A}_{10})\delta+2(6\delta\boldsymbol{A}_{01}+2\boldsymbol{A}_{11}) \\ 36\boldsymbol{A}_{00}\delta+12\boldsymbol{A}_{01} \end{bmatrix} \tag{1.115}$$

$$\begin{bmatrix} \frac{1}{\delta^3}\nabla^3 \boldsymbol{P}_{0,0} \\ \frac{1}{\delta^4}\nabla^3(\Delta \boldsymbol{P}_{0,0}) \\ \frac{1}{\delta^5}\nabla^3(\Delta^2 \boldsymbol{P}_{0,0}) \\ \frac{1}{\delta^6}\nabla^3(\Delta^3 \boldsymbol{P}_{0,0}) \end{bmatrix} = \begin{bmatrix} 6\boldsymbol{A}_{30} \\ 6(\delta^2\boldsymbol{A}_{00}+\delta\boldsymbol{A}_{10}+\boldsymbol{A}_{20}) \\ 6(6\delta\boldsymbol{A}_{00}+2\boldsymbol{A}_{10}) \\ 36\boldsymbol{A}_{00} \end{bmatrix}. \tag{1.116}$$

Let us name the 4×3 matrices on the right-hand sides of Eqs. (1.113), (1.114), (1.115) and (1.116), respectively as finite difference matrix A, finite difference matrix B, finite difference matrix C and finite difference matrix D (Fig. 1.28).

Calculations on the curve $\boldsymbol{P}(u, 0)$ in the u-direction at $w=0$ can be carried out by finite differences using matrix A, that is, Eq. (1.113). Next, to find the finite difference matrix to be used to calculate the curve $\boldsymbol{P}(u, \delta)$, as shown in Fig. (1.28) perform finite difference calculations for each element corresponding to matrices A, B, C and D. Therefore, to generate a family of curves extending along the u-direction separated by a parametric distance γ ($\delta=1/2^m$,

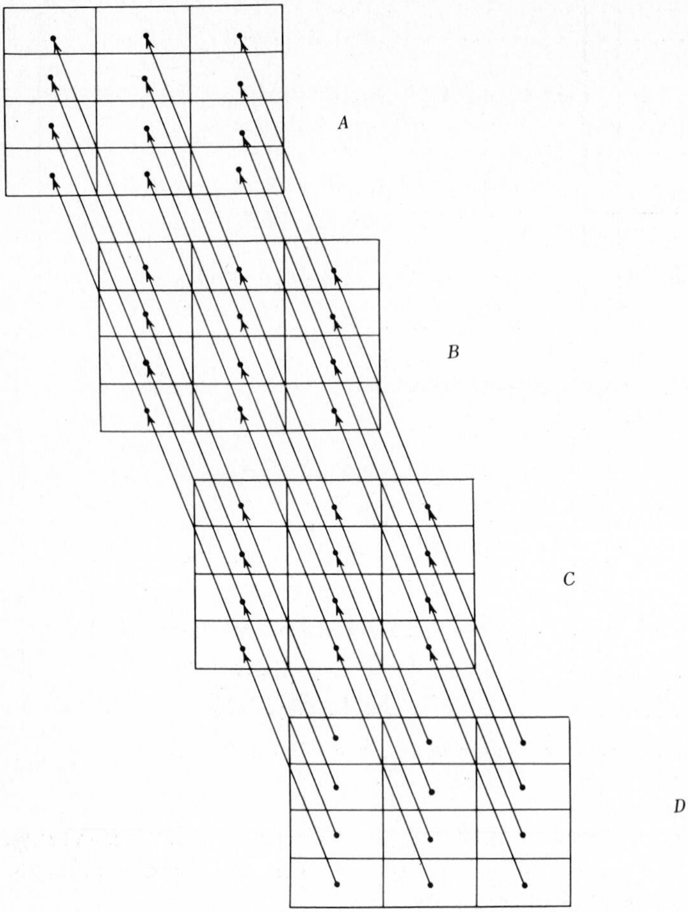

Fig. 1.28. Finite difference computations between elements of finite difference matrices A, B, C, D

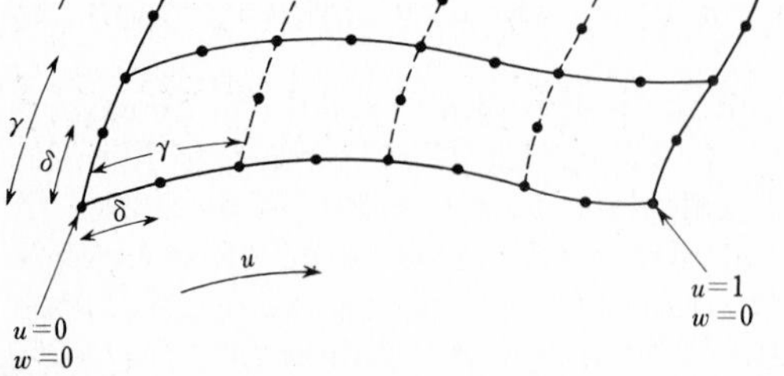

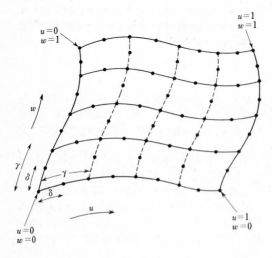

Fig. 1.29. Generation of points on a curved surface by finite difference computations

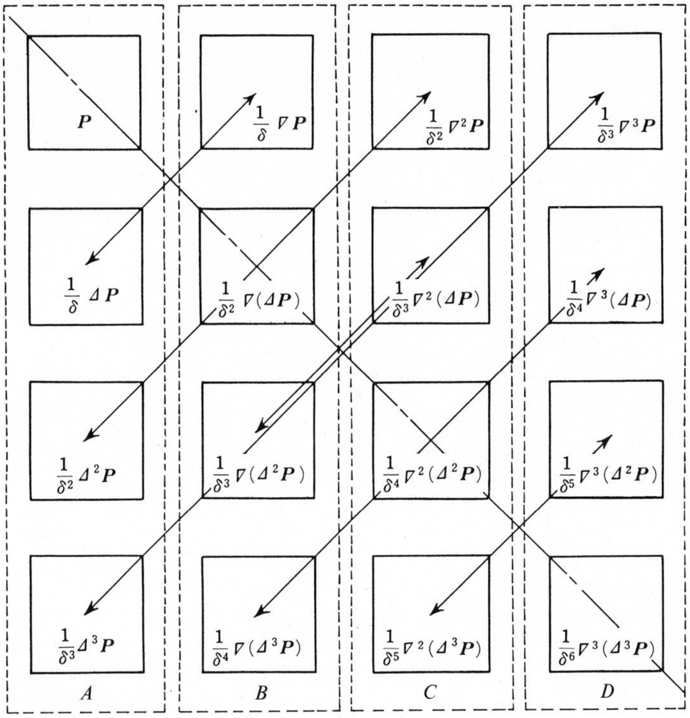

Fig. 1.30. Vector matrix transposition of finite difference matrices *A, B, C, D*

$\gamma = 1/2^n$, $m \geqq n$) in the w-direction, it is sufficient to perform 1 curve-generating finite difference operation in the u-direction for every 2^{m-n} finite difference operations between matrices (Fig. 1.29).

Next, to find a family of curves extending in the w-direction, as shown in Fig. 1.30 matrix transpositions are performed using the vectors of finite difference matrices *A, B, C* and *D*. In other words, the process described above can be performed with the order of u and w reversed; this corresponds to the transformations $A_{ij} \leftrightarrow A_{ji}$ in the finite difference matrices.

1.3.6 Subdivision of Surface Patches

Let us now divide up a curved surface patch, using a bi-cubic surface described by Eq. (1.111) (Fig. 1.31).

First, let us find equations to describe the surface patch $\boldsymbol{P}(u^*, w^*)$ $(0 \leqq u^* \leqq 1;\ 0 \leqq w^* \leqq 1)$ within the bi-cubic surface patch $\boldsymbol{P}(u, w)$ surrounded by $u = u_0$, $u = u_1$, $w = w_0$ and $w = w_1$. To do this the following parameter transformation must be performed in Eq. (1.111).

$$u = u_0 + u^*(u_1 - u_0) \tag{1.117}$$

$$w = w_0 + w^*(w_1 - w_0). \tag{1.118}$$

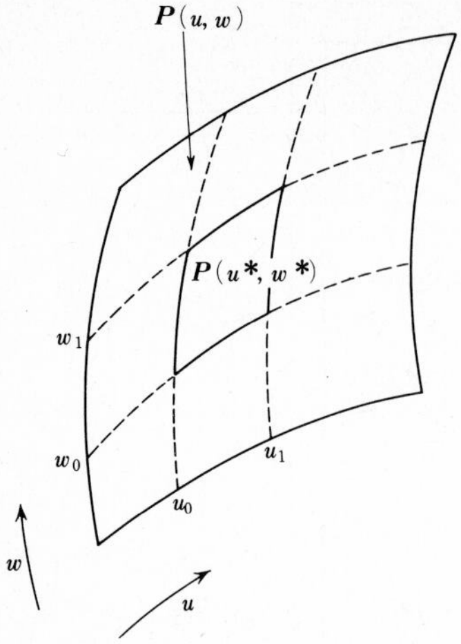

Fig. 1.31. Division of a surface patch (1)

Substituting (1.117) and (1.118) into (1.111), and using the result for parametric transformation of a curve (Eq. (1.55)), gives:

$$\boldsymbol{P}(u^*, w^*) = [u^{*3}\ u^{*2}\ u^*\ 1]$$

$$\times \begin{bmatrix} (u_1-u_0)^3 & 0 & 0 & 0 \\ 3(u_1-u_0)^2 u_0 & (u_1-u_0)^2 & 0 & 0 \\ 3(u_1-u_0) u_0^2 & 2(u_1-u_0) u_0 & u_1-u_0 & 0 \\ u_0^3 & u_0^2 & u_0 & 1 \end{bmatrix} \begin{bmatrix} \boldsymbol{A}_{00} & \boldsymbol{A}_{01} & \boldsymbol{A}_{02} & \boldsymbol{A}_{03} \\ \boldsymbol{A}_{10} & \boldsymbol{A}_{11} & \boldsymbol{A}_{12} & \boldsymbol{A}_{13} \\ \boldsymbol{A}_{20} & \boldsymbol{A}_{21} & \boldsymbol{A}_{22} & \boldsymbol{A}_{23} \\ \boldsymbol{A}_{30} & \boldsymbol{A}_{31} & \boldsymbol{A}_{32} & \boldsymbol{A}_{33} \end{bmatrix}$$

$$\times \begin{bmatrix} (w_1-w_0)^3 & 3(w_1-w_0)^2 w_0 & 3(w_1-w_0) w_0^2 & w_0^3 \\ 0 & (w_1-w_0)^2 & 2(w_1-w_0) w_0 & w_0^2 \\ 0 & 0 & w_1-w_0 & w_0 \\ 0 & 0 & 0 & 1 \end{bmatrix} \begin{bmatrix} w^{*3} \\ w^{*2} \\ w^* \\ 1 \end{bmatrix}$$

$$(0 \leqq u^* \leqq 1;\ 0 \leqq w^* \leqq 1). \tag{1.119}$$

Using Eq. (1.119), the curved surface formula for the case in which the surface patch $\boldsymbol{P}(u, w)$ in Fig. 1.32 is divided into 3 sections is obtained as follows:

- Formula for curved surface section A:
  in Eq. (1.119), set $u_0 = u_s$, $u_1 = 1$, $w_0 = 0$, $w_1 = 1$.
- Formula for curved surface section B:
  in Eq. (1.119), set $u_0 = 0$, $u_1 = u_s$, $w_0 = 0$, $w_1 = w_s$.
- Formula for curved surface section C:
  in Eq. (1.119), set $u_0 = 0$, $u_1 = u_s$, $w_0 = w_s$, $w_1 = 1$.

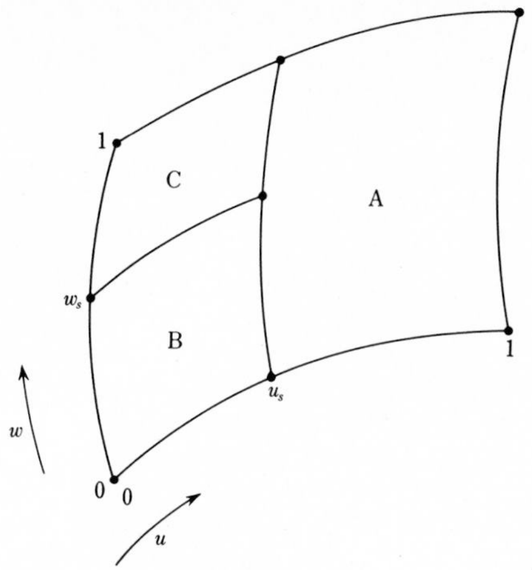

Fig. 1.32. Division of a surface patch (2)

1.3.7 Connection of Surface Patches

Let us consider the problem of connecting surface patch $\boldsymbol{P}_{\mathrm{II}}(u, w)$ along its curve $u=0$ continuously to surface patch $\boldsymbol{P}_{\mathrm{I}}(u, w)$ along its curve $u=1$ (Fig. 1.33).

First, the condition for the boundary curves of both patches to coincide is:

$$\boldsymbol{P}_{\mathrm{II}}(0, w)=\boldsymbol{P}_{\mathrm{I}}(1, w) \quad (0 \leqq w \leqq 1). \tag{1.120}$$

Next, on the boundary curve between both patches, in order for the slope to be continuous in the direction across the boundary curve it is necessary for

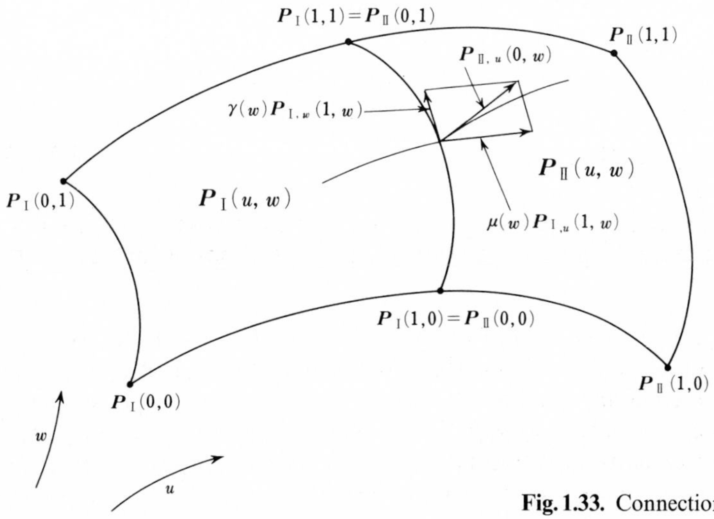

Fig. 1.33. Connection of surface patches

the tangent plane $\boldsymbol{P}_{\text{II}}(0, w)$ $(0 \leqq w \leqq 1)$ to coincide with the tangent plane $\boldsymbol{P}_{\text{I}}(1, w)$ $(0 \leqq w \leqq 1)$. In this case, the direction of the normal vector is continuous; that is, taking $\mu(w)$ to be a positive scalar function, we have:

$$\boldsymbol{P}_{\text{II},u}(0, w) \times \boldsymbol{P}_{\text{II},w}(0, w) = \mu(w)\,\boldsymbol{P}_{\text{I},u}(1, w) \times \boldsymbol{P}_{\text{I},w}(1, w) \ (0 \leqq w \leqq 1). \tag{1.121}$$

Since $\boldsymbol{P}_{\text{II},w}(0, w) = \boldsymbol{P}_{\text{I},w}(1, w)$, if $\gamma(w)$ is an arbitrary scalar function we have:

$$\boldsymbol{P}_{\text{II},u}(0, w) = \mu(w)\,\boldsymbol{P}_{\text{I},u}(1, w) + \gamma(w)\,\boldsymbol{P}_{\text{I},w}(1, w) \ \ (0 \leqq w \leqq 1). \tag{1.122}$$

The implication of Eq. (1.122) is that $\boldsymbol{P}_{\text{II},u}(0, w)$ of the second surface patch lies on the tangent plane of the first surface patch along the boundary curve.

Taking $\gamma(w) = 0$ as a special case, we obtain:

$$\boldsymbol{P}_{\text{II},u}(0, w) = \mu(w)\,\boldsymbol{P}_{\text{I},u}(1, w) \ \ (0 \leqq w \leqq 1). \tag{1.123}$$

In this case, the u-curve is smoothly connected across the boundary curve between the two surface patches (Fig. 1.34). This is a slope continuity condition and is a more severe condition than in the case of Eq. (1.122).

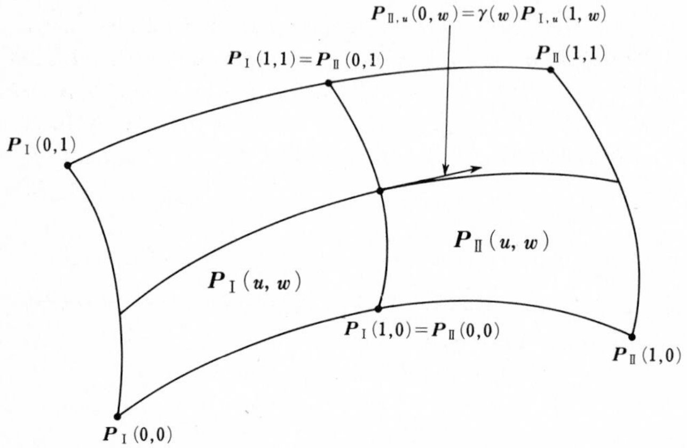

Fig. 1.34. Special case of connection by slope continuity (w-curves are continuous)

The condition expressed by Eq. (1.120) requires that on the boundary curve, the degree of w on $\boldsymbol{P}_{\text{II}}(u, w)$ must agree with the degree of w on $\boldsymbol{P}_{\text{I}}(u, w)$. This does not constrain the degree of w over all of the patch $\boldsymbol{P}_{\text{II}}(u, w)$. The degree of w on $\boldsymbol{P}_{\text{II}}(u, w)$ can be specified to have an arbitrary value n not lower than the degree of w on $\boldsymbol{P}_{\text{I}}(u, w)$ as long as it agrees with the degree on $\boldsymbol{P}_{\text{I}}(u, w)$ on the boundary curve.

In addition, surface patches can be connected as shown in Fig. 1.35 by means of a suitable parametric transformation.

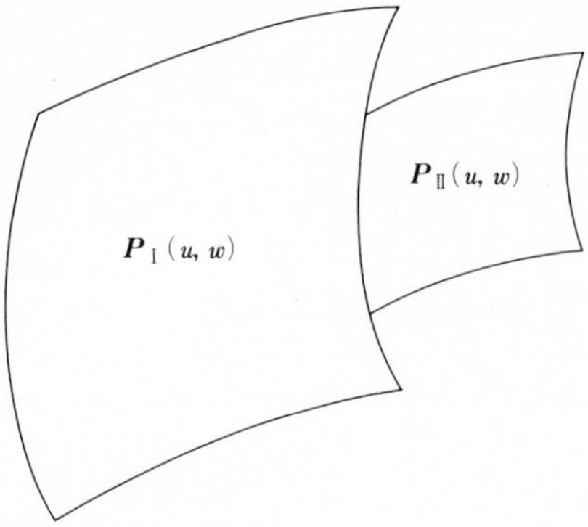

Fig. 1.35. Special case of patch connection

1.3.8 Degeneration of a Surface Patch

It is possible for 2 coincident vertices of the 4 corners of a surface patch to form a triangular surface patch (Fig. 1.36). Curve $\boldsymbol{P}(0, w)$ of this surface patch becomes a single point $\boldsymbol{D}$ regardless of the value which w takes from 0 to 1. In other words, the boundary curve degenerates to a point. When performing shape processing, for example in NC processing, when, for example, a tool path is calculated it is necessary to find the normal vector at every point on the surface. Normally, a normal vector on a surface is found as the vector product of tangent vectors in two parametric directions. However, at point $\boldsymbol{D}$ the vector product becomes:

$$\boldsymbol{P}_u(0, w) \times \boldsymbol{P}_w(0, w). \tag{1.124}$$

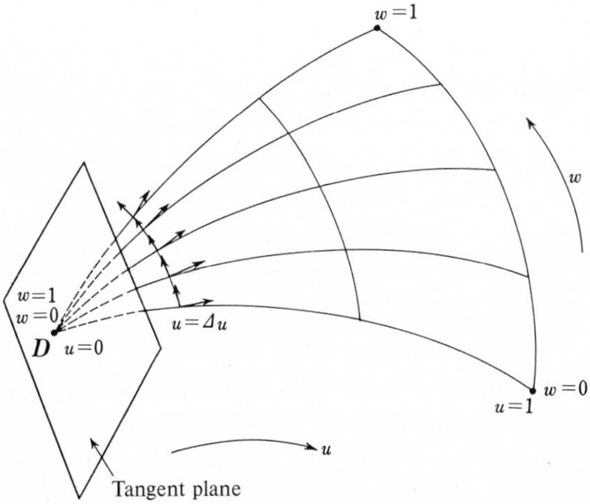

Fig. 1.36. A surface patch that has degenerated to a triangular shape

Since $\boldsymbol{P}_w(0, w)=\boldsymbol{0}$, the normal vector at point $\boldsymbol{D}$ cannot be determined by this method. In such a case, the normal vector is determined as a limit, as follows.

The unit normal vector $\boldsymbol{e}$ on the curve $u=\Delta u$ is:

$$\boldsymbol{e}=\frac{\boldsymbol{P}_u(\Delta u, w)\times\boldsymbol{P}_w(\Delta u, w)}{|\boldsymbol{P}_u(\Delta u, w)\times\boldsymbol{P}_w(\Delta u, w)|}. \tag{1.125}$$

We also have:

$$\boldsymbol{P}_u(\Delta u, w)\fallingdotseq\boldsymbol{P}_u(0, w)+\Delta u\,\boldsymbol{P}_{uu}(0, w) \tag{1.126}$$

$$\begin{aligned}\boldsymbol{P}_w(\Delta u, w)&\fallingdotseq\boldsymbol{P}_w(0, w)+\Delta u\,\boldsymbol{P}_{uw}(0, w)\\&=\Delta u\,\boldsymbol{P}_{uw}(0, w).\end{aligned} \tag{1.127}$$

Substituting the relations in Eqs. (1.126) and (1.127) into (1.125) and taking the limit gives, for the unit normal vector at point $\boldsymbol{D}$:

$$\begin{aligned}\boldsymbol{e}_D&=\lim_{\Delta u\to 0}\boldsymbol{e}=\lim_{\Delta u\to 0}\frac{\boldsymbol{P}_u(0, w)\times\boldsymbol{P}_{uw}(0, w)+\Delta u\,\boldsymbol{P}_{uu}(0, w)\times\boldsymbol{P}_{uw}(0, w)}{|\boldsymbol{P}_u(0, w)\times\boldsymbol{P}_{uw}(0, w)+\Delta u\,\boldsymbol{P}_{uu}(0, w)\times\boldsymbol{P}_{uw}(0, w)|}\\&=\frac{\boldsymbol{P}_u(0, w)\times\boldsymbol{P}_{uw}(0, w)}{|\boldsymbol{P}_u(0, w)\times\boldsymbol{P}_{uw}(0, w)|}.\end{aligned} \tag{1.128}$$

In order for the unit normal vector to be determined uniquely, formula (1.128) must have a unique value for any arbitrary w from 0 to 1.

In other words, when a triangular patch is expressed as an ordinary quadrangular patch, the unit normal vector at the degenerate point $\boldsymbol{D}$ cannot be found by normal means; it must be found using (1.128). And in order for formula (1.128) to be uniquely determined, special care must be taken (refer to Sects. 3.39 and 5.2.4).

1.3.9 Calculation of a Normal Vector on a Surface

A normal vector $\boldsymbol{N}$ at a point on a curved surface is calculated as follows:

$$\begin{aligned}\boldsymbol{N}&=\frac{\partial\boldsymbol{P}(u, w)}{\partial u}\times\frac{\partial\boldsymbol{P}(u, w)}{\partial w}\\&=\left[\frac{\partial(y, z)}{\partial(u, w)}\;\frac{\partial(z, x)}{\partial(u, w)}\;\frac{\partial(x, y)}{\partial(u, w)}\right]\\&\equiv[J_x\;J_y\;J_z].\end{aligned} \tag{1.129}$$

The Jacobians such as $\dfrac{\partial(y, z)}{\partial(u, w)}$ are defined as follows:

$$\frac{\partial(y,z)}{\partial(u,w)} \equiv J_x = \begin{vmatrix} \frac{\partial y}{\partial u} & \frac{\partial z}{\partial u} \\ \frac{\partial y}{\partial w} & \frac{\partial z}{\partial w} \end{vmatrix}.$$

In a special case, for example, when a triangular patch is represented as a quadrangular patch, one boundary line degenerates to a point. At this point the Jacobian becomes singular and the normal vector cannot be found by the above method (refer to Sect. 1.3.8).

The unit normal vector $\boldsymbol{e}$ is:

$$\boldsymbol{e} = \left[\frac{J_x}{\sqrt{J_x^2+J_y^2+J_z^2}} \;\; \frac{J_y}{\sqrt{J_x^2+J_y^2+J_z^2}} \;\; \frac{J_z}{\sqrt{J_x^2+J_y^2+J_z^2}} \right]. \quad (1.130)$$

1.3.10 Calculation of Surface Area and Volume of a Surface

To find the surface area of a curved surface, the first fundamental matrix can be used and the calculation performed with Eq. (1.93), but Jacobians can also be used, as follows.

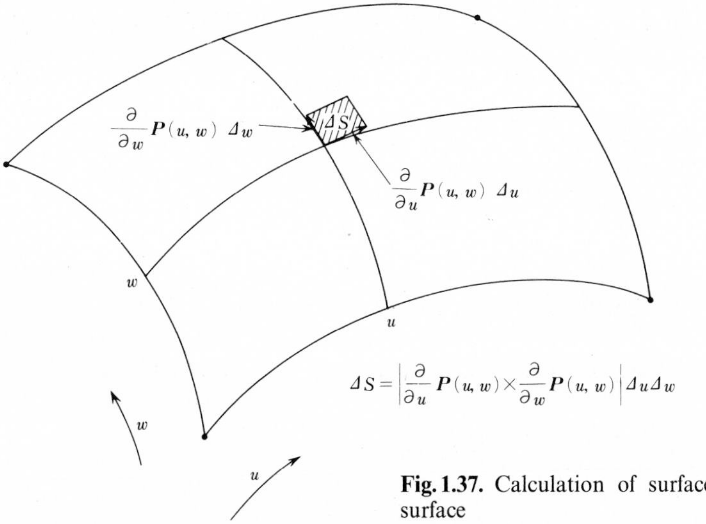

Fig. 1.37. Calculation of surface area of a curved surface

The surface area ΔS of an infinitesimal parallel quadrangle formed by the infinitesimal tangent vectors $\frac{\partial}{\partial u}\boldsymbol{P}(u, w)\,\Delta u$ and $\frac{\partial}{\partial w}\boldsymbol{P}(u, w)\,\Delta w$ in the u- and w-directions is (Fig. 1.37):

$$\Delta S = \left| \frac{\partial}{\partial u} \boldsymbol{P}(u, w) \times \frac{\partial}{\partial w} \boldsymbol{P}(u, w) \right| \Delta u \Delta w = |\boldsymbol{N}| \Delta u \Delta w$$
$$= \sqrt{J_x^2 + J_y^2 + J_z^2}\ \Delta u \Delta w. \tag{1.131}$$

The surface area S over the entire surface patch is then obtained as follows:

$$S = \int dS$$
$$= \int_0^1 \int_0^1 \sqrt{J_x^2 + J_y^2 + J_z^2}\ du dw. \tag{1.132}$$

Next, let us find the volume V of the space enclosed when a curved surface patch is projected onto the xy plane (Fig. 1.38). The projected area ΔS_z obtained by projecting an infinitesimal area on the xy plane is obtained by setting $J_x = J_y = 0$ in Eq. (1.131):

$$\Delta S_z = J_z \Delta u \Delta w.$$

This gives the following expression for the volume V enclosed by the column formed when ΔS is projected onto the xy plane:

$$\Delta V = z(u, w)\, \Delta S_z = z(u, w)\, J_z \Delta u \Delta w.$$

Accordingly, the total volume V becomes:

$$V = \int_0^1 \int_0^1 z(u, w)\, J_z du dw. \tag{1.133}$$

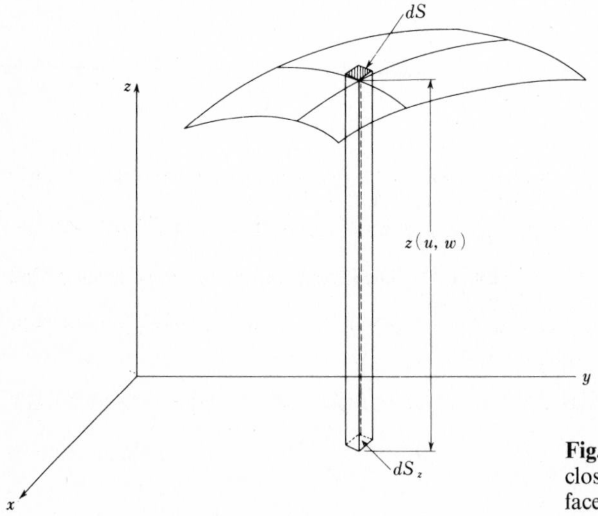

Fig. 1.38. Volume of the space enclosed by projecting a curved surface onto the xy plane

1.3.11 Offset Surfaces

The following equation can be used to generate a point on the curved surface $\boldsymbol{P}_0(u, w)$ that is offset a distance d with respect to the curved surface $\boldsymbol{P}(u, w)$.

$$\boldsymbol{P}_0(u, w) = \boldsymbol{P}(u, w) + \boldsymbol{e}(u, w)\, d. \tag{1.134}$$

In this equation, $\boldsymbol{e}(u, w)$ is the unit normal vector in the offset direction at point $\boldsymbol{P}(u, w)$.

To express $\boldsymbol{P}_0(u, w)$ as a bi-cubic surface in the format of Eq. (1.111), point $\boldsymbol{P}_0(u, w)$ can be calculated for the points corresponding to points $u, w = 0$, 1/3, 2/3, 1 on the curved surface $\boldsymbol{P}(u, w)$; then the bi-cubic surface passing through these 16 points can be found by solving simultaneous equations (Fig. 1.39). This curved surface is an approximation to the offset curved surface.

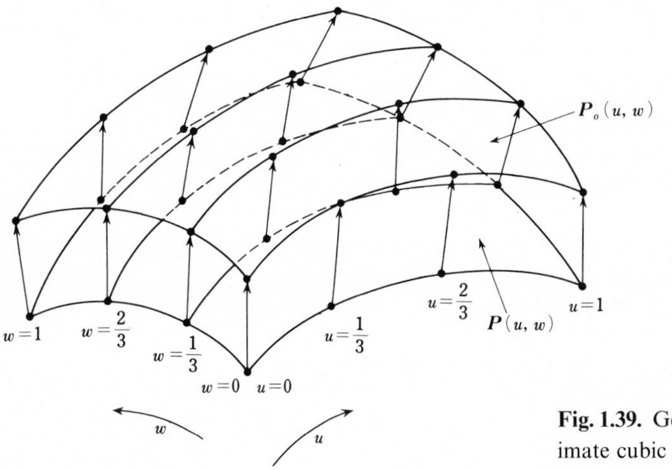

Fig. 1.39. Generation of an approximate cubic offset surface

References (Chap. 1)

4) Hosaka, Mamoru and Mitsuru Kuroda: On generation of curves and surfaces in CAD (in Japanese) *Information Processing* 17 (12) December, 1976.
5) Sasaki, Shigeo: *Differential Geometry* (in Japanese), Kyoritsu Shuppan, Basic Mathematics Textbooks 16, pp. 13—16.
6) Yamaguchi, Fujio: A method of designing free form surfaces by computer display (2nd report) (in Japanese), *Precision Machinery 43* (9).
7) Faux, I. D. and M. J. Pratt: *Computational Geometry for Design and Manufacture,* Ellis Horwood, Mathematics and its Applications.

2. Lagrange Interpolation

2.1 Lagrange Interpolation Curves

It is known that there exists one at most nth-order polynomial that connects the $(n+1)$ points (x_0, f_0), (x_1, f_1), ..., (x_n, f_n) having different abscissas [8].

First, let us use the following notation for the product of the differences of $x, x_0, \ldots, x_n$:

$$W(x) = (x-x_0)(x-x_1)\ldots(x-x_n) \equiv \prod_{j=0}^{n}(x-x_j). \tag{2.1}$$

Denote the above expression with the factor $(x-x_i)$ removed by $W_i(x)$:

$$W_i(x) = (x-x_0)(x-x_1)\ldots(x-x_{i-1})(x-x_{i+1})\ldots(x-x_n) \equiv \prod_{\substack{j=0 \\ j\neq i}}^{n}(x-x_j). \tag{2.2}$$

Then the polynomial $f_L(x)$ that connects a given sequence of points is given by the following formula (Fig. 2.1):

$$f_L(x) = \sum_{i=0}^{n} L_i(x) f_i \tag{2.3}$$

where:

$$L_i(x) = \frac{W_i(x)}{W_i(x_i)} = \prod_{\substack{j=0 \\ j\neq i}}^{n} \frac{(x-x_j)}{(x_i-x_j)}. \tag{2.4}$$

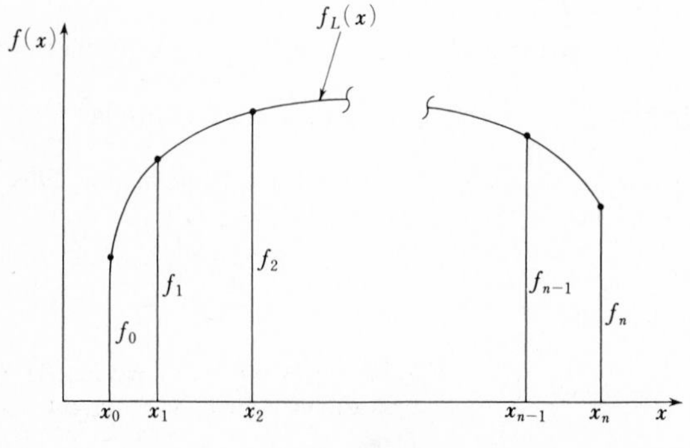

Fig. 2.1. Lagrange interpolation

Equation (2.3) is called Lagrange's formula, and $L_i(x)$ is called Lagrange's coefficient polynomial. $L_i(x)$ can also be expressed as follows:

$$L_i(x) = \frac{W(x)}{(x-x_i)\,W_i(x_i)} = \frac{W(x)}{(x-x_i)\,W'(x_i)}. \tag{2.5}$$

From Eq. (2.4) it is easy to show that the following relation holds for $L_i(x)$:

$$L_i(x_j) = \delta_{ij} = \begin{cases} 0 & (i \neq j) \\ 1 & (i = j) \end{cases} \tag{2.6}$$

showing that the polynomial interpolates these points. It is possible for $f_L(x)$ to be of degree less than n, depending on the coordinates of the points.

Lagrange's coefficient polynomial can be used to derive a parametric vector expression for a curve passing through the points $\boldsymbol{Q}_0, \boldsymbol{Q}_1, \boldsymbol{Q}_2, \ldots, \boldsymbol{Q}_n$:

$$\boldsymbol{P}(t) = \sum_{i=0}^{n} L_i(t)\,\boldsymbol{Q}_i \quad (0 \leqq t \leqq 1) \tag{2.7}$$

where:

$$L_i(t) = \prod_{\substack{j=0 \\ j \neq i}}^{n} \frac{(t-t_j)}{(t_i-t_j)}. \tag{2.8}$$

We also have:

$$t_0(=0) < t_1 < t_2 < \ldots < t_n(=1). \tag{2.9}$$

If the points $\boldsymbol{Q}_0, \boldsymbol{Q}_1, \ldots, \boldsymbol{Q}_n$ are nearly equidistant, we have:

$$t_i = \frac{i}{n}. \tag{2.10}$$

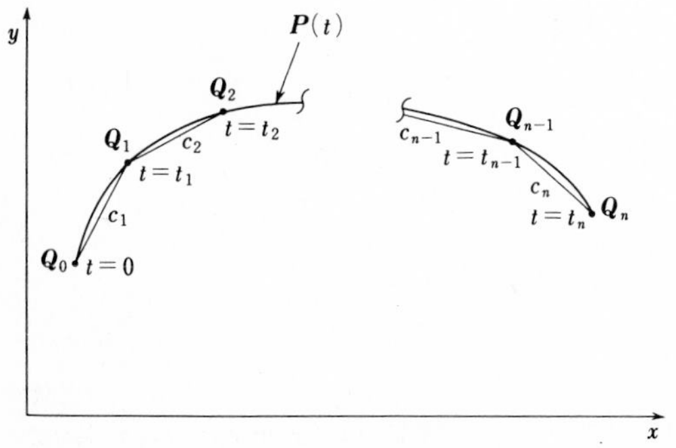

Fig. 2.2. Distribution of parameter values according to chord length

If the points are not approximately equidistant then a better result can be obtained by distributing the parameter t_i according to the chord length between points (refer to Fig. 2.2):

$$t_i = \frac{\sum_{j=1}^{i} c_j}{\sum_{j=1}^{n} c_j}. \tag{2.11}$$

If we have $n=3$ and $t_i = i/n$, then $L_i(t)$ become as follows:

$$\left.\begin{aligned} L_0(t) &= -\frac{9}{2}\left(t-\frac{1}{3}\right)\left(t-\frac{2}{3}\right)(t-1) \\ L_1(t) &= \frac{27}{2}\,t\left(t-\frac{2}{3}\right)(t-1) \\ L_2(t) &= -\frac{27}{2}\,t\left(t-\frac{1}{3}\right)(t-1) \\ L_3(t) &= \frac{9}{2}\,t\left(t-\frac{1}{3}\right)\left(t-\frac{2}{3}\right) \end{aligned}\right\}. \tag{2.12}$$

A graph of $L_i(t)$ is shown in Fig. 2.3, and an example of a curve generated by $L_i(t)$ in Fig. 2.4.

In general Lagrange curves work well up to $n=5$, that is, for up to 6 points, but when n becomes larger than this the high-degree polynomial effect causes fluctuations to occur, making it difficult to obtain a smooth curve.

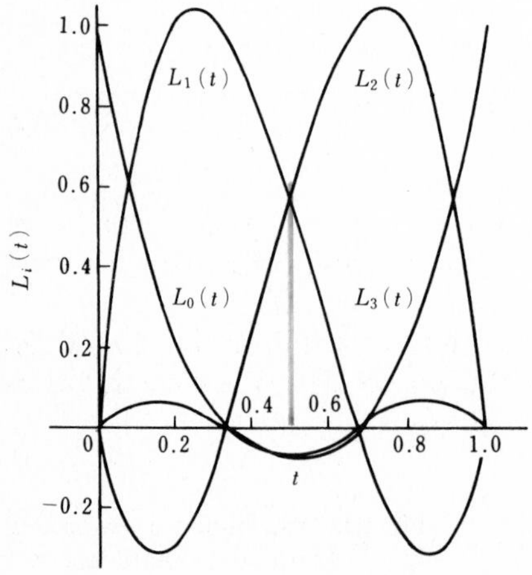

Fig. 2.3. Graph of Lagrange coefficient polynomial (for the case $n=3$)

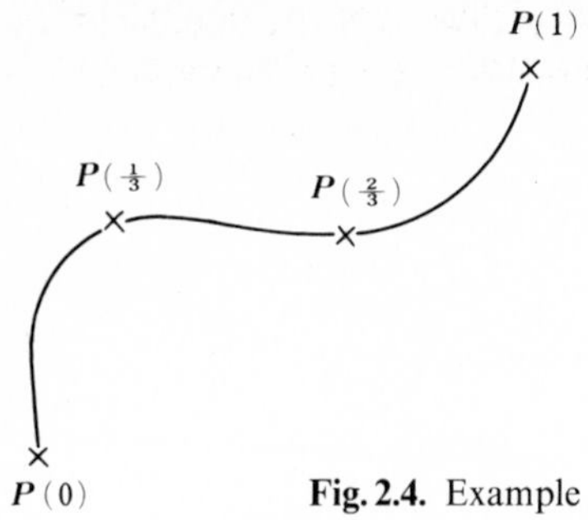

Fig. 2.4. Example of Lagrange curve

2.2 Expression in Terms of Divided Differences

Let f_0, f_1, ... be the function values at different abscissas x_0, x_1, ... The functional values do not necessarily need to be defined by an analytical formula; they could, for example, simply be given as values f_i at discrete data points x_i. In this case, the first-order divided difference $f[x_0, x_1]$ is given by the following formula:

$$f[x_0, x_1] = \frac{f_1 - f_0}{x_1 - x_0}. \tag{2.13}$$

Geometrically, these 1st-order divided differences give the slopes of the line segments connecting the data points (x_0, f_0) and (x_1, f_1). 2nd and higher order divided differences are defined as follows:

$$f[x_0, x_1, x_2] = \frac{f[x_1, x_2] - f[x_0, x_1]}{x_2 - x_0} \tag{2.14}$$

....

....

$$f[x_0, x_1, \ldots, x_n] = \frac{f[x_1, \ldots, x_n] - f[x_0, \ldots, x_{n-1}]}{x_n - x_0}. \tag{2.15}$$

Equation (2.15) gives the divided differences of arbitrary order in terms of the divided differences of one lower order. This equation is used to calculate the values of divided differences numerically.

The divided difference expression on the right-hand side of Eq. (2.15) can be applied repeatedly to express the divided difference of any order n in terms of linear combinations of $(n+1)$ function values $f_0, f_1, \ldots, f_n$:

$$f[x_0, x_1, \ldots, x_n] = \frac{f_0}{W_0(x_0)} + \frac{f_1}{W_1(x_1)} + \cdots + \frac{f_n}{W_n(x_n)} \tag{2.16}$$

$W_0(x_0)$, $W_1(x_1)$, ..., $W_n(x_n)$ in this equation are determined by Eq. (2.2). As can be seen from this equation, the divided differences are symmetrical with respect to their arguments x_0, x_1, ..., x_n, so the values of the divided differences are invariant no matter how the order of the arguments is interchanged.

Substituting x for x_0, f for f_0, x_0 for x_1 and f_0 for f_1 in Eq. (2.13) and solving for f gives:

$$f = f_0 + (x - x_0) f[x, x_0]. \tag{2.17}$$

Similarly, substituting x for x_0, f for f_0, x_0 for x_1, f_0 for f_1, x_1 for x_2 and f_1 for f_2 in Eq. (2.14) leads to:

$$f[x, x_0] = f[x_0, x_1] + (x - x_1) f[x, x_0, x_1]. \tag{2.18}$$

Substituting the right-hand side of this equation for $f[x, x_0]$ in Eq. (2.17) gives:

$$f = f_0 + (x - x_0) f[x_0, x_1] + (x - x_0)(x - x_1) f[x, x_0, x_1]. \tag{2.19}$$

Repeating the above substitutions in sequence leads to the following formula:

$$\begin{aligned} f = f_0 &+ (x - x_0) f[x_0, x_1] + (x - x_0)(x - x_1) f[x_0, x_1, x_2] + \dots \\ &+ (x - x_0)(x - x_1) \dots (x - x_{n-1}) f[x_0, \dots, x_n] \\ &+ (x - x_0) \dots (x - x_n) f[x, x_0, \dots, x_n]. \end{aligned} \tag{2.20}$$

This formula is called Newton's divided difference interpolation formula with remainder. The last term is the remainder. So the formula consists of the first $(n+1)$ terms and the remainder term. Let us focus attention on the first $(n+1)$ terms, the sum of which will be denoted by $\bar{f}$. $\bar{f}$ is of at most degree n.

Taking $x = x_0$ gives:

$$\bar{f}_{x=x_0} = f_0.$$

Then, taking $x = x_1$ gives:

$$\begin{aligned} \bar{f}_{x=x_1} &= f_0 + (x_1 - x_0) f[x_0, x_1] \\ &= f_0 + (x_1 - x_0) \frac{f_1 - f_0}{x_1 - x_0} \\ &= f_1. \end{aligned}$$

Then, taking $x = x_2$ gives:

$$\bar{f}_{x=x_2} = f_0 + (x_2 - x_0) f[x_0, x_1] + (x_2 - x_0)(x_2 - x_1) f[x_0, x_1, x_2] =$$

$$=f_0+(x_2-x_0)\frac{f_1-f_0}{x_1-x_0}+(x_2-x_0)(x_2-x_1)\frac{\dfrac{f_2-f_1}{x_2-x_1}-\dfrac{f_1-f_0}{x_1-x_0}}{x_2-x_0}$$

$$=f_2.$$

The same process can be continued, giving $\bar{f}_{x=x_n}=f_n$ when $x=x_n$. As was stated at the beginning of this chapter, there is only one polynomial that passes through the $(n+1)$ points (x_0, f_0), (x_1, f_1), ..., (x_n, f_n), so $\bar{f}$ is simply another expression for the Lagrange polynomial. This means that the Lagrange polynomial formula (2.3) can be expressed as:

$$f_L(x)\equiv\bar{f}=f_0+(x-x_0)f[x_0, x_1]+(x-x_0)(x-x_1)f[x_0, x_1, x_2]+\dots$$
$$+(x-x_0)(x-x_1)\dots(x-x_{n-1})f[x_0, x_1, \dots, x_n]. \tag{2.21}$$

Using this equation rather than Eq. (2.4) greatly reduces the number of time-consuming multiplications.

The discussion until now treats divided differences for scalar functions. Let us now consider divided differences for parametric vector functions.

Let $\boldsymbol{Q}_0$, $\boldsymbol{Q}_1$, ..., $\boldsymbol{Q}_n$ be position vectors corresponding to different parameters $t_0, t_1, \dots, t_n$. Then the 1st-order divided difference is:

$$\boldsymbol{Q}[t_0, t_1]=\frac{\boldsymbol{Q}_1-\boldsymbol{Q}_0}{t_1-t_0}. \tag{2.22}$$

The 2nd-order divided difference is:

$$\boldsymbol{Q}[t_0, t_1, t_2]=\frac{\boldsymbol{Q}[t_1, t_2]-\boldsymbol{Q}[t_0, t_1]}{t_2-t_0}. \tag{2.23}$$

Similarly, the nth-order divided difference is:

$$\boldsymbol{Q}[t_0, t_1, \dots, t_n]=\frac{\boldsymbol{Q}[t_1, \dots, t_n]-\boldsymbol{Q}[t_0, \dots, t_{n-1}]}{t_n-t_0}. \tag{2.24}$$

Defining the product of differences as:

$$W_i(t)=(t-t_0)(t-t_1)\dots(t-t_{i-1})(t-t_{i+1})\dots(t-t_n)\equiv\prod_{\substack{j=0\\ j\neq i}}^{n}(t-t_j) \tag{2.25}$$

gives, for an equation corresponding to Eq. (2.16):

$$\boldsymbol{Q}[t_0, t_1, \dots, t_n]=\frac{\boldsymbol{Q}_0}{W_0(t_0)}+\frac{\boldsymbol{Q}_1}{W_1(t_1)}+\dots+\frac{\boldsymbol{Q}_n}{W_n(t_n)}. \tag{2.26}$$

Then, the divided difference expression for the parametric Lagrange polynomial corresponding to Eq. (2.21) is:

$$\boldsymbol{P}(t)=\boldsymbol{Q}_0+(t-t_0)\boldsymbol{Q}[t_0,t_1]+(t-t_0)(t-t_1)\boldsymbol{Q}[t_0,t_1,t_2]+\dots$$
$$+(t-t_0)(t-t_1)\dots(t-t_{n-1})\boldsymbol{Q}[t_0,t_1,\dots,t_n]. \tag{2.27}$$

Problem 2.1. Using the divided difference expression for a parametric Lagrange polynomial, find the formula for the case $n=3$, $t_i=i/n$ and confirm that the coefficient polynomial formula agrees with formula (2.12).

Solution: First find the divided differences.

$$\boldsymbol{Q}[t_0,t_1]=\frac{\boldsymbol{Q}_1-\boldsymbol{Q}_0}{t_1-t_0}=\frac{\boldsymbol{Q}_1-\boldsymbol{Q}_0}{\frac{1}{3}-0}=3(\boldsymbol{Q}_1-\boldsymbol{Q}_0)$$

$$\boldsymbol{Q}[t_1,t_2]=\frac{\boldsymbol{Q}_2-\boldsymbol{Q}_1}{t_2-t_1}=\frac{\boldsymbol{Q}_2-\boldsymbol{Q}_1}{\frac{2}{3}-\frac{1}{3}}=3(\boldsymbol{Q}_2-\boldsymbol{Q}_1)$$

$$\boldsymbol{Q}[t_2,t_3]=\frac{\boldsymbol{Q}_3-\boldsymbol{Q}_2}{t_3-t_2}=\frac{\boldsymbol{Q}_3-\boldsymbol{Q}_2}{1-\frac{2}{3}}=3(\boldsymbol{Q}_3-\boldsymbol{Q}_2)$$

$$\boldsymbol{Q}[t_0,t_1,t_2]=\frac{\boldsymbol{Q}[t_1,t_2]-\boldsymbol{Q}[t_0,t_1]}{t_2-t_0}=\frac{3(\boldsymbol{Q}_2-\boldsymbol{Q}_1)-3(\boldsymbol{Q}_1-\boldsymbol{Q}_0)}{\frac{2}{3}-0}$$
$$=\frac{9}{2}(\boldsymbol{Q}_0-2\boldsymbol{Q}_1+\boldsymbol{Q}_2)$$

$$\boldsymbol{Q}[t_1,t_2,t_3]=\frac{\boldsymbol{Q}[t_2,t_3]-\boldsymbol{Q}[t_1,t_2]}{t_3-t_1}=\frac{3(\boldsymbol{Q}_3-\boldsymbol{Q}_2)-3(\boldsymbol{Q}_2-\boldsymbol{Q}_1)}{1-\frac{1}{3}}$$
$$=\frac{9}{2}(\boldsymbol{Q}_1-2\boldsymbol{Q}_2+\boldsymbol{Q}_3)$$

$$\boldsymbol{Q}[t_0,t_1,t_2,t_3]=\frac{\boldsymbol{Q}[t_1,t_2,t_3]-\boldsymbol{Q}[t_0,t_1,t_2]}{t_3-t_0}$$
$$=\frac{\frac{9}{2}(\boldsymbol{Q}_1-2\boldsymbol{Q}_2+\boldsymbol{Q}_3)-\frac{9}{2}(\boldsymbol{Q}_0-2\boldsymbol{Q}_1+\boldsymbol{Q}_2)}{1-0}$$
$$=\frac{9}{2}(-\boldsymbol{Q}_0+3\boldsymbol{Q}_1-3\boldsymbol{Q}_2+\boldsymbol{Q}_3).$$

Setting $n=3$ in formula (2.27), and substituting $\boldsymbol{Q}[t_0, t_1]$, $\boldsymbol{Q}_0[t_0, t_1, t_2]$ and $\boldsymbol{Q}[t_0, t_1, t_2, t_3]$ gives:

$$\begin{aligned}
\boldsymbol{P}(t) &= \boldsymbol{Q}_0 + (t-t_0)\boldsymbol{Q}[t_0, t_1] + (t-t_0)(t-t_1)\boldsymbol{Q}[t_0, t_1, t_2] \\
&\quad + (t-t_0)(t-t_1)(t-t_2)\boldsymbol{Q}[t_0, t_1, t_2, t_3] \\
&= \boldsymbol{Q}_0 + 3t(\boldsymbol{Q}_1 - \boldsymbol{Q}_0) + \frac{9}{2}t\left(t-\frac{1}{3}\right)(\boldsymbol{Q}_0 - 2\boldsymbol{Q}_1 + \boldsymbol{Q}_2) \\
&\quad + \frac{9}{2}t\left(t-\frac{1}{3}\right)\left(t-\frac{2}{3}\right)(-\boldsymbol{Q}_0 + 3\boldsymbol{Q}_1 - 3\boldsymbol{Q}_2 + \boldsymbol{Q}_3) \\
&= -\frac{9}{2}\left(t-\frac{1}{3}\right)\left(t-\frac{2}{3}\right)(t-1)\boldsymbol{Q}_0 + \frac{27}{2}t\left(t-\frac{2}{3}\right)(t-1)\boldsymbol{Q}_1 \\
&\quad - \frac{27}{2}t\left(t-\frac{1}{3}\right)(t-1)\boldsymbol{Q}_2 + \frac{9}{2}t\left(t-\frac{1}{3}\right)\left(t-\frac{2}{3}\right)\boldsymbol{Q}_3 \\
&\equiv L_0(t)\boldsymbol{Q}_0 + L_1(t)\boldsymbol{Q}_1 + L_2(t)\boldsymbol{Q}_2 + L_3(t)\boldsymbol{Q}_3 .
\end{aligned}$$

This confirms that the coefficient polynomial that has been obtained agrees with the Lagrange coefficient polynomial.

References (Chap. 2)

8) See for example Akira Sakurai, *Introduction to Spline Functions,* Tokyo Denki Daigaku Press, p. 14—15.
9) Forrest, A. R.: "Mathematical Principles for Curve and Surface Representation", in *Curved Surfaces in Engineering*. I. J. Brown (ed.), IPC Science and Technology Press Ltd., Guildford, Surrey, England, 1972, p. 5.

3. Hermite Interpolation

3.1 Hermite Interpolation

Hermite interpolation is a generalized form of Lagrange interpolation. Whereas Lagrange interpolation interpolates only between values of a function $f_0, f_1, \ldots, f_n$ at different abscissas $x_0, x_1, \ldots, x_n$, Hermite interpolation also interpolates between higher order derivatives (Fig. 3.1). The following discussion deals with Hermite interpolation of function values and slopes.

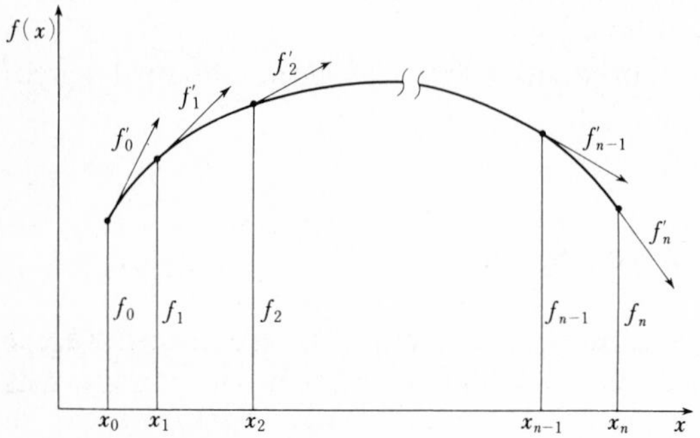

Fig. 3.1. Hermite interpolation

When function values $f_0, f_1, \ldots, f_n$ and slopes $f_0', f_1', \ldots, f_n'$ are given at different abscissas $x_0, x_1, \ldots, x_n$, the Hermite polynomial $f_H(x)$ used to interpolate between these data is given by the following formula[7]:

$$f_H(x) = \sum_{i=0}^{n} \sum_{r=0}^{1} H_{r,i}^*(x) f_i^{(r)} \tag{3.1}$$

where:

$$H_{0,i}^*(x) = [1 - 2L_i'(x_i)(x - x_i)] L_i^2(x) \tag{3.2}$$

$$H_{1,i}^*(x) = (x - x_i) L_i^2(x) \tag{3.3}$$

$L_i(x)$ is given by Eq. (2.4). From the fact that $L_i(x)$ is an nth-degree formula, we see that the polynomial in (3.1) is a $(2n+1)$-degree polynomial which satisfies $2(n+1)$ conditions.

The most common form of Hermite interpolation used in CAD interpolates not between $(n+1)$ points as in formula (3.1), but rather involves interpolation up to the kth-order derivative between 2 points x_0 and x_1:

$$f_H(x) = \sum_{i=0}^{1} \sum_{r=0}^{k} H_{r,i}(x) f_i^{(r)}. \tag{3.4}$$

This formula gives a $(2k+1)$-degree polynomial. $H_{r,i}(x)$ can be expressed in terms of Kronecker's delta as:

$$\frac{d^s}{dx^s} H_{r,i}(x)\bigg|_{x=xj} = \delta_{ij}\,\delta_{rs} \tag{3.5}$$

which means:

$$\frac{d^s}{dx^s} H_{r,i}(x_j) = \begin{cases} 1 & (i=j \quad \text{and} \quad r=s) \\ 0 & (i \neq j \quad \text{or} \quad r \neq s) \end{cases} \tag{3.6}$$

In this chapter, we will discuss Ferguson curves and surfaces and Coons surfaces based on Hermite interpolation.

3.2 Curves

3.2.1 Derivation of a Ferguson Curve Segment

As we found in Sect. 1.2.2, a parametric cubic curve is the lowest degree polynomial that can describe a space curve. A parametric cubic curve is given as follows:

$$\begin{aligned} \boldsymbol{P}(t) &= [t^3 \;\; t^2 \;\; t \;\; 1]\; [\boldsymbol{A} \;\; \boldsymbol{B} \;\; \boldsymbol{C} \;\; \boldsymbol{D}]^T \\ &\equiv [t^3 \;\; t^2 \;\; t \;\; 1]\, \boldsymbol{M} \end{aligned} \tag{3.7}$$

In this formula, $\boldsymbol{M}$ is a 4×3 matrix. Differentiating by t gives:

$$\dot{\boldsymbol{P}}(t) = [3t^2 \;\; 2t \;\; 1 \;\; 0]\, \boldsymbol{M}. \tag{3.8}$$

The curve is supposed to have a position $\boldsymbol{Q}_0$ and tangent vector $\dot{\boldsymbol{Q}}_0$ at $t=0$, and a position $\boldsymbol{Q}_1$ and tangent vector $\dot{\boldsymbol{Q}}_1$ at $t=1$ (Fig. 3.2). Substituting these conditions into formulas (3.7) and (3.8) and expressing the resulting relations in matrix form, we obtain:

$$\begin{bmatrix} \boldsymbol{Q}_0 \\ \boldsymbol{Q}_1 \\ \dot{\boldsymbol{Q}}_0 \\ \dot{\boldsymbol{Q}}_1 \end{bmatrix} = \begin{bmatrix} 0 & 0 & 0 & 1 \\ 1 & 1 & 1 & 1 \\ 0 & 0 & 1 & 0 \\ 3 & 2 & 1 & 0 \end{bmatrix} \boldsymbol{M}.$$

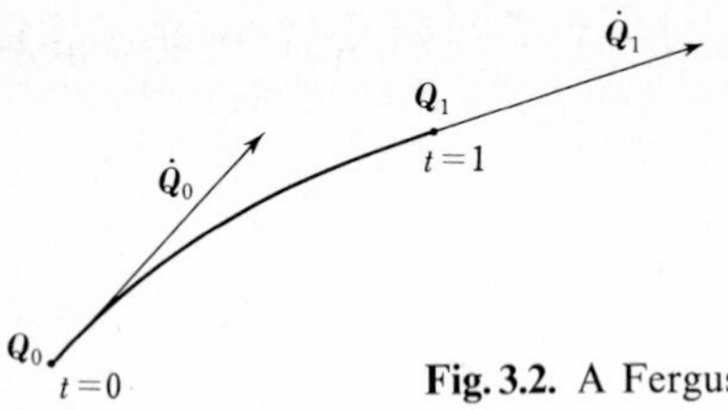

Fig. 3.2. A Ferguson curve segment

This equation can be inverted to find $\boldsymbol{M}$ as:

$$\boldsymbol{M}=\begin{bmatrix}0&0&0&1\\1&1&1&1\\0&0&1&0\\3&2&1&0\end{bmatrix}^{-1}\begin{bmatrix}\boldsymbol{Q}_0\\\boldsymbol{Q}_1\\\dot{\boldsymbol{Q}}_0\\\dot{\boldsymbol{Q}}_1\end{bmatrix}=\begin{bmatrix}2&-2&1&1\\-3&3&-2&-1\\0&0&1&0\\1&0&0&0\end{bmatrix}\begin{bmatrix}\boldsymbol{Q}_0\\\boldsymbol{Q}_1\\\dot{\boldsymbol{Q}}_0\\\dot{\boldsymbol{Q}}_1\end{bmatrix}.$$

Substituting this $\boldsymbol{M}$ into formula (3.7) gives the parametric cubic curve that satisfies the specified boundary conditions:

$$\boldsymbol{P}(t)=[t^3\ t^2\ t\ 1]\begin{bmatrix}2&-2&1&1\\-3&3&-2&-1\\0&0&1&0\\1&0&0&0\end{bmatrix}\begin{bmatrix}\boldsymbol{Q}_0\\\boldsymbol{Q}_1\\\dot{\boldsymbol{Q}}_0\\\dot{\boldsymbol{Q}}_1\end{bmatrix}\equiv[t^3\ t^2\ t\ 1]\boldsymbol{M}_c\begin{bmatrix}\boldsymbol{Q}_0\\\boldsymbol{Q}_1\\\dot{\boldsymbol{Q}}_0\\\dot{\boldsymbol{Q}}_1\end{bmatrix} \tag{3.9}$$

where:

$$\boldsymbol{M}_c=\begin{bmatrix}2&-2&1&1\\-3&3&-2&-1\\0&0&1&0\\1&0&0&0\end{bmatrix}. \tag{3.10}$$

The curve $\boldsymbol{P}(t)$ can also be written in the following form.

$$\boldsymbol{P}(t)=[H_{0,0}(t)\ H_{0,1}(t)\ H_{1,0}(t)\ H_{1,1}(t)]\begin{bmatrix}\boldsymbol{Q}_0\\\boldsymbol{Q}_1\\\dot{\boldsymbol{Q}}_0\\\dot{\boldsymbol{Q}}_1\end{bmatrix} \tag{3.11}$$

$$=H_{0,0}(t)\,\boldsymbol{Q}_0+H_{0,1}(t)\,\boldsymbol{Q}_1+H_{1,0}(t)\,\dot{\boldsymbol{Q}}_0+H_{1,1}(t)\,\dot{\boldsymbol{Q}}_1 \tag{3.12}$$

where:

$$\left.\begin{aligned}H_{0,0}(t)&=2t^3-3t^2+1=(t-1)^2(2t+1)\\H_{0,1}(t)&=-2t^3+3t^2\quad=t^2(3-2t)\\H_{1,0}(t)&=t^3-2t^2+t\quad=(t-1)^2t\\H_{1,1}(t)&=t^3-t^2\qquad\quad=(t-1)t^2\end{aligned}\right\}. \tag{3.13}$$

Since they will be referred to frequently, let us write out M_c^{-1} and the first and second derivatives of $H_{0,0}(t), \ldots, H_{1,1}(t)$:

$$M_c^{-1} = \begin{bmatrix} 0 & 0 & 0 & 1 \\ 1 & 1 & 1 & 1 \\ 0 & 0 & 1 & 0 \\ 3 & 2 & 1 & 0 \end{bmatrix}. \tag{3.14}$$

$$\left.\begin{aligned} \dot{H}_{0,0}(t) &= 6t^2 - 6t = 6t(t-1) \\ \dot{H}_{0,1}(t) &= -6t^2 + 6t = 6t(1-t) \\ \dot{H}_{1,0}(t) &= 3t^2 - 4t + 1 = (3t-1)(t-1) \\ \dot{H}_{1,1}(t) &= 3t^2 - 2t = t(3t-2) \end{aligned}\right\} \tag{3.15}$$

Table 3.1. Cubic Hermite interpolation functions

		$H_{0,0}(t)$	$H_{0,1}(t)$	$H_{1,0}(t)$	$H_{1,1}(t)$
Function value	$t=0$	1	0	0	0
	$t=1$	0	1	0	0
1st derivative	$t=0$	0	0	1	0
	$t=1$	0	0	0	1
2nd derivative	$t=0$	−6	6	−4	−2
	$t=1$	6	−6	2	4

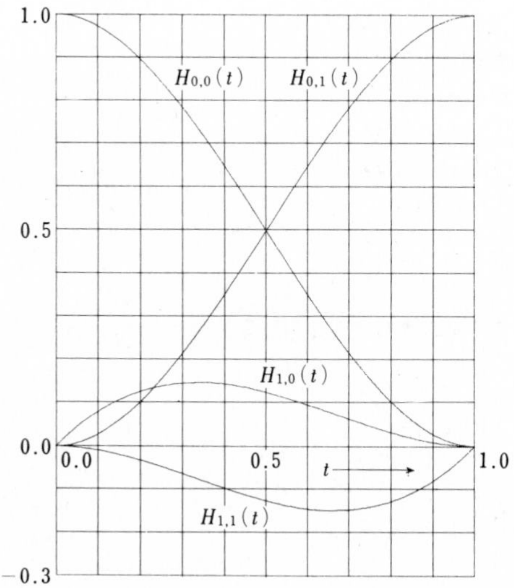

Fig. 3.3. Cubic Hermite interpolation functions

$$\left.\begin{aligned}\ddot{H}_{0,0}(t)&=12t-6=6(2t-1)\\ \ddot{H}_{0,1}(t)&=-12t+6=6(1-2t)\\ \ddot{H}_{1,0}(t)&=6t-4=2(3t-2)\\ \ddot{H}_{1,1}(t)&=6t-2=2(3t-1)\end{aligned}\right\}. \tag{3.16}$$

Since parametric cubic curves were first used by J.C. Ferguson and S.A. Coons, they are called Ferguson curves (or Coons curves). As shown in formula (3.12), these curves are expressed as linear combinations of vector data $\boldsymbol{Q}_0$, $\boldsymbol{Q}_1$, $\dot{\boldsymbol{Q}}_0$, $\dot{\boldsymbol{Q}}_1$. We can consider that input data are blended in linear combination with

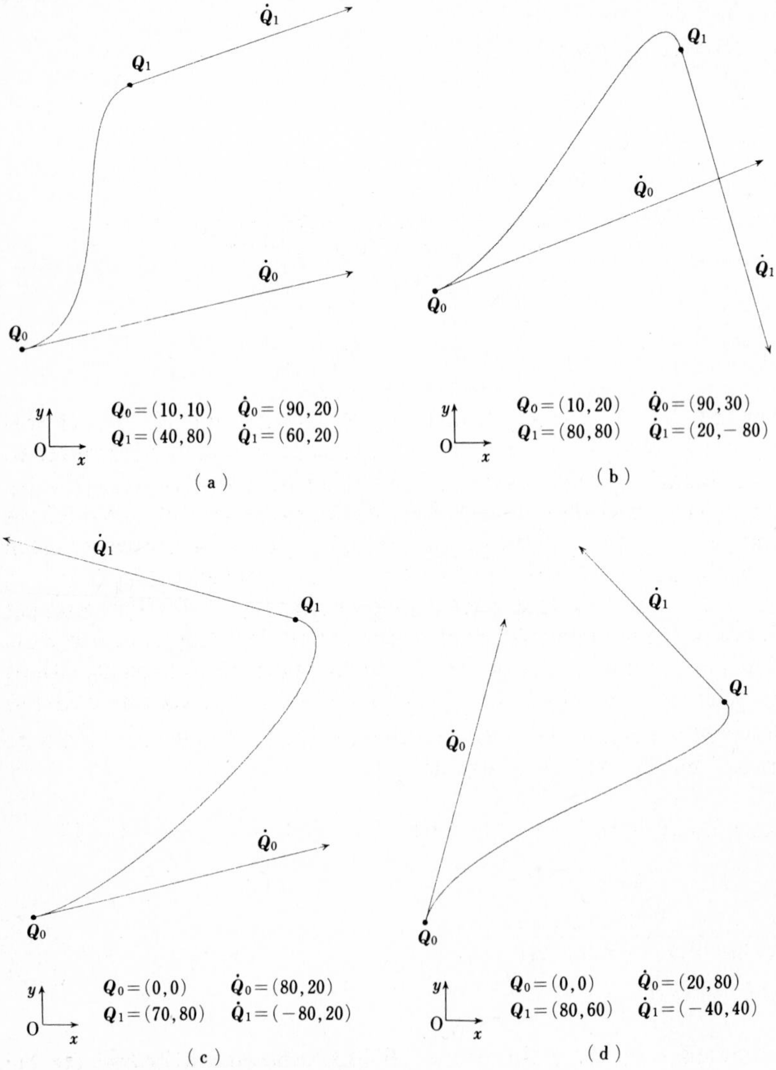

Fig. 3.4. Ferguson curve segments

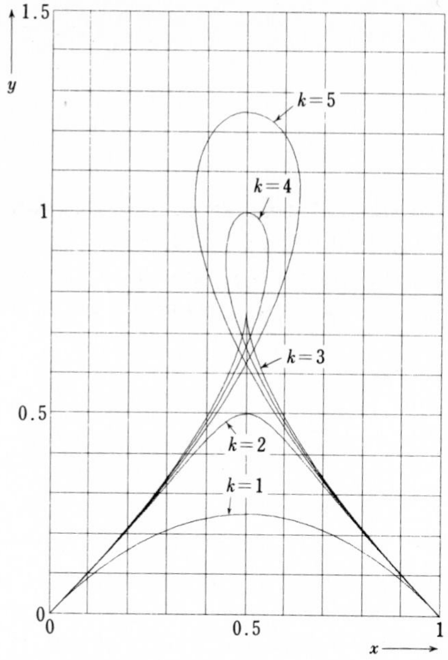

Fig. 3.5. The effect of tangent vector magnitude on curve shape[10)]

weights given by the coefficients $H_{0,0}(t)$, $H_{0,1}(t)$, $H_{1,0}(t)$ and $H_{1,1}(t)$. In this sense, a curve coefficient function expressed as a linear combination is referred to as a blending function. In particular, the blending functions $H_{0,0}(t)$, $H_{0,1}(t)$, $H_{1,0}(t)$ and $H_{1,1}(t)$ were used extensively by S. A. Coons, so they are called Coons blending functions. $H_{0,0}(t)$, $H_{0,1}(t)$, $H_{1,0}(t)$ and $H_{1,1}(t)$ have the properties given in Table 3.1 at $t=0$ and $t=1$. $H_{0,0}(t)$, ..., $H_{1,1}(t)$ are graphed in Fig. 3.3.

Examples of Ferguson curve segments are given in Fig. 3.4. The curve shapes change with not only the direction but also the magnitude of the tangent vectors at the starting and end points. In Fig. 3.5, the starting point is taken as $\boldsymbol{Q}_0=[0\ 0]$ and the end point as $\boldsymbol{Q}_1=[1\ 0]$. The tangent vectors at the starting and end points both have the magnitudes $\sqrt{2}k$. In this case, the Ferguson curve segment $\boldsymbol{P}(t)$ is expressed by the following formula:

$$\boldsymbol{P}(t)=H_{0,0}(t)\,[0\ 0]+H_{0,1}(t)\,[1\ 0]+H_{1,0}(t)\,[k\ k]+H_{1,1}(t)\,[k\ -k].$$

That is

$$\left.\begin{aligned} x&=2(k-1)t^3+3(1-k)t^2+kt \\ y&=k(-t^2+t) \end{aligned}\right\}. \tag{3.17}$$

To find the singular point of this curve, differentiate Eq. (3.17) and set the derivatives equal to 0:

$$\frac{dx}{dt} = 6(k-1)t^2 + 6(1-k)t + k = 0$$

$$\frac{dy}{dt} = k(-2t+1) = 0.$$

These equations can be solved to obtain $t=\frac{1}{2}$, $k=3$. The curve shapes obtained when k is varied are shown in Fig. 3.5. When k is near 1, a relatively natural curve shape is obtained between the two end points. As k increases above 1 the variation of the curve increases; at $k=3$ and $t=\frac{1}{2}$ a cusp appears. When k increases above 3 a loop appears.

3.2.2 Approximate Representation of a Circular Arc by a Ferguson Curve Segment

Let us consider how to approximate a circular arc of unit radius by a Ferguson curve segment (Fig. 3.6). Since the shape of a Ferguson curve segment is invariant under coordinate transformation, let us express it in terms of an $x'y'$ coordinate system with the x'-axis along the line that bisects the circular arc rather than the basic xy coordinate system.

The magnitudes of the tangent vectors to the curve at points $\boldsymbol{Q}_0$ and $\boldsymbol{Q}_1$ are both a. If we assume that the curve passes through (1, 0) at $t=\frac{1}{2}$, then we have:

$$\boldsymbol{P}\left(\frac{1}{2}\right) = \left[H_{0,0}\left(\frac{1}{2}\right) \; H_{0,1}\left(\frac{1}{2}\right) \; H_{1,0}\left(\frac{1}{2}\right) \; H_{1,1}\left(\frac{1}{2}\right)\right] \begin{bmatrix} \cos\theta & -\sin\theta \\ \cos\theta & \sin\theta \\ a\sin\theta & a\cos\theta \\ -a\sin\theta & a\cos\theta \end{bmatrix}$$

$$= \left[\cos\theta + \frac{a}{4}\sin\theta \;\; 0\right].$$

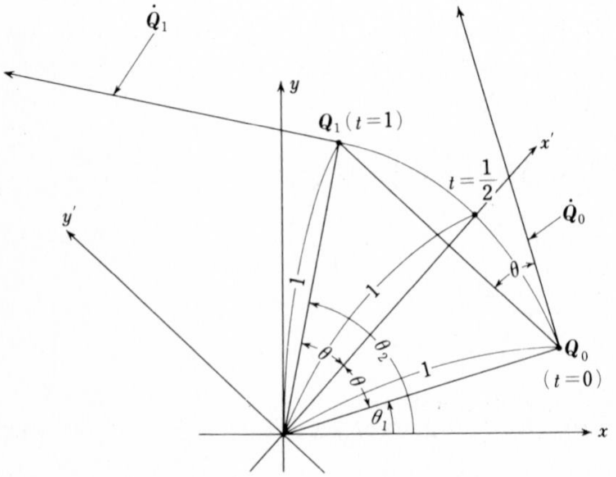

Fig. 3.6. Expression of a unit radius circular arc as a Ferguson curve segment

Consequently:

$$\cos\theta + \frac{a}{4}\sin\theta = 1 \quad \therefore\ a = \frac{4(1-\cos\theta)}{\sin\theta}.$$

The approximate formula for the circular arc becomes as follows:

$$P(t) = [H_{0,0}(t)\ H_{0,1}(t)\ H_{1,0}(t)\ H_{1,1}(t)]$$
$$\times \begin{bmatrix} \cos\theta & -\sin\theta \\ \cos\theta & \sin\theta \\ 4(1-\cos\theta) & 4(1-\cos\theta)/\tan\theta \\ -4(1-\cos\theta) & 4(1-\cos\theta)/\tan\theta \end{bmatrix} \quad (0 \leqq t \leqq 1) \qquad (3.18)$$

where:

$$\theta = (\theta_2 - \theta_1)/2.$$

To find the error with respect to a true circle, let us find the value of t which gives the extreme value of distance $r(t)$ from the origin to the curve:

$$r(t) = \sqrt{P^2(t)}. \qquad (3.19)$$

Differentiating this formula with respect to t and setting the derivative equal to 0 we obtain:

$$P(t) \cdot \dot{P}(t) = 0. \qquad (3.20)$$

Equation (3.20) is a 5th-degree equation in t. It can be easily seen that it has roots at $t=0$, $t=0.5$ and $t=1$, as follows:

$$P(0) = [\cos\theta\ \ -\sin\theta] \quad \dot{P}(0) = [4(1-\cos\theta)\ \ 4(1-\cos\theta)/\tan\theta]$$
$$\therefore P(0) \cdot \dot{P}(0) = 0$$
$$P\left(\frac{1}{2}\right) = [1\ \ 0] \quad \dot{P}\left(\frac{1}{2}\right) = [0\ \ 3\sin\theta - 2(1-\cos\theta)/\tan\theta]$$
$$\therefore P\left(\frac{1}{2}\right) \cdot \dot{P}\left(\frac{1}{2}\right) = 0$$
$$P(1) = [\cos\theta\ \ \sin\theta]$$
$$\dot{P}(1) = [-4(1-\cos\theta)\ \ 4(1-\cos\theta)/\tan\theta]$$
$$\therefore P(1) \cdot \dot{P}(1) = 0.$$

We find the roots of the remaining quadratic equation as follows:

$$t=\frac{1}{2}\pm\frac{1}{2}\sqrt{1-4(k_1/k_2)}\,. \tag{3.21}$$

Here, k_1 and k_2 are functions only of the angle ϕ. From the above equation we find that extreme values occur at two points symmetrically located with respect to $t=0.5$. From the value of t in Eq. (3.21) and from Eq. (3.19), it can be shown that the deviation from a true circle is always positive and that the extreme values of the deviation are equal[11] (refer to Fig. 3.7). Maximum deviations are given as functions of the center angle 2θ in Table 3.2.

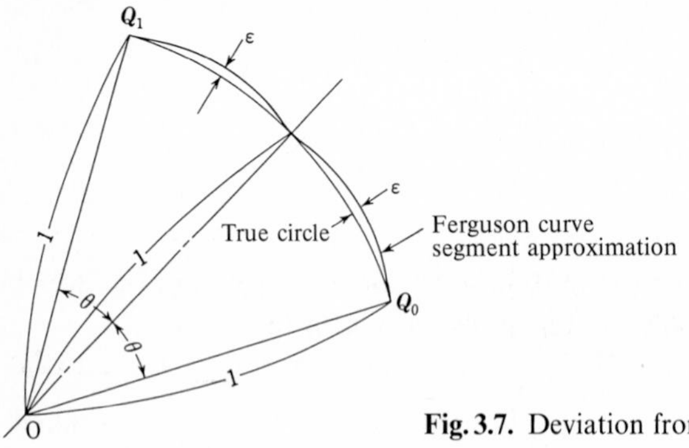

Fig. 3.7. Deviation from a true circle

Table 3.2. Error in approximating a circular arc as a Ferguson curve segment[11]

Center angle (2θ)	Deviation from a true circle (ε)
10°	5.0×10^{-10}
20°	3.3×10^{-8}
30°	3.7×10^{-7}
45°	4.0×10^{-6}
60°	2.4×10^{-5}
90°	2.7×10^{-4}
120°	1.5×10^{-3}
160°	8.9×10^{-3}
180°	1.5×10^{-2}

3.2.3 Hermite Interpolation Curves

By generalizing the method that we used to derive a Ferguson curve segment and specifying the positions and derivative vectors to order k at 2 points, a $(2k+1)$-degree polynomial curve can be obtained. A curve obtained by this method is called a Hermite interpolation curve. A Ferguson curve segment is a cubic Hermite interpolation curve. Cubic Hermite interpolation curves are very important in treating curves and surfaces in CAD.

The Hermite interpolation curve in the case $k=0$ is a straight line connecting 2 points:

$$\boldsymbol{P}(t)=(1-t)\boldsymbol{Q}_0+t\boldsymbol{Q}_1.$$

A 5th-degree Hermite interpolation curve can be derived by a method similar to that used to derive a Ferguson curve segment; the 5th-degree polynomial is given by the following formula.

$$\boldsymbol{P}(t)=[t^5\ t^4\ t^3\ t^2\ t\ 1]\times[\boldsymbol{A}\ \boldsymbol{B}\ \boldsymbol{C}\ \boldsymbol{D}\ \boldsymbol{E}\ \boldsymbol{F}]^T. \tag{3.22}$$

The 1st and 2nd order derivatives of formula (3.22) are:

$$\dot{\boldsymbol{P}}(t)=[5t^4\ 4t^3\ 3t^2\ 2t\ 1\ 0]\begin{bmatrix}\boldsymbol{A}\\ \boldsymbol{B}\\ \boldsymbol{C}\\ \boldsymbol{D}\\ \boldsymbol{E}\\ \boldsymbol{F}\end{bmatrix} \tag{3.23}$$

$$\ddot{\boldsymbol{P}}(t)=[20t^3\ 12t^2\ 6t\ 2\ 0\ 0]\begin{bmatrix}\boldsymbol{A}\\ \boldsymbol{B}\\ \boldsymbol{C}\\ \boldsymbol{D}\\ \boldsymbol{E}\\ \boldsymbol{F}\end{bmatrix}. \tag{3.24}$$

Assuming that the curve has a position $\boldsymbol{Q}_0$, tangent vector $\dot{\boldsymbol{Q}}_0$ and 2nd derivative vector $\ddot{\boldsymbol{Q}}_0$ at $t=0$ and a position $\boldsymbol{Q}_1$, tangent vector $\dot{\boldsymbol{Q}}_1$ and 2nd derivative vector $\ddot{\boldsymbol{Q}}_1$ at $t=1$ (Fig. 3.8), the following equation can be derived from (3.22), (3.23) and (3.24).

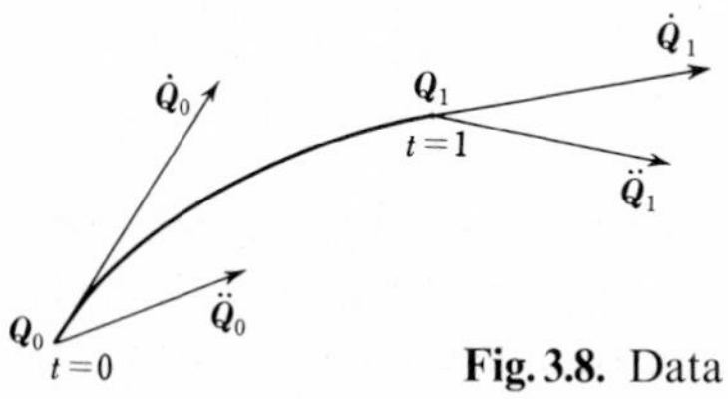

Fig. 3.8. Data for generating a 5th-degree Hermite interpolation curve

$$\begin{bmatrix} \boldsymbol{Q}_0 \\ \boldsymbol{Q}_1 \\ \dot{\boldsymbol{Q}}_0 \\ \dot{\boldsymbol{Q}}_1 \\ \ddot{\boldsymbol{Q}}_0 \\ \ddot{\boldsymbol{Q}}_1 \end{bmatrix} = \begin{bmatrix} 0 & 0 & 0 & 0 & 0 & 1 \\ 1 & 1 & 1 & 1 & 1 & 1 \\ 0 & 0 & 0 & 0 & 1 & 0 \\ 5 & 4 & 3 & 2 & 1 & 0 \\ 0 & 0 & 0 & 2 & 0 & 0 \\ 20 & 12 & 6 & 2 & 0 & 0 \end{bmatrix} \begin{bmatrix} \boldsymbol{A} \\ \boldsymbol{B} \\ \boldsymbol{C} \\ \boldsymbol{D} \\ \boldsymbol{E} \\ \boldsymbol{F} \end{bmatrix}$$

$$\begin{bmatrix} \boldsymbol{A} \\ \boldsymbol{B} \\ \boldsymbol{C} \\ \boldsymbol{D} \\ \boldsymbol{E} \\ \boldsymbol{F} \end{bmatrix} = \begin{bmatrix} 0 & 0 & 0 & 0 & 0 & 1 \\ 1 & 1 & 1 & 1 & 1 & 1 \\ 0 & 0 & 0 & 0 & 1 & 0 \\ 5 & 4 & 3 & 2 & 1 & 0 \\ 0 & 0 & 0 & 2 & 0 & 0 \\ 20 & 12 & 6 & 2 & 0 & 0 \end{bmatrix}^{-1} \begin{bmatrix} \boldsymbol{Q}_0 \\ \boldsymbol{Q}_1 \\ \dot{\boldsymbol{Q}}_0 \\ \dot{\boldsymbol{Q}}_1 \\ \ddot{\boldsymbol{Q}}_0 \\ \ddot{\boldsymbol{Q}}_1 \end{bmatrix}$$

$$= \begin{bmatrix} -6 & 6 & -3 & -3 & -\frac{1}{2} & \frac{1}{2} \\ 15 & -15 & 8 & 7 & \frac{3}{2} & -1 \\ -10 & 10 & -6 & -4 & -\frac{3}{2} & \frac{1}{2} \\ 0 & 0 & 0 & 0 & \frac{1}{2} & 0 \\ 0 & 0 & 1 & 0 & 0 & 0 \\ 1 & 0 & 0 & 0 & 0 & 0 \end{bmatrix} \begin{bmatrix} \boldsymbol{Q}_0 \\ \boldsymbol{Q}_1 \\ \dot{\boldsymbol{Q}}_0 \\ \dot{\boldsymbol{Q}}_1 \\ \ddot{\boldsymbol{Q}}_0 \\ \ddot{\boldsymbol{Q}}_1 \end{bmatrix}.$$

Substituting this relation into (3.22) gives:

$$\boldsymbol{P}(t) = [t^5 \ t^4 \ t^3 \ t^2 \ t \ 1] \times \begin{bmatrix} -6 & 6 & -3 & -3 & -\frac{1}{2} & \frac{1}{2} \\ 15 & -15 & 8 & 7 & \frac{3}{2} & -1 \\ -10 & 10 & -6 & -4 & -\frac{3}{2} & \frac{1}{2} \\ 0 & 0 & 0 & 0 & \frac{1}{2} & 0 \\ 0 & 0 & 1 & 0 & 0 & 0 \\ 1 & 0 & 0 & 0 & 0 & 0 \end{bmatrix} \begin{bmatrix} \boldsymbol{Q}_0 \\ \boldsymbol{Q}_1 \\ \dot{\boldsymbol{Q}}_0 \\ \dot{\boldsymbol{Q}}_1 \\ \ddot{\boldsymbol{Q}}_0 \\ \ddot{\boldsymbol{Q}}_1 \end{bmatrix} \tag{3.25}$$

$$= [K_{0,0}(t) \ K_{0,1}(t) \ K_{1,0}(t) \ K_{1,1}(t) \ K_{2,0}(t) \ K_{2,1}(t)] \times [\boldsymbol{Q}_0 \ \boldsymbol{Q}_1 \ \dot{\boldsymbol{Q}}_0 \ \dot{\boldsymbol{Q}}_1 \ \ddot{\boldsymbol{Q}}_0 \ \ddot{\boldsymbol{Q}}_1]^T \tag{3.26}$$

where:

$$\left.\begin{aligned} K_{0,0}(t) &= -6t^5+15t^4-10t^3+1 \\ K_{0,1}(t) &= 6t^5-15t^4+10t^3 \\ K_{1,0}(t) &= -3t^5+8t^4-6t^3+t \\ K_{1,1}(t) &= -3t^5+7t^4-4t^3 \\ K_{2,0}(t) &= -\frac{1}{2}t^5+\frac{3}{2}t^4-\frac{3}{2}t^3+\frac{1}{2}t^2 \\ K_{2,1}(t) &= \frac{1}{2}t^5-t^4+\frac{1}{2}t^3 \end{aligned}\right\} . \tag{3.27}$$

The properties of $K_{0,0}(t)$, ..., $K_{2,1}(t)$ at $t=0$ and $t=1$ are given in Table 3.3. Graphs of $K_{0,0}(t)$, ..., $K_{2,1}(t)$ are given in Fig. 3.9.

Table 3.3. 5th-degree Hermite interpolation functions

		$K_{0,0}(t)$	$K_{0,1}(t)$	$K_{1,0}(t)$	$K_{1,1}(t)$	$K_{2,0}(t)$	$K_{2,1}(t)$
Function value	$t=0$	1	0	0	0	0	0
	$t=1$	0	1	0	0	0	0
1st derivative	$t=0$	0	0	1	0	0	0
	$t=1$	0	0	0	1	0	0
2nd derivative	$t=0$	0	0	0	0	1	0
	$t=1$	0	0	0	0	0	1
3rd derivative	$t=0$	−60	60	−36	−24	−9	3
	$t=1$	−60	60	−24	−36	−3	9

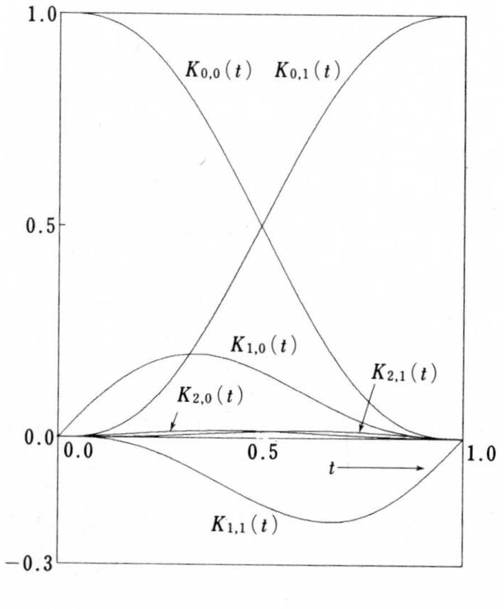

Fig. 3.9. A 5th-degree Hermite interpolation function

3.2.4 Partitioning of Ferguson Curve Segments

Let us now consider just the part of the Ferguson curve segment derived in Sect. 3.2.1 from $t=t_0$ to t_1 and express just this part of the curve in terms of a new parameter $t^*(0 \leqq t^* \leqq 1)$ (Fig. 3.10). From Eq. (1.53), the parameter transformation is:

$$t = t_0 + (t_1 - t_0)\, t^*.$$

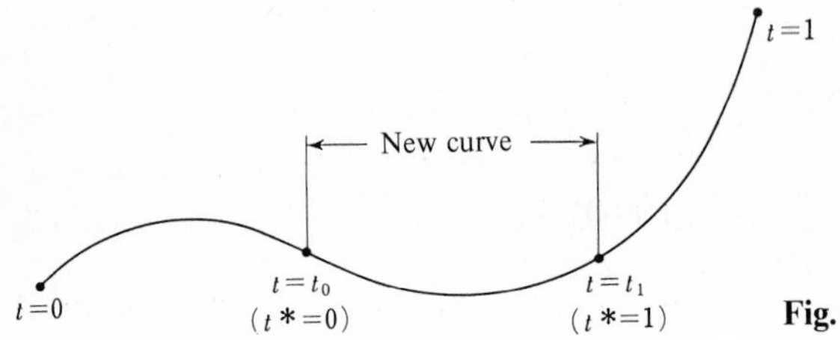

Fig. 3.10. Expression of part of a curve

Substituting this relation into (3.9) gives:

$$\begin{aligned}
\boldsymbol{P}(t^*) = [&(t_1-t_0)^3 t^{*3} + 3(t_1-t_0)^2 t_0 t^{*2} + 3(t_1-t_0) t_0^2 t^* + t_0^3 \\
&(t_1-t_0)^2 t^{*2} + 2(t_1-t_0) t_0 t^* + t_0^2 \quad (t_1-t_0) t^* + t_0 \quad 1]\ \boldsymbol{M}_c \begin{bmatrix} \boldsymbol{Q}_0 \\ \boldsymbol{Q}_1 \\ \dot{\boldsymbol{Q}}_0 \\ \dot{\boldsymbol{Q}}_1 \end{bmatrix}
\end{aligned}$$

$$= [t^{*3} \ \ t^{*2} \ \ t^* \ \ 1] \begin{bmatrix} (t_1-t_0)^3 & 0 & 0 & 0 \\ 3(t_1-t_0)^2 t_0 & (t_1-t_0)^2 & 0 & 0 \\ 3(t_1-t_0) t_0^2 & 2(t_1-t_0) t_0 & t_1-t_0 & 0 \\ t_0^3 & t_0^2 & t_0 & 1 \end{bmatrix} \boldsymbol{M}_c \begin{bmatrix} \boldsymbol{Q}_0 \\ \boldsymbol{Q}_1 \\ \dot{\boldsymbol{Q}}_0 \\ \dot{\boldsymbol{Q}}_1 \end{bmatrix}$$

$$= [t^{*3} \ \ t^{*2} \ \ t^* \ \ 1]\ \boldsymbol{M}_c$$

$$\times \begin{bmatrix} 2t_0^3 - 3t_0^2 + 1 & -2t_0^3 + 3t_0^2 & t_0^3 - 2t_0^2 + t_0 & t_0^3 - t_0^2 \\ 2t_1^3 - 3t_1^2 + 1 & -2t_1^3 + 3t_1^2 & t_1^3 - 2t_1^2 + t_1 & t_1^3 - t_1^2 \\ (t_1-t_0)(6t_0^2 - 6t_0) & (t_1-t_0)(-6t_0^2 + 6t_0) & (t_1-t_0)(3t_0^2 - 4t_0 + 1) & (t_1-t_0)(3t_0^2 - 2t_0) \\ (t_1-t_0)(6t_1^2 - 6t_1) & (t_1-t_0)(-6t_1^2 + 6t_1) & (t_1-t_0)(3t_1^2 - 4t_1 + 1) & (t_1-t_0)(3t_1^2 - 2t_1) \end{bmatrix}$$

$$\times \begin{bmatrix} \boldsymbol{Q}_0 \\ \boldsymbol{Q}_1 \\ \dot{\boldsymbol{Q}}_0 \\ \dot{\boldsymbol{Q}}_1 \end{bmatrix} = [t^{*3} \ \ t^{*2} \ \ t^* \ \ 1]\ \boldsymbol{M}_c$$

$$\times \begin{bmatrix} H_{0,0}(t_0) & H_{0,1}(t_0) & H_{1,0}(t_0) & H_{1,1}(t_0) \\ H_{0,0}(t_1) & H_{0,1}(t_1) & H_{1,0}(t_1) & H_{1,1}(t_1) \\ (t_1-t_0)\dot{H}_{0,0}(t_0) & (t_1-t_0)\dot{H}_{0,1}(t_0) & (t_1-t_0)\dot{H}_{1,0}(t_0) & (t_1-t_0)\dot{H}_{1,1}(t_0) \\ (t_1-t_0)\dot{H}_{0,0}(t_1) & (t_1-t_0)\dot{H}_{0,1}(t_1) & (t_1-t_0)\dot{H}_{1,0}(t_1) & (t_1-t_0)\dot{H}_{1,1}(t_1) \end{bmatrix} \begin{bmatrix} \boldsymbol{Q}_0 \\ \boldsymbol{Q}_1 \\ \dot{\boldsymbol{Q}}_0 \\ \dot{\boldsymbol{Q}}_1 \end{bmatrix} = \tag{3.28}$$

$$= [t^{*3} \ t^{*2} \ t^* \ 1] \ M_c \begin{bmatrix} P(t_0) \\ P(t_1) \\ (t_1 - t_0)\dot{P}(t_0) \\ (t_1 - t_0)\dot{P}(t_1) \end{bmatrix}. \tag{3.29}$$

In Eq. (3.29), the curve $P_1(t^*)$ obtained for $t_0 = 0$, $t_1 = t_s$ is the first half of the curve $P(t)$ divided at point $t = t_s$. This is:

$$P_1(t^*) = [t^{*3} \ t^{*2} \ t^* \ 1] \ M_c \begin{bmatrix} P(0) \\ P(t_s) \\ t_s\dot{P}(0) \\ t_s\dot{P}(t_s) \end{bmatrix} \quad (0 \leqq t^* \leqq 1). \tag{3.30}$$

The latter half curve $P_2(t^*)$ from $t = t_s$ to $t = 1$ can be obtained by setting $t_0 = t_s$ and $t_1 = 1$ in Eq. (3.29):

$$P_2(t^*) = [t^{*3} \ t^{*2} \ t^* \ 1] \ M_c \begin{bmatrix} P(t_s) \\ P(1) \\ (1 - t_s)\dot{P}(t_s) \\ (1 - t_s)\dot{P}(1) \end{bmatrix} \quad (0 \leqq t^* \leqq 1). \tag{3.31}$$

Thus the Ferguson curve segment $P(t)$ has been split into two Ferguson curve segments. This division does not change the shape of the curve.

3.2.5 Increase of Degree of a Ferguson Curve Segment

Let us now consider the problem of formally increasing the degree of a Ferguson curve segment to 5th order without changing its shape. The Ferguson curve segment can be formally transformed as follows.

$$P(t) = [t^3 \ t^2 \ t \ 1] \ M_c \begin{bmatrix} Q_0 \\ Q_1 \\ \dot{Q}_0 \\ \dot{Q}_1 \end{bmatrix}$$

$$= [t^5 \ t^4 \ t^3 \ t^2 \ t \ 1] \begin{bmatrix} 0 & 0 & 0 & 0 \\ 0 & 0 & 0 & 0 \\ 2 & -2 & 1 & 1 \\ -3 & 3 & -2 & -1 \\ 0 & 0 & 1 & 0 \\ 1 & 0 & 0 & 0 \end{bmatrix} \begin{bmatrix} Q_0 \\ Q_1 \\ \dot{Q}_0 \\ \dot{Q}_1 \end{bmatrix}. \tag{3.32}$$

A 5-th degree Hermite interpolation curve has the form given in Eq. (3.25). Taking the position vectors, tangent vectors and 2nd derivative vectors that define a 5th-degree Hermite interpolation curve having the same form as Eq. (3.32) to be $\boldsymbol{Q}_0^*$, $\boldsymbol{Q}_1^*$, $\dot{\boldsymbol{Q}}_0^*$, $\dot{\boldsymbol{Q}}_1^*$, $\ddot{\boldsymbol{Q}}_0^*$ and $\ddot{\boldsymbol{Q}}_1^*$, we obtain:

$$\begin{bmatrix} 0 & 0 & 0 & 0 \\ 0 & 0 & 0 & 0 \\ 2 & -2 & 1 & 1 \\ -3 & 3 & -2 & -1 \\ 0 & 0 & 1 & 0 \\ 1 & 0 & 0 & 0 \end{bmatrix} \begin{bmatrix} \boldsymbol{Q}_0 \\ \boldsymbol{Q}_1 \\ \dot{\boldsymbol{Q}}_0 \\ \dot{\boldsymbol{Q}}_1 \end{bmatrix} = \begin{bmatrix} -6 & 6 & -3 & -3 & -0.5 & 0.5 \\ 15 & -15 & 8 & 7 & 1.5 & -1 \\ -10 & 10 & -6 & -4 & -1.5 & 0.5 \\ 0 & 0 & 0 & 0 & 0.5 & 0 \\ 0 & 0 & 1 & 0 & 0 & 0 \\ 1 & 0 & 0 & 0 & 0 & 0 \end{bmatrix} \begin{bmatrix} \boldsymbol{Q}_0^* \\ \boldsymbol{Q}_1^* \\ \dot{\boldsymbol{Q}}_0^* \\ \dot{\boldsymbol{Q}}_1^* \\ \ddot{\boldsymbol{Q}}_0^* \\ \ddot{\boldsymbol{Q}}_1^* \end{bmatrix}$$

Consequently:

$$\begin{bmatrix} \boldsymbol{Q}_0^* \\ \boldsymbol{Q}_1^* \\ \dot{\boldsymbol{Q}}_0^* \\ \dot{\boldsymbol{Q}}_1^* \\ \ddot{\boldsymbol{Q}}_0^* \\ \ddot{\boldsymbol{Q}}_1^* \end{bmatrix} = \begin{bmatrix} -6 & 6 & -3 & -3 & -0.5 & 0.5 \\ 15 & -15 & 8 & 7 & 1.5 & -1 \\ -10 & 10 & -6 & -4 & -1.5 & 0.5 \\ 0 & 0 & 0 & 0 & 0.5 & 0 \\ 0 & 0 & 1 & 0 & 0 & 0 \\ 1 & 0 & 0 & 0 & 0 & 0 \end{bmatrix}^{-1} \begin{bmatrix} 0 & 0 & 0 & 0 \\ 0 & 0 & 0 & 0 \\ 2 & -2 & 1 & 1 \\ -3 & 3 & -2 & -1 \\ 0 & 0 & 1 & 0 \\ 1 & 0 & 0 & 0 \end{bmatrix} \begin{bmatrix} \boldsymbol{Q}_0 \\ \boldsymbol{Q}_1 \\ \dot{\boldsymbol{Q}}_0 \\ \dot{\boldsymbol{Q}}_1 \end{bmatrix}$$

$$= \begin{bmatrix} 0 & 0 & 0 & 0 & 0 & 1 \\ 1 & 1 & 1 & 1 & 1 & 1 \\ 0 & 0 & 0 & 0 & 1 & 0 \\ 5 & 4 & 3 & 2 & 1 & 0 \\ 0 & 0 & 0 & 2 & 0 & 0 \\ 20 & 12 & 6 & 2 & 0 & 0 \end{bmatrix} \begin{bmatrix} 0 & 0 & 0 & 0 \\ 0 & 0 & 0 & 0 \\ 2 & -2 & 1 & 1 \\ -3 & 3 & -2 & -1 \\ 0 & 0 & 1 & 0 \\ 1 & 0 & 0 & 0 \end{bmatrix} \begin{bmatrix} \boldsymbol{Q}_0 \\ \boldsymbol{Q}_1 \\ \dot{\boldsymbol{Q}}_0 \\ \dot{\boldsymbol{Q}}_1 \end{bmatrix}$$

$$= \begin{bmatrix} 1 & 0 & 0 & 0 \\ 0 & 1 & 0 & 0 \\ 0 & 0 & 1 & 0 \\ 0 & 0 & 0 & 1 \\ -6 & 6 & -4 & -2 \\ 6 & -6 & 2 & 4 \end{bmatrix} \begin{bmatrix} \boldsymbol{Q}_0 \\ \boldsymbol{Q}_1 \\ \dot{\boldsymbol{Q}}_0 \\ \dot{\boldsymbol{Q}}_1 \end{bmatrix} = \begin{bmatrix} \boldsymbol{Q}_0 \\ \boldsymbol{Q}_1 \\ \dot{\boldsymbol{Q}}_0 \\ \dot{\boldsymbol{Q}}_1 \\ 6(\boldsymbol{Q}_1 - \boldsymbol{Q}_0) - 2(2\dot{\boldsymbol{Q}}_0 + \dot{\boldsymbol{Q}}_1) \\ -6(\boldsymbol{Q}_1 - \boldsymbol{Q}_0) + 2(\dot{\boldsymbol{Q}}_0 + 2\dot{\boldsymbol{Q}}_1) \end{bmatrix}.$$

Therefore we have:

$$\left.\begin{aligned} &\boldsymbol{Q}_0^* = \boldsymbol{Q}_0, \ \boldsymbol{Q}_1^* = \boldsymbol{Q}_1, \ \dot{\boldsymbol{Q}}_0^* = \dot{\boldsymbol{Q}}_0, \ \dot{\boldsymbol{Q}}_1^* = \dot{\boldsymbol{Q}}_1 \\ &\ddot{\boldsymbol{Q}}_0^* = 6(\boldsymbol{Q}_1 - \boldsymbol{Q}_0) - 2(2\dot{\boldsymbol{Q}}_0 + \dot{\boldsymbol{Q}}_1), \ \ddot{\boldsymbol{Q}}_1^* = -6(\boldsymbol{Q}_1 - \boldsymbol{Q}_0) + 2(\dot{\boldsymbol{Q}}_0 + 2\dot{\boldsymbol{Q}}_1) \end{aligned}\right\}. \tag{3.33}$$

3.3 Surfaces

3.3.1 Ferguson Surface Patch

Ferguson conceived a method of generating a curved surface using the cubic Hermite interpolation curve (Ferguson curve segment) found in Sect. 3.2.1 [12)].

The data needed to define a surface "patch" are the position vectors at the 4 corners and tangent vectors in two parameter directions (Fig. 3.11).

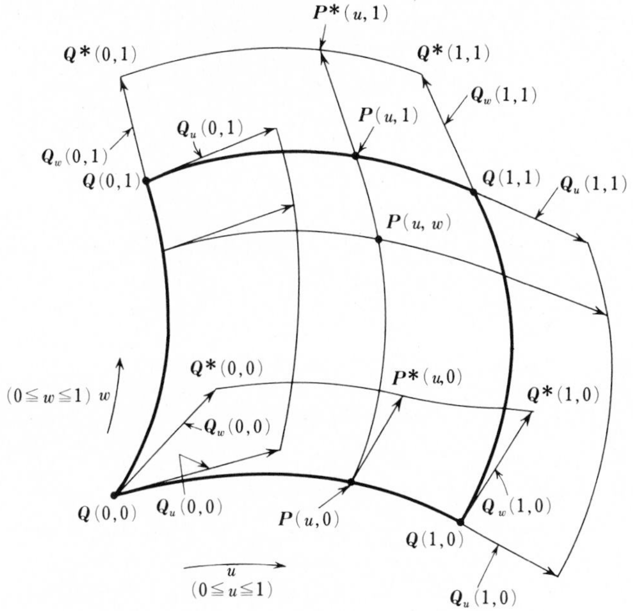

Fig. 3.11. Generation of a Ferguson surface patch

If these data are given, then the two boundary curves $\boldsymbol{P}(u,0)$ and $\boldsymbol{P}(u,1)$ in Fig. 3.11 can be expressed as Ferguson curve segments, by the following equations:

$$\left.\begin{aligned} \boldsymbol{P}(u,0) &= H_{0,0}(u)\,\boldsymbol{Q}(0,0) + H_{0,1}(u)\,\boldsymbol{Q}(1,0) \\ &\quad + H_{1,0}(u)\,\boldsymbol{Q}_u(0,0) + H_{1,1}(u)\,\boldsymbol{Q}_u(1,0) \\ \boldsymbol{P}(u,1) &= H_{0,0}(u)\,\boldsymbol{Q}(0,1) + H_{0,1}(u)\,\boldsymbol{Q}(1,1) \\ &\quad + H_{1,0}(u)\,\boldsymbol{Q}_u(0,1) + H_{1,1}(u)\,\boldsymbol{Q}_u(1,1) \end{aligned}\right\}. \tag{3.34}$$

Let $\boldsymbol{Q}^*(0,0)$, $\boldsymbol{Q}^*(1,0)$, $\boldsymbol{Q}^*(0,1)$ and $\boldsymbol{Q}^*(1,1)$ be the position vectors at the tips of the tangent vectors $\boldsymbol{Q}_w(0,0)$, $\boldsymbol{Q}_w(1,0)$, $\boldsymbol{Q}_w(0,1)$ and $\boldsymbol{Q}_w(1,1)$ in the w-direction at the 4 corner points. Next, create the two Ferguson curve segments connecting $\boldsymbol{Q}^*(0,0)$ and $\boldsymbol{Q}^*(1,0)$, and $\boldsymbol{Q}^*(0,1)$ and $\boldsymbol{Q}^*(1,1)$, respectively. For this purpose, assume that the tangent vectors in the u-direction are $\boldsymbol{Q}_u(0,0)$, $\boldsymbol{Q}_u(1,0)$, $\boldsymbol{Q}_u(0,1)$ and $\boldsymbol{Q}_u(1,1)$. At this time, the two curve segments are:

$$\left.\begin{aligned} \boldsymbol{P}^*(u,0) &= H_{0,0}(u)\,\boldsymbol{Q}^*(0,0) + H_{0,1}(u)\,\boldsymbol{Q}^*(1,0) \\ &\quad + H_{1,0}(u)\,\boldsymbol{Q}_u(0,0) + H_{1,1}(u)\,\boldsymbol{Q}_u(1,0) \\ \boldsymbol{P}^*(u,1) &= H_{0,0}(u)\,\boldsymbol{Q}^*(0,1) + H_{0,1}(u)\,\boldsymbol{Q}^*(1,1) \\ &\quad + H_{1,0}(u)\,\boldsymbol{Q}_u(0,1) + H_{1,1}(u)\,\boldsymbol{Q}_u(1,1) \end{aligned}\right\}. \tag{3.35}$$

When a certain value is specified for u, the 4 points are determined from Eq. (3.34) and (3.35). Assuming the tangent vectors in the w-direction at the 2 points $\boldsymbol{P}(u,0)$ and $\boldsymbol{P}(u,1)$ to be $\boldsymbol{P}^*(u,0) - \boldsymbol{P}(u,0)$ and $\boldsymbol{P}^*(u,1) - \boldsymbol{P}(u,1)$, respectively, the Ferguson curve segment connecting the 2 points $\boldsymbol{P}(u,0)$ and $\boldsymbol{P}(u,1)$ in the w-direction can be determined.

First, the tangent vector in the w-direction at the point $\boldsymbol{P}(u,0)$ is:

$$\begin{aligned} \boldsymbol{P}_w(u,0) &\equiv \boldsymbol{P}^*(u,0) - \boldsymbol{P}(u,0) \\ &= H_{0,0}(u)\,\boldsymbol{Q}^*(0,0) + H_{0,1}(u)\,\boldsymbol{Q}^*(1,0) + H_{1,0}(u)\,\boldsymbol{Q}_u(0,0) \\ &\quad + H_{1,1}(u)\,\boldsymbol{Q}_u(1,0) - (H_{0,0}(u)\,\boldsymbol{Q}(0,0) + H_{0,1}(u)\,\boldsymbol{Q}(1,0) \\ &\quad + H_{1,0}(u)\,\boldsymbol{Q}_u(0,0) + H_{1,1}(u)\,\boldsymbol{Q}_u(1,0)) \\ &= H_{0,0}(u)\,(\boldsymbol{Q}^*(0,0) - \boldsymbol{Q}(0,0)) + H_{0,1}(u)\,(\boldsymbol{Q}^*(1,0) - \boldsymbol{Q}(1,0)) \\ &= H_{0,0}(u)\,\boldsymbol{Q}_w(0,0) + H_{0,1}(u)\,\boldsymbol{Q}_w(1,0). \end{aligned} \tag{3.36}$$

Similarly, the tangent vector in the w-direction at the point $\boldsymbol{P}(u,1)$ is:

$$\boldsymbol{P}_w(u,1) \equiv \boldsymbol{P}^*(u,1) - \boldsymbol{P}(u,1) = H_{0,0}(u)\,\boldsymbol{Q}_w(0,1) + H_{0,1}(u)\,\boldsymbol{Q}_w(1,1). \tag{3.37}$$

Therefore, the Ferguson curve segment in the w-direction (that is, the surface patch) $\boldsymbol{P}(u,w)$ is:

$$\begin{aligned} \boldsymbol{P}(u,w) &= H_{0,0}(w)\,\boldsymbol{P}(u,0) + H_{0,1}(w)\,\boldsymbol{P}(u,1) + H_{1,0}(w)\,(\boldsymbol{P}^*(u,0) - \boldsymbol{P}(u,0)) \\ &\quad + H_{1,1}(w)\,(\boldsymbol{P}^*(u,1) - \boldsymbol{P}(u,1)) \\ &= [\boldsymbol{P}(u,0)\ \boldsymbol{P}(u,1)\ \boldsymbol{P}^*(u,0) - \boldsymbol{P}(u,0)\ \boldsymbol{P}^*(u,1) - \boldsymbol{P}(u,1)] \\ &\quad \times \begin{bmatrix} H_{0,0}(w) \\ H_{0,1}(w) \\ H_{1,0}(w) \\ H_{1,1}(w) \end{bmatrix} \end{aligned}$$

$$= \begin{bmatrix} H_{0,0}(u)\boldsymbol{Q}(0,0) + H_{0,1}(u)\boldsymbol{Q}(1,0) + H_{1,0}(u)\,\boldsymbol{Q}_u(0,0) + H_{1,1}(u)\,\boldsymbol{Q}_u(1,0) \\ H_{0,0}(u)\boldsymbol{Q}(0,1) + H_{0,1}(u)\boldsymbol{Q}(1,1) + H_{1,0}(u)\,\boldsymbol{Q}_u(0,1) + H_{1,1}(u)\,\boldsymbol{Q}_u(1,1) \\ H_{0,0}(u)\,\boldsymbol{Q}_w(0,0) + H_{0,1}(u)\,\boldsymbol{Q}_w(1,0) \\ H_{0,0}(u)\,\boldsymbol{Q}_w(0,1) + H_{0,1}(u)\,\boldsymbol{Q}_w(1,1) \end{bmatrix}^T \times$$

$$\times \begin{bmatrix} H_{0,0}(w) \\ H_{0,1}(w) \\ H_{1,0}(w) \\ H_{1,1}(w) \end{bmatrix}$$

$$= [H_{0,0}(u)\ H_{0,1}(u)\ H_{1,0}(u)\ H_{1,1}(u)]$$

$$\times \begin{bmatrix} \boldsymbol{Q}(0,0) & \boldsymbol{Q}(0,1) & \boldsymbol{Q}_w(0,0) & \boldsymbol{Q}_w(0,1) \\ \boldsymbol{Q}(1,0) & \boldsymbol{Q}(1,1) & \boldsymbol{Q}_w(1,0) & \boldsymbol{Q}_w(1,1) \\ \boldsymbol{Q}_u(0,0) & \boldsymbol{Q}_u(0,1) & \mathbf{0} & \mathbf{0} \\ \boldsymbol{Q}_u(1,0) & \boldsymbol{Q}_u(1,1) & \mathbf{0} & \mathbf{0} \end{bmatrix} \begin{bmatrix} H_{0,0}(w) \\ H_{0,1}(w) \\ H_{1,0}(w) \\ H_{1,1}(w) \end{bmatrix}. \tag{3.38}$$

If the order of operations is reversed so that first 2 curves are generated in the w-direction, then curves are generated in the u-direction, the same result (3.38) is obtained.

If a lattice of points is given, and 2 tangent vectors are given at each point as shown in Fig. 3.12, then Eq. (3.38) can be used to find a mathematical representation for each surface patch. At the boundary between neighboring surface patches, the boundary curve is shared and the tangent vector in the direction across the boundary curve are also shared, so the curved surface patches are connected continuously up to the slope. A surface given by Eq. (3.38) is called the Ferguson surface patch and was in practical use in the surface generating program FMILL used by the Boeing Company in the United States.

In the 4×4 matrix of the Ferguson surface patch, the first and second rows can be regarded as the patch boundary curves at $u=0$ and $u=1$, and the first and

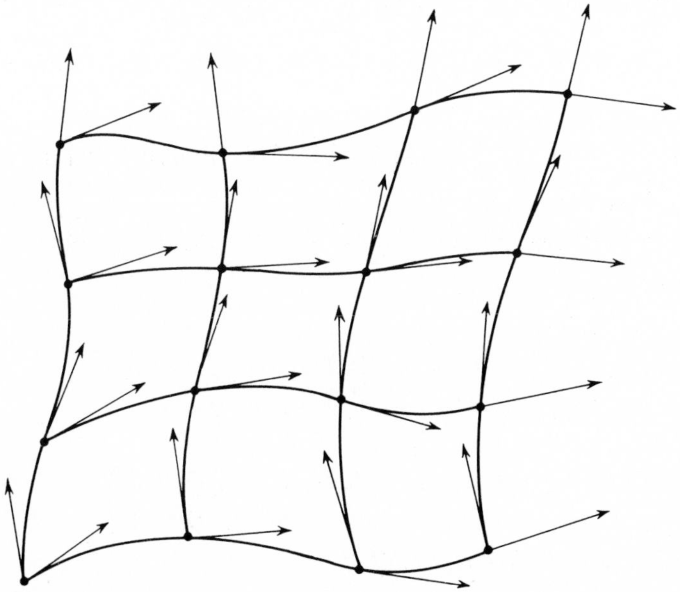

Fig. 3.12. Generation of a Ferguson surface by lattice point data

second columns as the patch boundary curves at $w=0$ and $w=1$, respectively. Note also that the third and fourth rows express the tangent vectors in the direction across the boundary curves at $u=0$ and $u=1$, and the third and fourth columns express the tangent vectors in the direction across the boundary curves at $w=0$ and $w=1$.

Let us consider the tangent vectors in the directions across a boundary curves. For example, the tangent vectors in the direction across the boundary curves at $w=0$ and $w=1$ are given by Eqs. (3.36) and (3.37), respectively. Noting that the relation

$$H_{0,0}(u)+H_{0,1}(u)\equiv 1$$

holds between the blending functions $H_{0,0}(u)$ and $H_{0,1}(u)$, we see that these tangent vectors become, respectively:

$$\left.\begin{aligned} \boldsymbol{P}_w(u,0)&\equiv \boldsymbol{P}^*(u,0)-\boldsymbol{P}(u,0)=(1-\alpha)\boldsymbol{Q}_w(0,0)+\alpha\boldsymbol{Q}_w(1,0)\\ \boldsymbol{P}_w(u,1)&\equiv \boldsymbol{P}^*(u,1)-\boldsymbol{P}(u,1)=(1-\alpha)\boldsymbol{Q}_w(0,1)+\alpha\boldsymbol{Q}_w(1,1)\end{aligned}\right\} \tag{3.39}$$
$$0\leqq\alpha=H_{0,1}(u)\leqq 1.$$

Relations similar to Eqs. (3.39) hold for the tangent vectors in the direction across the boundary curves at $u=0$ and $u=1$.

From the above discussion, it can be seen that on a Ferguson surface, the tangent vector in the direction across the boundary curve is obtained by interpolating linearly between the tangent vectors in the same direction at both end points.

In addition, we can find the cross partial derivative vector $\boldsymbol{P}_{uw}(u,w)$ at the four corners of the surface patch. From Eq. (3.38), we have:

$$\boldsymbol{P}_{uw}(u,w)=[\dot{H}_{0,0}(u)\ \dot{H}_{0,1}(u)\ \dot{H}_{1,0}(u)\ \dot{H}_{1,1}(u)]$$
$$\times\begin{bmatrix}\boldsymbol{Q}(0,0) & \boldsymbol{Q}(0,1) & \boldsymbol{Q}_w(0,0) & \boldsymbol{Q}_w(0,1)\\ \boldsymbol{Q}(1,0) & \boldsymbol{Q}(1,1) & \boldsymbol{Q}_w(1,0) & \boldsymbol{Q}_w(1,1)\\ \boldsymbol{Q}_u(0,0) & \boldsymbol{Q}_u(0,1) & \boldsymbol{0} & \boldsymbol{0}\\ \boldsymbol{Q}_u(1,0) & \boldsymbol{Q}_u(1,1) & \boldsymbol{0} & \boldsymbol{0}\end{bmatrix}\begin{bmatrix}\dot{H}_{0,0}(w)\\ \dot{H}_{0,1}(w)\\ \dot{H}_{1,0}(w)\\ \dot{H}_{1,1}(w)\end{bmatrix}$$

so if, for example, $u=w=0$, then:

$$\boldsymbol{P}_{uw}(0,0)=[0\ \ 0\ \ 1\ \ 0]\begin{bmatrix}\boldsymbol{Q}(0,0) & \boldsymbol{Q}(0,1) & \boldsymbol{Q}_w(0,0) & \boldsymbol{Q}_w(0,1)\\ \boldsymbol{Q}(1,0) & \boldsymbol{Q}(1,1) & \boldsymbol{Q}_w(1,0) & \boldsymbol{Q}_w(1,1)\\ \boldsymbol{Q}_u(0,0) & \boldsymbol{Q}_u(0,1) & \boldsymbol{0} & \boldsymbol{0}\\ \boldsymbol{Q}_u(1,0) & \boldsymbol{Q}_u(1,1) & \boldsymbol{0} & \boldsymbol{0}\end{bmatrix}\begin{bmatrix}0\\0\\1\\0\end{bmatrix}=\boldsymbol{0}.$$

A similar result is obtained at the other 3 points:

$$\boldsymbol{P}_{uw}(0,0)=\boldsymbol{P}_{uw}(1,0)=\boldsymbol{P}_{uw}(0,1)=\boldsymbol{P}_{uw}(1,1)=\boldsymbol{0}. \tag{3.40}$$

The cross partial derivative vector is also called the twist vector. The twist vector will be discussed in more detail below (refer to Sect. 3.3.4). On the Ferguson surface patch, in the vicinity of the 4 corners there is a tendency for the shape of the surface to become flattened (refer to Sect. 3.3.4).

3.3.2 The Coons Surface Patches (1964)

Against the background of the great interest in computer graphics centering on MIT (Massachusetts Institute of Technology) at the start of the 1960s, Associate Professor S.A. Coons of MIT announced a very powerful mathematical representation for defining the shape of a curved surface in 1964[13]. In 1967 he developed the method further into a more general curved surface representation. The curved surface representation which he proposed is called the Coons surface patch; it is widely known and has found a number of practical applications. Basic research has also been done on the Coons surface patch, making it clear that it is a very general surface representation.

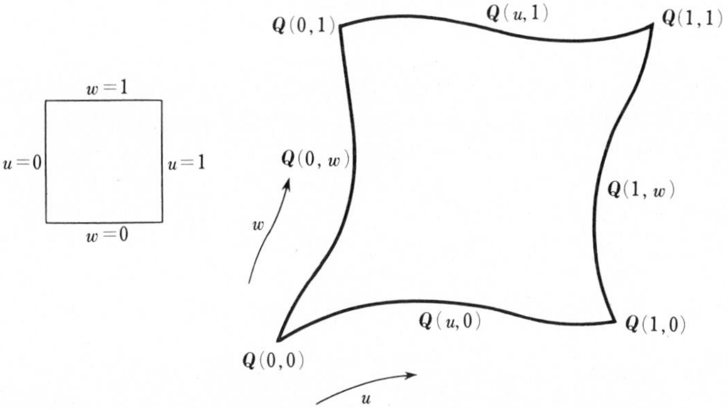

Fig. 3.13. A Coons surface patch (according to Coons' 1964 theory)

Consider a surface patch enclosed by four curves as shown in Fig. 3.13. Let u and w be two parameters used to express this surface patch; they vary within the ranges $0 \leqq u$ and $w \leqq 1$. Let $\boldsymbol{Q}(0,0)$, $\boldsymbol{Q}(0,1)$, $\boldsymbol{Q}(1,0)$ and $\boldsymbol{Q}(1,1)$ be the position vectors at the four corners, and denote the four boundary curves by $\boldsymbol{Q}(u,0)$, $\boldsymbol{Q}(u,1)$, $\boldsymbol{Q}(0,w)$ and $\boldsymbol{Q}(1,w)$. The formula for the Coons surface patch that was announced in 1964 is the following:

$$\begin{aligned}\boldsymbol{P}(u,w) = {} & C_{0,0}(u)\,\boldsymbol{Q}(0,w) + C_{0,1}(u)\,\boldsymbol{Q}(1,w) \\ & + C_{0,0}(w)\,\boldsymbol{Q}(u,0) + C_{0,1}(w)\,\boldsymbol{Q}(u,1) \\ & - C_{0,0}(u)\,C_{0,0}(w)\,\boldsymbol{Q}(0,0) - C_{0,0}(u)\,C_{0,1}(w)\,\boldsymbol{Q}(0,1) \\ & - C_{0,1}(u)\,C_{0,0}(w)\,\boldsymbol{Q}(1,0) - C_{0,1}(u)\,C_{0,1}(w)\,\boldsymbol{Q}(1,1). \qquad (3.41)\end{aligned}$$

This can be put into matrix form:

$$P(u,w) = -[-1 \;\; C_{0,0}(u) \;\; C_{0,1}(u)] \times \begin{bmatrix} \mathbf{0} & Q(u,0) & Q(u,1) \\ Q(0,w) & Q(0,0) & Q(0,1) \\ Q(1,w) & Q(1,0) & Q(1,1) \end{bmatrix} \begin{bmatrix} -1 \\ C_{0,0}(w) \\ C_{0,1}(w) \end{bmatrix}. \tag{3.42}$$

This can be decomposed into a sum of 3 terms:

$$P(u,w) = \underbrace{[C_{0,0}(u) \;\; C_{0,1}(u)] \begin{bmatrix} Q(0,w) \\ Q(1,w) \end{bmatrix}}_{P_A(u,w)} + \underbrace{[Q(u,0) \;\; Q(u,1)] \begin{bmatrix} C_{0,0}(w) \\ C_{0,1}(w) \end{bmatrix}}_{P_B(u,w)}$$
$$\underbrace{-[C_{0,0}(u) \;\; C_{0,1}(u)] \begin{bmatrix} Q(0,0) & Q(0,1) \\ Q(1,0) & Q(1,1) \end{bmatrix} \begin{bmatrix} C_{0,0}(w) \\ C_{0,1}(w) \end{bmatrix}}_{P_C(u,w)}. \tag{3.43}$$

In Eqs. (3.41) through (3.43), $C_{0,0}$ and $C_{0,1}$ are continuous, monotonic blending functions within the interval covered by the patch. It is clear that the surface patch $P(u,w)$ is expressed in terms of four boundary curves and the two scalar functions $C_{0,0}$ and $C_{0,1}$.

1) Apply the following conditions to the blending functions $C_{0,0}$ and $C_{0,1}$:

$$C_{0,0}(0) = 1, \;\; C_{0,0}(1) = 0, \;\; C_{0,1}(0) = 0, \;\; C_{0,1}(1) = 1. \tag{3.44}$$

Alternatively, these conditions can be expressed in terms of the Kronecker delta:

$$C_{0,i}(j) = \delta_{ij} \quad (i, j: 0, 1). \tag{3.45}$$

If this condition is satisfied, then the surface patch $P(u,w)$ has arbitrarily given curves as boundary curves. For example, in the surface patch formula (3.43), if we take $w=0$ then we have:

$$P(u,0) = [C_{0,0}(u) \;\; C_{0,1}(u)] \begin{bmatrix} Q(0,0) \\ Q(1,0) \end{bmatrix} + [Q(u,0) \;\; Q(u,1)] \begin{bmatrix} C_{0,0}(0) \\ C_{0,1}(0) \end{bmatrix}$$
$$-[C_{0,0}(u) \;\; C_{0,1}(u)] \begin{bmatrix} Q(0,0) & Q(0,1) \\ Q(1,0) & Q(1,1) \end{bmatrix} \begin{bmatrix} C_{0,0}(0) \\ C_{0,1}(0) \end{bmatrix}$$
$$= [C_{0,0}(u) \;\; C_{0,1}(u)] \begin{bmatrix} Q(0,0) \\ Q(1,0) \end{bmatrix} + Q(u,0)$$
$$-[C_{0,0}(u) \;\; C_{0,1}(u)] \begin{bmatrix} Q(0,0) \\ Q(1,0) \end{bmatrix}$$
$$= Q(u,0)$$

showing that at $w=0$ the surface patch $\boldsymbol{P}(u, w)$ coincides with an arbitrarily given boundary curve $\boldsymbol{Q}(u, 0)$. Similarly, for $w=1$, $u=0$ and $u=1$, it is easy to show that the surface patch coincides with arbitrarily given boundary curves.

To deepen our understanding of the Coons surface patch, let us show graphically what it means for $\boldsymbol{P}(u, w)$ to be enclosed by four boundary curves. The surface patch formula (3.43) contains three terms. All of the terms are expressions in terms of the two parameters u and w, so each term can be thought of as representing a surface. Let us denote these surfaces by $\boldsymbol{P}_A(u, w)$, $\boldsymbol{P}_B(u, w)$ and $\boldsymbol{P}_C(u, w)$, in that order. These are related by:

$$\boldsymbol{P}(u, w)=\boldsymbol{P}_A(u, w)+\boldsymbol{P}_B(u, w)-\boldsymbol{P}_C(u, w) \tag{3.46}$$

where:

$$\boldsymbol{P}_A(u, w)=[C_{0,0}(u) \;\; C_{0,1}(u)]\begin{bmatrix}\boldsymbol{Q}(0, w)\\ \boldsymbol{Q}(1, w)\end{bmatrix} \tag{3.47}$$

$$\boldsymbol{P}_B(u, w)=[\boldsymbol{Q}(u, 0) \;\; \boldsymbol{Q}(u, 1)]\begin{bmatrix}C_{0,0}(w)\\ C_{0,1}(w)\end{bmatrix} \tag{3.48}$$

$$\boldsymbol{P}_C(u, w)=[C_{0,0}(u) \;\; C_{0,1}(u)]\begin{bmatrix}\boldsymbol{Q}(0,0) & \boldsymbol{Q}(0, 1)\\ \boldsymbol{Q}(1,0) & \boldsymbol{Q}(1, 1)\end{bmatrix}\begin{bmatrix}C_{0,0}(w)\\ C_{0,1}(w)\end{bmatrix}. \tag{3.49}$$

The surface $\boldsymbol{P}_A(u, w)$ becomes $\boldsymbol{Q}(0, w)$ at $u=0$, and at $u=1$ $\boldsymbol{P}_A(1, w)=\boldsymbol{Q}(1, w)$, so, as shown in Fig. 3.14, $\boldsymbol{P}_A(u, w)$ is a curved surface such that the boundary curves on two sides are $\boldsymbol{Q}(0, w)$ and $\boldsymbol{Q}(1, w)$. The other two boundary curves of $\boldsymbol{P}_A(u, w)$, corresponding to $w=0$ and $w=1$, can be found by substituting $w=0$ and $w=1$ into Eq. (3.47):

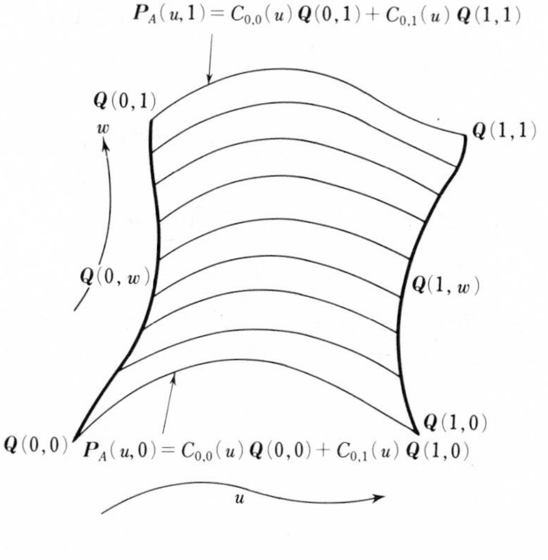

Fig. 3.14. The surface $\boldsymbol{P}_A(u, w)$

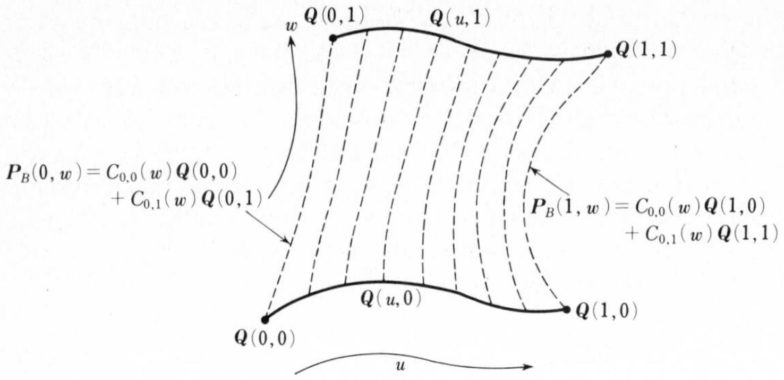

Fig. 3.15. The surface $P_B(u, w)$

$$\left.\begin{aligned} P_A(u, 0) &= C_{0,0}(u)\, Q(0,0) + C_{0,1}(u)\, Q(1,0) \\ P_A(u, 1) &= C_{0,0}(u)\, Q(0,1) + C_{0,1}(u)\, Q(1,1) \end{aligned}\right\}. \tag{3.50}$$ *)

Similarly, $P_B(u, w)$ is a curved surface which has $Q(u, 0)$ and $Q(u, 1)$ as two of its boundary curves (Fig. 3.15). The other two boundary curves of $P_B(u, w)$ are:

$$\left.\begin{aligned} P_B(0, w) &= C_{0,0}(w)\, Q(0,0) + C_{0,1}(w)\, Q(0,1) \\ P_B(1, w) &= C_{0,0}(w)\, Q(1,0) + C_{0,1}(w)\, Q(1,1) \end{aligned}\right\}. \tag{3.51}$$

Finally, let us discuss $P_C(u, w)$. To find the four boundary curves of this surface, set $u=0$, $u=1$, $w=0$ and $w=1$, respectively, to obtain:

$$\begin{aligned} P_C(0, w) &= [C_{0,0}(0)\;\; C_{0,1}(0)] \begin{bmatrix} Q(0,0) & Q(0,1) \\ Q(1,0) & Q(1,1) \end{bmatrix} \begin{bmatrix} C_{0,0}(w) \\ C_{0,1}(w) \end{bmatrix} \\ &= C_{0,0}(w)\, Q(0,0) + C_{0,1}(w)\, Q(0,1) \equiv P_B(0, w) \end{aligned} \tag{3.52}$$

$$P_C(1, w) = C_{0,0}(w)\, Q(1,0) + C_{0,1}(w)\, Q(1,1) \equiv P_B(1, w) \tag{3.53}$$

$$P_C(u, 0) = C_{0,0}(u)\, Q(0,0) + C_{0,1}(u)\, Q(1,0) \equiv P_A(u, 0) \tag{3.54}$$

$$P_C(u, 1) = C_{0,0}(u)\, Q(0,1) + C_{0,1}(u)\, Q(1,1) \equiv P_A(u, 1). \tag{3.55}$$

The surface $P_C(u, w)$ having the four curves $P_C(0, w)$, $P_C(1, w)$, $P_C(u, 0)$ and $P_C(u, 1)$ as boundaries is shown in Fig. 3.16.

From the above discussion, the Coons surface patch can be interpreted as follows. A Coons surface patch is composed of the three surfaces $P_A(u, w)$, $P_B(u, w)$ and $P_C(u, w)$. When surface $P_B(u, w)$ is added to curved surface $P_A(u, w)$, the curves

*) The next conditions are necessary for the blending functions $C_{0,0}$ and $C_{0,1}$, which are not explicitly written in the original paper (see Sect. 1.1.3):

$$C_{0,0}(t) + C_{0,1}(t) \equiv 1.$$

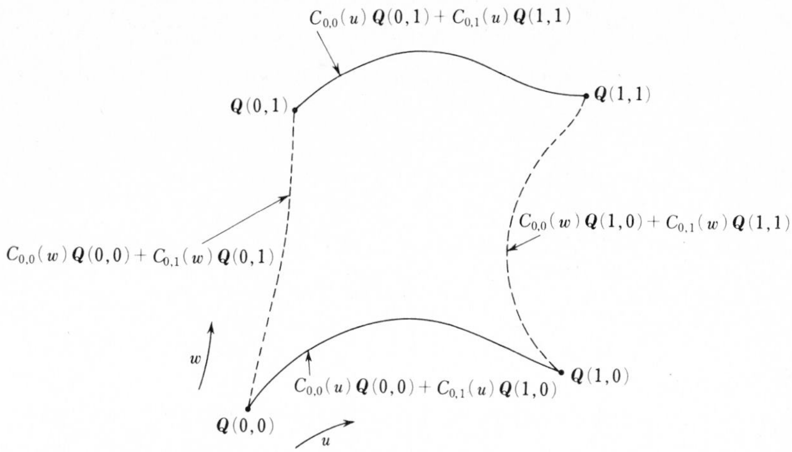

Fig. 3.16. The surface $\boldsymbol{P}_C(u, w)$

corresponding to $u=0$, $u=1$, $w=0$ and $w=1$ include some excess curves in addition to the boundary curves $\boldsymbol{Q}(0, w)$, $\boldsymbol{Q}(1, w)$, $\boldsymbol{Q}(u, 0)$ and $\boldsymbol{Q}(u, 1)$. These excess curves are $\boldsymbol{P}_B(0, w)$, $\boldsymbol{P}_B(1, w)$, $\boldsymbol{P}_A(u, 0)$ and $\boldsymbol{P}_A(u, 1)$. As can be seen from Eqs. (3.52) through (3.55), the third surface $\boldsymbol{P}_C(u, w)$ has these excess curves as its four boundary curves. Consequently, a Coons surface patch can be interpreted as expressing a surface that is bounded by the four arbitrarily specified boundary curves $\boldsymbol{Q}(u, 0)$, $\boldsymbol{Q}(u, 1)$, $\boldsymbol{Q}(0, w)$ and $\boldsymbol{Q}(1, w)$ (Fig. 3.17).

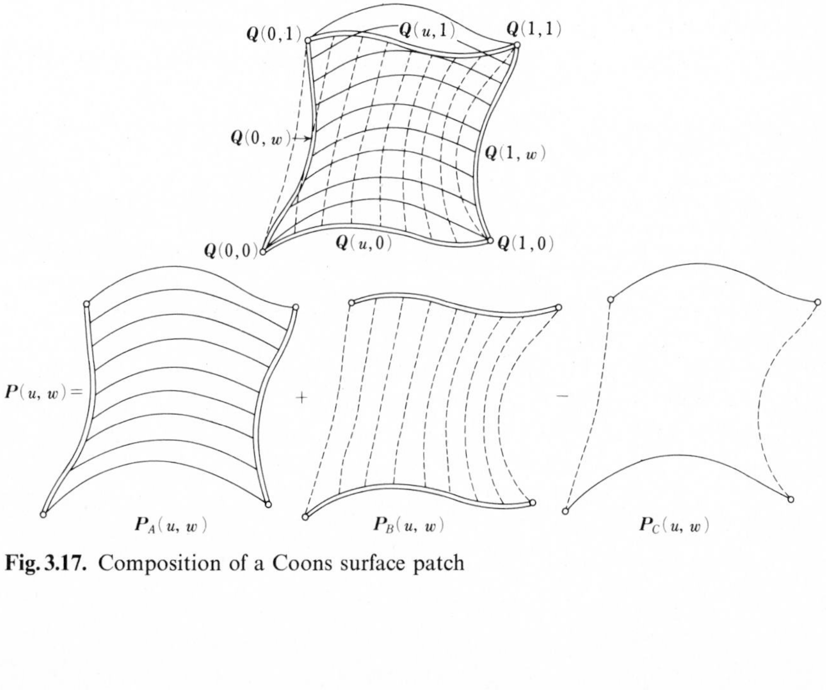

Fig. 3.17. Composition of a Coons surface patch

2) Add the following conditions for the blending functions to satisfy.

$$\dot{C}_{0,0}(0)=0,\ \dot{C}_{0,0}(1)=0,\ \dot{C}_{0,1}(0)=0,\ \dot{C}_{0,1}(1)=0 \tag{3.56}$$

that is:

$$\dot{C}_{0,i}(j)=0 \quad (i, j:0, 1). \tag{3.57}$$

In this case, we will investigate the tangent vectors in the direction across the boundary curves of the surface patch. For example, to find the tangent vector $\boldsymbol{P}_u(0, w)$ that crosses the boundary curve $\boldsymbol{P}(0, w)=\boldsymbol{Q}(0, w)$, first differentiate Eq. (3.43) by u:

$$\begin{aligned}\boldsymbol{P}_u(u, w) &= [\dot{C}_{0,0}(u)\ \ \dot{C}_{0,1}(u)]\begin{bmatrix}\boldsymbol{Q}(0, w)\\ \boldsymbol{Q}(1, w)\end{bmatrix}+[\boldsymbol{Q}_u(u,0)\ \ \boldsymbol{Q}_u(u, 1)]\begin{bmatrix}C_{0,0}(w)\\ C_{0,1}(w)\end{bmatrix}\\ &\quad -[\dot{C}_{0,0}(u)\ \ \dot{C}_{0,1}(u)]\begin{bmatrix}\boldsymbol{Q}(0,0) & \boldsymbol{Q}(0,1)\\ \boldsymbol{Q}(1,0) & \boldsymbol{Q}(1,1)\end{bmatrix}\begin{bmatrix}C_{0,0}(w)\\ C_{0,1}(w)\end{bmatrix}.\end{aligned} \tag{3.58}$$

Then set $u=0$:

$$\begin{aligned}\boldsymbol{P}_u(0, w) &= [\dot{C}_{0,0}(0)\ \ \dot{C}_{0,1}(0)]\begin{bmatrix}\boldsymbol{Q}(0, w)\\ \boldsymbol{Q}(1, w)\end{bmatrix}+[\boldsymbol{Q}_u(0,0)\ \ \boldsymbol{Q}_u(0, 1)]\begin{bmatrix}C_{0,0}(w)\\ C_{0,1}(w)\end{bmatrix}\\ &\quad -[\dot{C}_{0,0}(0)\ \ \dot{C}_{0,1}(0)]\begin{bmatrix}\boldsymbol{Q}(0,0) & \boldsymbol{Q}(0,1)\\ \boldsymbol{Q}(1,0) & \boldsymbol{Q}(1,1)\end{bmatrix}\begin{bmatrix}C_{0,0}(w)\\ C_{0,1}(w)\end{bmatrix}\\ &= C_{0,0}(w)\,\boldsymbol{Q}_u(0, 0)+C_{0,1}(w)\,\boldsymbol{Q}_u(0, 1).\end{aligned} \tag{3.59}$$

This equation shows that the tangent vectors in the direction across the boundary curve $\boldsymbol{P}(0, w)=\boldsymbol{Q}(0, w)$ are expressed only in terms of the tangent vectors in that direction at both ends of the boundary curve and the blending

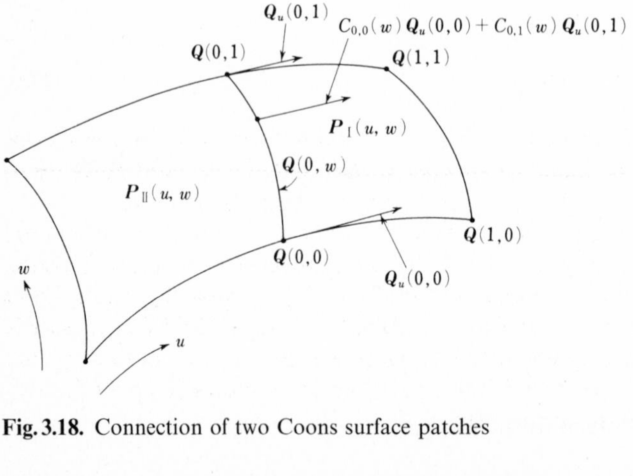

Fig. 3.18. Connection of two Coons surface patches

functions, and are independent of the shape of the boundary curve. Similarly, it can be seen that for the other boundary curves as well, the tangent vectors in the direction across the boundary curve are expressed by a formula similar to (3.59) in terms of the tangent vectors in that direction at both ends and the blending functions.

Consequently, if two such surface patches $\boldsymbol{P}_{\mathrm{I}}(u, w)$ and $\boldsymbol{P}_{\mathrm{II}}(u, w)$ share the same boundary curves, and have the same tangent vectors in the direction across the boundary curve at both ends of the boundary curve, then those two surface patches are connected continuously to the slope along that boundary curve (Fig. 3.18).

3) Add the following conditions to those to be satisfied by the blending functions.

$$\ddot{C}_{0,0}(0)=0, \ \ddot{C}_{0,0}(1)=0, \ \ddot{C}_{0,1}(0)=0, \ \ddot{C}_{0,1}(1)=0 \tag{3.60}$$

or equivalently:

$$\ddot{C}_{0,i}(j)=0 \quad (i, j: 0, 1). \tag{3.61}$$

Differentiating Eq. (3.58) by u once more and setting $u=0$ gives:

$$\boldsymbol{P}_{uu}(0, w)=C_{0,0}(w)\,\boldsymbol{Q}_{uu}(0, 0)+C_{0,1}(w)\,\boldsymbol{Q}_{uu}(0, 1). \tag{3.62}$$

This equation shows that along the boundary curve $\boldsymbol{P}(0, w)=\boldsymbol{Q}(0, w)$, the 2nd derivative vectors in the direction across the curve are determined solely by the 2nd derivative vectors in that direction at both ends of the curve. That is, along the boundary curve another surface patch can be connected with continuity up to the curvature.

Similarly, we have:

$$C_{0,0}^{(n)}(0)=0, \ C_{0,0}^{(n)}(1)=0, \ C_{0,1}^{(n)}(0)=0, \ C_{0,1}^{(n)}(1)=0 \tag{3.63}$$

or equivalently:

$$C_{0,i}^{(n)}(j)=0 \quad (i, j: 0, 1). \tag{3.64}$$

If these conditions are satisfied, then along a boundary curve of the Coons surface, connection can be made to another surface patch with continuity up to the nth-derivative vector in the direction across the boundary curve.

The conditions which must be satisfied by the blending functions which are needed to provide continuity up to the 2nd derivative vector in the direction across a boundary curve are summarized in Table 3.4. Comparing Table 3.4 to Tables 3.1 and 3.3, respectively, we see that functions $H_{0,0}(t)$ and $H_{0,1}(t)$ satisfy the conditions for continuity up to the derivative vector (slope continuity) and functions $K_{0,0}(t)$ and $K_{0,1}(t)$ satisfy the conditions for continuity up to the 2nd derivative vector (curvature continuity).

		$C_{0,0}(t)$	$C_{0,1}(t)$
Function value	$t=0$	1	0
	$t=1$	0	1
1st derivative	$t=0$	0	0
	$t=1$	0	0
2nd derivative	$t=0$	0	0
	$t=1$	0	0

Table 3.4. Conditions on the functions $C_{0,0}(t)$ and $C_{0,1}(t)$ so that the connection will be continuous up to the 2nd derivative vector

4) Taking the partial derivative of $\boldsymbol{P}_u(u, w)$ in Eq. (3.58) with respect to w and finding the cross partial derivative vector or twist vector $\boldsymbol{P}_{uw}(u, w)$ we obtain:

$$\boldsymbol{P}_{uw}(u,w)=[\dot{C}_{0,0}(u)\ \dot{C}_{0,1}(u)]\begin{bmatrix}\boldsymbol{Q}_w(0,w)\\ \boldsymbol{Q}_w(1,w)\end{bmatrix}+[\boldsymbol{Q}_u(u,0)\ \ \boldsymbol{Q}_u(u,1)]\begin{bmatrix}\dot{C}_{0,0}(w)\\ \dot{C}_{0,1}(w)\end{bmatrix}$$
$$-[\dot{C}_{0,0}(u)\ \dot{C}_{0,1}(u)]\begin{bmatrix}\boldsymbol{Q}(0,0) & \boldsymbol{Q}(0,1)\\ \boldsymbol{Q}(1,0) & \boldsymbol{Q}(1,1)\end{bmatrix}\begin{bmatrix}\dot{C}_{0,0}(w)\\ \dot{C}_{0,1}(w)\end{bmatrix}. \qquad (3.65)$$

Taking, for example, $u=w=0$ gives:

$$\boldsymbol{P}_{uw}(0,0)=[\dot{C}_{0,0}(0)\ \dot{C}_{0,1}(0)]\begin{bmatrix}\boldsymbol{Q}_w(0,0)\\ \boldsymbol{Q}_w(1,0)\end{bmatrix}+[\boldsymbol{Q}_u(0,0)\ \ \boldsymbol{Q}_u(0,1)]\begin{bmatrix}\dot{C}_{0,0}(0)\\ \dot{C}_{0,1}(0)\end{bmatrix}$$
$$-[\dot{C}_{0,0}(0)\ \dot{C}_{0,1}(0)]\begin{bmatrix}\boldsymbol{Q}(0,0) & \boldsymbol{Q}(0,1)\\ \boldsymbol{Q}(1,0) & \boldsymbol{Q}(1,1)\end{bmatrix}\begin{bmatrix}\dot{C}_{0,0}(0)\\ \dot{C}_{0,1}(0)\end{bmatrix}$$
$$=\boldsymbol{0}.$$

Similarly, the twist vector can be shown to be $\boldsymbol{0}$ at the other 3 points:

$$\boldsymbol{P}_{uw}(0,0)=\boldsymbol{P}_{uw}(1,0)=\boldsymbol{P}_{uw}(0,1)=\boldsymbol{P}_{uw}(1,1)=\boldsymbol{0}. \qquad (3.66)$$

The relations (3.66) also apply to a Ferguson surface patch. For a discussion of the effect of the twist vector on the surface shape, refer to Sect. 3.3.4.

5) On the Coons surface patch, the boundary curves can be given arbitrarily. Let us suppose the boundary curves $\boldsymbol{Q}(0, w)$ and $\boldsymbol{Q}(1, w)$ are given by the following equations:

$$\left.\begin{aligned}\boldsymbol{Q}(0,w)&=C_{0,0}(w)\,\boldsymbol{Q}(0,0)+C_{0,1}(w)\,\boldsymbol{Q}(0,1)\\ \boldsymbol{Q}(1,w)&=C_{0,0}(w)\,\boldsymbol{Q}(1,0)+C_{0,1}(w)\,\boldsymbol{Q}(1,1)\end{aligned}\right\}. \qquad (3.67)$$

In this case, the surface $\boldsymbol{P}_A(u,w)$ is:

$$\begin{aligned}\boldsymbol{P}_A(u,w)&=[C_{0,0}(u)\;\;C_{0,1}(u)]\begin{bmatrix}\boldsymbol{Q}(0,w)\\ \boldsymbol{Q}(1,w)\end{bmatrix}\\ &=[C_{0,0}(u)\;\;C_{0,1}(u)]\begin{bmatrix}C_{0,0}(w)\,\boldsymbol{Q}(0,0)+C_{0,1}(w)\,\boldsymbol{Q}(0,1)\\ C_{0,0}(w)\,\boldsymbol{Q}(1,0)+C_{0,1}(w)\,\boldsymbol{Q}(1,1)\end{bmatrix}\\ &=[C_{0,0}(u)\;\;C_{0,1}(u)]\begin{bmatrix}\boldsymbol{Q}(0,0) & \boldsymbol{Q}(0,1)\\ \boldsymbol{Q}(1,0) & \boldsymbol{Q}(1,1)\end{bmatrix}\begin{bmatrix}C_{0,0}(w)\\ C_{0,1}(w)\end{bmatrix}\\ &\equiv \boldsymbol{P}_C(u,w)\end{aligned}$$

so that, as a result, the Coons surface patch becomes:

$$\boldsymbol{P}(u,w)=\boldsymbol{P}_B(u,w).$$

Meanwhile, let us suppose the boundary curves $\boldsymbol{Q}(u,0)$ and $\boldsymbol{Q}(u,1)$ are given by the following equations:

$$\left.\begin{aligned}\boldsymbol{Q}(u,0)&=C_{0,0}(u)\,\boldsymbol{Q}(0,0)+C_{0,1}(u)\,\boldsymbol{Q}(1,0)\\ \boldsymbol{Q}(u,1)&=C_{0,0}(u)\,\boldsymbol{Q}(0,1)+C_{0,1}(u)\,\boldsymbol{Q}(1,1)\end{aligned}\right\}. \tag{3.68}$$

In this case, the surface $\boldsymbol{P}_B(u,w)$ becomes:

$$\boldsymbol{P}_B(u,w)\equiv \boldsymbol{P}_C(u,w)$$

so that the Coons surface patch becomes:

$$\boldsymbol{P}(u,w)=\boldsymbol{P}_A(u,w).$$

Consequently, if all four boundary curves $\boldsymbol{Q}(0,w)$, $\boldsymbol{Q}(1,w)$, $\boldsymbol{Q}(u,0)$ and $\boldsymbol{Q}(u,1)$ are given by Eqs. (3.67) and (3.68), then the Coons surface patch $\boldsymbol{P}(u,w)$ becomes:

$$\begin{aligned}\boldsymbol{P}(u,w)&=[C_{0,0}(u)\;\;C_{0,1}(u)]\begin{bmatrix}\boldsymbol{Q}(0,0) & \boldsymbol{Q}(0,1)\\ \boldsymbol{Q}(1,0) & \boldsymbol{Q}(1,1)\end{bmatrix}\begin{bmatrix}C_{0,0}(w)\\ C_{0,1}(w)\end{bmatrix}\\ &\equiv \boldsymbol{P}_C(u,w).\end{aligned} \tag{3.69}$$

6) The lowest order polynomial which satisfies the condition equations (3.44) for the Coons blending functions is:

$$C_{0,0}(t)=1-t,\qquad C_{0,1}(t)=t. \tag{3.70}$$

In this case, the Coons surface becomes:

$$P(u,w)=[1-u \ \ u]\begin{bmatrix} Q(0,w) \\ Q(1,w) \end{bmatrix}+[Q(u,0) \ \ Q(u,1)]\begin{bmatrix} 1-w \\ w \end{bmatrix}$$
$$-[1-u \ \ u]\begin{bmatrix} Q(0,0) & Q(0,1) \\ Q(1,0) & Q(1,1) \end{bmatrix}\begin{bmatrix} 1-w \\ w \end{bmatrix}. \quad (3.71)$$

This is the formula for a surface patch of C^0 continuity.

As was discussed in section 5), if the four boundary curves may be assumed to have the following form:

$$Q(0,w)=(1-w)\,Q(0,0)+w\,Q(0,1)$$
$$Q(1,w)=(1-w)\,Q(1,0)+w\,Q(1,1)$$
$$Q(u,0)=(1-u)\,Q(0,0)+u\,Q(1,0)$$
$$Q(u,1)=(1-u)\,Q(0,1)+u\,Q(1,1)$$

then the Coons surface patch is a bilinear surface patch such as the following:

$$P(u,w)=[1-u \ \ u]\begin{bmatrix} Q(0,0) & Q(0,1) \\ Q(1,0) & Q(1,1) \end{bmatrix}\begin{bmatrix} 1-w \\ w \end{bmatrix}. \quad (3.72)$$

7) The simplest polynomials which satisfy both conditions (3.44) and (3.56) for Coons blending functions are $H_{0,0}(t)$ and $H_{0,1}(t)$. In this case, the Coons surface is:

$$P(u,w)=[H_{0,0}(u) \ \ H_{0,1}(u)]\begin{bmatrix} Q(0,w) \\ Q(1,w) \end{bmatrix}+[Q(u,0) \ \ Q(u,1)]\begin{bmatrix} H_{0,0}(w) \\ H_{0,1}(w) \end{bmatrix}$$
$$-[H_{0,0}(u) \ \ H_{0,1}(u)]\begin{bmatrix} Q(0,0) & Q(0,1) \\ Q(1,0) & Q(1,1) \end{bmatrix}\begin{bmatrix} H_{0,0}(w) \\ H_{0,1}(w) \end{bmatrix}. \quad (3.73)$$

This is the equation for a surface patch that has class C^1 continuity.

Let us assume that the four boundary curves are Ferguson curve segments (Eq. (3.12)):

$$Q(0,w)=H_{0,0}(w)\,Q(0,0)+H_{0,1}(w)\,Q(0,1)+H_{1,0}(w)\,Q_w(0,0)$$
$$+H_{1,1}(w)\,Q_w(0,1)$$
$$Q(1,w)=H_{0,0}(w)\,Q(1,0)+H_{0,1}(w)\,Q(1,1)+H_{1,0}(w)\,Q_w(1,0)$$
$$+H_{1,1}(w)\,Q_w(1,1)$$
$$Q(u,0)=H_{0,0}(u)\,Q(0,0)+H_{0,1}(u)\,Q(1,0)+H_{1,0}(u)\,Q_u(0,0)$$
$$+H_{1,1}(u)\,Q_u(1,0)$$
$$Q(u,1)=H_{0,0}(u)\,Q(0,1)+H_{0,1}(u)\,Q(1,1)+H_{1,0}(u)\,Q_u(0,1)$$
$$+H_{1,1}(u)\,Q_u(1,1).$$

In this case, $\boldsymbol{P}_A(u, w)$ in the Coons surface patch equation is:

$$\boldsymbol{P}_A(u,w) = [H_{0,0}(u)\ H_{0,1}(u)] \begin{bmatrix} \boldsymbol{Q}(0,w) \\ \boldsymbol{Q}(1,w) \end{bmatrix}$$

$$= [H_{0,0}(u)\ H_{0,1}(u)] \begin{bmatrix} \boldsymbol{Q}(0,0) & \boldsymbol{Q}(0,1) & \boldsymbol{Q}_w(0,0) & \boldsymbol{Q}_w(0,1) \\ \boldsymbol{Q}(1,0) & \boldsymbol{Q}(1,1) & \boldsymbol{Q}_w(1,0) & \boldsymbol{Q}_w(1,1) \end{bmatrix} \begin{bmatrix} H_{0,0}(w) \\ H_{0,1}(w) \\ H_{1,0}(w) \\ H_{1,1}(w) \end{bmatrix}$$

$$= [H_{0,0}(u)\ H_{0,1}(u)\ H_{1,0}(u)\ H_{1,1}(u)] \times \begin{bmatrix} \boldsymbol{Q}(0,0) & \boldsymbol{Q}(0,1) & \boldsymbol{Q}_w(0,0) & \boldsymbol{Q}_w(0,1) \\ \boldsymbol{Q}(1,0) & \boldsymbol{Q}(1,1) & \boldsymbol{Q}_w(1,0) & \boldsymbol{Q}_w(1,1) \\ \boldsymbol{0} & \boldsymbol{0} & \boldsymbol{0} & \boldsymbol{0} \\ \boldsymbol{0} & \boldsymbol{0} & \boldsymbol{0} & \boldsymbol{0} \end{bmatrix} \begin{bmatrix} H_{0,0}(w) \\ H_{0,1}(w) \\ H_{1,0}(w) \\ H_{1,1}(w) \end{bmatrix}. \tag{3.74}$$

The $\boldsymbol{P}_B(u, w)$ surface is:

$$\boldsymbol{P}_B(u,w) = [H_{0,0}(u)\ H_{0,1}(u)\ H_{1,0}(u)\ H_{1,1}(u)] \times \begin{bmatrix} \boldsymbol{Q}(0,0) & \boldsymbol{Q}(0,1) & \boldsymbol{0} & \boldsymbol{0} \\ \boldsymbol{Q}(1,0) & \boldsymbol{Q}(1,1) & \boldsymbol{0} & \boldsymbol{0} \\ \boldsymbol{Q}_u(0,0) & \boldsymbol{Q}_u(0,1) & \boldsymbol{0} & \boldsymbol{0} \\ \boldsymbol{Q}_u(1,0) & \boldsymbol{Q}_u(1,1) & \boldsymbol{0} & \boldsymbol{0} \end{bmatrix} \begin{bmatrix} H_{0,0}(w) \\ H_{0,1}(w) \\ H_{1,0}(w) \\ H_{1,1}(w) \end{bmatrix} \tag{3.75}$$

In addition, the $\boldsymbol{P}_C(u, w)$ surface can be deformed as follows:

$$\boldsymbol{P}_C(u,w) = [H_{0,0}(u)\ H_{0,1}(u)\ H_{1,0}(u)\ H_{1,1}(u)] \times \begin{bmatrix} \boldsymbol{Q}(0,0) & \boldsymbol{Q}(0,1) & \boldsymbol{0} & \boldsymbol{0} \\ \boldsymbol{Q}(1,0) & \boldsymbol{Q}(1,1) & \boldsymbol{0} & \boldsymbol{0} \\ \boldsymbol{0} & \boldsymbol{0} & \boldsymbol{0} & \boldsymbol{0} \\ \boldsymbol{0} & \boldsymbol{0} & \boldsymbol{0} & \boldsymbol{0} \end{bmatrix} \begin{bmatrix} H_{0,0}(w) \\ H_{0,1}(w) \\ H_{1,0}(w) \\ H_{1,1}(w) \end{bmatrix}. \tag{3.76}$$

From Eqs. (3.74), (3.75) and (3.76), the Coons surface patch $\boldsymbol{P}(u, w)$ is:

$$\boldsymbol{P}(u,w) = [H_{0,0}(u)\ H_{0,1}(u)\ H_{1,0}(u)\ H_{1,1}(u)] \times \begin{bmatrix} \boldsymbol{Q}(0,0) & \boldsymbol{Q}(0,1) & \boldsymbol{Q}_w(0,0) & \boldsymbol{Q}_w(0,1) \\ \boldsymbol{Q}(1,0) & \boldsymbol{Q}(1,1) & \boldsymbol{Q}_w(1,0) & \boldsymbol{Q}_w(1,1) \\ \boldsymbol{Q}_u(0,0) & \boldsymbol{Q}_u(0,1) & \boldsymbol{0} & \boldsymbol{0} \\ \boldsymbol{Q}_u(1,0) & \boldsymbol{Q}_u(1,1) & \boldsymbol{0} & \boldsymbol{0} \end{bmatrix} \begin{bmatrix} H_{0,0}(w) \\ H_{0,1}(w) \\ H_{1,0}(w) \\ H_{1,1}(w) \end{bmatrix} \tag{3.77}$$

so that the Coons surface patch agrees with the Ferguson surface patch. In fact, the Ferguson surface patch is a special case of the Coons surface patch proposed in 1964, in which the boundary curves are cubic Hermite interpolation curves (Ferguson curve segments).

3.3.3 The Coons Surface Patches (1967)

In 1967 Coons announced a generalization of the surface patch formula which he first announced in 1964, with higher degree terms added. This generalized surface has the following form:

$$\begin{aligned} \boldsymbol{P}(u,w) = & \sum_{i=0}^{1}\sum_{r=0}^{n} C_{r,i}(u)\,\boldsymbol{Q}^{r,0}(i,w) + \sum_{j=0}^{1}\sum_{s=0}^{n} C_{s,j}(w)\;\boldsymbol{Q}^{0,s}(u,j) \\ & - \sum_{i=0}^{1}\sum_{j=0}^{1}\sum_{r=0}^{n}\sum_{s=0}^{n} C_{r,i}(u)\,C_{s,j}(w)\,\boldsymbol{Q}^{r,s}(i,j) \end{aligned} \tag{3.78}$$

where:

$$\boldsymbol{Q}^{a,b}(i,j) = \frac{\partial^{a+b}}{\partial u^a \partial w^b}\boldsymbol{Q}(u,w)\bigg|_{\substack{u=i\\w=j}} . \tag{3.79}$$

Equation (3.78) can be expressed in matrix form as:

$$\begin{aligned} \boldsymbol{P}(u,w) = & [C_{0,0}(u)\;\; C_{0,1}(u)\;\; C_{1,0}(u)\;\; C_{1,1}(u)\;\ldots] \begin{bmatrix} \boldsymbol{Q}(0,w) \\ \boldsymbol{Q}(1,w) \\ \boldsymbol{Q}_u(0,w) \\ \boldsymbol{Q}_u(1,w) \\ \vdots \end{bmatrix} \\ & + [\boldsymbol{Q}(u,0)\;\; \boldsymbol{Q}(u,1)\;\; \boldsymbol{Q}_w(u,0)\;\; \boldsymbol{Q}_w(u,1)\;\ldots] \begin{bmatrix} C_{0,0}(w) \\ C_{0,1}(w) \\ C_{1,0}(w) \\ C_{1,1}(w) \\ \vdots \end{bmatrix} \\ & - [C_{0,0}(u)\;\; C_{0,1}(u)\;\; C_{1,0}(u)\;\; C_{1,1}(u)\;\ldots] \\ & \times \begin{bmatrix} \boldsymbol{Q}(0,0) & \boldsymbol{Q}(0,1) & \boldsymbol{Q}_w(0,0) & \boldsymbol{Q}_w(0,1) & \ldots \\ \boldsymbol{Q}(1,0) & \boldsymbol{Q}(1,1) & \boldsymbol{Q}_w(1,0) & \boldsymbol{Q}_w(1,1) & \ldots \\ \boldsymbol{Q}_u(0,0) & \boldsymbol{Q}_u(0,1) & \boldsymbol{Q}_{uw}(0,0) & \boldsymbol{Q}_{uw}(0,1) & \ldots \\ \boldsymbol{Q}_u(1,0) & \boldsymbol{Q}_u(1,1) & \boldsymbol{Q}_{uw}(1,0) & \boldsymbol{Q}_{uw}(1,1) & \ldots \\ \ldots & \ldots & \ldots & \ldots & \ldots \\ \ldots & \ldots & \ldots & \ldots & \ldots \end{bmatrix} \begin{bmatrix} C_{0,0}(w) \\ C_{0,1}(w) \\ C_{1,0}(w) \\ C_{1,1}(w) \\ \vdots \\ \vdots \end{bmatrix} . \end{aligned} \tag{3.80}$$

From a practical point of view, it is sufficient to use the following equation in place of Eq. (3.78):

$$\boldsymbol{P}(u,w)=\sum_{i=0}^{1}\sum_{r=0}^{1} C_{r,i}(u)\,\boldsymbol{Q}^{r,0}(i,w)+\sum_{j=0}^{1}\sum_{s=0}^{1} C_{s,j}(w)\,\boldsymbol{Q}^{0,s}(u,j)$$

$$-\sum_{i=0}^{1}\sum_{j=0}^{1}\sum_{r=0}^{1}\sum_{s=0}^{1} C_{r,i}(u)\,C_{s,j}(w)\,\boldsymbol{Q}^{r,s}(i,j). \tag{3.81}$$

Equation (3.81) can be expressed in matrix form as:

$$\boldsymbol{P}(u,w)=-[-1\;\; C_{0,0}(u)\;\; C_{0,1}(u)\;\; C_{1,0}(u)\;\; C_{1,1}(u)]$$

$$\times\begin{bmatrix} \boldsymbol{0} & \boldsymbol{Q}(u,0) & \boldsymbol{Q}(u,1) & \boldsymbol{Q}_w(u,0) & \boldsymbol{Q}_w(u,1)\\ \boldsymbol{Q}(0,w) & \boldsymbol{Q}(0,0) & \boldsymbol{Q}(0,1) & \boldsymbol{Q}_w(0,0) & \boldsymbol{Q}_w(0,1)\\ \boldsymbol{Q}(1,w) & \boldsymbol{Q}(1,0) & \boldsymbol{Q}(1,1) & \boldsymbol{Q}_w(1,0) & \boldsymbol{Q}_w(1,1)\\ \boldsymbol{Q}_u(0,w) & \boldsymbol{Q}_u(0,0) & \boldsymbol{Q}_u(0,1) & \boldsymbol{Q}_{uw}(0,0) & \boldsymbol{Q}_{uw}(0,1)\\ \boldsymbol{Q}_u(1,w) & \boldsymbol{Q}_u(1,0) & \boldsymbol{Q}_u(1,1) & \boldsymbol{Q}_{uw}(1,0) & \boldsymbol{Q}_{uw}(1,1) \end{bmatrix}\begin{bmatrix} -1\\ C_{0,0}(w)\\ C_{0,1}(w)\\ C_{1,0}(w)\\ C_{1,1}(w)\end{bmatrix} \tag{3.82}$$

or, alternatively:

$$\boldsymbol{P}(u,w)=\underbrace{[C_{0,0}(u)\;\; C_{0,1}(u)\;\; C_{1,0}(u)\;\; C_{1,1}(u)]\begin{bmatrix}\boldsymbol{Q}(0,w)\\ \boldsymbol{Q}(1,w)\\ \boldsymbol{Q}_u(0,w)\\ \boldsymbol{Q}_u(1,w)\end{bmatrix}}_{\boldsymbol{P}_A(u,w)}.$$

$$+\underbrace{[\boldsymbol{Q}(u,0)\;\; \boldsymbol{Q}(u,1)\;\; \boldsymbol{Q}_w(u,0)\;\; \boldsymbol{Q}_w(u,1)]\begin{bmatrix}C_{0,0}(w)\\ C_{0,1}(w)\\ C_{1,0}(w)\\ C_{1,1}(w)\end{bmatrix}}_{\boldsymbol{P}_B(u,w)}$$

$$-[C_{0,0}(u)\;\; C_{0,1}(u)\;\; C_{1,0}(u)\;\; C_{1,1}(u)]$$

$$\underbrace{\times\begin{bmatrix} \boldsymbol{Q}(0,0) & \boldsymbol{Q}(0,1) & \boldsymbol{Q}_w(0,0) & \boldsymbol{Q}_w(0,1)\\ \boldsymbol{Q}(1,0) & \boldsymbol{Q}(1,1) & \boldsymbol{Q}_w(1,0) & \boldsymbol{Q}_w(1,1)\\ \boldsymbol{Q}_u(0,0) & \boldsymbol{Q}_u(0,1) & \boldsymbol{Q}_{uw}(0,0) & \boldsymbol{Q}_{uw}(0,1)\\ \boldsymbol{Q}_u(1,0) & \boldsymbol{Q}_u(1,1) & \boldsymbol{Q}_{uw}(1,0) & \boldsymbol{Q}_{uw}(1,1) \end{bmatrix}\begin{bmatrix}C_{0,0}(w)\\ C_{0,1}(w)\\ C_{1,0}(w)\\ C_{1,1}(w)\end{bmatrix}}_{\boldsymbol{P}_C(u,w)} \tag{3.83}$$

The surface patch formula (3.81) (or (3.82) or (3.83)) shows that one surface patch is defined by the four boundary curves $\boldsymbol{Q}(0,w)$, $\boldsymbol{Q}(1,w)$, $\boldsymbol{Q}(u,0)$ and $\boldsymbol{Q}(u,1)$; the functions $\boldsymbol{Q}_u(0,w)$, $\boldsymbol{Q}_u(1,w)$, $\boldsymbol{Q}_w(u,0)$ and $\boldsymbol{Q}_w(u,1)$ of the tangent

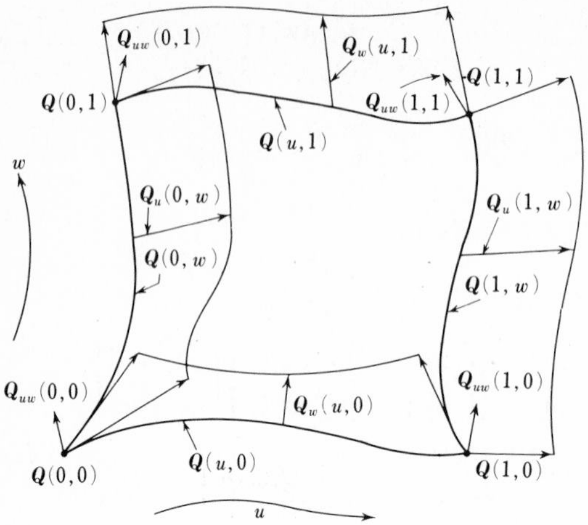

Fig. 3.19. A Coons surface patch (according to Coons' 1967 theory)

vectors in the directions across those boundary curves, respectively, the tangent vectors and the twist vectors at the 4 corner points (Fig. 3.19). $C_{0,0}(t)$, $C_{0,1}(t)$, $C_{1,0}(t)$ and $C_{1,1}(t)$ are monotonic, continuous scalar functions; the conditions on $C_{0,0}(t)$ and $C_{0,1}(t)$ have already been given (refer to Table 3.4).

The surface patch announced in 1967, like that of 1964, is expressed in terms of sums and differences of 3 surfaces.

1) The following conditions are imposed on the blending functions:

$$C_{0,0}(0)=1,\ C_{0,0}(1)=0,\ C_{0,1}(0)=0,\ C_{0,1}(1)=1 \tag{3.84}$$

$$C_{1,0}(0)=0,\ C_{1,0}(1)=0,\ C_{1,1}(0)=0,\ C_{1,1}(1)=0. \tag{3.85}$$

These conditions can be expressed in terms of the Kronecker delta:

$$C_{0,i}(j)=\delta_{ij},\ \ C_{1,i}(j)=0 \quad (i, j: 0, 1). \tag{3.86}$$

Let us show that in this case, the surface patch $\boldsymbol{P}(u, w)$ has the specified curves $\boldsymbol{Q}(u, 0)$, $\boldsymbol{Q}(0, w)$, $\boldsymbol{Q}(u, 1)$ and $\boldsymbol{Q}(1, w)$ as boundary curves. Let us consider, for example, the surface formula (3.83) with $w=0$:

$$\boldsymbol{P}(u,0)=[C_{0,0}(u)\ C_{0,1}(u)\ C_{1,0}(u)\ C_{1,1}(u)]\begin{bmatrix}\boldsymbol{Q}(0,0)\\ \boldsymbol{Q}(1,0)\\ \boldsymbol{Q}_u(0,0)\\ \boldsymbol{Q}_u(1,0)\end{bmatrix}$$

$$+[\boldsymbol{Q}(u,0)\ \boldsymbol{Q}(u,1)\ \boldsymbol{Q}_w(u,0)\ \boldsymbol{Q}_w(u,1)]\begin{bmatrix}1\\0\\0\\0\end{bmatrix}-$$

$$-[C_{0,0}(u)\ C_{0,1}(u)\ C_{1,0}(u)\ C_{1,1}(u)]$$

$$\times\begin{bmatrix} \boldsymbol{Q}(0,0) & \boldsymbol{Q}(0,1) & \boldsymbol{Q}_w(0,0) & \boldsymbol{Q}_w(0,1) \\ \boldsymbol{Q}(1,0) & \boldsymbol{Q}(1,1) & \boldsymbol{Q}_w(1,0) & \boldsymbol{Q}_w(1,1) \\ \boldsymbol{Q}_u(0,0) & \boldsymbol{Q}_u(0,1) & \boldsymbol{Q}_{uw}(0,0) & \boldsymbol{Q}_{uw}(0,1) \\ \boldsymbol{Q}_u(1,0) & \boldsymbol{Q}_u(1,1) & \boldsymbol{Q}_{uw}(1,0) & \boldsymbol{Q}_{uw}(1,1) \end{bmatrix}\begin{bmatrix}1\\0\\0\\0\end{bmatrix}$$

$$=[C_{0,0}(u)\ C_{0,1}(u)\ C_{1,0}(u)\ C_{1,1}(u)]\begin{bmatrix}\boldsymbol{Q}(0,0)\\ \boldsymbol{Q}(1,0)\\ \boldsymbol{Q}_u(0,0)\\ \boldsymbol{Q}_u(1,0)\end{bmatrix}$$

$$+\boldsymbol{Q}(u,0)$$

$$-[C_{0,0}(u)\ C_{0,1}(u)\ C_{1,0}(u)\ C_{1,1}(u)]\begin{bmatrix}\boldsymbol{Q}(0,0)\\ \boldsymbol{Q}(1,0)\\ \boldsymbol{Q}_u(0,0)\\ \boldsymbol{Q}_u(1,0)\end{bmatrix}$$

$$=\boldsymbol{Q}(u,0).$$

We can then repeat the same procedure for the other cases to show that the surface patch expressed by (3.83) has the specified curves as boundary curves.

2) Next, apply the following additional conditions to the blending functions:

$$\dot{C}_{0,0}(0)=0,\ \dot{C}_{0,0}(1)=0,\ \dot{C}_{0,1}(0)=0,\ \dot{C}_{0,1}(1)=0 \tag{3.87}$$

$$\dot{C}_{1,0}(0)=1,\ \dot{C}_{1,0}(1)=0,\ \dot{C}_{1,1}(0)=0,\ \dot{C}_{1,1}(1)=1 \tag{3.88}$$

Conditions (3.87) and (3.88) can be expressed in terms of the Kronecker delta:

$$\dot{C}_{0,i}(j)=0,\ \dot{C}_{1,i}(j)=\delta_{ij}\quad (i,j:0,1). \tag{3.89}$$

Let us show that in this case, on the boundary curves of the surface patch $\boldsymbol{P}(u,w)$ the tangent vectors in the directions across those curves agree with the specified tangent vector functions. We can show this as follows. First, differentiating the surface formula with respect to u, we obtain:

$$\boldsymbol{P}_u(u,w)=[\dot{C}_{0,0}(u)\ \dot{C}_{0,1}(u)\ \dot{C}_{1,0}(u)\ \dot{C}_{1,1}(u)]\begin{bmatrix}\boldsymbol{Q}(0,w)\\ \boldsymbol{Q}(1,w)\\ \boldsymbol{Q}_u(0,w)\\ \boldsymbol{Q}_u(1,w)\end{bmatrix}$$

$$+[\boldsymbol{Q}_u(u,0)\ \boldsymbol{Q}_u(u,1)\ \boldsymbol{Q}_{wu}(u,0)\ \boldsymbol{Q}_{wu}(u,1)]\begin{bmatrix}C_{0,0}(w)\\ C_{0,1}(w)\\ C_{1,0}(w)\\ C_{1,1}(w)\end{bmatrix}-$$

$$-[\dot{C}_{0,0}(u)\ \dot{C}_{0,1}(u)\ \dot{C}_{1,0}(u)\ \dot{C}_{1,1}(u)]$$

$$\times\begin{bmatrix} \boldsymbol{Q}(0,0) & \boldsymbol{Q}(0,1) & \boldsymbol{Q}_w(0,0) & \boldsymbol{Q}_w(0,1) \\ \boldsymbol{Q}(1,0) & \boldsymbol{Q}(1,1) & \boldsymbol{Q}_w(1,0) & \boldsymbol{Q}_w(1,1) \\ \boldsymbol{Q}_u(0,0) & \boldsymbol{Q}_u(0,1) & \boldsymbol{Q}_{uw}(0,0) & \boldsymbol{Q}_{uw}(0,1) \\ \boldsymbol{Q}_u(1,0) & \boldsymbol{Q}_u(1,1) & \boldsymbol{Q}_{uw}(1,0) & \boldsymbol{Q}_{uw}(1,1) \end{bmatrix} \begin{bmatrix} C_{0,0}(w) \\ C_{0,1}(w) \\ C_{1,0}(w) \\ C_{1,1}(w) \end{bmatrix} \tag{3.90}$$

Setting $u=0$ in this expression gives:

$$\boldsymbol{P}_u(0,w)=[0\ \ 0\ \ 1\ \ 0]\begin{bmatrix} \boldsymbol{Q}(0,w) \\ \boldsymbol{Q}(1,w) \\ \boldsymbol{Q}_u(0,w) \\ \boldsymbol{Q}_u(1,w) \end{bmatrix}$$

$$+[\boldsymbol{Q}_u(0,0)\ \boldsymbol{Q}_u(0,1)\ \boldsymbol{Q}_{wu}(0,0)\ \boldsymbol{Q}_{wu}(0,1)]\begin{bmatrix} C_{0,0}(w) \\ C_{0,1}(w) \\ C_{1,0}(w) \\ C_{1,1}(w) \end{bmatrix}$$

$$-[0\ \ 0\ \ 1\ \ 0]\begin{bmatrix} \boldsymbol{Q}(0,0) & \boldsymbol{Q}(0,1) & \boldsymbol{Q}_w(0,0) & \boldsymbol{Q}_w(0,1) \\ \boldsymbol{Q}(1,0) & \boldsymbol{Q}(1,1) & \boldsymbol{Q}_w(1,0) & \boldsymbol{Q}_w(1,1) \\ \boldsymbol{Q}_u(0,0) & \boldsymbol{Q}_u(0,1) & \boldsymbol{Q}_{uw}(0,0) & \boldsymbol{Q}_{uw}(0,1) \\ \boldsymbol{Q}_u(1,0) & \boldsymbol{Q}_u(1,1) & \boldsymbol{Q}_{uw}(1,0) & \boldsymbol{Q}_{uw}(1,1) \end{bmatrix} \begin{bmatrix} C_{0,0}(w) \\ C_{0,1}(w) \\ C_{1,0}(w) \\ C_{1,1}(w) \end{bmatrix}$$

$$=\boldsymbol{Q}_u(0,w)+C_{1,0}(w)(\boldsymbol{Q}_{wu}(0,0)-\boldsymbol{Q}_{uw}(0,0))+C_{1,1}(w)(\boldsymbol{Q}_{wu}(0,1)-\boldsymbol{Q}_{uw}(0,1)).$$

If $\boldsymbol{Q}_{wu}(0,0)=\boldsymbol{Q}_{uw}(0,0)$ and $\boldsymbol{Q}_{wu}(0,1)=\boldsymbol{Q}_{uw}(0,1)$, then we have

$$\boldsymbol{P}_u(0,w)=\boldsymbol{Q}_u(0,w)\,^{*)}.$$

Next, let us consider the cross partial derivative vectors, that is, the twist vectors, at the 4 corners. Differentiating Eq. (3.90) again with respect to w gives:

$$\boldsymbol{P}_{uw}(u,w)=[\dot{C}_{0,0}(u)\ \dot{C}_{0,1}(u)\ \dot{C}_{1,0}(u)\ \dot{C}_{1,1}(u)]\begin{bmatrix} \boldsymbol{Q}_w(0,w) \\ \boldsymbol{Q}_w(1,w) \\ \boldsymbol{Q}_{uw}(0,w) \\ \boldsymbol{Q}_{uw}(1,w) \end{bmatrix}$$

$$+[\boldsymbol{Q}_u(u,0)\ \boldsymbol{Q}_u(u,1)\ \boldsymbol{Q}_{wu}(u,0)\ \boldsymbol{Q}_{wu}(u,1)]\begin{bmatrix} \dot{C}_{0,0}(w) \\ \dot{C}_{0,1}(w) \\ \dot{C}_{1,0}(w) \\ \dot{C}_{1,1}(w) \end{bmatrix}-$$

*) Although this is not stated explicitly in Coons' paper, it is necessary to assume that on a Coons surface patch the cross partial derivative vectors are uniquely determined at the 4 corners. This implies that $\boldsymbol{Q}_{uw}(0,0)=\boldsymbol{Q}_{wu}(0,0)$, $\boldsymbol{Q}_{uw}(1,0)=\boldsymbol{Q}_{wu}(1,0)$, $\boldsymbol{Q}_{uw}(0,1)=\boldsymbol{Q}_{wu}(0,1)$ and $\boldsymbol{Q}_{uw}(1,1)=\boldsymbol{Q}_{wu}(1,1)$.

$$-[\dot{C}_{0,0}(u)\ \dot{C}_{0,1}(u)\ \dot{C}_{1,0}(u)\ \dot{C}_{1,1}(u)]$$

$$\times\begin{bmatrix} \boldsymbol{Q}(0,0) & \boldsymbol{Q}(0,1) & \boldsymbol{Q}_w(0,0) & \boldsymbol{Q}_w(0,1) \\ \boldsymbol{Q}(1,0) & \boldsymbol{Q}(1,1) & \boldsymbol{Q}_w(1,0) & \boldsymbol{Q}_w(1,1) \\ \boldsymbol{Q}_u(0,0) & \boldsymbol{Q}_u(0,1) & \boldsymbol{Q}_{uw}(0,0) & \boldsymbol{Q}_{uw}(0,1) \\ \boldsymbol{Q}_u(1,0) & \boldsymbol{Q}_u(1,1) & \boldsymbol{Q}_{uw}(1,0) & \boldsymbol{Q}_{uw}(1,1) \end{bmatrix}\begin{bmatrix} \dot{C}_{0,0}(w) \\ \dot{C}_{0,1}(w) \\ \dot{C}_{1,0}(w) \\ \dot{C}_{1,1}(w) \end{bmatrix}.$$

Setting, for example, $u=w=0$ in the above equation gives:

$$\boldsymbol{P}_{uw}(0,0)=[0\ 0\ 1\ 0]\begin{bmatrix} \boldsymbol{Q}_w(0,0) \\ \boldsymbol{Q}_w(1,0) \\ \boldsymbol{Q}_{uw}(0,0) \\ \boldsymbol{Q}_{uw}(1,0) \end{bmatrix}$$

$$+[\boldsymbol{Q}_u(0,0)\ \boldsymbol{Q}_u(0,1)\ \boldsymbol{Q}_{wu}(0,0)\ \boldsymbol{Q}_{wu}(0,1)]\begin{bmatrix} 0 \\ 0 \\ 1 \\ 0 \end{bmatrix}$$

$$-[0\ 0\ 1\ 0]\begin{bmatrix} \boldsymbol{Q}(0,0) & \boldsymbol{Q}(0,1) & \boldsymbol{Q}_w(0,0) & \boldsymbol{Q}_w(0,1) \\ \boldsymbol{Q}(1,0) & \boldsymbol{Q}(1,1) & \boldsymbol{Q}_w(1,0) & \boldsymbol{Q}_w(1,1) \\ \boldsymbol{Q}_u(0,0) & \boldsymbol{Q}_u(0,1) & \boldsymbol{Q}_{uw}(0,0) & \boldsymbol{Q}_{uw}(0,1) \\ \boldsymbol{Q}_u(1,0) & \boldsymbol{Q}_u(1,1) & \boldsymbol{Q}_{uw}(1,0) & \boldsymbol{Q}_{uw}(1,1) \end{bmatrix}\begin{bmatrix} 0 \\ 0 \\ 1 \\ 0 \end{bmatrix}$$

$$=\boldsymbol{Q}_{uw}(0,0)+\boldsymbol{Q}_{wu}(0,0)-\boldsymbol{Q}_{uw}(0,0)$$

$$=\boldsymbol{Q}_{wu}(0,0).$$

If $\boldsymbol{Q}_{wu}(0,0)=\boldsymbol{Q}_{uw}(0,0)$, then $\boldsymbol{P}_{uw}(0,0)=\boldsymbol{Q}_{uw}(0,0)$. Similarly, it can be shown that at the other 3 points if $\boldsymbol{Q}_{wu}(1,0)=\boldsymbol{Q}_{uw}(1,0)$, $\boldsymbol{Q}_{wu}(0,1)=\boldsymbol{Q}_{uw}(0,1)$ and $\boldsymbol{Q}_{wu}(1,1)=\boldsymbol{Q}_{uw}(1,1)$, then $\boldsymbol{P}_{uw}(1,0)=\boldsymbol{Q}_{uw}(1,0)$, $\boldsymbol{P}_{uw}(0,1)=\boldsymbol{Q}_{uw}(0,1)$ and $\boldsymbol{P}_{uw}(1,1)=\boldsymbol{Q}_{uw}(1,1)$, respectively.

3) Now add the following conditions to also be applied to the blending functions.

$$\left.\begin{matrix} \ddot{C}_{0,0}(0)=0, & \ddot{C}_{0,0}(1)=0, & \ddot{C}_{0,1}(0)=0, & \ddot{C}_{0,1}(1)=0 \\ \ddot{C}_{1,0}(0)=0, & \ddot{C}_{1,0}(1)=0, & \ddot{C}_{1,1}(0)=0, & \ddot{C}_{1,1}(1)=0 \end{matrix}\right\}. \tag{3.91}$$

In a compact notation these become:

$$\ddot{C}_{0,i}(j)=\ddot{C}_{1,i}(j)=0 \quad (i,j:0,1). \tag{3.92}$$

Differentiating Eq. (3.90) again by u gives:

$$P_{uu}(u,w)=[\ddot{C}_{0,0}(u)\ \ddot{C}_{0,1}(u)\ \ddot{C}_{1,0}(u)\ \ddot{C}_{1,1}(u)]\begin{bmatrix}Q(0,w)\\Q(1,w)\\Q_u(0,w)\\Q_u(1,w)\end{bmatrix}$$

$$+[Q_{uu}(u,0)\ Q_{uu}(u,1)\ Q_{wuu}(u,0)\ Q_{wuu}(u,1)]\begin{bmatrix}C_{0,0}(w)\\C_{0,1}(w)\\C_{1,0}(w)\\C_{1,1}(w)\end{bmatrix}$$

$$-[\ddot{C}_{0,0}(u)\ \ddot{C}_{0,1}(u)\ \ddot{C}_{1,0}(u)\ \ddot{C}_{1,1}(u)]$$

$$\times\begin{bmatrix}Q(0,0) & Q(0,1) & Q_w(0,0) & Q_w(0,1)\\Q(1,0) & Q(1,1) & Q_w(1,0) & Q_w(1,1)\\Q_u(0,0) & Q_u(0,1) & Q_{uw}(0,0) & Q_{uw}(0,1)\\Q_u(1,0) & Q_u(1,1) & Q_{uw}(1,0) & Q_{uw}(1,1)\end{bmatrix}\begin{bmatrix}C_{0,0}(w)\\C_{0,1}(w)\\C_{1,0}(w)\\C_{1,1}(w)\end{bmatrix}.$$

Setting $u=0$ in this equation gives:

$$P_{uu}(0,w)=C_{0,0}(w)\,Q_{uu}(0,0)+C_{0,1}(w)\,Q_{uu}(0,1)+C_{1,0}(w)\,Q_{wuu}(0,0)$$
$$+C_{1,1}(w)\,Q_{wuu}(0,1)\,. \tag{3.93}$$

This means that along the boundary curve corresponding to $u=0$, the 2nd derivative functions with respect to the parameter directions across those curves can be expressed in terms of only the vectors at the end points, $Q_{uu}(0,0)$, $Q_{uu}(0,1)$, $Q_{wuu}(0,0)$ and $Q_{wuu}(0,1)$. Consequently, if two neighboring patches have the same vectors at both ends of the boundary curve, then they are connected with continuity up to the 2nd derivative vectors (curvature continuity).

The conditions applied to the blending functions are summarized in Table 3.5. As can be seen from Tables 3.1 and 3.3, $H_{0,0}(t)$, $H_{0,1}(t)$, $H_{1,0}(t)$ and $H_{1,1}(t)$ satisfy Coons' blending function conditions up to the 1st derivatives, while $K_{0,0}(t)$, $K_{0,1}(t)$, $K_{1,0}(t)$ and $K_{1,1}(t)$ satisfy them up to the 2nd derivatives.

In general, on a Coons surface patch the condition which must be applied to the blending functions in order to make it possible to independently specify the nth-derivative vectors in the directions across the boundary curves is:

$$C_{q,i}^{(p)}(j)=\delta_{pq}\,\delta_{ij} \tag{3.94}$$
$$i,j:0,1$$
$$p,q:0,1,\ldots,n.$$

Table 3.5. Conditions on the functions $C_{0,0}(t)$, $C_{0,1}(t)$, $C_{1,0}(t)$ and $C_{1,1}(t)$ so that connection will be continuous up to the 2nd derivative vectors

		$C_{0,0}(t)$	$C_{0,1}(t)$	$C_{1,0}(t)$	$C_{1,1}(t)$
Function value	$t=0$	1	0	0	0
	$t=1$	0	1	0	0
1st derivative	$t=0$	0	0	1	0
	$t=1$	0	0	0	1
2nd derivative	$t=0$	0	0	0	0
	$t=1$	0	0	0	0

4) Let us assume that for the surface patch of Eq. (3.83), the four boundary curves can be expressed in the following form:

$$\left.\begin{aligned} \boldsymbol{Q}(0,w) &= C_{0,0}(w)\,\boldsymbol{Q}(0,0)+C_{0,1}(w)\,\boldsymbol{Q}(0,1)\\ &\quad +C_{1,0}(w)\,\boldsymbol{Q}_w(0,0)+C_{1,1}(w)\,\boldsymbol{Q}_w(0,1)\\ \boldsymbol{Q}(1,w) &= C_{0,0}(w)\,\boldsymbol{Q}(1,0)+C_{0,1}(w)\,\boldsymbol{Q}(1,1)\\ &\quad +C_{1,0}(w)\,\boldsymbol{Q}_w(1,0)+C_{1,1}(w)\,\boldsymbol{Q}_w(1,1)\\ \boldsymbol{Q}(u,0) &= C_{0,0}(u)\,\boldsymbol{Q}(0,0)+C_{0,1}(u)\,\boldsymbol{Q}(1,0)\\ &\quad +C_{1,0}(u)\,\boldsymbol{Q}_u(0,0)+C_{1,1}(u)\,\boldsymbol{Q}_u(1,0)\\ \boldsymbol{Q}(u,1) &= C_{0,0}(u)\,\boldsymbol{Q}(0,1)+C_{0,1}(u)\,\boldsymbol{Q}(1,1)\\ &\quad +C_{1,0}(u)\,\boldsymbol{Q}_u(0,1)+C_{1,1}(u)\,\boldsymbol{Q}_u(1,1) \end{aligned}\right\}. \tag{3.95}$$

These imply that:

$$\begin{aligned} \boldsymbol{P}(u,w) &= [C_{0,0}(u)\;\; C_{0,1}(u)]\begin{bmatrix}\boldsymbol{Q}(0,0) & \boldsymbol{Q}(0,1)\\ \boldsymbol{Q}(1,0) & \boldsymbol{Q}(1,1)\end{bmatrix}\begin{bmatrix}C_{0,0}(w)\\ C_{0,1}(w)\end{bmatrix}\\ &\quad -[-1\;\; C_{1,0}(u)\;\; C_{1,1}(u)]\\ &\quad \times\begin{bmatrix}\boldsymbol{0} & \boldsymbol{Q}_w(u,0) & \boldsymbol{Q}_w(u,1)\\ \boldsymbol{Q}_u(0,w) & \boldsymbol{Q}_{uw}(0,0) & \boldsymbol{Q}_{uw}(0,1)\\ \boldsymbol{Q}_u(1,w) & \boldsymbol{Q}_{uw}(1,0) & \boldsymbol{Q}_{uw}(1,1)\end{bmatrix}\begin{bmatrix}-1\\ C_{1,0}(w)\\ C_{1,1}(w)\end{bmatrix}. \end{aligned} \tag{3.96}$$

5) Next, let us assume that the following equations apply to the tangent vector functions in the directions across the boundary curves:

$$\left.\begin{aligned}
\boldsymbol{Q}_u(0,w) &= C_{0,0}(w)\,\boldsymbol{Q}_u(0,0)+C_{0,1}(w)\,\boldsymbol{Q}_u(0,1)\\
&\quad +C_{1,0}(w)\,\boldsymbol{Q}_{uw}(0,0)+C_{1,1}(w)\,\boldsymbol{Q}_{uw}(0,1)\\
\boldsymbol{Q}_u(1,w) &= C_{0,0}(w)\,\boldsymbol{Q}_u(1,0)+C_{0,1}(w)\,\boldsymbol{Q}_u(1,1)\\
&\quad +C_{1,0}(w)\,\boldsymbol{Q}_{uw}(1,0)+C_{1,1}(w)\,\boldsymbol{Q}_{uw}(1,1)\\
\boldsymbol{Q}_w(u,0) &= C_{0,0}(u)\,\boldsymbol{Q}_w(0,0)+C_{0,1}(u)\,\boldsymbol{Q}_w(1,0)\\
&\quad +C_{1,0}(u)\,\boldsymbol{Q}_{uw}(0,0)+C_{1,1}(u)\,\boldsymbol{Q}_{uw}(1,0)\\
\boldsymbol{Q}_w(u,1) &= C_{0,0}(u)\,\boldsymbol{Q}_w(0,1)+C_{0,1}(u)\,\boldsymbol{Q}_w(1,1)\\
&\quad +C_{1,0}(u)\,\boldsymbol{Q}_{uw}(0,1)+C_{1,1}(u)\,\boldsymbol{Q}_{uw}(1,1)
\end{aligned}\right\}. \tag{3.97}$$

These imply that:

$$\begin{aligned}
\boldsymbol{P}(u,w) &= [C_{1,0}(u)\;\; C_{1,1}(u)]\begin{bmatrix}\boldsymbol{Q}_{uw}(0,0) & \boldsymbol{Q}_{uw}(0,1)\\ \boldsymbol{Q}_{uw}(1,0) & \boldsymbol{Q}_{uw}(1,1)\end{bmatrix}\begin{bmatrix}C_{1,0}(w)\\ C_{1,1}(w)\end{bmatrix}\\
&\quad -[-1\;\; C_{0,0}(u)\;\; C_{0,1}(u)]\\
&\quad \times\begin{bmatrix}\boldsymbol{0} & \boldsymbol{Q}(u,0) & \boldsymbol{Q}(u,1)\\ \boldsymbol{Q}(0,w) & \boldsymbol{Q}(0,0) & \boldsymbol{Q}(0,1)\\ \boldsymbol{Q}(1,w) & \boldsymbol{Q}(1,0) & \boldsymbol{Q}(1,1)\end{bmatrix}\begin{bmatrix}-1\\ C_{0,0}(w)\\ C_{0,1}(w)\end{bmatrix}.
\end{aligned} \tag{3.98}$$

6) Denote the 1st, 2nd and 3rd surfaces in Eq. (3.83) by $\boldsymbol{P}_A(u,w)$, $\boldsymbol{P}_B(u,w)$ and $\boldsymbol{P}_C(u,w)$, respectively.

Assuming that Eqs. (3.95) and (3.97) apply, we obtain:

$$\boldsymbol{P}_A(u,w)=\boldsymbol{P}_B(u,w)=\boldsymbol{P}_C(u,w)$$

so that:

$$\begin{aligned}
\boldsymbol{P}(u,w) &= \boldsymbol{P}_C(u,w)\\
&= [C_{0,0}(u)\;\; C_{0,1}(u)\;\; C_{1,0}(u)\;\; C_{1,1}(u)]\\
&\quad \times\begin{bmatrix}\boldsymbol{Q}(0,0) & \boldsymbol{Q}(0,1) & \boldsymbol{Q}_w(0,0) & \boldsymbol{Q}_w(0,1)\\ \boldsymbol{Q}(1,0) & \boldsymbol{Q}(1,1) & \boldsymbol{Q}_w(1,0) & \boldsymbol{Q}_w(1,1)\\ \boldsymbol{Q}_u(0,0) & \boldsymbol{Q}_u(0,1) & \boldsymbol{Q}_{uw}(0,0) & \boldsymbol{Q}_{uw}(0,1)\\ \boldsymbol{Q}_u(1,0) & \boldsymbol{Q}_u(1,1) & \boldsymbol{Q}_{uw}(1,0) & \boldsymbol{Q}_{uw}(1,1)\end{bmatrix}\begin{bmatrix}C_{0,0}(w)\\ C_{0,1}(w)\\ C_{1;0}(w)\\ C_{1,1}(w)\end{bmatrix}.
\end{aligned} \tag{3.99}$$

In Eq. (3.99), if we use the cubic Hermite interpolation functions $H_{0,0}(t)$, $H_{0,1}(t)$, $H_{1,0}(t)$ and $H_{1,1}(t)$ as blending functions, we obtain:

$$P(u,w)=[H_{0,0}(u)\ H_{0,1}(u)\ H_{1,0}(u)\ H_{1,1}(u)]$$
$$\times\begin{bmatrix} Q(0,0) & Q(0,1) & Q_w(0,0) & Q_w(0,1)\\ Q(1,0) & Q(1,1) & Q_w(1,0) & Q_w(1,1)\\ Q_u(0,0) & Q_u(0,1) & Q_{uw}(0,0) & Q_{uw}(0,1)\\ Q_u(1,0) & Q_u(1,1) & Q_{uw}(1,0) & Q_{uw}(1,1)\end{bmatrix}\begin{bmatrix}H_{0,0}(w)\\ H_{0,1}(w)\\ H_{1,0}(w)\\ H_{1,1}(w)\end{bmatrix}. \quad (3.100)$$

Using simpler notation, this becomes:

$$P(u,w)=[H_{0,0}(u)\ H_{0,1}(u)\ H_{1,0}(u)\ H_{1,1}(u)]$$
$$\times B_c[H_{0,0}(w)\ H_{0,1}(w)\ H_{1,0}(w)\ H_{1,1}(w)]^T \quad (3.101)$$
$$=UM_cB_cM_c^T W^T \quad (3.102)$$

where M_c is the matrix defined in (3.10). We also have:

$$B_c=\begin{bmatrix} Q(0,0) & Q(0,1) & Q_w(0,0) & Q_w(0,1)\\ Q(1,0) & Q(1,1) & Q_w(1,0) & Q_w(1,1)\\ Q_u(0,0) & Q_u(0,1) & Q_{uw}(0,0) & Q_{uw}(0,1)\\ Q_u(1,0) & Q_u(1,1) & Q_{uw}(1,0) & Q_{uw}(1,1)\end{bmatrix} \quad (3.103)$$

$$U=[u^3\quad u^2\quad u\quad 1]$$
$$W=[w^3\quad w^2\quad w\quad 1].$$

Since B_c is a matrix which expresses the surface patch boundary conditions, it is sometimes called the boundary condition matrix (Fig. 3.20).

Since, in the surface patch equation (3.100), $H_{0,0}$, $H_{0,1}$, $H_{1,0}$ and $H_{1,1}$ are expressed as cubic polynomials in the respective parameters, they are expressed

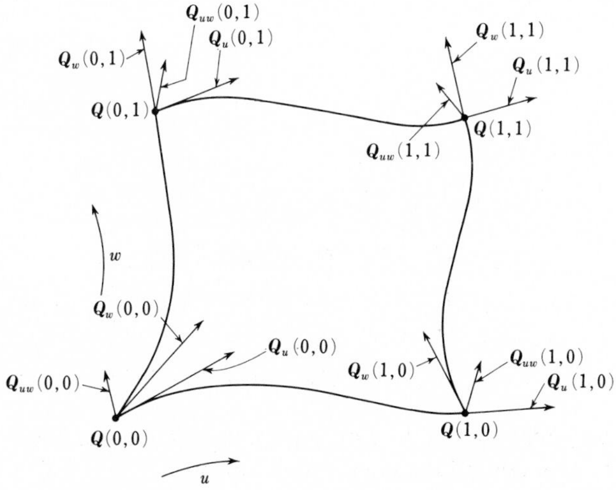

Fig. 3.20. Vectors that define a Coons bi-cubic surface patch

in the bi-cubic polynomial form of equation (1.109). Consequently, the surface described by Eqs. (3.100), (3.101) and (3.102) are called bi-cubic Coons patches.

The curved surface having the form (3.100) is a special case of a Coons surface; since it has a very simple form it is easy to use and is a standard type of Coons surface.

3.3.4 Twist Vectors and Surface Shapes

Cross partial derivative vectors, that is, twist vectors, do not occur in studies of curves but do come up in treatment of curved surfaces. In the Ferguson surface patch and the 1964 Coons surface patch, the twist vectors are **0** at all 4 corners. As we mentioned in discussing the description of the Ferguson surface patch, a tangent vector in the direction across a boundary curve can be found by simply interpolating between the tangent vectors in that direction at the 2 ends of the boundary curve using the functions $H_{0,0}$ and $H_{0,1}$. For example, $\boldsymbol{P}_u(0, w)$ is obtained as:

$$\boldsymbol{P}_u(0, w) = H_{0,0}(w)\,\boldsymbol{P}_u(0, 0) + H_{0,1}(w)\,\boldsymbol{P}_u(0, 1). \qquad (3.104)$$

In this interpolation, the rate of increase of $\boldsymbol{P}_u(0, w)$ in the w-direction is forcibly held to **0** at both ends (Fig. 3.21(a)). At both ends, the vectors showing the rate of tangent vector increase in the direction across the boundary curve are also specified; applying the cubic Hermite interpolation that was applied to the position vectors to the tangent vectors in the direction across the boundary curve gives a smoother interpolation. To do this, instead of equation (3.104), use:

$$\begin{aligned}\boldsymbol{P}_u(0, w) = {} & H_{0,0}(w)\,\boldsymbol{P}_u(0, 0) + H_{0,1}(w)\,\boldsymbol{P}_u(0, 1) + H_{1,0}(w)\,\boldsymbol{P}_{uw}(0, 0) \\ & + H_{1,1}(w)\,\boldsymbol{P}_{uw}(0, 1).\end{aligned} \qquad (3.105)$$

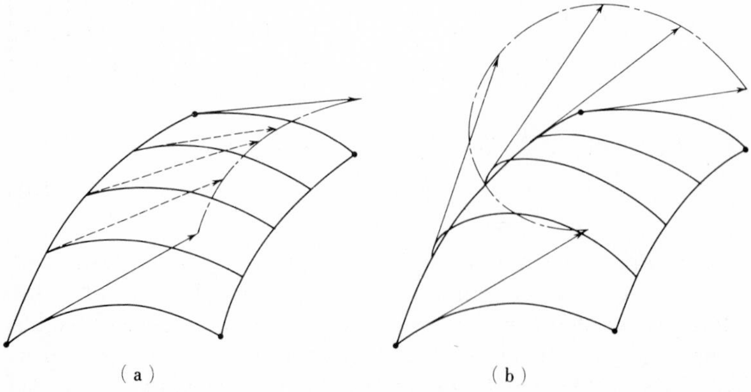

Fig. 3.21. Effect of twist vectors on surface shape. (a) Case in which twist vectors are zero vectors; (b) case in which twist vectors are specified to be nonzero vectors

If the tangent vector in the direction across the boundary curve is determined according to Eq. (3.105) in the course of deriving Ferguson's surface formula, then surface patch formula (3.100), which is a special case of Coons' 1967 surface, is obtained. If the twist vector is $\mathbf{0}$, that part will be flattened, but if the twist vector is nonzero then, from Fig. 3.21(b), it can be expected that the shape will become rounded. Since the effect of the twist vector on the surface shape is rather subtle, it is normally difficult to recognize that effect from a static image. It is easier to understand by moving the curved surface slowly on the display, or test-cutting the shape with an NC machine.

In the following discussion we will take Coons' bi-cubic surface patch as an example and see what we can learn from the formulas about the effect of the twist vector on the surface shape. On Coons' bi-cubic surface patch, let us find the position of a point separated by a small distance Δu, Δw from the corner at $\boldsymbol{Q}(0,0)$. Taking $t=\Delta t$ in $H_{0,0}(t)$, $H_{0,1}(t)$, $H_{1,0}(t)$ and $H_{1,1}(t)$, and neglecting terms of quadratic or higher order which are small compared to 1, we obtain:

$$\begin{aligned}
H_{0,0}(\Delta t) &= 2\Delta t^3 - 3\Delta t^2 + 1 \fallingdotseq 1\\
H_{0,1}(\Delta t) &= -2\Delta t^3 + 3\Delta t^2 \fallingdotseq 0\\
H_{1,0}(\Delta t) &= \Delta t^3 - 2\Delta t^2 + \Delta t \fallingdotseq \Delta t\\
H_{1,1}(\Delta t) &= \Delta t^3 - \Delta t^2 \fallingdotseq 0.
\end{aligned} \tag{3.106}$$

Substituting these relations into surface patch formula (3.100), we find, for a point on the surface:

$$\begin{aligned}
\boldsymbol{P}(\Delta u, \Delta w) &= [H_{0,0}(\Delta u)\ H_{0,1}(\Delta u)\ H_{1,0}(\Delta u)\ H_{1,1}(\Delta u)]\\
&\quad \times \boldsymbol{B}_c [H_{0,0}(\Delta w)\ H_{0,1}(\Delta w)\ H_{1,0}(\Delta w)\ H_{1,1}(\Delta w)]^T\\
&\fallingdotseq [1\ \ 0\ \ \Delta u\ \ 0] \begin{bmatrix} \boldsymbol{Q}(0,0) & \boldsymbol{Q}(0,1) & \boldsymbol{Q}_w(0,0) & \boldsymbol{Q}_w(0,1)\\ \boldsymbol{Q}(1,0) & \boldsymbol{Q}(1,1) & \boldsymbol{Q}_w(1,0) & \boldsymbol{Q}_w(1,1)\\ \boldsymbol{Q}_u(0,0) & \boldsymbol{Q}_u(0,1) & \boldsymbol{Q}_{uw}(0,0) & \boldsymbol{Q}_{uw}(0,1)\\ \boldsymbol{Q}_u(1,0) & \boldsymbol{Q}_u(1,1) & \boldsymbol{Q}_{uw}(1,0) & \boldsymbol{Q}_{uw}(1,1) \end{bmatrix} \begin{bmatrix} 1\\ 0\\ \Delta w\\ 0 \end{bmatrix}\\
&= \boldsymbol{Q}(0,0) + \boldsymbol{Q}_u(0,0)\,\Delta u + \boldsymbol{Q}_w(0,0)\,\Delta w + \boldsymbol{Q}_{uw}(0,0)\,\Delta u\,\Delta w.
\end{aligned} \tag{3.107}$$

From Eq. (3.107), we see that what we are essentially doing is to construct a vector by adding to the corner position vector $\boldsymbol{Q}(0,0)$, a diagonal vector of the parallelogram made by the two small tangent vectors in the u- and w-direction and the very small twist vector (Fig. 3.22). Therefore, if the twist vector is $\mathbf{0}$ at the corner $\boldsymbol{Q}(0,0)$ of the surface patch, we see that the surface will be flat in the vicinity of that corner. Next, let us investigate the effect of the twist vector at a point other than $\boldsymbol{Q}(0,0)$, using the similar method.

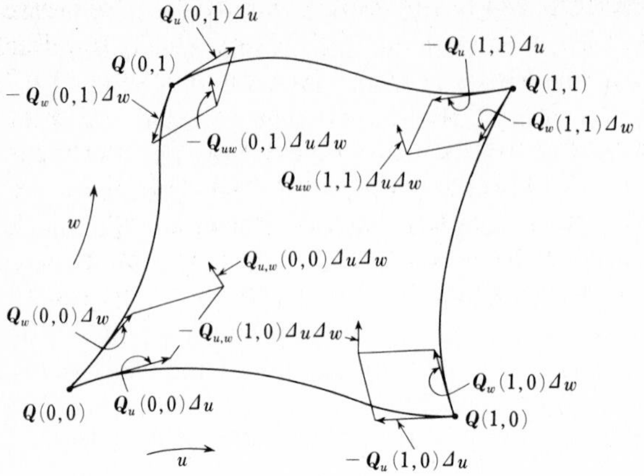

Fig. 3.22. Points on a surface separated by Δu, Δw from the 4 corners

We have:

$$\begin{aligned}
H_{0,0}(1-\Delta t) &= 2(1-\Delta t)^3 - 3(1-\Delta t)^2 + 1 \fallingdotseq 0\\
H_{0,1}(1-\Delta t) &= -2(1-\Delta t)^3 + 3(1-\Delta t)^2 \fallingdotseq 1\\
H_{1,0}(1-\Delta t) &= (1-\Delta t)^3 - 2(1-\Delta t)^2 + 1 - \Delta t \fallingdotseq 0\\
H_{1,1}(1-\Delta t) &= (1-\Delta t)^3 - (1-\Delta t)^2 \fallingdotseq -\Delta t
\end{aligned} \tag{3.108}$$

so, in the vicinity of $\boldsymbol{Q}(1,0)$:

$$\begin{aligned}
\boldsymbol{P}(1-\Delta u, \Delta w) &\fallingdotseq [0 \ \ 1 \ \ 0 - \Delta u] \, \boldsymbol{B}_c \begin{bmatrix} 1 \\ 0 \\ \Delta w \\ 0 \end{bmatrix}\\
&= \boldsymbol{Q}(1,0) - \boldsymbol{Q}_u(1,0)\,\Delta u + \boldsymbol{Q}_w(1,0)\,\Delta w - \boldsymbol{Q}_{uw}(1,0)\,\Delta u\,\Delta w.
\end{aligned} \tag{3.109}$$

Similarly, in the vicinity of $\boldsymbol{Q}(0,1)$:

$$\begin{aligned}
\boldsymbol{P}(\Delta u, 1-\Delta w) &\fallingdotseq [1 \ \ 0 \ \ \Delta u \ \ 0] \, \boldsymbol{B}_c \begin{bmatrix} 0 \\ 1 \\ 0 \\ -\Delta w \end{bmatrix}\\
&= \boldsymbol{Q}(0,1) + \boldsymbol{Q}_u(0,1)\,\Delta u - \boldsymbol{Q}_w(0,1)\,\Delta w - \boldsymbol{Q}_{uw}(0,1)\,\Delta u\,\Delta w.
\end{aligned} \tag{3.110}$$

Finally, in the vicinity of $\boldsymbol{Q}(1,1)$, we have:

$$\begin{aligned}
\boldsymbol{P}(1-\Delta u, 1-\Delta w) &\fallingdotseq [0 \ \ 1 \ \ 0 \ \ -\Delta u] \, \boldsymbol{B}_c \begin{bmatrix} 0 \\ 1 \\ 0 \\ -\Delta w \end{bmatrix}\\
&= \boldsymbol{Q}(1,1) - \boldsymbol{Q}_u(1,1)\,\Delta u - \boldsymbol{Q}_w(1,1)\,\Delta w + \boldsymbol{Q}_{uw}(1,1)\,\Delta u\,\Delta w.
\end{aligned} \tag{3.111}$$

The twist vector relations described above are shown in Fig. 3.22. Particular attention should be paid to the fact that at points $\boldsymbol{Q}(0,0)$ and $\boldsymbol{Q}(1,1)$ the effect of the twist vector is $+\boldsymbol{Q}_{uw}(0,0)\Delta u\Delta w$ or $+\boldsymbol{Q}_{uw}(1,1)\Delta u\Delta w$; while at $\boldsymbol{Q}(1,0)$ and $\boldsymbol{Q}(0,1)$ it is $-\boldsymbol{Q}_{uw}(1,0)\Delta u\Delta w$ or $-\boldsymbol{Q}_{uw}(0,1)\Delta u\Delta w$, respectively. The direction of the effect is opposite at alternative corners. Therefore, a twist vector that pushes the surface upward at $\boldsymbol{Q}(0,0)$ will also push it upward at $\boldsymbol{Q}(1,1)$, but will push it downward, tending to create an indentation, at $\boldsymbol{Q}(1,0)$ and $\boldsymbol{Q}(0,1)$. In the example shown in Fig. 3.23, the twist vector is $\mathbf{0}$, but if vertical twist of equal magnitude are added downward at the 4 corners, and their magnitude is

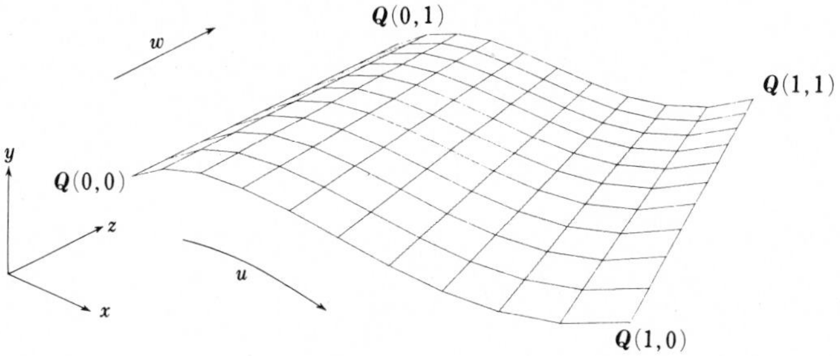

$\boldsymbol{Q}(0,0)$: (0,0,0)	$\boldsymbol{Q}(0,1)$: (0,0,1000)	$\boldsymbol{Q}_w(0,0)$: (0,0,1000)	$\boldsymbol{Q}_w(0,1)$: (0,0,1000)
$\boldsymbol{Q}(1,0)$: (1000,0,0)	$\boldsymbol{Q}(1,1)$: (1000,0,1000)	$\boldsymbol{Q}_w(1,0)$: (0,0,1000)	$\boldsymbol{Q}_w(1,1)$: (0,0,1000)
$\boldsymbol{Q}_u(0,0)$: (1000,500,0)	$\boldsymbol{Q}_u(0,1)$: (1000,500,0)	$\boldsymbol{Q}_{uw}(0,0)$: (0,0,0)	$\boldsymbol{Q}_{uw}(0,1)$: (0,0,0)
$\boldsymbol{Q}_u(1,0)$: (1000,500,0)	$\boldsymbol{Q}_u(1,1)$: (1000,500,0)	$\boldsymbol{Q}_{uw}(1,0)$: (0,0,0)	$\boldsymbol{Q}_{uw}(1,1)$: (0,0,0)

Fig. 3.23. A surface with zero twist vectors

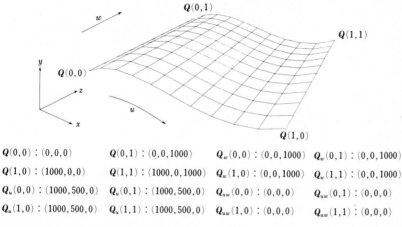

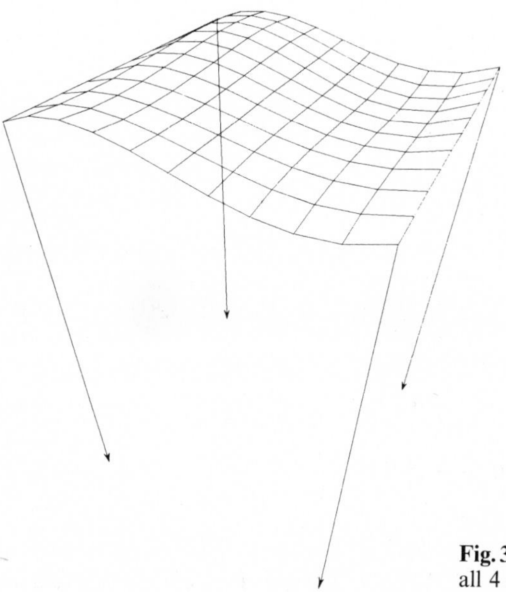

Fig. 3.24. A surface with the twist vectors at all 4 corners taken to be (0, −1500, 0)

gradually increased, the surface shape changes as shown in Figs. 3.24 and 3.25. If the same twist vectors are added pointing vertically upward instead of downward, the result is as shown in Fig. 3.26.

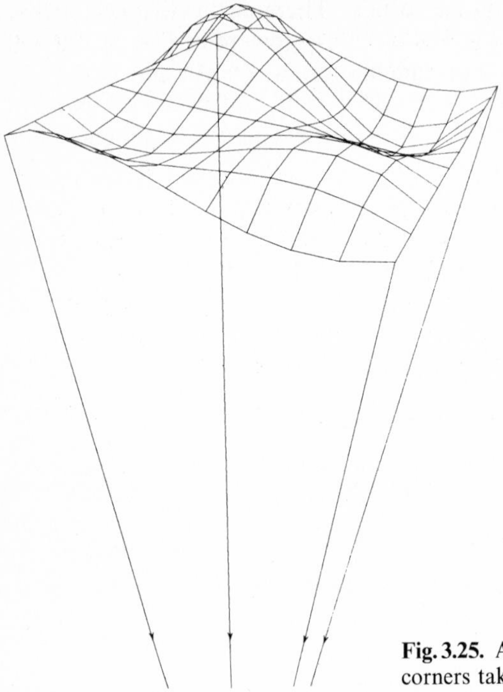

Fig. 3.25. A surface with the twist vectors at all 4 corners taken to be (0, −12000, 0)

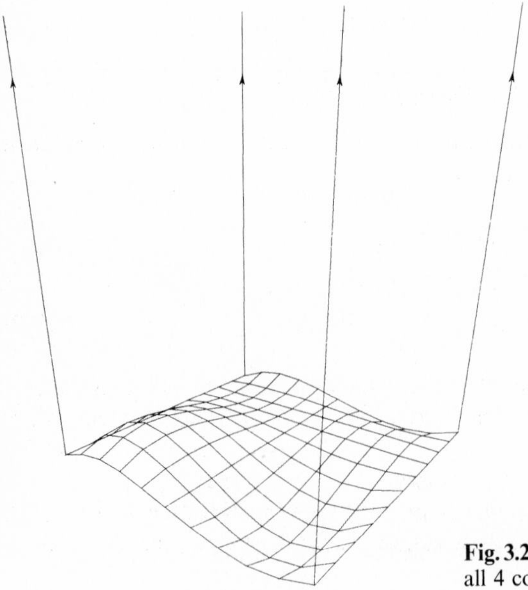

Fig. 3.26. A surface with the twist vectors at all 4 corners taken to be (0, +5500, 0)

3.3.5 Methods of Determining Twist Vectors

Since it is very difficult for a human being to determine suitable twist vectors, it is desirable to have a program that will do it automatically according to a rational scheme.

Several methods of automatically determining twist vectors will now be discussed.

(1) Forrest's Method [15]

If one knows not only the 4 corners of a surface patch but also 12 other points through which the surface patch passes, then the twist vector can be determined using Coons' bi-cubic surface formula (3.100) (Fig. 3.27).

Normally, however, one can not expect that these other 12 points, especially the 4 interior points, will be given. Normally only the 4 boundary curves are given, and we must determine the twist vector from these alone.

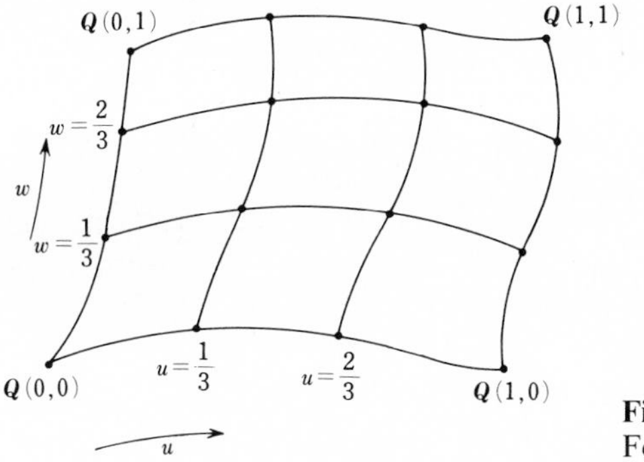

Fig. 3.27. Determination of a twist vector by Forrest's method

(2) Method Using Bi-linear Surface Formula [15]

Differentiating the bi-linear surface formula (3.72) partially with respect to u and w and finding the twist vector gives:

$$\begin{aligned} \boldsymbol{P}_{uw}(u, w) &= [-1 \;\; 1] \begin{bmatrix} \boldsymbol{Q}(0,0) & \boldsymbol{Q}(0,1) \\ \boldsymbol{Q}(1,0) & \boldsymbol{Q}(1,1) \end{bmatrix} \begin{bmatrix} -1 \\ 1 \end{bmatrix} \\ &= \boldsymbol{Q}(0,0) - \boldsymbol{Q}(1,0) - \boldsymbol{Q}(0,1) + \boldsymbol{Q}(1,1). \end{aligned} \tag{3.112}$$

It can be seen from this equation that for a bi-linear surface, the twist vector $\boldsymbol{P}_{uw}(u, w)$ is fixed at all points of the surface patch. The following algorithm permits this twist vector to be used to determine the twist vector for Coons' bi-cubic surface formula.

Let us consider four surface patches joined together as shown in Fig. 3.28. The twist vectors for each surface patch can be found as $\boldsymbol{w}_0$, $\boldsymbol{w}_1$, $\boldsymbol{w}_2$ and $\boldsymbol{w}_3$ from Eq. (3.112).

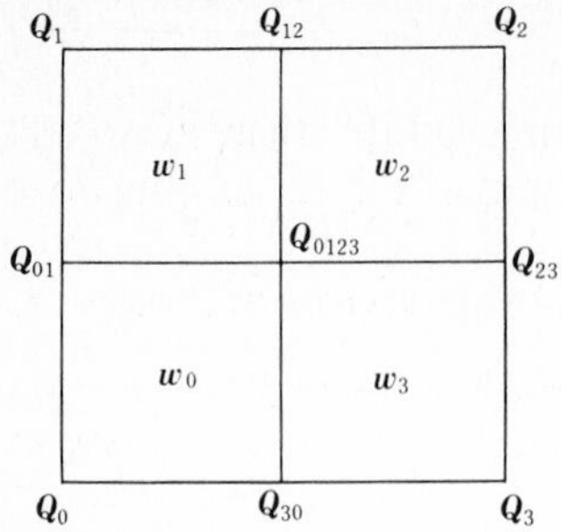

Fig. 3.28. Determination of a twist vector using a bi-linear surface formula

1) Let $\boldsymbol{w}_i$ be the twist vector at point $\boldsymbol{Q}_i$.
2) Take the twist vector at a point $\boldsymbol{Q}_{ij}$ to be the average of the twist vectors of the two surface patches which share $\boldsymbol{Q}_{ij}$: $\frac{1}{2}(\boldsymbol{w}_i+\boldsymbol{w}_j)$.
3) The twist vector at point $\boldsymbol{Q}_{0123}$ is the average of the twist vectors of the four surface patches which share $\boldsymbol{Q}_{0123}$: $\frac{1}{4}(\boldsymbol{w}_0+\boldsymbol{w}_1+\boldsymbol{w}_2+\boldsymbol{w}_3)$.

This algorithm is very simple and gives a better result than simply forcing the twist vector to be $\mathbf{0}$.

(3) Adini's Method[15)]

This is a bit more complicated than method (2), but it uses more data and gives a better result. It uses the class C^0 Coons surface patch formula (3.71). In this surface formula, assume that the boundary curves are Ferguson curve segments:

$$\left.\begin{aligned}
\boldsymbol{Q}(0,w)&=H_{0,0}(w)\,\boldsymbol{Q}(0,0)+H_{0,1}(w)\,\boldsymbol{Q}(0,1)\\
&\quad+H_{1,0}(w)\,\boldsymbol{Q}_w(0,0)+H_{1,1}(w)\,\boldsymbol{Q}_w(0,1)\\
\boldsymbol{Q}(1,w)&=H_{0,0}(w)\,\boldsymbol{Q}(1,0)+H_{0,1}(w)\,\boldsymbol{Q}(1,1)\\
&\quad+H_{1,0}(w)\,\boldsymbol{Q}_w(1,0)+H_{1,1}(w)\,\boldsymbol{Q}_w(1,1)\\
\boldsymbol{Q}(u,0)&=H_{0,0}(u)\,\boldsymbol{Q}(0,0)+H_{0,1}(u)\,\boldsymbol{Q}(1,0)\\
&\quad+H_{1,0}(u)\,\boldsymbol{Q}_u(0,0)+H_{1,1}(u)\,\boldsymbol{Q}_u(1,0)\\
\boldsymbol{Q}(u,1)&=H_{0,0}(u)\,\boldsymbol{Q}(0,1)+H_{0,1}(u)\,\boldsymbol{Q}(1,1)\\
&\quad+H_{1,0}(u)\,\boldsymbol{Q}_u(0,1)+H_{1,1}(u)\,\boldsymbol{Q}_u(1,1)
\end{aligned}\right\}. \tag{3.113}$$

In this case, the cross partial derivative vector for the surface patch formula (3.71) is:

$$\begin{aligned}
\boldsymbol{P}_{uw}(u,w)&=[-1\;\;1]\begin{bmatrix}\boldsymbol{Q}_w(0,w)\\ \boldsymbol{Q}_w(1,w)\end{bmatrix}+[\boldsymbol{Q}_u(u,0)\;\;\boldsymbol{Q}_u(u,1)]\begin{bmatrix}-1\\ 1\end{bmatrix}\\
&\quad-[-1\;\;1]\begin{bmatrix}\boldsymbol{Q}(0,0) & \boldsymbol{Q}(0,1)\\ \boldsymbol{Q}(1,0) & \boldsymbol{Q}(1,1)\end{bmatrix}\begin{bmatrix}-1\\ 1\end{bmatrix}.
\end{aligned} \tag{3.114}$$

Using equations (3.113) gives:

$$\begin{aligned}
P_{uw}(0,0) &= -Q_w(0,0)+Q_w(1,0)-Q_u(0,0)+Q_u(0,1)-C \equiv w(0,0)\\
P_{uw}(0,1) &= -Q_w(0,1)+Q_w(1,1)-Q_u(0,0)+Q_u(0,1)-C \equiv w(0,1)\\
P_{uw}(1,0) &= -Q_w(0,0)+Q_w(1,0)-Q_u(1,0)+Q_u(1,1)-C \equiv w(1,0)\\
P_{uw}(1,1) &= -Q_w(0,1)+Q_w(1,1)-Q_u(1,0)+Q_u(1,1)-C \equiv w(1,1)
\end{aligned} \tag{3.115}$$

where:

$$C = Q(0,0)-Q(1,0)-Q(0,1)+Q(1,1).$$

In this case, the twist vectors will in general be different at the 4 corners of the surface patch. When the curved surface patches are connected, the twist vectors at each point are determined as follows.

When 4 surface patches are connected as shown in Fig. 3.29, the twist vectors at the 4 corners of each patch can be found as follows.

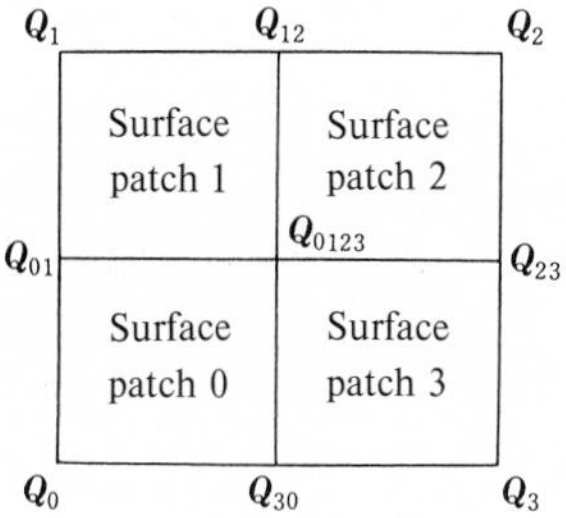

Fig. 3.29. Determination of a twist vector by Adini's method

1) The twist vector at a point Q_i is $w_i(Q_i)$*).
2) The twist vector at a point Q_{ij} is $\frac{1}{2}(w_i(Q_{ij})+w_j(Q_{ij}))$.
3) The twist vector at point Q_{0123} is $\frac{1}{4}(w_0(Q_{0123})+w_1(Q_{0123})+w_2(Q_{0123})+w_3(Q_{0123}))$.

This method of determining the twist vector is called Adini's method.

(4) Method Using the Twist Vector of a Spline Surface (refer to Chap. 4)

Assume that position vectors Q_{ij} which determine the 4 corners of surface patches are given as a lattice (Fig. 3.30):

$$Q_{ij} \quad (i=0, 1, \ldots, m; j=0, 1, \ldots, n).$$

When the interpolating spline method described in Chap. 4 is used, the tangent vectors $Q_{u,i,j}(i=0, 1, \ldots, m)$ which define a curve that passes through a

*) We assume this notation denotes the twist vector at Q_i in the i-th patch.

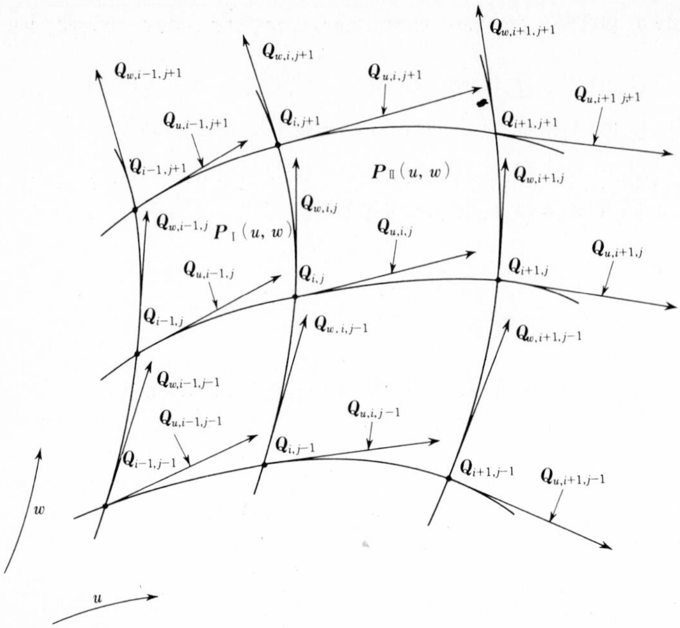

Fig. 3.30. Method using the twist vectors of a spline curved surface

sequence of points in the u-direction and is continuous up to the curvature vector can be obtained by solving the following condition equation:

$$Q_{u,i-1,j}+4Q_{u,i,j}+Q_{u,i+1,j}=3(Q_{i+1,j}-Q_{i-1,j})^{*)} \quad (i=1, 2, \ldots, m-1). \tag{3.116}$$

Similarly, the tangent vectors $Q_{w,i,j}(j=0, 1, \ldots, n)$ which define a curve that passes through a sequence of points in the w-direction and is continuous up to the curvature vector can be obtained by solving the following condition equation:

$$Q_{w,i,j-1}+4Q_{w,i,j}+Q_{w,i,j+1}=3(Q_{i,j+1}-Q_{i,j-1}) \quad (j=1, 2, \ldots, n-1). \tag{3.117}$$

Thus, the lattice of points are connected by curves which are continuous up to the curvature vector. These curves are supposed to be the boundary curves which separate neighboring bi-cubic Coons surface patches (Fig. 3.30):

$$P_{\mathrm{I}}(u, w)=UM_cB_{c,\mathrm{I}}M_c^TW^T \tag{3.118}$$

$$P_{\mathrm{II}}(u, w)=UM_cB_{c,\mathrm{II}}M_c^TW^T. \tag{3.119}$$

---

*) Here we use Eq. (4.40) for simplicity. Ideally Eq. (4.38) is better.

Since two neighboring surface patches share the same boundary curve, we have:

$$P_{\mathrm{I}}(1,w)=P_{\mathrm{II}}(0,w).$$

If we apply the condition that the 2nd derivative vectors in the direction across the boundary curve are continuous, we obtain:

$$[6\ 2\ 0\ 0]\,M_c B_{c,\mathrm{I}}=[0\ 2\ 0\ 0]\,M_c B_{c,\mathrm{II}}.$$

Therefore:

$$[6\ -6\ 2\ 4]\begin{bmatrix} Q_{i-1,j} & Q_{i-1,j+1} & Q_{w,i-1,j} & Q_{w,i-1,j+1} \\ Q_{i,j} & Q_{i,j+1} & Q_{w,i,j} & Q_{w,i,j+1} \\ Q_{u,i-1,j} & Q_{u,i-1,j+1} & Q_{uw,i-1,j} & Q_{uw,i-1,j+1} \\ Q_{u,i,j} & Q_{u,i,j+1} & Q_{uw,i,j} & Q_{uw,i,j+1} \end{bmatrix}$$

$$=[-6\ 6\ -4\ -2]\begin{bmatrix} Q_{i,j} & Q_{i,j+1} & Q_{w,i,j} & Q_{w,i,j+1} \\ Q_{i+1,j} & Q_{i+1,j+1} & Q_{w,i+1,j} & Q_{w,i+1,j+1} \\ Q_{u,i,j} & Q_{u,i,j+1} & Q_{uw,i,j} & Q_{uw,i,j+1} \\ Q_{u,i+1,j} & Q_{u,i+1,j+1} & Q_{uw,i+1,j} & Q_{uw,i+1,j+1} \end{bmatrix}.$$

From the above matrix equation, the following four condition equations are obtained:

$$6Q_{i-1,j}-6Q_{i,j}+2Q_{u,i-1,j}+4Q_{u,i,j}$$
$$=-6Q_{i,j}+6Q_{i+1,j}-4Q_{u,i,j}-2Q_{u,i+1,j} \tag{3.120}$$

$$6Q_{i-1,j+1}-6Q_{i,j+1}+2Q_{u,i-1,j+1}+4Q_{u,i,j+1}$$
$$=-6Q_{i,j+1}+6Q_{i+1,j+1}-4Q_{u,i,j+1}-2Q_{u,i+1,j+1} \tag{3.121}$$

$$6Q_{w,i-1,j}-6Q_{w,i,j}+2Q_{uw,i-1,j}+4Q_{uw,i,j}$$
$$=-6Q_{w,i,j}+6Q_{w,i+1,j}-4Q_{uw,i,j}-2Q_{uw,i+1,j} \tag{3.122}$$

$$6Q_{w,i-1,j+1}-6Q_{w,i,j+1}+2Q_{uw,i-1,j+1}+4Q_{uw,i,j+1}$$
$$=-6Q_{w,i,j+1}+6Q_{w,i+1,j+1}-4Q_{uw,i,j+1}-2Q_{uw,i+1,j+1}. \tag{3.123}$$

The first two condition equations, (3.120) and (3.121), are essentially the same as condition equation (3.116). The last two equations, which are essentially the same as each other, express the following condition:

$$Q_{uw,i-1,j}+4Q_{uw,i,j}+Q_{uw,i+1,j}=3(Q_{w,i+1,j}-Q_{w,i-1,j})\quad (i=1,2,\ldots,m-1). \tag{3.124}$$

Equation (3.124) expresses the condition that the twist vector is determined so as to make the 2nd derivative vectors in the direction across a boundary curve in the w-direction equal. The right-hand side of Eq. (3.124) can be found from Eq. (3.117), so, by adding two more conditions on $\boldsymbol{Q}_{uw,i,j}$, Eq. (3.124) can be solved to determine $\boldsymbol{Q}_{uw,i,j}(i=0, 1, \ldots, m; j=0, 1, \ldots, n)$.

By the above procedure we have determined a twist vector subject to the condition that the 2nd derivative vectors in the direction across a boundary curve in the w-direction are equal. An equation corresponding to (3.124) can be derived for a twist vector subject to the condition that the 2nd derivative vectors in the direction across a boundary curve in the u-direction are equal. It can be shown that the twist vectors found by these two methods are the same[7]. Therefore, by using a twist vector derived in this way, we can produce a surface such that the 2nd derivative vectors in the directions across curves in both the u- and w-directions are continuous.

3.3.6 Partial Surface Representation of the Coons Bi-cubic Surface Patch

On the bi-cubic Coons surface patch described by:

$$\boldsymbol{P}(u, w) = \boldsymbol{U}\boldsymbol{M}_c\boldsymbol{B}_c\boldsymbol{M}_c^T\boldsymbol{W}^T$$

let us try to express the part of the surface in the range $u_0 \leqq u \leqq u_1$, $w_0 \leqq w \leqq w_1$ in terms of the new parameters u^* and w^* $(0 \leqq u^*, w^* \leqq 1)$ (Fig. 3.31). To do this we perform the following parameter transformation:

$$u = u_0 + u^*(u_1 - u_0)$$
$$w = w_0 + w^*(w_1 - w_0).$$

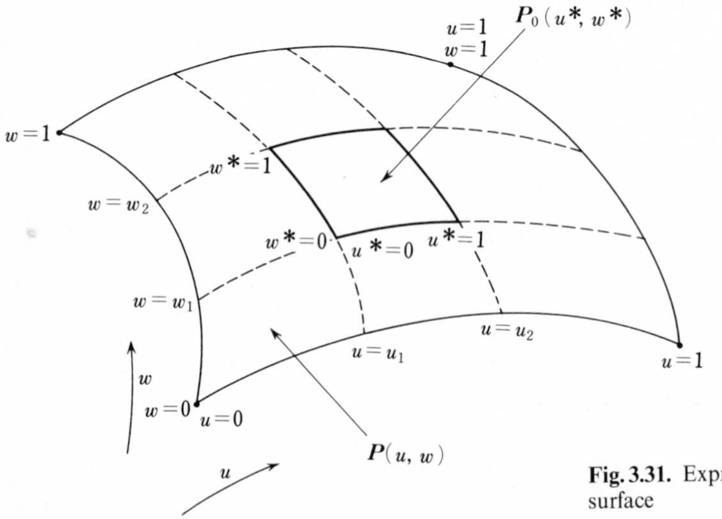

Fig. 3.31. Expression of a partial surface

Applying the result of this parameter transformation for the case of a curve (Eq. (3.28)) gives:

$$\boldsymbol{P}(u^*, w^*) = [u^{*3} \;\; u^{*2} \;\; u^* \;\; 1]\,\boldsymbol{M}_c$$

$$\times \begin{bmatrix} H_{0,0}(u_0) & H_{0,1}(u_0) & H_{1,0}(u_0) & H_{1,1}(u_0) \\ H_{0,0}(u_1) & H_{0,1}(u_1) & H_{1,0}(u_1) & H_{1,1}(u_1) \\ (u_1-u_0)\dot{H}_{0,0}(u_0) & (u_1-u_0)\dot{H}_{0,1}(u_0) & (u_1-u_0)\dot{H}_{1,0}(u_0) & (u_1-u_0)\dot{H}_{1,1}(u_0) \\ (u_1-u_0)\dot{H}_{0,0}(u_1) & (u_1-u_0)\dot{H}_{0,1}(u_1) & (u_1-u_0)\dot{H}_{1,0}(u_1) & (u_1-u_0)\dot{H}_{1,1}(u_1) \end{bmatrix} \boldsymbol{B}_c$$

$$\times \begin{bmatrix} H_{0,0}(w_0) & H_{0,0}(w_1) & (w_1-w_0)\dot{H}_{0,0}(w_0) & (w_1-w_0)\dot{H}_{0,0}(w_1) \\ H_{0,1}(w_0) & H_{0,1}(w_1) & (w_1-w_0)\dot{H}_{0,1}(w_0) & (w_1-w_0)\dot{H}_{0,1}(w_1) \\ H_{1,0}(w_0) & H_{1,0}(w_1) & (w_1-w_0)\dot{H}_{1,0}(w_0) & (w_1-w_0)\dot{H}_{1,0}(w_1) \\ H_{1,1}(w_0) & H_{1,1}(w_1) & (w_1-w_0)\dot{H}_{1,1}(w_0) & (w_1-w_0)\dot{H}_{1,1}(w_1) \end{bmatrix} \boldsymbol{M}_c^T \begin{bmatrix} w^{*3} \\ w^{*2} \\ w^* \\ 1 \end{bmatrix}$$

$$= [u^{*3} \;\; u^{*2} \;\; u^* \;\; 1]\,\boldsymbol{M}_c$$

$$\times \begin{bmatrix} \boldsymbol{P}(u_0,0) & \boldsymbol{P}(u_0,1) & \boldsymbol{P}_w(u_0,0) & \boldsymbol{P}_w(u_0,1) \\ \boldsymbol{P}(u_1,0) & \boldsymbol{P}(u_1,1) & \boldsymbol{P}_w(u_1,0) & \boldsymbol{P}_w(u_1,1) \\ (u_1-u_0)\boldsymbol{P}_u(u_0,0) & (u_1-u_0)\boldsymbol{P}_u(u_0,1) & (u_1-u_0)\boldsymbol{P}_{uw}(u_0,0) & (u_1-u_0)\boldsymbol{P}_{uw}(u_0,1) \\ (u_1-u_0)\boldsymbol{P}_u(u_1,0) & (u_1-u_0)\boldsymbol{P}_u(u_1,1) & (u_1-u_0)\boldsymbol{P}_{uw}(u_1,0) & (u_1-u_0)\boldsymbol{P}_{uw}(u_1,1) \end{bmatrix}$$

$$\times \begin{bmatrix} H_{0,0}(w_0) & H_{0,0}(w_1) & (w_1-w_0)\dot{H}_{0,0}(w_0) & (w_1-w_0)\dot{H}_{0,0}(w_1) \\ H_{0,1}(w_0) & H_{0,1}(w_1) & (w_1-w_0)\dot{H}_{0,1}(w_0) & (w_1-w_0)\dot{H}_{0,1}(w_1) \\ H_{1,0}(w_0) & H_{1,0}(w_1) & (w_1-w_0)\dot{H}_{1,0}(w_0) & (w_1-w_0)\dot{H}_{1,0}(w_1) \\ H_{1,1}(w_0) & H_{1,1}(w_1) & (w_1-w_0)\dot{H}_{1,1}(w_0) & (w_1-w_0)\dot{H}_{1,1}(w_1) \end{bmatrix} \boldsymbol{M}_c^T \begin{bmatrix} w^{*3} \\ w^{*2} \\ w^* \\ 1 \end{bmatrix}$$

$$= [u^{*3} \;\; u^{*2} \;\; u^* \;\; 1]\,\boldsymbol{M}_c$$

$$\times \begin{bmatrix} \boldsymbol{P}(u_0,w_0) & \boldsymbol{P}(u_0,w_1) & (w_1-w_0)\boldsymbol{P}_w(u_0,w_0) & (w_1-w_0)\boldsymbol{P}_w(u_0,w_1) \\ \boldsymbol{P}(u_1,w_0) & \boldsymbol{P}(u_1,w_1) & (w_1-w_0)\boldsymbol{P}_w(u_1,w_0) & (w_1-w_0)\boldsymbol{P}_w(u_1,w_1) \\ (u_1-u_0)\boldsymbol{P}_u(u_0,w_0) & (u_1-u_0)\boldsymbol{P}_u(u_0,w_1) & (u_1-u_0)(w_1-w_0)\boldsymbol{P}_{uw}(u_0,w_0) & (u_1-u_0)(w_1-w_0)\boldsymbol{P}_{uw}(u_0,w_1) \\ (u_1-u_0)\boldsymbol{P}_u(u_1,w_0) & (u_1-u_0)\boldsymbol{P}_u(u_1,w_1) & (u_1-u_0)(w_1-w_0)\boldsymbol{P}_{uw}(u_1,w_0) & (u_1-u_0)(w_1-w_0)\boldsymbol{P}_{uw}(u_1,w_1) \end{bmatrix} \boldsymbol{M}_c^T \begin{bmatrix} w^{*3} \\ w^{*2} \\ w^* \\ 1 \end{bmatrix}. \tag{3.125}$$

3.3.7 Connection of the Coons Bi-cubic Surface Patches

Let us now consider the problem of finding a simple method by which Coons bi-cubic surface patches can be connected (refer to Sect. 1.3.7).

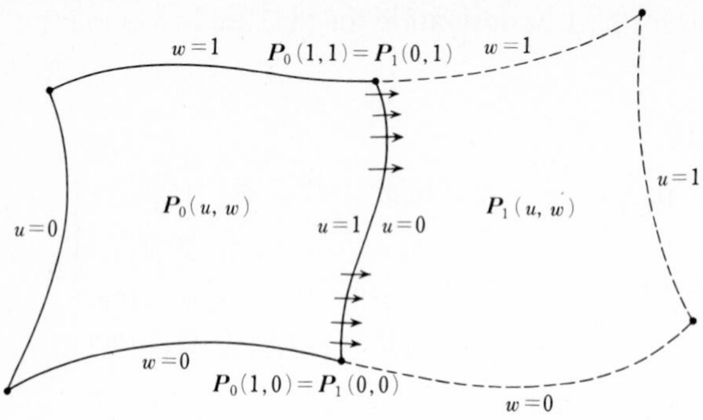

Fig. 3.32. Connection of Coons bi-cubic surface patches (case 1)

Case 1

Let us connect two surface patches $\boldsymbol{P}_0(u, w)$ and $\boldsymbol{P}_1(u, w)$ together with continuity up to the slope (class C^1) as shown in Fig. 3.32. Figure 3.32 shows the case in which the $\boldsymbol{P}_0(1, w)$ boundary curve of $\boldsymbol{P}_0(u, w)$ is connected to the $\boldsymbol{P}_1(0, w)$ boundary curve of $\boldsymbol{P}_1(u, w)$.

First let us consider the condition that the two boundary curves must be equal:

$$\boldsymbol{P}_1(0, w) = \boldsymbol{P}_0(1, w).$$

The boundary curve $\boldsymbol{P}_0(1, w)$ is expressed by the 2nd row of matrix $\boldsymbol{B}_c(\boldsymbol{B}_{c0})$. Since the boundary curve $\boldsymbol{P}_1(0, w)$ of $\boldsymbol{P}_1(u, w)$ is expressed by the 1st row of the $\boldsymbol{B}_c$ matrix of $\boldsymbol{P}_1(u, w)$ $(\boldsymbol{B}_{c1})$, we have, for corresponding elements, the condition that each element in the 1st row of $\boldsymbol{B}_{c1}$ must be equal to the corresponding element in the 2nd row of $\boldsymbol{B}_{c0}$.

Next, let us consider the continuity of slope in the direction across the connecting curve. Along the connecting curve $\boldsymbol{P}_0(1, w)$ of $\boldsymbol{P}_0(u, w)$, the tangent vector function $\boldsymbol{P}_{0,u}(1, w)$ in the direction across the boundary curve is given by the following formula:

$$\begin{aligned}\boldsymbol{P}_{0,u}(1, w) = & H_{0,0}(w)\,\boldsymbol{Q}_u(1,0) + H_{0,1}(w)\,\boldsymbol{Q}_u(1,1)\\ & + H_{1,0}(w)\,\boldsymbol{Q}_{uw}(1,0) + H_{1,1}(w)\,\boldsymbol{Q}_{uw}(1,1).\end{aligned} \tag{3.126}$$

From this formula we see that the 4th row of the $\boldsymbol{B}_c$ matrix of $\boldsymbol{P}_0(u, w)$ expresses the tangent vector function in the direction across the connection curve. In addition, along the connencting curve $\boldsymbol{P}_1(0, w)$ of the surface patch $\boldsymbol{P}_1(u, w)$, the tangent vector in the direction across the connecting curve is expressed by the 3rd row of the $\boldsymbol{B}_c$ matrix of $\boldsymbol{P}_1(u, w)$. Therefore, in the direction across the connecting curve, in order for the slope to be continuous we have the condition that each 3rd row element of matrix $\boldsymbol{B}_{c1}$ must be equal to the corresponding 4th row element of the matrix $\boldsymbol{B}_{c0}$ times r, where r is an arbitrary positive scalar. This relation is shown in Fig. 3.33. Blank elements in matrix $\boldsymbol{B}_{c1}$ are not related to the connection to $\boldsymbol{P}_0(u, w)$; as far as the connection is concerned those vectors can be chosen freely.

(a)

$Q(1,0)$	$Q(1,1)$	$Q_w(1,0)$	$Q_w(1,1)$
$Q_u(1,0)$	$Q_u(1,1)$	$Q_{uw}(1,0)$	$Q_{uw}(1,1)$

(b)

$Q(1,0)$	$Q(1,1)$	$Q_w(1,0)$	$Q_w(1,1)$
$rQ_u(1,0)$	$rQ_u(1,1)$	$rQ_{uw}(1,0)$	$rQ_{uw}(1,1)$

Fig. 3.33. B_c matrix relations for the connection in case 1. (a) matrix B_{c0}; (b) matrix B_{c1}

Case 2

Next, let us consider the case in which the previously defined surface patch $P_0(u, w)$ is connected along both sides to the surface patches $P_1(u, w)$ and $P_2(u, w)$, as shown in Fig. 3.34, with continuity up to the slope. In this case, it is sufficient to repeat the procedure of Case 1 twice. The B_c matrix elements of $P_1(u, w)$ and $P_2(u, w)$ become as shown in Fig. 3.35. Elements which are left blank in matrices B_{c1} and B_{c2} are unrelated to the connection to $P_0(u, w)$; as far as the connections are concerned those vectors can be chosen freely.

Case 3

Now consider the case in which the previously defined surface patch $P_0(u, w)$ is connected to 8 other surface patches around it, with continuity up to the slope, as shown in Fig. 3.36.

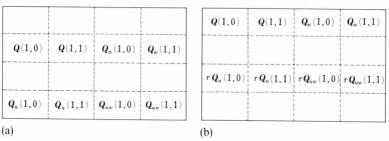

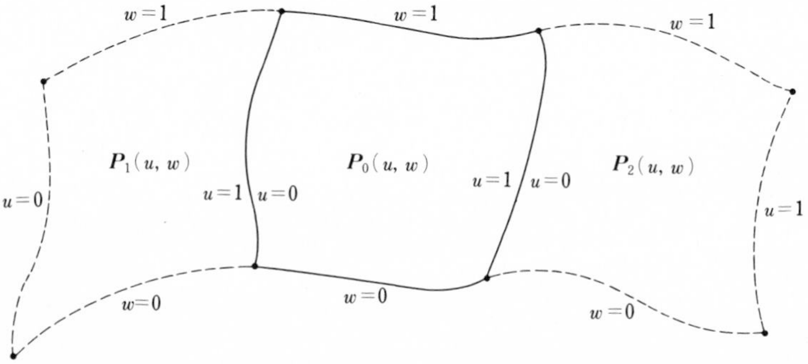

Fig. 3.34. Connection of Coons bi-cubic surface patches (case 2)

(a)

$Q(0,0)$	$Q(0,1)$	$Q_w(0,0)$	$Q_w(0,1)$
$r_1Q_u(0,0)$	$r_1Q_u(0,1)$	$r_1Q_{uw}(0,0)$	$r_1Q_{uw}(0,1)$

(b)

$Q(0,0)$	$Q(0,1)$	$Q_w(0,0)$	$Q_w(0,1)$
$Q(1,0)$	$Q(1,1)$	$Q_w(1,0)$	$Q_w(1,1)$
$Q_u(0,0)$	$Q_u(0,1)$	$Q_{uw}(0,0)$	$Q_{uw}(0,1)$
$Q_u(1,0)$	$Q_u(1,1)$	$Q_{uw}(1,0)$	$Q_{uw}(1,1)$

(c)

$Q(1,0)$	$Q(1,1)$	$Q_w(1,0)$	$Q_w(1,1)$
$r_2Q_u(1,0)$	$r_2Q_u(1,1)$	$r_2Q_{uw}(1,0)$	$r_2Q_{uw}(1,1)$

Fig. 3.35. B_c matrix relations for the connection in case 2 (a) matrix B_{c1}; (b) matrix B_{c0}; (c) matrix B_{c2}

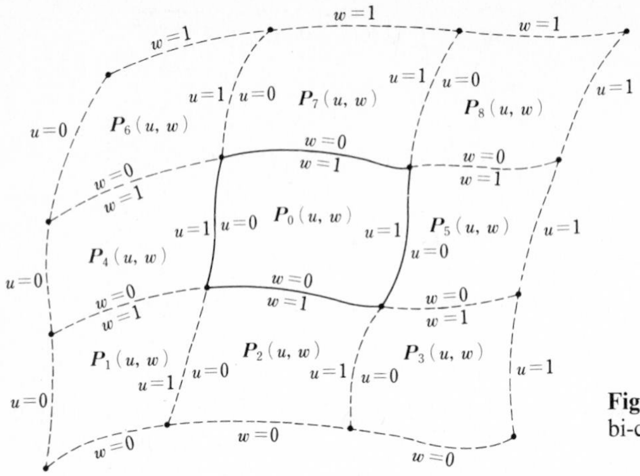

Fig. 3.36. Connection of Coons bi-cubic surface patches (case 3)

First, the B_c matrices of $P_4(u,w)$ and $P_5(u,w)$ are found in exactly the same way as in Case 2. Next, consider the B_c matrices of $P_2(u,w)$ and $P_7(u,w)$. In matrix B_{c0}, the 1st and 2nd columns express the $w=0$ and $w=1$ boundary curves, while the 3rd and 4th columns express the tangent vector functions in the direction across the boundary curves along the boundary curves. From this we see that the elements of the 2nd and 4th columns of B_{c2} and of the 1st and 3rd columns of B_{c7} are determined by the slope continuity condition. Next, for the remaining 4 surface patches, $P_1(u,w)$, $P_3(u,w)$, $P_6(u,w)$ and $P_8(u,w)$, the conditions on the B_c matrix elements are determined by the connection conditions to the two neighboring surface patches in each case. The results are shown in Fig. 3.37. In this figure, the elements that are left blank in each B_c matrix can be freely selected as far as the connections are concerned. However, if one of the blank elements is determined, then some of the others will be determined by their relationship to it. For example, if element (2,2) of matrix B_{c7} is V, then element (1,2) of matrix B_{c8} is also V.

3.3.8 Shape Control of the Coons Bi-cubic Surface Patch

The Coons bi-cubic surface patch is defined by only 16 vectors: the position vectors $Q(0,0)$, $Q(1,0)$, $Q(0,1)$ and $Q(1,1)$ at the 4 corners of the patch, the tangent vectors in the u- and w-directions $Q_u(0,0)$, $Q_w(0,0)$, $Q_u(1,0)$, $Q_w(1,0)$, $Q_u(0,1)$, $Q_w(0,1)$, $Q_u(1,1)$ and $Q_w(1,1)$, and the twist vectors $Q_{uw}(0,0)$, $Q_{uw}(1,0)$, $Q_{uw}(0,1)$ and $Q_{uw}(1,1)$. For neighboring surface patches, as explained in the preceeding section if these vectors are suitably determined, the connection will be continuous up to the slope in the direction across the boundary curve. In Fig. 3.38, this continuity condition is maintained as 9 surface patches are connected together. The 2 tangent vectors at each of the 4 corners of each surface patch are shown. In the figure, these vectors are shown lying in a plane. The twist vectors at the 4 corners of each patch are all set to be $\mathbf{0}$, so they are not shown in the figure. The magnitudes of the tangent vectors at those points are set equal to the sides of the grid rectangles.

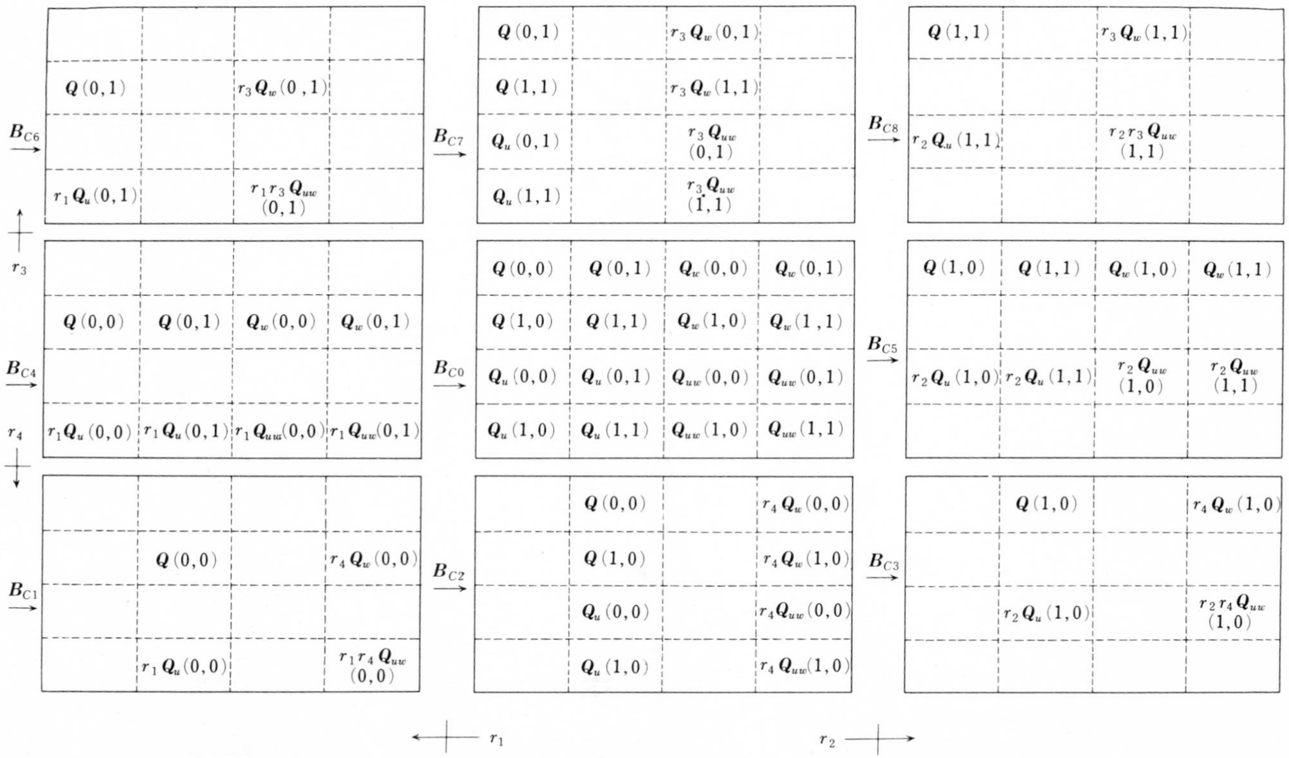

Fig. 3.37. $\boldsymbol{B}_c$ matrix relations for the connection in case 3

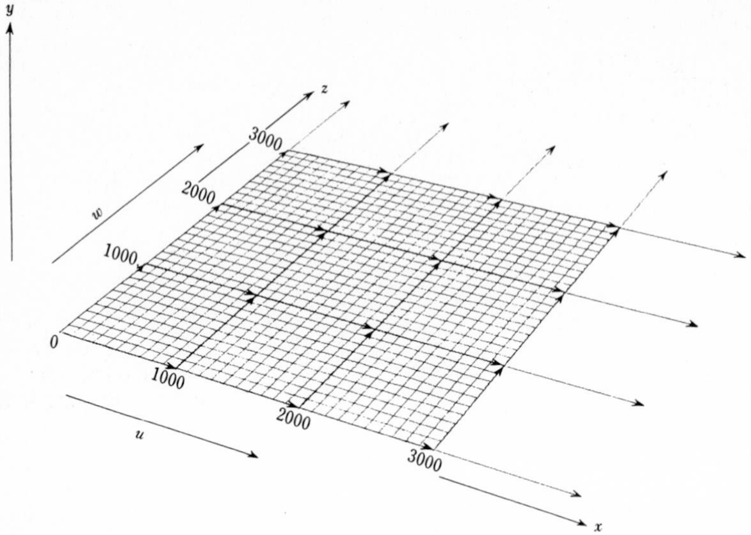

Fig. 3.38. Nine surface patches lying on a plane. Vectors at corners of all 9 surface patches: tangent vector in u-direction: $(+1000, 0, 0)$; tangent vector in w-direction: $(0, 0, +1000)$; twist vector: $(0, 0, 0)$

(1) Control of the Position Vectors

If the y-coordinate (in the vertical direction) of only one of the four corner points in the center patch is varied, the surface bulges up so as to pass through that point (Fig. 3.39). This shape variation occurs in all four of the surface patches which share the point which was varied; it does not extend to the other surface patches. If all 4 of the corner points of the center surface patch are moved exactly the same distance in the y-direction the center patch remains flat as it moves (Fig. 3.40).

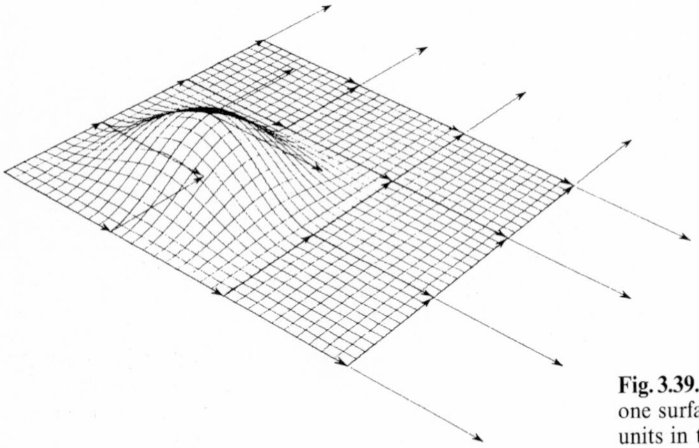

Fig. 3.39. Movement of only one surface patch corner 500 units in the y-direction

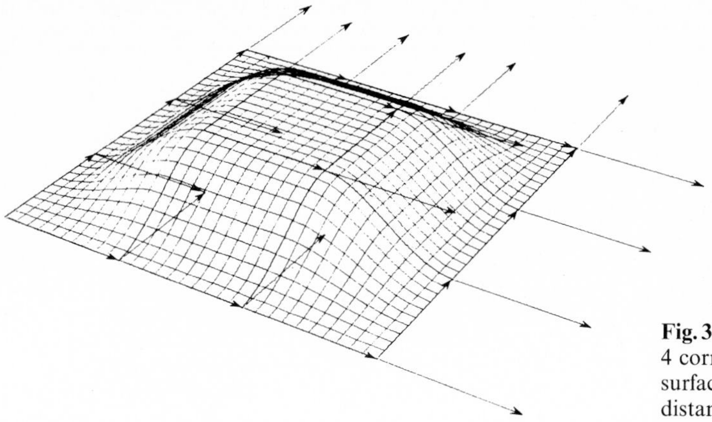

Fig. 3.40. Movement of all 4 corners of the center surface patch the same distance in the y-direction

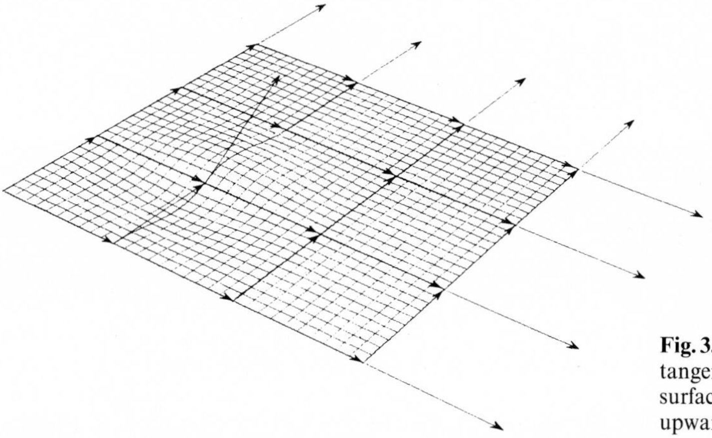

Fig. 3.41. Rotation of the tangent vector at a surface patch corner 30° upward

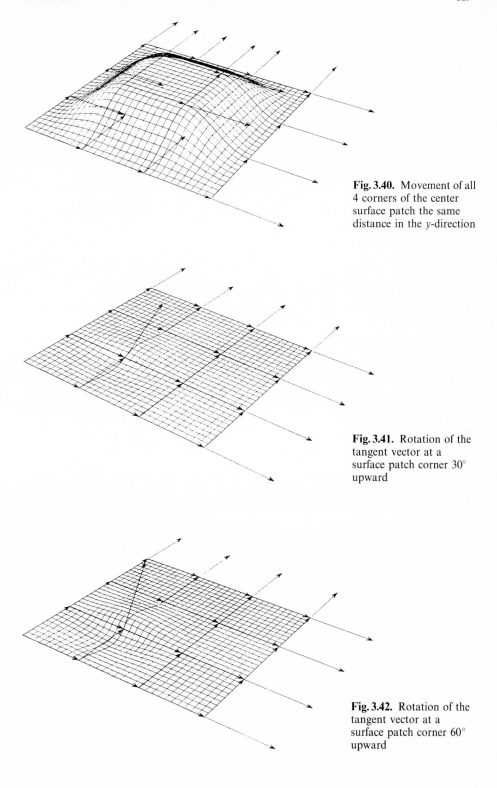

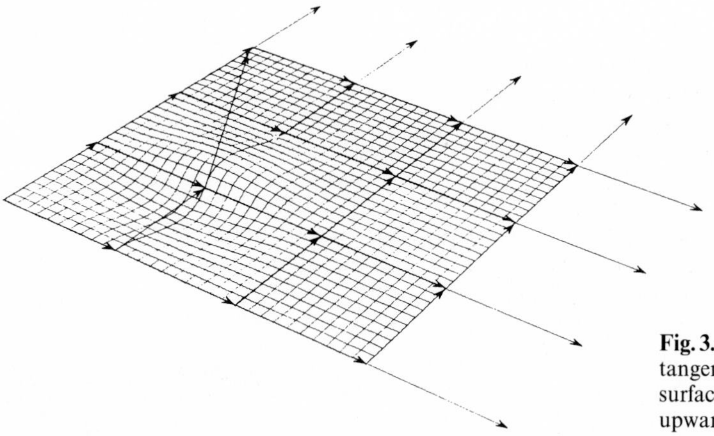

Fig. 3.42. Rotation of the tangent vector at a surface patch corner 60° upward

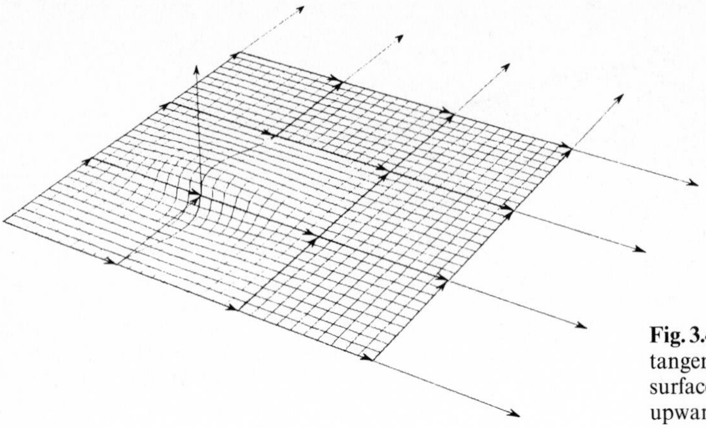

Fig. 3.43. Rotation of the tangent vector at a surface patch corner 90° upward

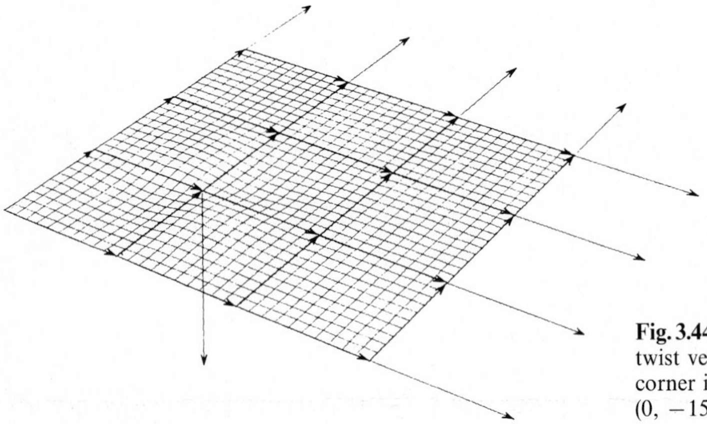

Fig. 3.44. Surface when the twist vector at a patch corner is taken to be (0, −1500, 0)

(2) Control of Tangent Vectors

Starting from the condition shown in Fig. 3.38, rotate one of the tangent vectors at one corner of a surface patch while holding its magnitude fixed. The result of rotation by 30° is shown in Fig. 3.41, and of rotation by 60° in Fig. 3.42. If it is rotated another 30° so that it points straight up, the surface shape becomes as shown in Fig. 3.43. Varying a tangent vector changes the shapes of the four surface patches which share the point at which it is defined.

(3) Control of Twist Vectors

Starting from the condition shown in Fig. 3.38, suppose that a vertical twist vector is added downward at one corner of a surface patch, as shown in Fig. 3.44. The effect of the twist vector on the surface shape can be easily seen by increasing its magnitude, as shown in Fig. 3.45. The effect of the twist vector is to cause parts of the surface to bulge up or become indented near the point where it is added. The direction of the effect is opposite on neighboring surface patches. If the twist vector points up instead of down, the direction of its effect on the surface is reversed.

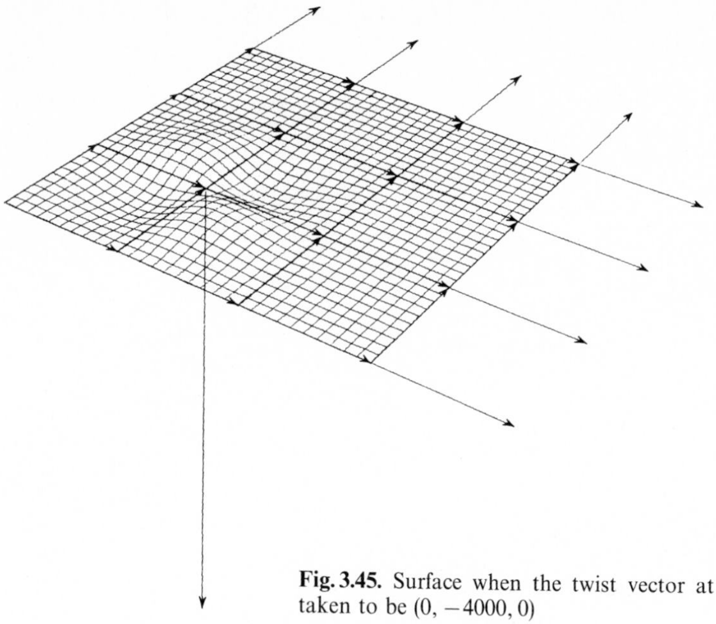

Fig. 3.45. Surface when the twist vector at a patch corner is taken to be (0, −4000, 0)

3.3.9 Triangular Patches Formed by Degeneration

Let us consider the problem of expressing a degerate triangular patch by a bicubic Coons surface (refer to Sect. 1.3.8).

We can find $\boldsymbol{P}_u(0, w)$ from Eq. (3.100) as:

$$\begin{aligned}\boldsymbol{P}_u(0,w) &= [0\ \ 0\ \ 1\ \ 0]\,\boldsymbol{M}_c\boldsymbol{B}_c\boldsymbol{M}_c^T\,[w^3\ \ w^2\ \ w\ \ 1]^T \\ &= H_{0,0}(w)\,\boldsymbol{Q}_u(0,0)+H_{0,1}(w)\,\boldsymbol{Q}_u(0,1) \\ &\quad +H_{1,0}(w)\,\boldsymbol{Q}_{uw}(0,0)+H_{1,1}(w)\,\boldsymbol{Q}_{uw}(0,1)\,. \end{aligned} \tag{3.127}$$

We also have, for $\boldsymbol{P}_{uw}(0, w)$:

$$\begin{aligned}\boldsymbol{P}_{uw}(0,w) &= \dot{H}_{0,0}(w)\,\boldsymbol{Q}_u(0,0)+\dot{H}_{0,1}(w)\,\boldsymbol{Q}_u(0,1) \\ &\quad +\dot{H}_{1,0}(w)\,\boldsymbol{Q}_{uw}(0,0)+\dot{H}_{1,1}(w)\,\boldsymbol{Q}_{uw}(0,1)\,. \end{aligned} \tag{3.128}$$

Both formulas (3.127) and (3.128) are expressed as linear combinations of the four vectors $\boldsymbol{Q}_u(0,0)$, $\boldsymbol{Q}_u(0,1)$, $\boldsymbol{Q}_{uw}(0,0)$ and $\boldsymbol{Q}_{uw}(0,1)$. The tangent plane at the degenerate point must agree with the plane defined by the two tangent vectors $\boldsymbol{Q}_u(0,0)$ and $\boldsymbol{Q}_u(0,1)$. Consequently, if the other two vectors, $\boldsymbol{Q}_{uw}(0,0)$ and $\boldsymbol{Q}_{uw}(0,1)$, are both in this same plane, the unit tangent vector described by (3.128) is uniquely determined for an arbitrary value of w in the range $0 \leq w \leq 1$.

3.3.10 Decomposition of Coons Surface Patches and 3 Types in Constructing Surfaces

As can be seen from the discussion up to this point, the Coons surface patch can be decomposed into 3 surfaces, given by the general formula (3.78) for the Coons surface patch:

$$\boldsymbol{P}_A(u,w)=\sum_{i=0}^{1}\sum_{r=0}^{n} C_{r,i}(u)\,\boldsymbol{Q}^{r,0}(i,w) \tag{3.129}$$

$$\boldsymbol{P}_B(u,w)=\sum_{j=0}^{1}\sum_{s=0}^{n} C_{s,j}(w)\,\boldsymbol{Q}^{0,s}(u,j) \tag{3.130}$$

$$\boldsymbol{P}_C(u,w)=\sum_{i=0}^{1}\sum_{j=0}^{1}\sum_{r=0}^{n}\sum_{s=0}^{n} C_{r,i}(u)\,C_{s,j}(w)\,\boldsymbol{Q}^{r,s}(i,j). \tag{3.131}$$

These are combined to form the Coons surface patch according to:

$$\boldsymbol{P}(u,w)=\boldsymbol{P}_A(u,w)+\boldsymbol{P}_B(u,w)-\boldsymbol{P}_C(u,w). \tag{3.132}$$

The surface $\boldsymbol{P}_A(u,w)$ is generated by the two curves $\boldsymbol{Q}(0,w)$ and $\boldsymbol{Q}(1,w)$ in the w-direction and the ith derivative vectors $\boldsymbol{Q}^{i,0}(0,w)$ and $\boldsymbol{Q}^{i,0}(1,w)$ in the u-direction along those curves up to $i=n$. Surface $\boldsymbol{P}_B(u,w)$ forms a pair with the surface $\boldsymbol{P}_A(u,w)$. $\boldsymbol{P}_B(u,w)$ is the surface generated by giving the two curves $\boldsymbol{Q}(u,0)$ and $\boldsymbol{Q}(u,1)$ in the u-direction and the ith derivative vectors $\boldsymbol{Q}^{0,i}(u,0)$ and $\boldsymbol{Q}^{0,i}(u,1)$ in the w-direction along those two curves up to $i=n$. A surface created in this way by giving data with respect to curves in one direction is called a loft surface (Fig. 3.46).

Surface $\boldsymbol{P}_C(u,w)$ is defined by specifying positions with respect to given lattice points and higher order derivative vectors. Such a surface is called a Cartesian product surface or a tensor product surface (Fig. 3.47).

A Coons surface patch consists of loft surfaces in both directions and a Cartesian product surface (Fig. 3.48). Therefore, it is defined by specifying data for curves in both directions and data with respect to lattice points. We have already pointed out, in Sects. 3.3.2 and 3.3.3, that a Coons surface patch

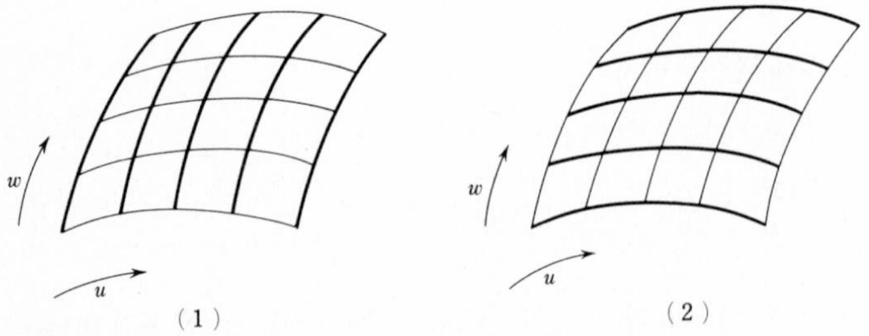

Fig. 3.46. Loft surfaces generated by specifying a group of curves in (1) the w-direction, (2) the u-direction

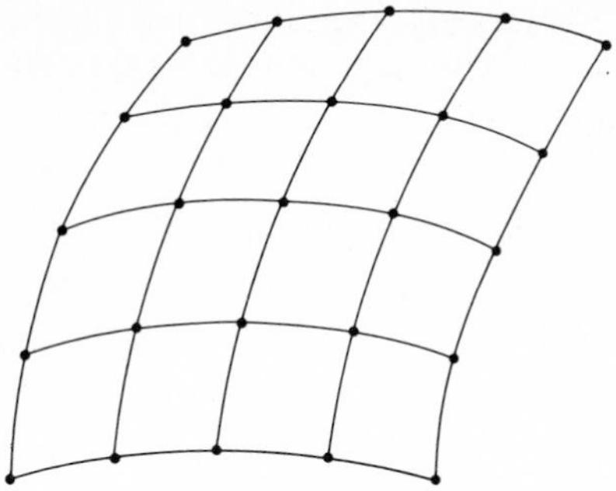

Fig. 3.47. A Cartesian product surface (tensor product surface)

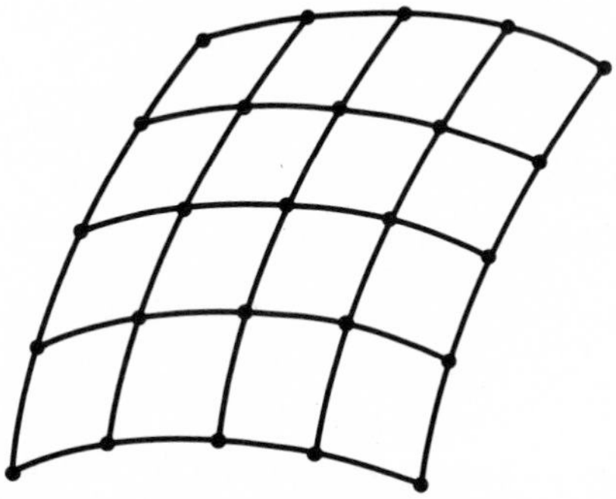

Fig. 3.48. A Coons curved surface patch

includes the loft surfaces and a Cartesian product surface as special cases. Thus, we see that the Coons surface patch is a very general surface formula. The Coons surface patch is often called a Boolean sum type surface.

3.3.11 Some Considerations on Hermite Interpolation Curves and Surfaces

Curves and surfaces based on Hermite interpolation are generated by specifying the positions of points through which they pass and higher order derivative vectors at those points. As can be guessed from the discussion thus far, they present a number of problems when a person tries to change, control or design such a shape.

1) When a curve is defined, it is necessary to define not only the position vectors at points through which the curve passes, but also the tangent vectors and higher order derivative vectors at those points. In the case of a surface, the position vectors must be specified at lattice points through which the surface passes, along with tangent vectors in 2 directions, twist vectors, etc. at those points. In general, there are many points at which data must be given, so a large quantity of date must be input to generate a curve or surface.

2) Since it is hard to predict the effect of a tangent vector or twist vector on a curve or surface shape in some cases it is very hard for a person to determine suitable values for them. In addition, although curve segments and surface

patches can be mathematically connected by making the directions of the tangent vectors equal at the contact points, if their magnitudes change, the shapes will change. How to choose the optimum magnitude is not made clear in the papers of Ferguson and Coons.

3) To control a shape, it is necessary to control different types of vectors, specifically position vectors, tangent vectors and twist vectors. In general, depending on the type of vectors, the magnitudes of the vectors are of different orders, so control tends to be difficult.

4) The effects of tangent vectors and twist vectors on curves and surfaces are difficult to tell at a glance and differ from point to point, which causes complications in shape control (refer to Sect. 3.3.4).

References (Chap. 3)

10) Forrest, A.R.: "Curves and Surfaces for Computer Aided Design", Ph. D. thesis, Univ. Cambridge, July 1968, p. 19.
11) Barnhill, R.E. and R.F. Riesenfeld (eds.): *Computer Aided Geometric Design,* Academic, New York, 1975, p. 272—274.
12) Gellert, G.O.: "Geometric Computing-Electronic Geometry for Semiautomated Design", Machine Design, March 1965.
13) Coons, S.A.: "Surfaces for computer-aided design of space figures", M.I.T. ESL 9442-M-139, January 1964.
14) Coons, S.A.: "Surfaces for Computer Aided Design of Space Forms", *MIT Project MAC,* TR-41, June 1967.
15) Barnhill, R.E., J.H. Brown, and I.M. Klucewicz: "A New Twist in Computer Aided Geometric Design", *Comput. Graphics Image Processing, 8* (1): 78—91, August 1978.

4. Spline Interpolation

4.1 Splines

When a smooth curve passing through a specified sequence of points is generated, use of the shape of a curve produced by a long narrow elastic band such as a steel band has long been used in the design of, for example, ships and automobiles. An elastic band used for such a purpose is called a spline. The spline can be made to assume the shape of a smooth curve passing through the specified points by attaching a suitable number of weights, called *weights* or *ducks* (Fig. 4.1).

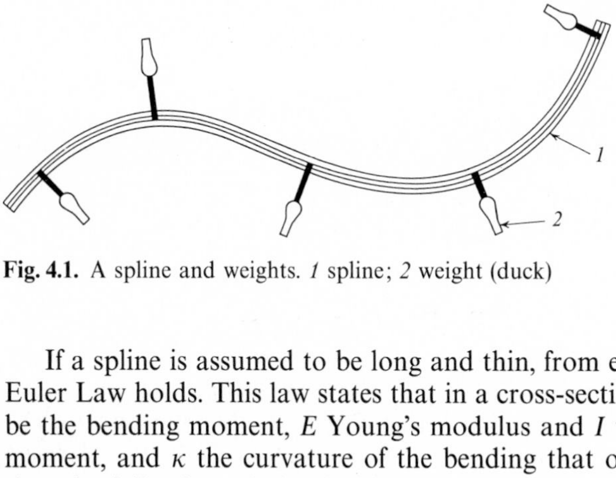

Fig. 4.1. A spline and weights. *1* spline; *2* weight (duck)

If a spline is assumed to be long and thin, from elastic theory the Bernoulli-Euler Law holds. This law states that in a cross-section of the spline, if we let M be the bending moment, E Young's modulus and I the cross-section 2nd order moment, and κ the curvature of the bending that occurs in that cross-section, then the following relation holds:

$$M = EI\kappa. \tag{4.1}$$

Here the deformation of the spline is assumed to be sufficiently small. That is, in an $x-y$ Cartesian coordinate system, the spline deformation curve (where y represents the deflection) can be expressed as follows, assuming that $dy/dx \ll 1$:

$$\kappa(x) = \frac{\dfrac{d^2y}{dx^2}}{\left\{1+\left(\dfrac{dy}{dx}\right)^2\right\}^{\frac{3}{2}}} \fallingdotseq \frac{d^2y}{dx^2}.$$

From Eq. (4.1), we can write this as:

$$\frac{d^2y}{dx^2} = \frac{1}{EI} M(x). \tag{4.2}$$

Since the weights act as simple support points for the spline, the variation of the bending moment M between the weights is linear. Therefore, if we apply some constraints to both ends of the spline and solve Eq. (4.2), the elastic curve is determined. This curve consists of separate cubic segments between each pair of weights which are connected together; not only the slopes but the curvatures are continuous at the weight positions. However, the rate of change of curvature is discontinuous at the weight positions.

According to elasticity theory, the shape of the elastic curve minimizes the total bending energy stored in the spline:

$$\text{total bending energy} = \frac{EI}{2} \int_0^l \kappa^2(s)\, ds \Rightarrow \text{minimum}. \tag{4.3}$$

Here $\kappa(s)$ is the spline curvature, EI is the spline flexural rigidity, l is the total spline length and s is the parameter along the spline length.

4.2 Spline Functions

In the preceding section we explained that the term "spline" comes from the name of an elastic ruler that is free to bend, and discussed the properties of this freely bending ruler from a physical point of view. Mathematically speaking, a spline is a generalization of the physical spline which has presently become a very important technique in numerical analysis and CAD. Let us now consider the mathematical aspects of splines. We noted in the preceding section that a physical spline bends in such a way as to minimize the total bending energy. The following discussion explores the mathematical analogue of this physical property.

First, let us define a spline function.

Definition 4.1. A (polynomial) spline function of order M and degree $m = M - 1$, having an increasing sequence of real numbers:

$$x_{-1} < x_0 < x_1 < \cdots < x_{n-1} < x_n < x_{n+1} \text{ *)}$$

as knots, is a function $S(x)$ which satisfies the following two conditions ① and ②.

*) In this chapter, the knots are all taken to be different. However, in general they can be allowed to coincide. In Chap. 6, a more general knot sequence in which the knots are allowed to coincide is used.

① $S(x)$ is a polynomial of degree m or less within each interval (x_{i-1}, x_i) $(i=0, 1, 2, \ldots, n+1; x_{-1}=-\infty, x_{n+1}=+\infty)$.

② $S(x)$ and its 1-st, 2-nd, ..., $(m-1)$-st derivatives are continuous in (x_{-1}, x_{n+1}).

Thus, a spline function consists of a number of polynomials which are defined separately in small intervals and connected together as smoothly as possible. A spline function of degree m has an m-th derivative that is a step function; the $(m-1)$-st and lower-order derivatives are continuous. From the above definition, an m-th degree polynomial can be regarded as a special case of an m-th degree spline. Note that a 1-st degree spline function is a bent line. In this book, a spline function will often be abbreviated as simply "spline".

4.3 Mathematical Representation of Spline Functions

Let $S(x)$ be an m-th degree spline function. $S(x)$ is continuous to the $(m-1)$-st order derivative at knots x_i $(i=0, 1, \ldots, n)$. Letting $P_{m,i}$ be an m-th degree polynomial in the interval (x_{i-1}, x_i), we have:

$$P_{m,i}^{(r)}(x_i) = P_{m,i+1}^{(r)}(x_i) \quad (r=0, 1, \ldots, m-1; i=0, 1, \ldots, n).$$

Since the m-th degree polynomial $g(x)=P_{m,i+1}(x)-P_{m,i}(x)$ and its derivatives up to order $m-1$ have roots at $x=x_i$, we have:

$$g(x)=P_{m,i+1}(x)-P_{m,i}(x)=c_i(x-x_i)^m$$

where c_i is a constant. This equation can be rewritten as:

$$P_{m,i+1}(x)=P_{m,i}(x)+c_i(x-x_i)^m.$$

When an m-th degree polynomial $P_{m,0}(x)$ is given in the interval (x_{-1}, x_0), $P_{m,1}(x)$ is related to it in the interval (x_0, x_1) by:

$$P_{m,1}(x)=P_{m,0}(x)+c_0(x-x_0)^m.$$

The polynomial $P_{m,2}(x)$ in the interval (x_1, x_2) becomes:

$$P_{m,2}(x)=P_{m,0}(x)+c_0(x-x_0)^m+c_1(x-x_1)^m.$$

Similarly, the polynomial $P_{m,i+1}(x)$ in the interval (x_i, x_{i+1}) can be written as:

$$P_{m,i+1}(x)=P_{m,0}(x)+\sum_{j=0}^{i} c_j(x-x_j)^m.$$

Rewriting $P_{m,i+1}(x)$ in the spline notation $S(x)$, we have:

$$S(x)=P_{m,0}(x)+\sum_{j=0}^{i} c_j(x-x_j)^m \quad (x_i \leqq x \leqq x_{i+1}). \tag{4.4}$$

Using the truncated power function:

$$(x-x_i)_+^m=\begin{cases}(x-x_i)^m & (x>x_i)\\ 0 & (x\leqq x_i)\end{cases}$$

we have, over the entire domain of x:

$$S(x)=P_{m,0}(x)+\sum_{i=0}^{n} c_i(x-x_i)_+^m. \tag{4.5}$$

In Eq. (4.5), the c_i's are unknown. The next step is to find them. Since the m-th order derivative of an m-th degree spline function is a step function, $S^{(m)}(x)$ becomes discontinuous at the knot points. Let us look at the m-th order difference between the spline functions on both sides of the knot x_i. Using Eq. (4.4) and differentiating gives:

$$S^{(m)}(x_i+0)-S^{(m)}(x_i-0)=m!\,c_i. \tag{4.6}$$

Solving for c_i gives:

$$c_i=\frac{1}{m!}\left[S^{(m)}(x_i+0)-S^{(m)}(x_i-0)\right]. \tag{4.7}$$

4.4 Natural Splines

Among spline functions, the natural spline defined below is highly significant in that it has the property of minimum interpolation.

Definition 4.2. A spline which is at most of degree $k-1$ in the two end intervals (x_{-1}, x_0) and (n_n, x_{n+1}) and $2k-1$ in the other intervals is called a natural spline *).

From the results in the preceding section, it is clear that a natural spline can be expressed in the following form, from Eq. (4.5):

$$S(x)=P_{k-1,0}(x)+\sum_{i=0}^{n} c_i(x-x_i)_+^{2k-1}. \tag{4.8}$$

---

*) In the case $k=2$, for $x_i \leqq x \leqq x_{i+1}$ $(i=0, 1, \ldots, n-1)$ the spline is cubic; for $x_{-1}<x<x_0$ and $x_n<x<x_{n+1}$ it is linear; since this is the actual shape of a physical spline, the name "natural spline" is given to it.

Let us find the conditions on the coefficients c_i. For $x \geqq x_n$, $S(x)$ can be written as in the above equation without the subscript "+":

$$S(x) = P_{k-1,0}(x) + \sum_{i=0}^{n} c_i (x - x_i)^{2k-1} \quad (x \geqq x_n). \tag{4.9}$$

In this equation, the k-th derivative of $S(x)$ is:

$$S^{(k)}(x) = (2k-1)(2k-2)\ldots(k+1)\,k \sum_{i=0}^{n} c_i (x - x_i)^{k-1} \quad (x \geqq x_n).$$

This equation can be rewritten as follows:

$$\begin{aligned} S^{(k)}(x) &= (2k-1)(2k-2)\ldots(k+1)\,k \sum_{i=0}^{n} c_i (x - x_i)^{k-1} \\ &= (2k-1)(2k-2)\ldots(k+1)\,k \sum_{i=0}^{n} c_i \sum_{r=0}^{k-1} \binom{k-1}{r} x^{k-1-r} (-1)^r x_i^r \\ &= (2k-1)(2k-2)\ldots(k+1)\,k \sum_{i=0}^{k-1} \left[(-1)^r \binom{k-1}{r} x^{k-1-r} \sum_{i=0}^{n} c_i x_i^r \right] \\ &\quad (x \geqq x_n). \end{aligned} \tag{4.10}$$

Since $S(x)$ is a natural spline, for $x \geqq x_n$ it is of at most degree $k-1$, so the k-th derivative must be identically 0. Since the right side of Eq. (4.10) is a polynomial of degree $(k-1)$ in x, for the entire expression to be identically 0 the coefficient of each power of x must be 0:

$$\sum_{i=0}^{n} c_i x_i^r = 0 \quad (r = 0, 1, 2, \ldots, k-1). \tag{4.11}$$

It is easy to see that the derivative of a spline is also a spline, of one lower degree, having the same knots. Since a natural spline has k, $(k+1)$, ..., $(2k+2)$-order derivatives equal to 0 in both end intervals (x_{-1}, x_0) and (x_n, x_{n+1}), the respective derivatives must be 0 at the knots x_0 and x_n.

Definition 4.3. A spline of degree k that is identically 0 in both end intervals (x_{-1}, x_0) and (x_n, x_{n+1}) is called a C-spline.

From this definition, we see that the k, $(k+1)$, ..., $(2k-2)$-order derivatives of a natural spline of degree $(2k-1)$ are C-splines.

4.5 Natural Splines and the Minimum Interpolation Property

A natural spline has the property corresponding to minimizing potential energy, that is, Eq. (4.3). Let us now give some auxiliary theorems needed to prove this.

Auxiliary Theorem 4.1. Given a C-spline with knots $a<x_0<x_1<\ldots<x_n<b$, with $C(x)$ expressed in the form:

$$C(x)=\sum_{i=0}^{n} b_i(x-x_i)_+^{k-1}$$

and let $f(x)$ be a function that is continuously differentiable m times in the interval (a,b). Then the following holds:

$$\int_a^b C(x)f^{(k)}(x)\,dx=(-1)^k(k-1)!\sum_{i=0}^{n} b_i f(x_i). \tag{4.12}$$

Proof: Partially integrating the left-hand side of Eq. (4.12) gives:

$$\int_a^b C(x)f^{(k)}(x)\,dx=\left[C(x)f^{(k-1)}(x)\right]_a^b-\int_a^b C'(x)f^{(k-1)}(x)\,dx.$$

From the definition of a C-spline, at $x=a$ and $x=b$ $C(x)$ and all of its derivatives are identically 0, which implies:

$$\int_a^b C(x)f^{(k)}(x)\,dx=(-1)\int_a^b C'(x)f^{(k-1)}(x)\,dx.$$

Partially integrating the right-hand side again and using the fact that the derivatives of $C(x)$ are 0 at $x=a$ and $x=b$ gives:

$$\begin{aligned}\int_a^b C(x)f^{(k)}(x)\,dx&=(-1)\left[C'(x)f^{(k-2)}(x)\right]_a^b-(-1)\int_a^b C''(x)f^{(k-2)}(x)\,dx\\&=(-1)^2\int_a^b C''(x)f^{(k-2)}(x)\,dx.\end{aligned}$$

Repeating this procedure gives:

$$\int_a^b C(x)f^{(k)}(x)\,dx=(-1)^{k-1}\int_a^b C^{(k-1)}(x)f'(x)\,dx. \tag{4.13}$$

Since $C(x)$ is of degree $(k-1)$, $C^{(k-1)}(x)$ is a constant $\eta_i(x_i<x<x_{i+1})$. Therefore, the integral on the left-hand side of Eq. (4.13) is, in each interval:

$$(-1)^{k-1}\int_{x_i}^{x_{i+1}} C^{(k-1)}(x)f'(x)\,dx=(-1)^{k-1}\eta_i[f(x_{i+1})-f(x_i)]$$

which implies:

$$\int_a^b C(x)f^{(k)}(x)\,dx=(-1)^{k-1}\sum_{i=0}^{n-1}\eta_i[f(x_{i+1})-f(x_i)]$$
$$=(-1)^k\sum_{i=0}^{n}(\eta_i-\eta_{i-1})f(x_i) \qquad (4.14)$$

where $\eta_{-1}=\eta_n=0$ in Eq. (4.14).

From Eq. (4.6), we have:

$$\eta_i-\eta_{i-1}=C^{(k-1)}(x_i+0)-C^{(k-1)}(x_i-0)$$
$$=(k-1)!\,b_i.$$

Substituting this relation into Eq. (4.14) gives:

$$\int_a^b C(x)f^{(k)}(x)\,dx=(-1)^k(k-1)!\sum_{i=0}^{n}b_if(x_i)$$

Q.E.D.

The purposes of this section are to show that, when $(n+1)$ data points (x_0,y_0), (x_1,y_1), ..., (x_n,y_n) are given (such that $a<x_0<x_1<\ldots<x_n<b$) and $k(<n+1)$ is given, with $x_0,x_1,\ldots,x_n$ as knots, then the natural spline of degree $(2k-1)$ that interpolates between the y values is uniquely determined; and that among all possible functions that can interpolate between the given data points, this natural interpolating spline is the function that minimizes:

$$\int_a^b [f^{(k)}(x)]^2\,dx. \qquad (4.15)$$

Theorem 4.1. There exists a unique natural spline of degree $(2k-1)$ $(k<n+1)$ that interpolates between the data points (x_0,y_0), (x_1,y_1), ..., (x_n,y_n), with knots $x_0<x_1<\ldots<x_n$.

Proof: Since a natural spline of degree $(2k-1)$ is given by Eq. (4.9), the condition that it interpolates between the specified data points is expressed mathematically as follows:

$$P_{k-1,0}(x_j)+\sum_{i=0}^{n}c_i(x_j-x_i)_+^{2k-1}=y_j \quad (j=0,1,2,\ldots,n). \qquad (4.16)$$

The coefficients c_i must satisfy the condition of Eq. (4.11):

$$\sum_{i=0}^{n} c_i x_i^r = 0 \quad (r = 0, 1, 2, \ldots, k-1). \tag{4.17}$$

In Eq. (4.16), the first term includes k unknown coefficients and the second term includes $(n+1)$ c_i's, so there are a total of $(k+n+1)$ unknown coefficients. Therefore, Eqs. (4.16) and (4.17) constitute a system of $(k+n+1)$ simultaneous linear equations.

Next, let us show that these simultaneous linear equations have a unique set of solutions. This is equivalent to showing that the homogeneous equations with $y_j = 0$ $(j = 0, 1, 2, \ldots, n)$ have only zero solutions; with data points:

$$(x_0, 0), (x_1, 0), (x_2, 0), \ldots, (x_n, 0) \tag{4.18}$$

and that the natural spline of degree $(2k-1)$ that interpolates between these data points can only be identically zero.

Let $f(x)$ be the natural spline of degree $(2k-1)$ that interpolates between the sequence of points (4.18). As was stated in the preceding section, $f^{(k)}(x)$ is a C-spline of degree $(k-1)$ that has the knots $a < x_0 < x_1 < \ldots < x_n < b$. Since $f(x)$ is 0 at these knots, then, from Auxiliary Theorem 4.1, if we take $C(x) = f^{(k)}(x)$, then we have:

$$\int_a^b C(x) f^{(k)}(x)\, dx = \int_a^b [f^{(k)}(x)]^2\, dx = 0.$$

This implies that $f^{(k)}(x) = 0$ $(a \leqq x \leqq b)$, so that $f(x)$ is a polynomial of degree not more than $(k-1)$. Since $f(x)$ is 0 at k or more different points (the $(n+1)$ points $x_0, x_1, x_2, \ldots, x_n$), $f(x)$ must be identically 0 *). This shows that the solutions of the homogeneous equations must be 0. Therefore, the coefficients of the natural spline given by Eq. (4.9) are determined uniquely by Eqs. (4.16) and (4.17). Q.E.D.

Theorem 4.2. Let $S(x)$ be a natural spline of degree $(2k-1)$ which has knots $a < x_0 < x_1 < \ldots < x_n < b$ and interpolates between the $(n+1)$ data points (x_0, y_0), (x_1, y_1), $\ldots$, (x_n, y_n). We assume that $k < n+1$. In addition, let $f(x)$ be an arbitrary function that interpolates between the above data points and that its derivative functions up to order k are continuous. Then the following relation holds:

$$\int_a^b [S^{(k)}(x)]^2\, dx \leqq \int_a^b [f^{(k)}(x)]^2\, dx.$$

Equality only occurs in the case $f(x) = S(x)$.

---

*) This follows from an algebraic theorem: If a polynomial of degree n is 0 at more than n different points, then that polynomial is identically 0.

Proof: Start with the identity:

$$f^{(k)}(x) = S^{(k)}(x) + [f^{(k)}(x) - S^{(k)}(x)].$$

Squaring both sides gives:

$$[f^{(k)}(x)]^2 = [S^{(k)}(x)]^2 + [f^{(k)}(x) - S^{(k)}(x)]^2 + 2\,S^{(k)}(x)\,[f^{(k)}(x) - S^{(k)}(x)].$$

Integrating both sides of this equation, we obtain:

$$\int_a^b [f^{(k)}(x)]^2\,dx = \int_a^b [S^{(k)}(x)]^2\,dx + \int_a^b [f^{(k)}(x) - S^{(k)}(x)]^2\,dx$$
$$+ 2\int_a^b S^{(k)}(x)\,[f^{(k)}(x) - S^{(k)}(x)]\,dx. \qquad (4.19)$$

Auxiliary Theorem 4.1 can be applied to the third term on the right-hand side. Since $S(x)$ is a natural spline of degree $(2k-1)$, $S^{(k)}(x)$ is clearly, from the discussion in the preceding section, a C-spline of degree $(k-1)$ which has the same knots as $S(x)$. Since, by assumption, $f(x)$ and $S(x)$ interpolate between the same data points, we have at each knot:

$$f(x_i) - S(x_i) = 0 \quad (i = 0, 1, 2, \ldots, n).$$

Substituting $S^{(k)}(x)$ for $C(x)$ and $f(x) - S(x)$ for $f(x)$ in the auxiliary theorem, we obtain:

$$\int_a^b S^{(k)}(x)\,[f^{(k)}(x) - S^{(k)}(x)]\,dx = 0.$$

The first and second terms on the right-hand side of Eq. (4.19) are non-negative. Therefore we have:

$$\int_a^b [S^{(k)}(x)]^2\,dx \leqq \int_a^b [f^{(k)}(x)]^2\,dx.$$

Equality holds only when the second term on the right-hand side of Eq. (4.19) is 0, that is, when $f^{(k)}(x) - S^{(k)}(x)$ is identically 0 in the interval $a \leqq x \leqq b$. Therefore, $f(x) - S(x)$ is a polynomial of degree not more than $(k-1)$. However, we know that this polynomial is 0 at $(n+1) > k$ different values of x $x_0, x_1, \ldots, x_n$.

Therefore, $f(x) - S(x)$ is identically 0; equality holds only when $f(x) = S(x)$. Q.E.D.

The above discussion shows that a function which interpolates between the $(n+1)$ data points (x_0, y_0), (x_1, y_1), $\ldots$, (x_n, y_n) and minimizes the expression (4.15) is a natural spline of degree $(2k-1)$, where $k < n+1$. This is called the minimal interpolation property of a natural spline. In this sense it can be said that a natural spline is the smoothest possible interpolation function. To find a natural spline, it is necessary to solve the condition equations (4.16) and (4.17).

4.6 Smoothing Splines

In the preceding section, it was shown that a natural spline that passes through the specified points (x_0, y_0), (x_1, y_1), ..., (x_n, y_n) has the minimal interpolation property. In this section we will consider the case in which the data at these points include a certain amount of error, and find the function $f(x)$ that ① minimizes the "deviation" from the data values and ② is itself as smooth as possible (Fig. 4.2). The first condition involves minimizing the sum of the squares of errors:

$$E_1 = \sum_{i=0}^{n} [y_i - f(x_i)]^2$$

while the second condition involves minimization in the sense of the preceding section, that is, we must minimize:

$$E_2 = \int_a^b [f^{(k)}(x)]^2 \, dx .$$

Schoenberg[16)] showed that when $x_0 < x_1 < \ldots < x_n$ and w_i and g are arbitrary positive numbers, there exists a unique function that minimizes:

$$\sigma = \sum_{i=0}^{n} w_i [y_i - f(x_i)]^2 + g \int_a^b [f^{(k)}(x)]^2 \, dx \tag{4.20}$$

and that function is a natural spline of degree $(2k-1)$ with knots $x_0, x_1, \ldots, x_n$. In this case, the intervals between knots are arbitrary. Such a spline is called a smoothing spline.

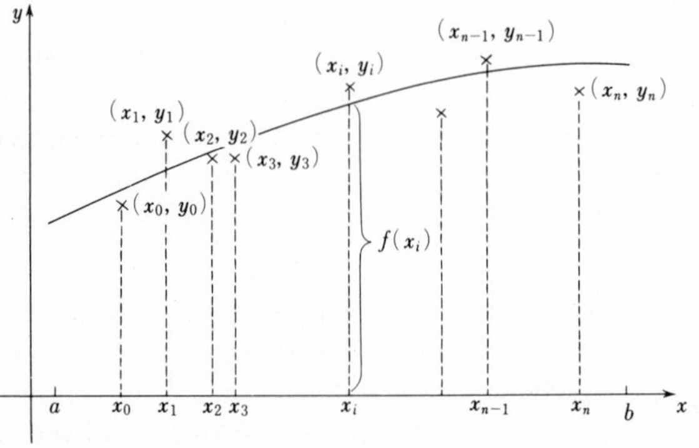

Fig. 4.2. A smoothing spline

4.7 Parametric Spline Curves

When the points $\boldsymbol{Q}_0, \boldsymbol{Q}_1, \ldots, \boldsymbol{Q}_n$ through which a curve passes are given, suppose that a curve is generated by connecting n parametric curve segments $\boldsymbol{P}_i (i=1,2,\ldots,n)$ with continuity up to the curvature (Fig. 4.3).

First, consider the two segments $\boldsymbol{P}_1$ and $\boldsymbol{P}_2$ having the three points $\boldsymbol{Q}_0$, $\boldsymbol{Q}_1$, $\boldsymbol{Q}_2$ as their end points (Fig. 4.4). These curves are expressed by cubic formulas with the curve length s as a parameter:

$$\boldsymbol{P}_1(s) = [s^3 \ \ s^2 \ \ s \ \ 1]\, \boldsymbol{M}_1 \tag{4.21}$$

$$\boldsymbol{P}_2(s) = [s^3 \ \ s^2 \ \ s \ \ 1]\, \boldsymbol{M}_2 \tag{4.22}$$

where $\boldsymbol{M}_1$ and $\boldsymbol{M}_2$ are 4×3 matrices. Differentiating these two equations with respect to s gives:

$$\boldsymbol{P}_1'(s) = [3s^2 \ \ 2s \ \ 1 \ \ 0]\, \boldsymbol{M}_1 \tag{4.23}$$

$$\boldsymbol{P}_2'(s) = [3s^2 \ \ 2s \ \ 1 \ \ 0]\, \boldsymbol{M}_2 . \tag{4.24}$$

Let s_0, s_1 and s_2 be the values of the parameter s at points $\boldsymbol{Q}_0$, $\boldsymbol{Q}_1$ and $\boldsymbol{Q}_2$, respectively, and denote the unit tangent vectors at those same points by $\boldsymbol{Q}_0'$, $\boldsymbol{Q}_1'$ and $\boldsymbol{Q}_2'$. Then, from Eqs. (4.21), (4.22), (4.23) and (4.24) the following two equations hold:

$$\begin{bmatrix} \boldsymbol{Q}_0 \\ \boldsymbol{Q}_1 \\ \boldsymbol{Q}_0' \\ \boldsymbol{Q}_1' \end{bmatrix} = \begin{bmatrix} s_0^3 & s_0^2 & s_0 & 1 \\ s_1^3 & s_1^2 & s_1 & 1 \\ 3s_0^2 & 2s_0 & 1 & 0 \\ 3s_1^2 & 2s_1 & 1 & 0 \end{bmatrix} \boldsymbol{M}_1 \equiv \boldsymbol{N}_1 \boldsymbol{M}_1 \tag{4.25}$$

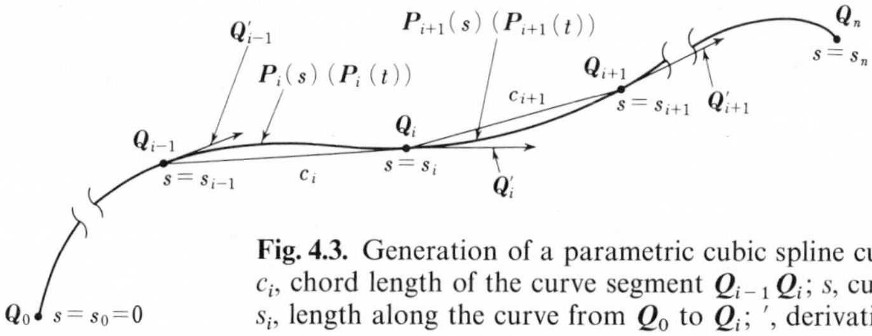

Fig. 4.3. Generation of a parametric cubic spline curve.
c_i, chord length of the curve segment $\boldsymbol{Q}_{i-1}\boldsymbol{Q}_i$; s, curve length parameter; s_i, length along the curve from $\boldsymbol{Q}_0$ to $\boldsymbol{Q}_i$; ′, derivative with respect to s

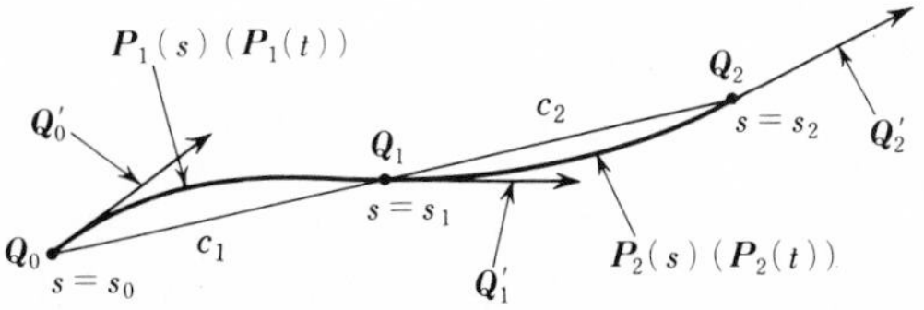

Fig. 4.4. The first two curve segments

$$\begin{bmatrix} \boldsymbol{Q}_1 \\ \boldsymbol{Q}_2 \\ \boldsymbol{Q}'_1 \\ \boldsymbol{Q}'_2 \end{bmatrix} = \begin{bmatrix} s_1^3 & s_1^2 & s_1 & 1 \\ s_2^3 & s_2^2 & s_2 & 1 \\ 3s_1^2 & 2s_1 & 1 & 0 \\ 3s_2^2 & 2s_2 & 1 & 0 \end{bmatrix} \boldsymbol{M}_2 \equiv \boldsymbol{N}_2 \boldsymbol{M}_2. \tag{4.26}$$

Finding $\boldsymbol{M}_1$ and $\boldsymbol{M}_2$ from Eqs. (4.25) and (4.26) and substituting into Eqs. (4.21) and (4.22) gives:

$$\boldsymbol{P}_1(s) = [s^3 \;\; s^2 \;\; s \;\; 1]\, \boldsymbol{N}_1^{-1} \begin{bmatrix} \boldsymbol{Q}_0 \\ \boldsymbol{Q}_1 \\ \boldsymbol{Q}'_0 \\ \boldsymbol{Q}'_1 \end{bmatrix} \tag{4.27}$$

$$\boldsymbol{P}_2(s) = [s^3 \;\; s^2 \;\; s \;\; 1]\, \boldsymbol{N}_2^{-1} \begin{bmatrix} \boldsymbol{Q}_1 \\ \boldsymbol{Q}_2 \\ \boldsymbol{Q}'_1 \\ \boldsymbol{Q}'_2 \end{bmatrix} \tag{4.28}$$

where $\boldsymbol{N}_1^{-1}$ is given by:

$$\boldsymbol{N}_1^{-1} = \frac{1}{(s_1 - s_0)^3} \times \begin{bmatrix} 2 & -2 & s_1 - s_0 & s_1 - s_0 \\ -3(s_0 + s_1) & 3(s_0 + s_1) & -(s_0 + 2s_1)(s_1 - s_0) & -(2s_0 + s_1)(s_1 - s_0) \\ 6s_0 s_1 & -6s_0 s_1 & s_1(2s_0 + s_1)(s_1 - s_0) & s_0(s_0 + 2s_1)(s_1 - s_0) \\ s_1^2(s_1 - 3s_0) & s_0^2(3s_1 - s_0) & -s_0 s_1^2(s_1 - s_0) & -s_0^2 s_1(s_1 - s_0) \end{bmatrix} \tag{4.29}$$

and $\boldsymbol{N}_2^{-1}$ is obtained by increasing all of the subscripts in the above formula by 1.

Differentiating Eqs. (4.23) and (4.24) again to obtain the second derivatives gives:

$$\boldsymbol{P}''_1(s) = [6s \;\; 2 \;\; 0 \;\; 0]\, \boldsymbol{N}_1^{-1} \begin{bmatrix} \boldsymbol{Q}_0 \\ \boldsymbol{Q}_1 \\ \boldsymbol{Q}'_0 \\ \boldsymbol{Q}'_1 \end{bmatrix} \tag{4.30}$$

$$\boldsymbol{P}''_2(s) = [6s \;\; 2 \;\; 0 \;\; 0]\, \boldsymbol{N}_2^{-1} \begin{bmatrix} \boldsymbol{Q}_1 \\ \boldsymbol{Q}_2 \\ \boldsymbol{Q}'_1 \\ \boldsymbol{Q}'_2 \end{bmatrix}. \tag{4.31}$$

As was stated in Sect. 1.2.2, the second derivative vectors $\boldsymbol{P}_1''(s)$ and $\boldsymbol{P}_2''(s)$ with respect to length along the curve s have the curvature as their magnitude and point toward the center of curvature.

Calculating the curvature vector of the curve $\boldsymbol{P}_1(s)$ at $s=s_1$, we obtain:

$$\boldsymbol{P}_1''(s_1)=[6s_1 \;\; 2 \;\; 0 \;\; 0]\,\boldsymbol{N}_1^{-1}\begin{bmatrix}\boldsymbol{Q}_0\\ \boldsymbol{Q}_1\\ \boldsymbol{Q}_0'\\ \boldsymbol{Q}_1'\end{bmatrix}$$

$$=\frac{2}{(s_1-s_0)^3}[3s_1 \;\; 1 \;\; 0 \;\; 0]$$

$$\times\begin{bmatrix}2 & -2 & s_1-s_0 & s_1-s_0\\ -3(s_0+s_1) & 3(s_0+s_1) & -(s_0+2s_1)(s_1-s_0) & -(2s_0+s_1)(s_1-s_0)\\ 6s_0s_1 & -6s_0s_1 & s_1(2s_0+s_1)(s_1-s_0) & s_0(s_0+2s_1)(s_1-s_0)\\ s_1^2(s_1-3s_0) & s_0^2(3s_1-s_0) & -s_0s_1^2(s_1-s_0) & -s_0^2s_1(s_1-s_0)\end{bmatrix}\begin{bmatrix}\boldsymbol{Q}_0\\ \boldsymbol{Q}_1\\ \boldsymbol{Q}_0'\\ \boldsymbol{Q}_1'\end{bmatrix}$$

$$=\frac{2}{(s_1-s_0)^2}[3 \;\; -3 \;\; s_1-s_0 \;\; 2(s_1-s_0)]\begin{bmatrix}\boldsymbol{Q}_0\\ \boldsymbol{Q}_1\\ \boldsymbol{Q}_0'\\ \boldsymbol{Q}_1'\end{bmatrix}. \tag{4.32}$$

Similarly, calculating the curvature vector of the curve $\boldsymbol{P}_2(s)$ at $s=s_1$, we obtain:

$$\boldsymbol{P}_2''(s_1)=\frac{2}{(s_2-s_1)^2}[-3 \;\; 3 \;\; -2(s_2-s_1) \;\; -(s_2-s_1)]\begin{bmatrix}\boldsymbol{Q}_1\\ \boldsymbol{Q}_2\\ \boldsymbol{Q}_1'\\ \boldsymbol{Q}_2'\end{bmatrix}. \tag{4.33}$$

If we assume that the curves $\boldsymbol{P}_1(s)$ and $\boldsymbol{P}_2(s)$ have the same curvature vectors at $s=s_1$, then from Eqs. (4.32) and (4.33) the following condition holds:

$$(s_2-s_1)\boldsymbol{Q}_0'+2(s_2-s_0)\boldsymbol{Q}_1'+(s_1-s_0)\boldsymbol{Q}_2'$$
$$=\frac{3}{(s_1-s_0)(s_2-s_1)}\{(s_1-s_0)^2(\boldsymbol{Q}_2-\boldsymbol{Q}_1)+(s_2-s_1)^2(\boldsymbol{Q}_1-\boldsymbol{Q}_0)\}. \tag{4.34}$$

Equation (4.34) expresses the condition that the two curve segments $\boldsymbol{P}_1(s)$ and $\boldsymbol{P}_2(s)$ connecting the first three points $\boldsymbol{Q}_0$, $\boldsymbol{Q}_1$ and $\boldsymbol{Q}_2$ are connected with continuity up to the curvature vector at the connection points. This can be generalized to the condition that the two curve segments $\boldsymbol{P}_i(s)$ and $\boldsymbol{P}_{i+1}(s)$ connecting any three points $\boldsymbol{Q}_{i-1}$, $\boldsymbol{Q}_i$ and $\boldsymbol{Q}_{i+1}$ are connected with continuity up to the curvature vector at $\boldsymbol{Q}_i$ by making the following substitutions in Eq. (4.34):

$$\left.\begin{array}{l} 0 \to i-1 \\ 1 \to i \\ 2 \to i+1 \end{array}\right\}. \tag{4.35}$$

Therefore we have:

$$\begin{aligned}(s_{i+1}-s_i)\,\boldsymbol{Q}'_{i-1}&+2(s_{i+1}-s_{i-1})\,\boldsymbol{Q}'_i+(s_i-s_{i-1})\,\boldsymbol{Q}'_{i+1}\\ &=\frac{3}{(s_i-s_{i-1})(s_{i+1}-s_i)}\{(s_i-s_{i-1})^2(\boldsymbol{Q}_{i+1}-\boldsymbol{Q}_i)+(s_{i+1}-s_i)^2(\boldsymbol{Q}_i-\boldsymbol{Q}_{i-1})\}\\ &(1\leqq i\leqq n-1).\end{aligned} \tag{4.36}$$

If the curve length is approximated by chord length:

$$\left.\begin{array}{l} s_i-s_{i-1} \fallingdotseq c_i \\ s_{i+1}-s_i \fallingdotseq c_{i+1} \end{array}\right\} \tag{4.37}$$

then Eq. (4.36) becomes:

$$\begin{aligned}c_{i+1}\boldsymbol{Q}'_{i-1}&+2(c_{i+1}+c_i)\,\boldsymbol{Q}'_i+c_i\boldsymbol{Q}'_{i+1}\\ &=\frac{3}{c_i c_{i+1}}\{c_i^2(\boldsymbol{Q}_{i+1}-\boldsymbol{Q}_i)+c_{i+1}^2(\boldsymbol{Q}_i-\boldsymbol{Q}_{i-1})\}\\ &(1\leqq i\leqq n-1).\end{aligned} \tag{4.38}$$

As a special case, consider a given sequence of points separated by approximately equal intervals:

$$c_{i-1}=c_i=c_{i+1}=c\,.$$

Then Eq. (4.38) becomes:

$$c\boldsymbol{Q}'_{i-1}+4c\boldsymbol{Q}'_i+c\boldsymbol{Q}'_{i+1}=3(\boldsymbol{Q}_{i+1}-\boldsymbol{Q}_{i-1}). \tag{4.39}$$

Expressed in terms of the tangent vectors this becomes:

$$\dot{\boldsymbol{Q}}_{i-1}+4\dot{\boldsymbol{Q}}_i+\dot{\boldsymbol{Q}}_{i+1}=3(\boldsymbol{Q}_{i+1}-\boldsymbol{Q}_{i-1}). \tag{4.40}$$

A formula for the i-th curve segment can be obtained by performing the subscript substitution (4.35) in (4.27):

$$\boldsymbol{P}_i(s)=[s^3\ \ s^2\ \ s\ \ 1]\,\boldsymbol{N}_i^{-1}\begin{bmatrix}\boldsymbol{Q}_{i-1}\\ \boldsymbol{Q}_i\\ \boldsymbol{Q}'_{i-1}\\ \boldsymbol{Q}'_i\end{bmatrix}. \tag{4.41}$$

Next, performing a parameter transformation for the i-th curve segment $\boldsymbol{P}_i(s)$, $0 \leqq t \leqq 1$ corresponds to $s_{i-1} \leqq s \leqq s_i$, if the transformation is of the form:

$$s = s_{i-1} + (s_i - s_{i-1})t. \tag{4.42}$$

Substituting this relation into (4.41) gives:

$$\begin{aligned}
\boldsymbol{P}_i(t) &= [t^3 \; t^2 \; t \; 1] \\
&\times \begin{bmatrix} (s_i - s_{i-1})^3 & 0 & 0 & 0 \\ 3s_{i-1}(s_i - s_{i-1})^2 & (s_i - s_{i-1})^2 & 0 & 0 \\ 3s_{i-1}^2(s_i - s_{i-1}) & 2s_{i-1}(s_i - s_{i-1}) & s_i - s_{i-1} & 0 \\ s_{i-1}^3 & s_{i-1}^2 & s_i & 1 \end{bmatrix} \boldsymbol{N}_i^{-1} \begin{bmatrix} \boldsymbol{Q}_{i-1} \\ \boldsymbol{Q}_i \\ \boldsymbol{Q}'_{i-1} \\ \boldsymbol{Q}'_i \end{bmatrix} \\
&= [t^3 \; t^2 \; t \; 1] \begin{bmatrix} 2 & -2 & s_i - s_{i-1} & s_i - s_{i-1} \\ -3 & 3 & -2(s_i - s_{i-1}) & -(s_i - s_{i-1}) \\ 0 & 0 & s_i - s_{i-1} & 0 \\ 1 & 0 & 0 & 0 \end{bmatrix} \begin{bmatrix} \boldsymbol{Q}_{i-1} \\ \boldsymbol{Q}_i \\ \boldsymbol{Q}'_{i-1} \\ \boldsymbol{Q}'_i \end{bmatrix}
\end{aligned}$$

$$= [t^3 \; t^2 \; t \; 1] \begin{bmatrix} 2 & -2 & 1 & 1 \\ -3 & 3 & -2 & -1 \\ 0 & 0 & 1 & 0 \\ 1 & 0 & 0 & 0 \end{bmatrix} \begin{bmatrix} \boldsymbol{Q}_{i-1} \\ \boldsymbol{Q}_i \\ (s_i - s_{i-1})\boldsymbol{Q}'_{i-1} \\ (s_i - s_{i-1})\boldsymbol{Q}'_i \end{bmatrix} \tag{4.43}$$

$$= [H_{0,0}(t) \; H_{0,1}(t) \; H_{1,0}(t) \; H_{1,1}(t)] \begin{bmatrix} \boldsymbol{Q}_{i-1} \\ \boldsymbol{Q}_i \\ (s_i - s_{i-1})\boldsymbol{Q}'_{i-1} \\ (s_i - s_{i-1})\boldsymbol{Q}'_i \end{bmatrix} \tag{4.44}$$

$$\fallingdotseq [H_{0,0}(t) \; H_{0,1}(t) \; H_{1,0}(t) \; H_{1,1}(t)] \begin{bmatrix} \boldsymbol{Q}_{i-1} \\ \boldsymbol{Q}_i \\ c_i \boldsymbol{Q}'_{i-1} \\ c_i \boldsymbol{Q}'_i \end{bmatrix} \tag{4.45}$$

$$\equiv [H_{0,0}(t) \; H_{0,1}(t) \; H_{1,0}(t) \; H_{1,1}(t)] \begin{bmatrix} \boldsymbol{Q}_{i-1} \\ \boldsymbol{Q}_i \\ \dot{\boldsymbol{Q}}_{i-1} \\ \dot{\boldsymbol{Q}}_i \end{bmatrix}. \tag{4.46}$$

As can be seen from Eq. (4.45), to determine all curve segments $\boldsymbol{P}_i(t)$ $(i = 1, 2, \ldots, n)$ it is necessary that the $(n+1)$ unit tangent vectors $\boldsymbol{Q}'_0, \boldsymbol{Q}'_1, \ldots, \boldsymbol{Q}'_n$ be determined. However, since $(n-1)$ conditions are given by Eq. (4.38) which expresses the condition that the curvature vectors be continuous, it is necessary to add two more conditions to (4.38) so that all of the unit tangent vectors are determined and, hence, all of the curve segments are determined.

Let us now take a closer look at the connection points between two curve segments $\boldsymbol{P}_i(t)$ and $\boldsymbol{P}_{i+1}(t)$ which have been determined in the above manner. The formulas for curve segments $\boldsymbol{P}_i(t)$ and $\mathrm{P}_{i+1}(t)$ are:

$$\boldsymbol{P}_i(t)=[H_{0,0}(t)\ H_{0,1}(t)\ H_{1,0}(t)\ H_{1,1}(t)]\begin{bmatrix}\boldsymbol{Q}_{i-1}\\ \boldsymbol{Q}_i\\ c_i\boldsymbol{Q}'_{i-1}\\ c_i\boldsymbol{Q}'_i\end{bmatrix} \tag{4.47}$$

$$\boldsymbol{P}_{i+1}(t)=[H_{0,0}(t)\ H_{0,1}(t)\ H_{1,0}(t)\ H_{1,1}(t)]\begin{bmatrix}\boldsymbol{Q}_i\\ \boldsymbol{Q}_{i+1}\\ c_{i+1}\boldsymbol{Q}'_i\\ c_{i+1}\boldsymbol{Q}'_{i+1}\end{bmatrix}. \tag{4.48}$$

The tangent vectors at the connection point are found by differentiating (4.47) and (4.48) and setting $t=1$ and $t=0$, respectively:

$$\dot{\boldsymbol{P}}_i(1)=c_i\boldsymbol{Q}'_i$$
$$\dot{\boldsymbol{P}}_{i+1}(0)=c_{i+1}\boldsymbol{Q}'_i.$$

These give the relation:

$$\dot{\boldsymbol{P}}_{i+1}(0)=\frac{c_{i+1}}{c_i}\dot{\boldsymbol{P}}_i(1)\equiv\lambda_i\dot{\boldsymbol{P}}_i(1)$$

where:

$$\lambda_i=\frac{c_{i+1}}{c_i}. \tag{4.49}$$

Next, let us find the relation between the second derivative vectors $\ddot{\boldsymbol{P}}_i(1)$ and $\ddot{\boldsymbol{P}}_{i+1}(0)$. Using Eqs. (3.16) for the second derivatives of $H_{0,0}(t), \ldots, H_{1,1}(t)$ gives:

$$\ddot{\boldsymbol{P}}_i(1)=-6(\boldsymbol{Q}_i-\boldsymbol{Q}_{i-1})+2c_i(2\boldsymbol{Q}'_i+\boldsymbol{Q}'_{i-1}) \tag{4.50}$$

$$\ddot{\boldsymbol{P}}_{i+1}(0)=6(\boldsymbol{Q}_{i+1}-\boldsymbol{Q}_i)-2c_{i+1}(2\boldsymbol{Q}'_i+\boldsymbol{Q}'_{i+1}). \tag{4.51}$$

Substituting $\boldsymbol{Q}_i-\boldsymbol{Q}_{i-1}$ and $\boldsymbol{Q}_{i+1}-\boldsymbol{Q}_i$ in (4.50) and (4.51) into (4.38) and rearranging terms gives:

$$\ddot{\boldsymbol{P}}_{i+1}(0)=\left(\frac{c_{i+1}}{c_i}\right)^2\ddot{\boldsymbol{P}}_i(1)\equiv\lambda_i^2\ddot{\boldsymbol{P}}_i(1) \tag{4.52}$$
$$\lambda_i=c_{i+1}/c_i.$$

The curvature vector continuity condition (4.52) is equivalent to the general case (1.52) with $\beta=0$.

We now use curvature vector equation (1.24) to determine whether or not the curvature vector is continuous on both sides of the connection point when relations (4.49) and (4.52) hold.

For the segments on both sides of the connection point $\boldsymbol{Q}_i$, first, on the curve segment $\boldsymbol{P}_i(t)$ side, we have:

$$\boldsymbol{P}_i'' = \frac{(\dot{\boldsymbol{P}}_i(1) \times \ddot{\boldsymbol{P}}_i(1)) \times \dot{\boldsymbol{P}}_i(1)}{\dot{\boldsymbol{P}}_i(1)^4} \equiv \kappa_i \boldsymbol{n}_i .$$

On the curve segment $\boldsymbol{P}_{i+1}(t)$ side we have:

$$\begin{aligned}\boldsymbol{P}_{i+1}'' &= \frac{(\dot{\boldsymbol{P}}_{i+1}(0) \times \ddot{\boldsymbol{P}}_{i+1}(0)) \times \dot{\boldsymbol{P}}_{i+1}(0)}{\dot{\boldsymbol{P}}_{i+1}(0)^4} \\ &= \frac{(\lambda_i \dot{\boldsymbol{P}}_i(1) \times \lambda_i^2 \ddot{\boldsymbol{P}}_i(1)) \times \lambda_i \dot{\boldsymbol{P}}_i(1)}{(\lambda_i \dot{\boldsymbol{P}}_i(1))^4} \\ &= \frac{(\dot{\boldsymbol{P}}_i(1) \times \ddot{\boldsymbol{P}}_i(1)) \times \dot{\boldsymbol{P}}_i(1)}{\dot{\boldsymbol{P}}_i(1)^4} \\ &= \kappa_i \boldsymbol{n}_i .\end{aligned}$$

This confirms that we have generated a curve which, at the connection point, is continuous not only to the slope, but to the curvature vector.

4.8 End Conditions on a Spline Curve

Equation (4.38) holds for $1 \leqq i \leqq n-1$. This relation can be expressed in matrix form as follows:

$$\bar{\boldsymbol{M}} \begin{bmatrix} \boldsymbol{Q}_0' \\ \boldsymbol{Q}_1' \\ \vdots \\ \boldsymbol{Q}_n' \end{bmatrix} = \bar{\boldsymbol{B}} \tag{4.53}$$

where:

$$\bar{\boldsymbol{M}} = \begin{bmatrix} c_2 & 2(c_2+c_1) & c_1 & 0 & 0 & \dots & \dots & 0 \\ 0 & c_3 & 2(c_3+c_2) & c_2 & 0 & \dots & \dots & 0 \\ 0 & \dots & \dots & \dots & \dots & \dots & \dots & \dots \\ \dots & \dots & \dots & \dots & \dots & \dots & \dots & \dots \\ \dots & \dots & \dots & \dots & \dots & \dots & \dots & \dots \\ 0 & \dots & \dots & \dots & \dots & c_n & 2(c_n+c_{n-1}) & c_{n-1} \end{bmatrix}$$

$$\bar{\boldsymbol{B}} = \begin{bmatrix} \frac{3}{c_1 c_2}\{c_1^2(\boldsymbol{Q}_2-\boldsymbol{Q}_1)+c_2^2(\boldsymbol{Q}_1-\boldsymbol{Q}_0)\} \\ \frac{3}{c_2 c_3}\{c_2^2(\boldsymbol{Q}_3-\boldsymbol{Q}_2)+c_3^2(\boldsymbol{Q}_2-\boldsymbol{Q}_1)\} \\ \cdots\cdots\cdots \\ \cdots\cdots\cdots \\ \frac{3}{c_{n-1} c_n}\{c_{n-1}^2(\boldsymbol{Q}_n-\boldsymbol{Q}_{n-1})+c_n^2(\boldsymbol{Q}_{n-1}-\boldsymbol{Q}_{n-2})\} \end{bmatrix}.$$

Matrix $\bar{\boldsymbol{M}}$ is $(n-1)\times(n+1)$; it is not square, so an inverse matrix does not exist. If we express conditions at the ends $\boldsymbol{Q}_0$ and $\boldsymbol{Q}_n$ of the curve by some means, we can replace $\bar{\boldsymbol{M}}$ and $\bar{\boldsymbol{B}}$ by an $(n+1)\times(n+1)$ matrix $\boldsymbol{M}$ and an $(n+1)\times 1$ matrix $\boldsymbol{B}$; then Eq. (4.53) is replaced by the following:

$$\boldsymbol{M}\begin{bmatrix} \boldsymbol{Q}_0' \\ \boldsymbol{Q}_1' \\ \vdots \\ \boldsymbol{Q}_n' \end{bmatrix} = \boldsymbol{B}. \tag{4.54}$$

From this we obtain:

$$\begin{bmatrix} \mathrm{Q}_0' \\ \boldsymbol{Q}_1' \\ \vdots \\ \boldsymbol{Q}_n' \end{bmatrix} = \boldsymbol{M}^{-1}\boldsymbol{B}.$$

$\boldsymbol{Q}_0'$, $\boldsymbol{Q}_1'$, ..., $\boldsymbol{Q}_n'$ are all determined. Substituting these into Eq. (4.45) determines all of the curve segments $\boldsymbol{P}_i(t)$ $(i=1, 2, \ldots, n)$.

(1) Case in which unit tangent vectors are specified at both end points $\boldsymbol{Q}_0$ and $\boldsymbol{Q}_n$

Specify the unit tangent vectors $\boldsymbol{Q}_0'$ and $\boldsymbol{Q}_n'$ at the end points $\boldsymbol{Q}_0$ and $\boldsymbol{Q}_n$ as follows:

$$\boldsymbol{Q}_0' = \boldsymbol{t}_0$$
$$\boldsymbol{Q}_n' = \boldsymbol{t}_n.$$

Then the matrices $\boldsymbol{M}$ and $\boldsymbol{B}$ become:

$$
\boldsymbol{M}=\begin{bmatrix}
1 & 0 & \cdots & \cdots & \cdots & \cdots & 0\\
c_2 & 2(c_2+c_1) & c_1 & 0 & \cdots & \cdots & \cdots\\
0 & c_3 & 2(c_3+c_2) & c_2 & \cdots & \cdots & \cdots\\
\cdots & \cdots & \cdots & \cdots & \cdots & \cdots & \cdots\\
\cdots & \cdots & \cdots & \cdots & \cdots & \cdots & \cdots\\
\cdots & \cdots & \cdots & \cdots & c_n & 2(c_n+c_{n-1}) & c_{n-1}\\
0 & \cdots & \cdots & \cdots & 0 & 0 & 1
\end{bmatrix}
$$

$$
\boldsymbol{B}=\begin{bmatrix}
\boldsymbol{t}_0\\
\dfrac{3}{c_1c_2}\{c_1^2(\boldsymbol{Q}_2-\boldsymbol{Q}_1)+c_2^2(\boldsymbol{Q}_1-\boldsymbol{Q}_0)\}\\
\dfrac{3}{c_2c_3}\{c_2^2(\boldsymbol{Q}_3-\boldsymbol{Q}_2)+c_3^2(\boldsymbol{Q}_2-\boldsymbol{Q}_1)\}\\
\cdots\\
\cdots\\
\dfrac{3}{c_{n-1}c_n}\{c_{n-1}^2(\boldsymbol{Q}_n-\boldsymbol{Q}_{n-1})+c_n^2(\boldsymbol{Q}_{n-1}-\boldsymbol{Q}_{n-2})\}\\
\boldsymbol{t}_n
\end{bmatrix}.
$$

Examples of curves generated in this case are shown in Figs. 4.5, 4.6 and 4.7.

(2) Case in which the curvature is 0 at both end points $\boldsymbol{Q}_0$ and $\boldsymbol{Q}_n$

The 2nd derivative of (4.47) with respect to s is, regarding dt/ds is approximately constant

$$
\boldsymbol{P}_i''=\left(\frac{dt}{ds}\right)^2[\ddot{H}_{0,0}(t)\ \ddot{H}_{0,1}(t)\ \ddot{H}_{1,0}(t)\ \ddot{H}_{1,1}(t)]\begin{bmatrix}\boldsymbol{Q}_{i-1}\\ \boldsymbol{Q}_i\\ c_i\boldsymbol{Q}_{i-1}'\\ c_i\boldsymbol{Q}_i'\end{bmatrix}. \tag{4.55}
$$

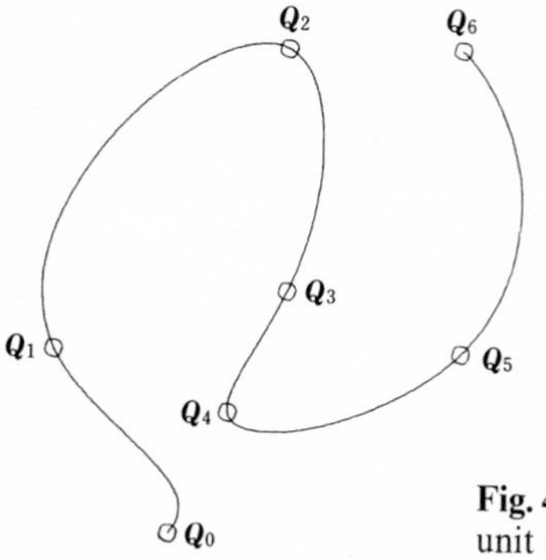

Fig. 4.5. Example of spline curve generation (case in which unit tangent vectors at both ends are specified ——(1))

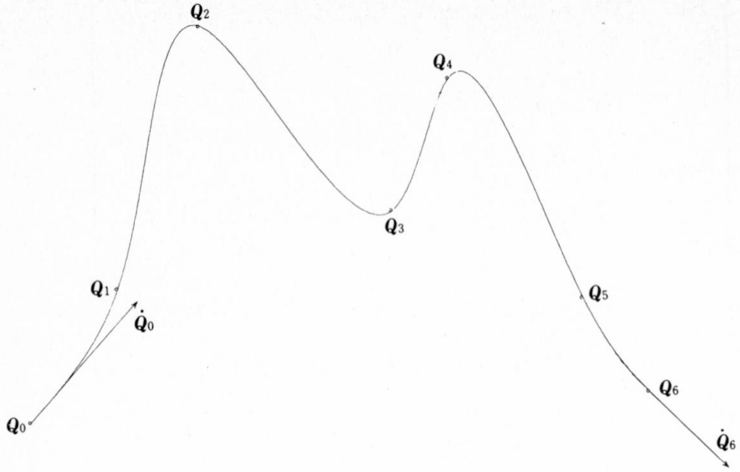

Fig. 4.6. Example of spline curve generation (case in which the unit tangent vectors at both ends are specified ——(2))

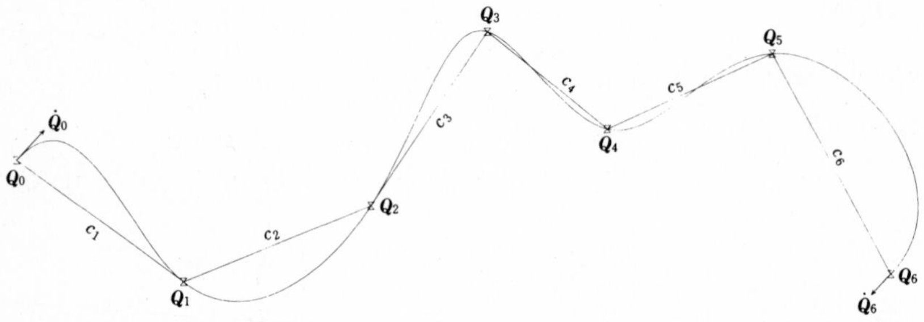

Fig. 4.7. Example of spline curve generation (case in which the unit tangent vectors at both end points are specified ——(3))

Setting $i=1$, $t=0$ and $\boldsymbol{P}_1''=\boldsymbol{0}$ and using relation (3.16) gives:

$$\boldsymbol{0}=[-6\ \ 6\ \ -4\ \ -2]\begin{bmatrix}\boldsymbol{Q}_0\\ \boldsymbol{Q}_1\\ c_1\boldsymbol{Q}_0'\\ c_1\boldsymbol{Q}_1'\end{bmatrix}. \tag{4.56}$$

This relation can be simplified to:

$$2c_1\boldsymbol{Q}_0'+c_1\boldsymbol{Q}_1'=-3\boldsymbol{Q}_0+3\boldsymbol{Q}_1. \tag{4.57}$$

Setting $i=n$, $t=1$ and $\boldsymbol{P}_n''=\boldsymbol{0}$ in (4.55) and using relation (3.16) gives:

$$\boldsymbol{0}=[6\ \ -6\ \ 2\ \ 4]\begin{bmatrix}\boldsymbol{Q}_{n-1}\\ \boldsymbol{Q}_n\\ c_n\boldsymbol{Q}'_{n-1}\\ c_n\boldsymbol{Q}'_n\end{bmatrix}. \tag{4.58}$$

This can be simplified to:

$$c_n\boldsymbol{Q}'_{n-1}+2c_n\boldsymbol{Q}'_n=-3\boldsymbol{Q}_{n-1}+3\boldsymbol{Q}_n. \tag{4.59}$$

From conditions (4.57) and (4.59), matrices $\boldsymbol{M}$ and $\boldsymbol{B}$ can be expressed as follows:

$$\boldsymbol{M}=\begin{bmatrix}2c_1 & c_1 & 0 & \cdots & \cdots & \cdots & \cdots\\ c_2 & 2(c_2+c_1) & c_1 & 0 & \cdots & \cdots & \cdots\\ 0 & c_3 & 2(c_3+c_2) & c_2 & \cdots & \cdots & \cdots\\ \cdots & \cdots & \cdots & \cdots & \cdots & \cdots & \cdots\\ \cdots & \cdots & \cdots & \cdots & \cdots & \cdots & \cdots\\ 0 & \cdots & \cdots & c_{n-1} & 2(c_{n-1}+c_{n-2}) & c_{n-2} & 0\\ 0 & \cdots & \cdots & \cdots & c_n & 2(c_n+c_{n-1}) & c_{n-1}\\ 0 & \cdots & \cdots & \cdots & 0 & c_n & 2c_n\end{bmatrix}$$

$$\boldsymbol{B}=\begin{bmatrix}3(\boldsymbol{Q}_1-\boldsymbol{Q}_0)\\ \dfrac{3}{c_1c_2}\{c_1^2(\boldsymbol{Q}_2-\boldsymbol{Q}_1)+c_2^2(\boldsymbol{Q}_1-\boldsymbol{Q}_0)\}\\ \dfrac{3}{c_2c_3}\{c_2^2(\boldsymbol{Q}_3-\boldsymbol{Q}_2)+c_3^2(\boldsymbol{Q}_2-\boldsymbol{Q}_1)\}\\ \cdots\\ \cdots\\ \dfrac{3}{c_{n-1}c_n}\{c_{n-1}^2(\boldsymbol{Q}_n-\boldsymbol{Q}_{n-1})+c_n^2(\boldsymbol{Q}_{n-1}-\boldsymbol{Q}_{n-2})\}\\ 3(\boldsymbol{Q}_n-\boldsymbol{Q}_{n-1})\end{bmatrix}.$$

Examples of curves generated in this case are shown in Figs. 4.8, 4.9 and 4.10.

(3) Case in which the slopes and curvature vectors are equal at both end points $\boldsymbol{Q}_0$ and $\boldsymbol{Q}_n$

Consider the conditions that the slopes and curvature vectors are equal at both ends of a spline curve:

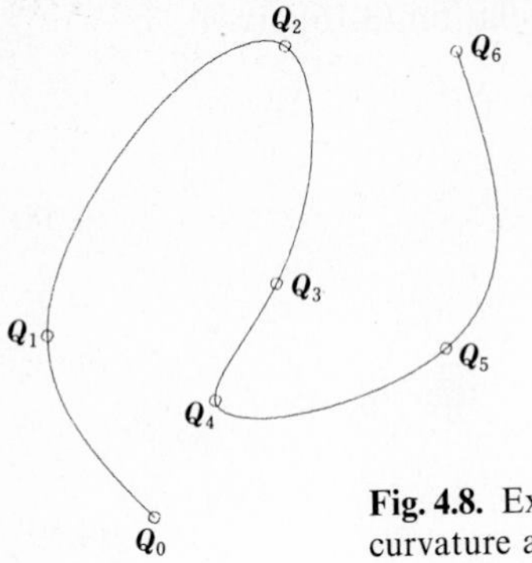

Fig. 4.8. Example of spline curve generation (case in which the curvature at both end points is taken to be 0 ——(1))

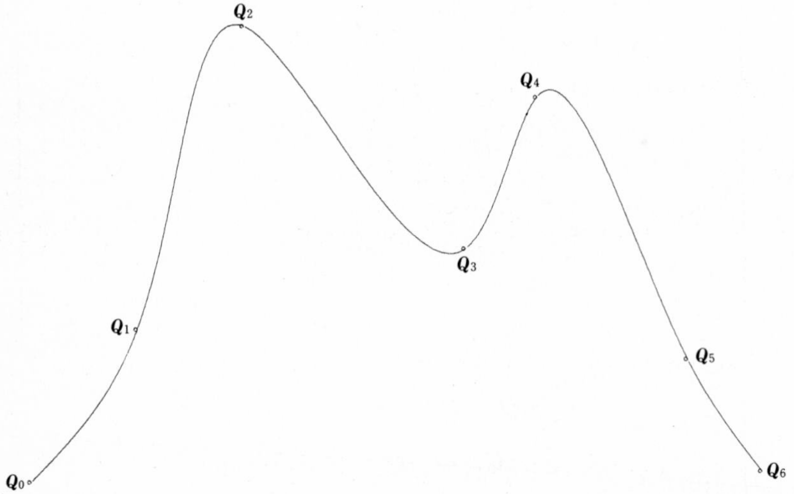

Fig. 4.9. Example of spline curve generation (case in which the curvature at both end points is taken to be 0 ——(2))

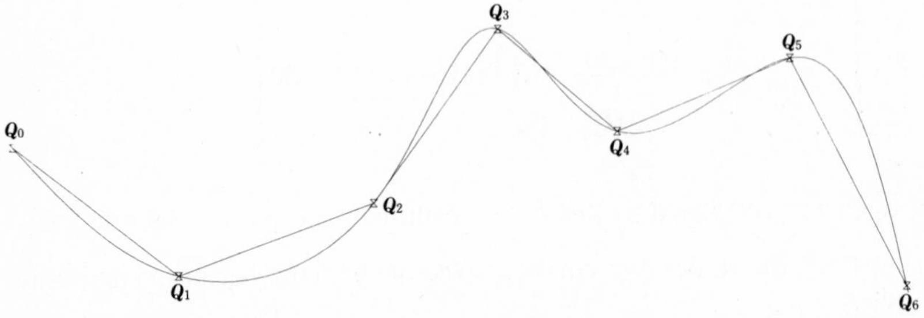

Fig. 4.10. Example of spline curve generation (case in which the curvature at both end points is taken to be 0 ——(3))

$$P'_{1,t=0} = P'_{n,t=1} \tag{4.60}$$

$$P''_{1,t=0} = P''_{n,t=1}\,. \tag{4.61}$$

It is clear that condition (4.60) implies that:

$$Q'_0 = Q'_n\,. \tag{4.62}$$

Setting $ds/dt \risingdotseq c_1$ in Eq. (4.55) gives:

$$P''_{1,t=0} = \frac{1}{c_1^2}\,[-6\;\;6\;\;-4\;\;-2]\begin{bmatrix} Q_0 \\ Q_1 \\ c_1 Q'_0 \\ c_1 Q'_1 \end{bmatrix}. \tag{4.63}$$

Similarly, setting $ds/dt \risingdotseq c_n$ in Eq. (4.55) gives:

$$P''_{n,t=1} = \frac{1}{c_n^2}\,[6\;\;-6\;\;2\;\;4]\begin{bmatrix} Q_{n-1} \\ Q_n \\ c_n Q'_{n-1} \\ c_n Q'_n \end{bmatrix}. \tag{4.64}$$

Applying relation (4.61) to Eqs. (4.63) and (4.64) and using relation (4.62) gives the following relation:

$$c_n Q'_1 + c_1 Q'_{n-1} + 2(c_n + c_1)\,Q'_n = \frac{3}{c_n c_1}\,\{c_n^2(Q_1 - Q_0) + c_1^2(Q_n - Q_{n-1})\}\,. \tag{4.65}$$

From conditions (4.62) and (4.65), we see that matrices $\boldsymbol{M}$ and $\boldsymbol{B}$ can be expressed as follows:

$$\boldsymbol{M} = \begin{bmatrix} 1 & 0 & \cdots & \cdots & \cdots & \cdots & -1 \\ c_2 & 2(c_2+c_1) & c_1 & 0 & \cdots & \cdots & 0 \\ 0 & c_3 & 2(c_3+c_2) & c_2 & \cdots & \cdots & \cdots \\ \cdots & \cdots & \cdots & \cdots & \cdots & \cdots & \cdots \\ \cdots & \cdots & \cdots & \cdots & \cdots & \cdots & \cdots \\ 0 & \cdots & \cdots & c_{n-1} & 2(c_{n-1}+c_{n-2}) & c_{n-2} & 0 \\ 0 & \cdots & \cdots & \cdots & c_n & 2(c_n+c_{n-1}) & c_{n-1} \\ 0 & c_n & 0 & \cdots & \cdots & c_1 & 2(c_n+c_1) \end{bmatrix}$$

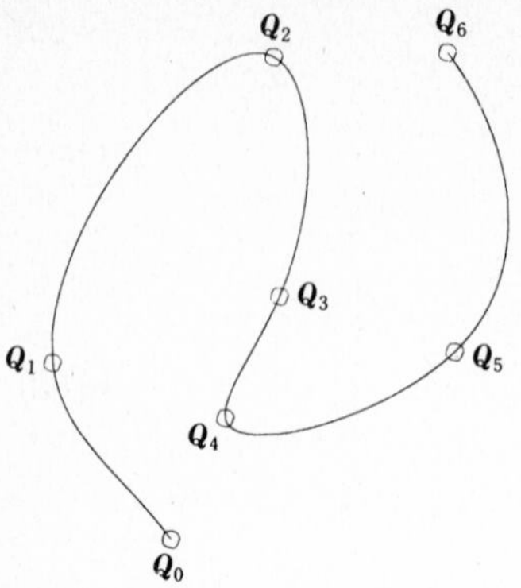

Fig. 4.11. Example of spline curve generation (case in which slope and curvature vectors at both end points are taken to be equal)

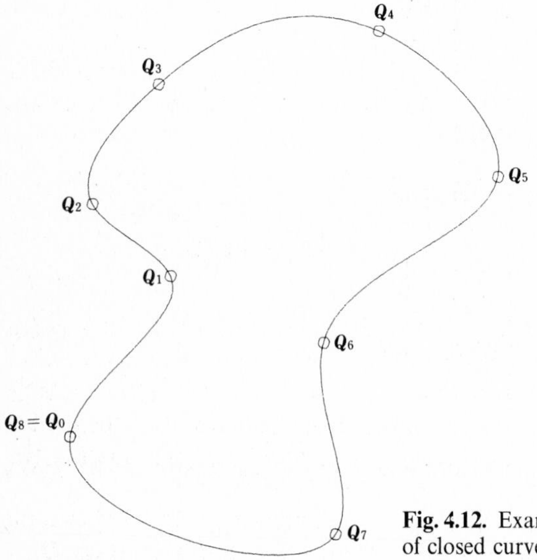

Fig. 4.12. Example of spline curve generation (case of closed curve ——(1))

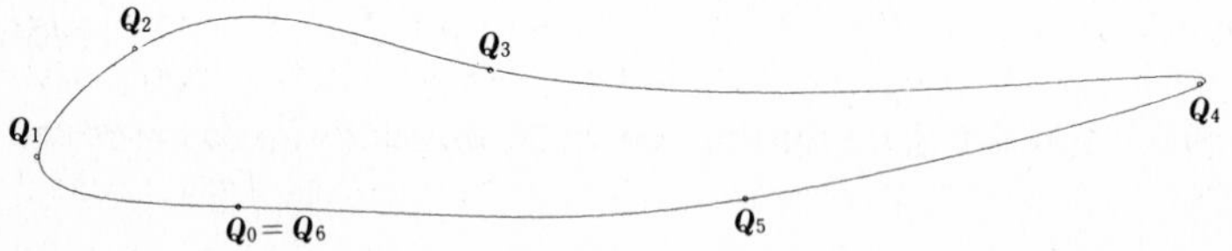

Fig. 4.13. Example of spline curve generation (case of closed curve ——(2))

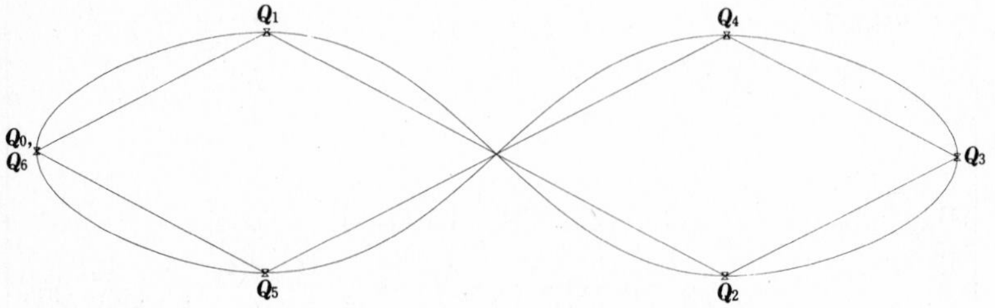

Fig. 4.14. Example of spline curve generation (case of closed curve ——(3))

$$B=\begin{bmatrix} \mathbf{0} \\ \dfrac{3}{c_1c_2}\{c_1^2(Q_2-Q_1)+c_2^2(Q_1-Q_0)\} \\ \dfrac{3}{c_2c_3}\{c_2^2(Q_3-Q_2)+c_3^2(Q_2-Q_1)\} \\ \cdots \\ \cdots \\ \dfrac{3}{c_{n-1}c_n}\{c_{n-1}^2(Q_n-Q_{n-1})+c_n^2(Q_{n-1}-Q_{n-2})\} \\ \dfrac{3}{c_nc_1}\{c_n^2(Q_1-Q_0)+c_1^2(Q_n-Q_{n-1})\} \end{bmatrix}.$$

A curve generated by this method is called a periodic spline curve. To make it a closed curve, it is sufficient to take $Q_n = Q_0$. Examples of these spline curves are shown in Figs. 4.11, 4.12, 4.13 and 4.14.

4.9 Cubic Spline Curves Using Circular Arc Length

In the spline curves discussed so far, in their derivation the length of a curve between 2 points has been approximated by the chord length. This assumption is appropriate when the variation of the curve between specified points is relatively slight, but when a fairly large variation is anticipated, in general a curve closer to the actual spline curve will be obtained by approximating the curve length by circular arc length.

The radius of the circle passing through the 3 points Q_{i-2}, Q_{i-1}, Q_i can be found from Eq. (A.12) in Appendix A: the arc lengths a'_{i-1} and a'_i between Q_{i-2} and Q_{i-1} and Q_{i-1} and Q_i, respectively can be found from Eq. (A.13) (Fig. 4.15). Advancing one point to consider the 3 points Q_{i-1}, Q_i and Q_{i+1}, letting a''_i and a''_{i+1} be the arc lengths between points Q_{i-1} and Q_i and Q_i and Q_{i+1}, the curve length between points Q_{i-1} and Q_i is approximated by $a_i = (a'_i + a''_i)/2$. A similar procedure is repeated to generate circular arc lengths through the specified

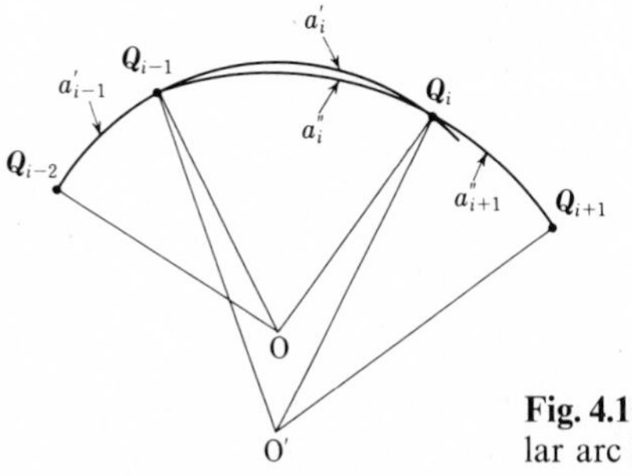

Fig. 4.15. Approximation of curve length by average circular arc length

sequence of points $Q_0, Q_1, \ldots, Q_{n-1}, Q_n$, thus giving approximate curve lengths for all of the spans except the two end spans. The arc lengths of the two end spans are found by constructing the circles passing through the points Q_0, Q_1 and Q_2 and Q_{n-2}, Q_{n-1} and Q_n, and using the circular arc lengths between Q_0 and Q_1 and between Q_{n-1} and Q_n. In this case, to find the unit tangent vector Q_i' $(1 \leqq i \leqq (n-1))$, c is replaced by a in Eq. (4.38):

$$a_{i+1} Q'_{i-1} + 2(a_{i+1} + a_i) Q'_i + a_i Q'_{i+1} = \frac{3}{a_i a_{i+1}} \{a_i^2 (Q_{i+1} - Q_i) + a_{i+1}^2 (Q_i - Q_{i-1})\} \tag{4.66}$$

and also in the various end conditions discussed in Sect. 4.8.

For the unit tangent vectors Q_i' $(i=0, 1, \ldots, n)$ which are obtained, the curve segment $P_i(t)$ in the i-th span can be expressed as follows:

$$\begin{aligned} P_i(t) &= [H_{0,0}(t)\ H_{0,1}(t)\ H_{1,0}(t)\ H_{1,1}(t)] \begin{bmatrix} Q_{i-1} \\ Q_i \\ a_i Q'_{i-1} \\ a_i Q'_i \end{bmatrix} \\ &= [H_{0,0}(t)\ H_{0,1}(t)\ H_{1,0}(t)\ H_{1,1}(t)] \begin{bmatrix} Q_{i-1} \\ Q_i \\ \dot{Q}_{i-1} \\ \dot{Q}_i \end{bmatrix} \quad (1 \leqq i \leqq n). \end{aligned} \tag{4.67}$$

4.10 *B*-Splines

A spline function can be determined which satisfies the following conditions between the knots x_j, x_{j+1}, x_{j+2}, x_{j+3}.

Condition ① is that the function S satisfies the following equations at the two end knots:

$$S(x_j) = \dot{S}(x_j) = 0$$
$$S(x_{j+3}) = \dot{S}(x_{j+3}) = 0.$$

Condition ② is that at intermediate knots, the function S takes a specified value, for example:

$$S(x_{j+1}) = h.$$

A function which satisfies these conditions is a C-spline of degree 2 (Fig. 4.16(a)).

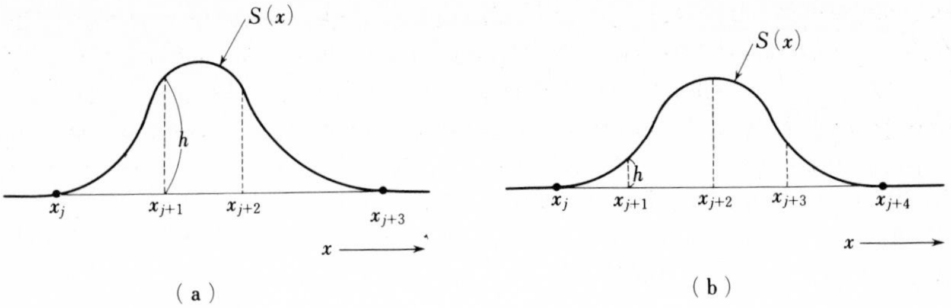

Fig. 4.16. *B*-splines

Similarly, a spline function can be determined which satisfies the following conditions between the knots x_j, x_{j+1}, x_{j+2}, x_{j+3} and x_{j+4}.

Condition ① is that the function *S* satisfies the following equations at the two end knots:

$$S(x_j)=\dot{S}(x_j)=\ddot{S}(x_j)=0$$
$$S(x_{j+4})=\dot{S}(x_{j+4})=\ddot{S}(x_{j+4})=0.$$

Condition ② is that at intermediate knots, the function *S* takes a specified value, for example:

$$S(x_{j+1})=h.$$

A function which satisfies these conditions is a *C*-spline of degree 3 (Fig. 4.16(b)).

Let us now derive a polynomial spline function of degree 2 which satisfies the first condition.

$\boldsymbol{x_j \leqq x \leqq x_{j+1}}$

The following quadratic $S(x)$ satisfies $S(x_j)=\dot{S}(x_j)=0$ and $S(x_{j+1})=h$.

$$S(x)=\frac{h}{(x_{j+1}-x_j)^2}(x-x_j)^2. \tag{4.68}$$

$\boldsymbol{x_{j+1} \leqq x \leqq x_{j+2}}$

From spline Eq. (4.5), calling the unknown constant b_1 we can write:

$$S(x)=\frac{h}{(x_{j+1}-x_j)^2}(x-x_j)^2+b_1(x-x_{j+1})^2. \tag{4.69}$$

$\boldsymbol{x_{j+2} \leqq x \leqq x_{j+3}}$

Calling the unknown constant b_2, a quadratic expression that satisfies the condition $S(x_{j+3})=\dot{S}(x_{j+3})=0$ can be written as follows:

$$S(x)=b_2(x-x_{j+3})^2. \tag{4.70}$$

By adding the condition that the curves in the 2-nd and 3-rd spans must have both position and slope continuity at $x=x_{j+2}$, the unknown constants b_1 and b_2 can be determined as follows:

$$b_1=\frac{-(x_{j+2}-x_j)\,(x_{j+3}-x_j)}{(x_{j+1}-x_j)^2\,(x_{j+2}-x_{j+1})\,(x_{j+3}-x_{j+1})}\,h$$

$$b_2=\frac{x_{j+2}-x_j}{(x_{j+1}-x_j)\,(x_{j+3}-x_{j+1})\,(x_{j+3}-x_{j+2})}\,h.$$

Then the curves in each span can be found from (4.71), (4.72) and (4.73).

$\boldsymbol{x_j \leqq x \leqq x_{j+1}}$

$$S(x)=\frac{h}{(x_{j+1}-x_j)^2}\,(x-x_j)^2 \tag{4.71}$$

$\boldsymbol{x_{j+1} \leqq x \leqq x_{j+2}}$

$$S(x)=\frac{h}{(x_{j+1}-x_j)^2}\left[(x-x_j)^2-\frac{(x_{j+2}-x_j)\,(x_{j+3}-x_j)}{(x_{j+2}-x_{j+1})\,(x_{j+3}-x_{j+1})}\,(x-x_{j+1})^2\right] \tag{4.72}$$

$\boldsymbol{x_{j+2} \leqq x \leqq x_{j+3}}$

$$S(x)=\frac{(x_{j+2}-x_j)\,h}{(x_{j+1}-x_j)\,(x_{j+3}-x_{j+1})\,(x_{j+3}-x_{j+2})}\,(x-x_{j+3})^2. \tag{4.73}$$

There is another method in which, as a condition for determining the function, in place of specifying the value h the area enclosed by the curve and the x-axis is specified. Since the area is:

$$\int_{x_j}^{x_{j+3}} S(x)\,dx=\frac{h}{3}\,\frac{(x_{j+2}-x_j)\,(x_{j+3}-x_j)}{x_{j+1}-x_j}.$$

Schoenberg used a method in which he specified a unit area. In this case, we have:

$$h=\frac{3\,(x_{j+1}-x_j)}{(x_{j+2}-x_j)\,(x_{j+3}-x_j)}.$$

Cox and De Boor specified the area as follows so that Cauchy's relation holds (refer to Eq. (1.4)) and normalized:

$$\frac{h}{3}\,\frac{(x_{j+2}-x_j)\,(x_{j+3}-x_j)}{x_{j+1}-x_j}=\frac{x_{j+3}-x_j}{3}. \tag{4.74}$$

In this case we have:

$$h=\frac{x_{j+1}-x_j}{x_{j+2}-x_j}. \tag{4.75}$$

In this case, the spline function is expressed as follows in each interval:

$\boldsymbol{x_j \leqq x \leqq x_{j+1}}$

$$S(x)=\frac{1}{(x_{j+1}-x_j)\,(x_{j+2}-x_j)}\,(x-x_j)^2 \tag{4.76}$$

$\boldsymbol{x_{j+1} \leqq x \leqq x_{j+2}}$

$$S(x)=\frac{1}{(x_{j+1}-x_j)\,(x_{j+2}-x_j)} \times\left[(x-x_j)^2-\frac{(x_{j+2}-x_j)\,(x_{j+3}-x_j)}{(x_{j+2}-x_{j+1})\,(x_{j+3}-x_{j+1})}\,(x-x_{j+1})^2\right] \tag{4.77}$$

$\boldsymbol{x_{j+2} \leqq x \leqq x_{j+3}}$

$$S(x)=\frac{1}{(x_{j+3}-x_{j+1})\,(x_{j+3}-x_{j+2})}\,(x-x_{j+3})^2. \tag{4.78}$$

(Equations (4.76) to (4.78) agree with Eq. (6.90) to be derived in Chap. 6). The function defined by Eqs. (4.76) to (4.78), with spans continued indefinitely to the left and right and all but 3 spans having functions values of 0, as shown in Fig. 4.16, is called a *B*-spline (Basis spline) or fundamental spline of degree 2. A normalized *B*-spline of degree 3 can be derived by a similar method. In the case of degree 3, in the right-hand side of Eq. (4.74) the area is specified as $(x_{j+4}-x_j)/4$. A *B*-spline is a special case of a *C*-spline. For further details on *B*-splines refer to Chap. 6.

4.11 Generation of Spline Surfaces

The method used to generate spline curves can be used to generate spline surfaces that are continuous up to the curvature.

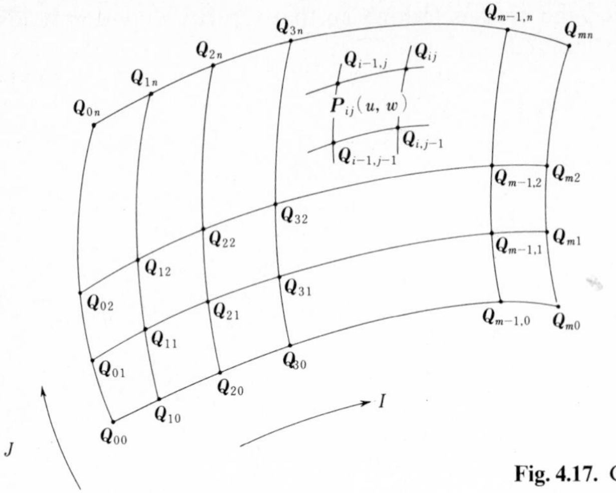

Fig. 4.17. Generation of a spline surface

Assume that a lattice of points $\boldsymbol{Q}_{ij} (i=0, 1, \ldots, m; j=0, 1, \ldots, n)$ is given (Fig. 4.17).

① In Fig. 4.17 there are $(n+1)$ rows of points in the I-direction. If conditions are given for the unit tangent vectors in the I-direction at the start and end points of each of these rows of points, then, letting h_i be the chord length in the I-direction*), the $(n+1)$ spline curves in the I-direction are determined by solving the following condition equations:

$$h_{i+1}\boldsymbol{Q}_{I,i-1,j}+2(h_{i+1}+h_i)\boldsymbol{Q}_{I,i,j}+h_i\boldsymbol{Q}_{I,i+1,j}$$
$$=\frac{3}{h_i h_{i+1}}\{h_i^2(\boldsymbol{Q}_{i+1,j}-\boldsymbol{Q}_{i,j})+h_{i+1}^2(\boldsymbol{Q}_{i,j}-\boldsymbol{Q}_{i-1,j})\}^{**)} \tag{4.79}$$
$$(i=1, 2, \ldots, m-1;\ j=0, 1, \ldots, n).$$

In this case, if normalization is done in each interval using a parameter u, then the tangent vectors with respect to the parameter u are:

$$\left.\begin{array}{ll}\cdot\ \boldsymbol{Q}_{u,i-1,j-1}=h_i\boldsymbol{Q}_{I,i-1,j-1} & \cdot\ \boldsymbol{Q}_{u,i,j-1}=h_i\boldsymbol{Q}_{I,i,j-1}\\ \cdot\ \boldsymbol{Q}_{u,i-1,j}=h_i\boldsymbol{Q}_{I,i-1,j} & \cdot\ \boldsymbol{Q}_{u,i,j}=h_i\boldsymbol{Q}_{I,i,j}\end{array}\right\}. \tag{4.80}$$

(Refer to equations (4.45) and (4.46).)

② Following a similar procedure to the above, for the $(m+1)$ rows of points in the J-direction, the following condition equations hold:

---

*) For simplicity, h_i is assumed to not vary in the J-direction.

**) $\boldsymbol{Q}_{I,i,j}$ is the derivative vector at point $\boldsymbol{Q}_{i,j}$ with respect to curve length in the I-direction.

$$k_{j+1}\boldsymbol{Q}_{J,i,j-1}+2(k_{j+1}+k_j)\boldsymbol{Q}_{J,i,j}+k_j\boldsymbol{Q}_{J,i,j+1}$$
$$=\frac{3}{k_j k_{j+1}}\{k_j^2(\boldsymbol{Q}_{i,j+1}-\boldsymbol{Q}_{i,j})+k_{j+1}^2(\boldsymbol{Q}_{i,j}-\boldsymbol{Q}_{i,j-1})\} \tag{4.81}$$
$$(i=0,1,\ldots,m;\ j=1,2,\ldots,n-1).$$

k_j is the chord length in the J-direction, and is assumed to not vary in the I-direction. For each of the rows of points, if conditions on the unit tangent vectors in the J-direction at the starting and end points are given, Eqs. (4.81) can be solved. In this case, if normalization is done in each interval with a parameter w, then the tangent vectors with respect to w are:

$$\left.\begin{array}{ll}\dot{\boldsymbol{Q}}_{w,i-1,j-1}=k_j\boldsymbol{Q}_{J,i-1,j-1} & \dot{\boldsymbol{Q}}_{w,i,j-1}=k_j\boldsymbol{Q}_{J,i,j-1}\\ \dot{\boldsymbol{Q}}_{w,i-1,j}=k_j\boldsymbol{Q}_{J,i-1,j} & \dot{\boldsymbol{Q}}_{w,i,j}=k_j\boldsymbol{Q}_{J,i,j}\end{array}\right\}. \tag{4.82}$$

(Refer to Eqs. (4.45) and (4.46).)

Procedures ① and ② above generate a net of spline curves passing through the given lattice points.

③ In Fig. 4.18, focus attention on the previously determined group of unit tangent vectors in the I-direction, $\boldsymbol{Q}_{I,0j}$, $\boldsymbol{Q}_{I,mj}$ $(j=0,1,\ldots,n)$. First let us explain the significance of $\boldsymbol{Q}_{I,0j}$ $(j=0,1,\ldots,n)$. If the two end vectors $\boldsymbol{Q}_{IJ,00}$ and $\boldsymbol{Q}_{IJ,0n}$ are suitably specified, the $\boldsymbol{Q}_{IJ,0j}$ $(j=1,2,\ldots,n-1)$ that will smoothly interpolate between the $\boldsymbol{Q}_{I,0j}$ in the J-direction can be found from the following equations, corresponding to (4.81)*):

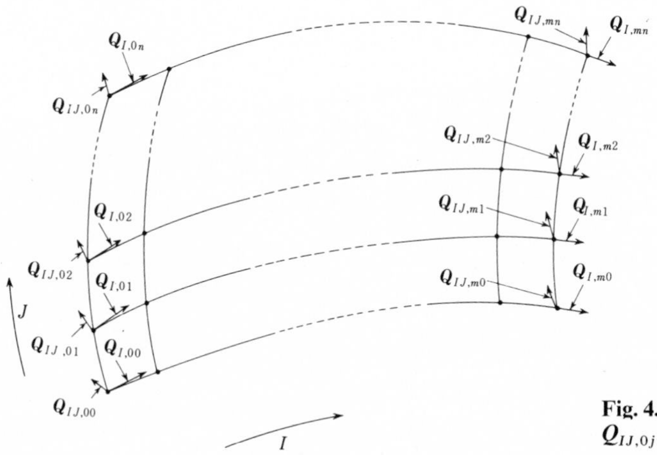

Fig. 4.18. Determination of $\boldsymbol{Q}_{IJ,0j}$, $\boldsymbol{Q}_{IJ,mj}$ $(j=0,1,\ldots,n)$

*) $\boldsymbol{Q}_{IJ,ij}$ is the cross partial derivative vector with respect to curve lengths in both the I- and J-directions at point $\boldsymbol{Q}_{ij}$.

$$k_{j+1}\boldsymbol{Q}_{IJ,0,j-1}+2(k_{j+1}+k_j)\boldsymbol{Q}_{IJ,0,j}+k_j\boldsymbol{Q}_{IJ,0,j+1}$$
$$=\frac{3}{k_jk_{j+1}}\{k_j^2(\boldsymbol{Q}_{I,0,j+1}-\boldsymbol{Q}_{I,0,j})+k_{j+1}^2(\boldsymbol{Q}_{I,0,j}-\boldsymbol{Q}_{I,0,j-1})\} \tag{4.83}$$
$$(j=1,2,\ldots,n-1).$$

A similar method can be used to determine the $\boldsymbol{Q}_{IJ,mj}$ $(j=1,2,\ldots,n-1)$ that will interpolate smoothly between the $\boldsymbol{Q}_{I,mj}(j=0,1,\ldots,n)$. In many cases $\boldsymbol{Q}_{IJ,00}$, $\boldsymbol{Q}_{IJ,0n}$, $\boldsymbol{Q}_{IJ,m0}$ and $\boldsymbol{Q}_{IJ,mn}$ are simply taken to be zero vectors.

④ Next, focus attention on the groups of unit tangent vectors $\boldsymbol{Q}_{J,ij}$ $(i=0,1,\ldots,m;\ j=0,1,\ldots,n)$ in the J-direction, previously determined at each point, which extend in the I direction (Fig. 4.19). There are $(n+1)$ such groups, for $j=0,1,\ldots,n$. For each group, the first and last vectors $\boldsymbol{Q}_{IJ,0j}$ and $\boldsymbol{Q}_{IJ,mj}$ were already found in procedure ③. By using the equation corresponding to Eq. (4.83), the $\boldsymbol{Q}_{IJ,ij}$ $(i=1,\ldots,m-1)$ that will interpolate smoothly between the $\boldsymbol{Q}_{J,ij}$ $(i=0,1,\ldots,m)$ can be determined. The condition equations are as follows:

$$h_{i+1}\boldsymbol{Q}_{IJ,i-1,j}+2(h_{i+1}+h_i)\boldsymbol{Q}_{IJ,i,j}+h_i\boldsymbol{Q}_{IJ,i+1,j}$$
$$=\frac{3}{h_ih_{i+1}}[h_i^2(\boldsymbol{Q}_{J,i+1,j}-\boldsymbol{Q}_{J,i,j})+h_{i+1}^2(\boldsymbol{Q}_{J,i,j}-\boldsymbol{Q}_{J,i-1,j})] \tag{4.84}$$
$$(i=1,2,\ldots,m-1;\ j=0,1,\ldots,n).$$

By solving these equations, $\boldsymbol{Q}_{IJ,ij}$ are determined at all points. $\boldsymbol{Q}_{IJ,ij}$ is the cross partial derivative vector with respect to curve length s. Concerning the parameters u, w $(0\leqq u, w\leqq 1)$, the cross partial derivative vector, for example at point $\boldsymbol{Q}_{ij}$, is, from Eq. (4.80):

$$\boldsymbol{Q}_{u,i,j}=h_i\boldsymbol{Q}_{I,i,j}$$

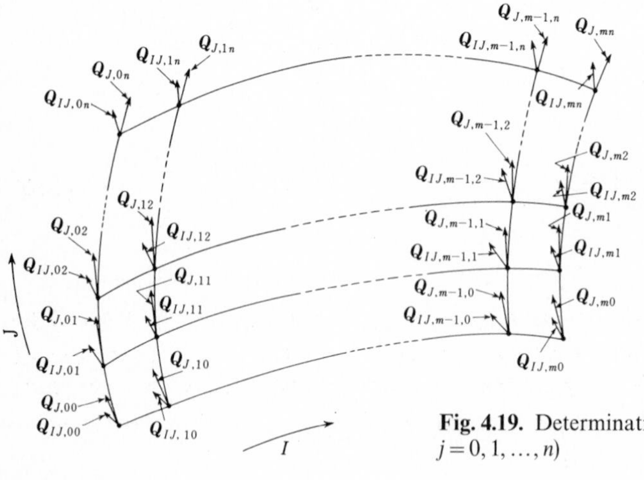

Fig. 4.19. Determination of $\boldsymbol{Q}_{IJ,ij}(i=0,1,\ldots,m;\ j=0,1,\ldots,n)$

so, if s is curve length in the J-direction, we have:

$$\begin{aligned}\boldsymbol{Q}_{uw,i,j} &= h_i \boldsymbol{Q}_{IJ,i,j} \frac{\partial s}{\partial w} \\ &\fallingdotseq h_i k_j \boldsymbol{Q}_{IJ,i,j}.\end{aligned}$$

Then, at each point we have:

$$\begin{array}{ll} \cdot\, \boldsymbol{Q}_{uw,i-1,j-1} = h_i k_j \boldsymbol{Q}_{IJ,i-1,j-1} & \cdot\, \boldsymbol{Q}_{uw,i,j-1} = h_i k_j \boldsymbol{Q}_{IJ,i,j-1} \\ \cdot\, \boldsymbol{Q}_{uw,i-1,j} = h_i k_j \boldsymbol{Q}_{IJ,i-1,j} & \cdot\, \boldsymbol{Q}_{uw,i,j} = h_i k_j \boldsymbol{Q}_{IJ,i,j} \end{array}. \tag{4.85}$$

Using Eqs. (4.80), (4.82) and (4.85), from Eq. (3.102) we obtain $\boldsymbol{P}_{ij}(u, w)$ as:

$$\begin{aligned}\boldsymbol{P}_{ij}(u,w) &= \boldsymbol{U}\boldsymbol{M}_c \\ &\times \begin{bmatrix} \boldsymbol{Q}_{i-1,j-1} & \boldsymbol{Q}_{i-1,j} & \boldsymbol{Q}_{w,i-1,j-1} & \boldsymbol{Q}_{w,i-1,j} \\ \boldsymbol{Q}_{i,j-1} & \boldsymbol{Q}_{i,j} & \boldsymbol{Q}_{w,i,j-1} & \boldsymbol{Q}_{w,i,j} \\ \boldsymbol{Q}_{u,i-1,j-1} & \boldsymbol{Q}_{u,i-1,j} & \boldsymbol{Q}_{uw,i-1,j-1} & \boldsymbol{Q}_{uw,i-1,j} \\ \boldsymbol{Q}_{u,i,j-1} & \boldsymbol{Q}_{u,i,j} & \boldsymbol{Q}_{uw,i,j-1} & \boldsymbol{Q}_{uw,i,j} \end{bmatrix} \boldsymbol{M}_c^T \boldsymbol{W}^T \\ &= \boldsymbol{U}\boldsymbol{M}_c \\ &\times \begin{bmatrix} \boldsymbol{Q}_{i-1,j-1} & \boldsymbol{Q}_{i-1,j} & k_j \boldsymbol{Q}_{J,i-1,j-1} & k_j \boldsymbol{Q}_{J,i-1,j} \\ \boldsymbol{Q}_{i,j-1} & \boldsymbol{Q}_{i,j} & k_j \boldsymbol{Q}_{J,i,j-1} & k_j \boldsymbol{Q}_{J,i,j} \\ h_i \boldsymbol{Q}_{I,i-1,j-1} & h_i \boldsymbol{Q}_{I,i-1,j} & h_i k_j \boldsymbol{Q}_{IJ,i-1,j-1} & h_i k_j \boldsymbol{Q}_{IJ,i-1,j} \\ h_i \boldsymbol{Q}_{I,i,j-1} & h_i \boldsymbol{Q}_{I,i,j} & h_i k_j \boldsymbol{Q}_{IJ,i,j-1} & h_i k_j \boldsymbol{Q}_{IJ,i,j} \end{bmatrix} \boldsymbol{M}_c^T \boldsymbol{W}^T \\ &= \boldsymbol{U}\boldsymbol{M}_c \\ &\times \begin{bmatrix} 1 & 0 & 0 & 0 \\ 0 & 1 & 0 & 0 \\ 0 & 0 & h_i & 0 \\ 0 & 0 & 0 & h_i \end{bmatrix} \begin{bmatrix} \boldsymbol{Q}_{i-1,j-1} & \boldsymbol{Q}_{i-1,j} & \boldsymbol{Q}_{J,i-1,j-1} & \boldsymbol{Q}_{J,i-1,j} \\ \boldsymbol{Q}_{i,j-1} & \boldsymbol{Q}_{i,j} & \boldsymbol{Q}_{J,i,j-1} & \boldsymbol{Q}_{J,i,j} \\ \boldsymbol{Q}_{I,i-1,j-1} & \boldsymbol{Q}_{I,i-1,j} & \boldsymbol{Q}_{IJ,i-1,j-1} & \boldsymbol{Q}_{IJ,i-1,j} \\ \boldsymbol{Q}_{I,i,j-1} & \boldsymbol{Q}_{I,i,j} & \boldsymbol{Q}_{IJ,i,j-1} & \boldsymbol{Q}_{IJ,i,j} \end{bmatrix} \\ &\times \begin{bmatrix} 1 & 0 & 0 & 0 \\ 0 & 1 & 0 & 0 \\ 0 & 0 & k_j & 0 \\ 0 & 0 & 0 & k_j \end{bmatrix} \boldsymbol{M}_c^T \boldsymbol{W}^T. \end{aligned} \tag{4.86}$$

In procedures ③ and ④, spline surfaces can be generated even if the order of the procedures for the I- and J-directions is opposite to that given here. It has been shown that following the procedures in either order produces the same surface.

References (Chap. 4)

16) Schoenberg, I. J.: "Spline functions and the problem of graduation, proceedings of the National Academy of Sciences of the U.S.A., 52 (1964), 947—950.

5. The Bernstein Approximation

5.1 Curves

5.1.1 Modification of Ferguson Curve Segments

As explained in Chap. 3, curves and surfaces based on Hermite interpolation position vectors of 2 points $\boldsymbol{Q}_0$ and $\boldsymbol{Q}_1$ and the tangent vectors at those points $\dot{\boldsymbol{Q}}_0$ and $\dot{\boldsymbol{Q}}_1$ (Chap. 3):

$$\boldsymbol{P}(t)=[t^3 \; t^2 \; t \; 1]\begin{bmatrix} 2 & -2 & 1 & 1 \\ -3 & 3 & -2 & -1 \\ 0 & 0 & 1 & 0 \\ 1 & 0 & 0 & 0 \end{bmatrix}\begin{bmatrix} \boldsymbol{Q}_0 \\ \boldsymbol{Q}_1 \\ \dot{\boldsymbol{Q}}_0 \\ \dot{\boldsymbol{Q}}_1 \end{bmatrix}. \tag{5.1}$$

As explained in Chap. 3, curves and surfaces based on Hermite interpolation have the following problems:

① At the end points, the effects of the tangent vectors and twist vectors on shapes of curves and surfaces are in opposite directions, causing confusion when a person tries to control the shape.

② To control a curve shape, two different types of vectors, position vectors and and tangent vectors, must be manipulated. To control a surface shape, these two plus a third type, twist vectors, must be manipulated. In particular, the relation between the twist vector and surface shape is difficult to grasp intuitively; also, the orders of magnitude of the different vectors are different, so human manipulation is difficult.

One way to solve these problems is shown in Fig. 5.1. On a Ferguson curve segment, a vector is shown at point $\boldsymbol{Q}_1$ in the reverse direction of the

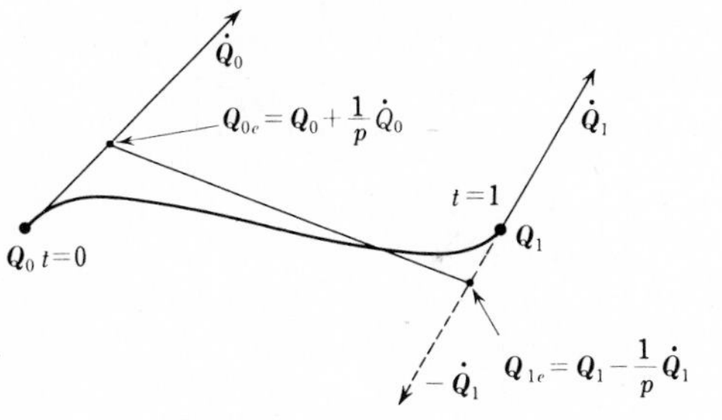

Fig. 5.1. Modification of a Ferguson curve segment

tangent vector $\dot{\boldsymbol{Q}}_1$. A cubic curve is supposed to be expressed by the two points $\boldsymbol{Q}_0$ and $\boldsymbol{Q}_1$ and 2 appropriate points on the tangent vectors which are shown, as $\boldsymbol{Q}_{0e}$ and $\boldsymbol{Q}_{1e}$ [18].

For the last 2 points, let us use the point $\boldsymbol{Q}_{0e}$ at a distance $1/p$ times the tangent vector from point $\boldsymbol{Q}_0$, and point $\boldsymbol{Q}_{1e}$ a distance $1/p$ times the tangent vector, direction reversed, from point $\boldsymbol{Q}_1$:

$$\boldsymbol{Q}_{0e} = \boldsymbol{Q}_0 + \frac{1}{p}\dot{\boldsymbol{Q}}_0 \tag{5.2}$$

$$\boldsymbol{Q}_{1e} = \boldsymbol{Q}_1 - \frac{1}{p}\dot{\boldsymbol{Q}}_1 . \tag{5.3}$$

Solving Eqs. (5.2) and (5.3) for $\boldsymbol{Q}_0$ and $\boldsymbol{Q}_1$, substituting in Eq. (5.1) and rearranging gives:

$$\boldsymbol{P}(t) = [t^3\ t^2\ t\ 1]\begin{bmatrix} 2 & -2 & 1 & 1 \\ -3 & 3 & -2 & -1 \\ 0 & 0 & 1 & 0 \\ 1 & 0 & 0 & 0 \end{bmatrix}\begin{bmatrix} \boldsymbol{Q}_0 \\ \boldsymbol{Q}_1 \\ p(\boldsymbol{Q}_{0e}-\boldsymbol{Q}_0) \\ -p(\boldsymbol{Q}_{1e}-\boldsymbol{Q}_1) \end{bmatrix}$$

$$= [t^3\ t^2\ t\ 1]\begin{bmatrix} 2 & -2 & 1 & 1 \\ -3 & 3 & -2 & -1 \\ 0 & 0 & 1 & 0 \\ 1 & 0 & 0 & 0 \end{bmatrix}\begin{bmatrix} 1 & 0 & 0 & 0 \\ 0 & 1 & 0 & 0 \\ -p & 0 & p & 0 \\ 0 & p & 0 & -p \end{bmatrix}\begin{bmatrix} \boldsymbol{Q}_0 \\ \boldsymbol{Q}_1 \\ \boldsymbol{Q}_{0e} \\ \boldsymbol{Q}_{1e} \end{bmatrix}$$

$$= [t^3\ t^2\ t\ 1]\begin{bmatrix} 2-p & -2+p & p & -p \\ -3+2p & 3-p & -2p & p \\ -p & 0 & p & 0 \\ 1 & 0 & 0 & 0 \end{bmatrix}\begin{bmatrix} \boldsymbol{Q}_0 \\ \boldsymbol{Q}_1 \\ \boldsymbol{Q}_{0e} \\ \boldsymbol{Q}_{1e} \end{bmatrix}$$

$$= [t^3\ t^2\ t\ 1]\begin{bmatrix} 2-p & p & -p & -2+p \\ -3+2p & -2p & p & 3-p \\ -p & p & 0 & 0 \\ 1 & 0 & 0 & 0 \end{bmatrix}\begin{bmatrix} \boldsymbol{Q}_0 \\ \boldsymbol{Q}_{0e} \\ \boldsymbol{Q}_{1e} \\ \boldsymbol{Q}_1 \end{bmatrix}$$

$$\equiv [X_0(t)\ X_1(t)\ X_2(t)\ X_3(t)]\begin{bmatrix} \boldsymbol{Q}_0 \\ \boldsymbol{Q}_{0e} \\ \boldsymbol{Q}_{1e} \\ \boldsymbol{Q}_1 \end{bmatrix} \tag{5.4}$$

where:

$$\left.\begin{aligned} X_0(t) &= (1-t)^2\{1+(2-p)t\} \\ X_1(t) &= pt(1-t)^2 \\ X_2(t) &= pt^2(1-t) \\ X_3(t) &= t^2\{3-p+(-2+p)t\} \end{aligned}\right\} . \tag{5.5}$$

It is known that a Cauchy relation (Eq. (1.4)) holds among $X_0(t)$, $X_1(t)$, $X_2(t)$ and $X_3(t)$, regardless of the value of p:

$$X_0(t)+X_1(t)+X_2(t)+X_3(t) \equiv 1. \tag{5.6}$$

It can be shown that the following are the conditions for a curve to always exist inside the convex polygon formed by $\boldsymbol{Q}_0$, $\boldsymbol{Q}_{0e}$, $\boldsymbol{Q}_{1e}$ and $\boldsymbol{Q}_1$:

$$X_0(t) \geqq 0,\ X_1(t) \geqq 0,\ X_2(t) \geqq 0,\ X_3(t) \geqq 0 \quad (0 \leqq t \leqq 1). \tag{5.7}$$

From these inequalities it follows that the curve is inside the polygon when:

$$0 \leqq p \leqq 3. \tag{5.8}$$

In the case $p=0$, the curve $\boldsymbol{P}(t)$ reduces to a straight line connecting the 2 points $\boldsymbol{Q}_0$ and $\boldsymbol{Q}_1$. As p is increased, the curve approaches closer to the polygon $\boldsymbol{Q}_0$, $\boldsymbol{Q}_{0e}$, $\boldsymbol{Q}_{1e}$, $\boldsymbol{Q}_1$ (Fig. 5.2). If p exceeds the limit of condition (5.8), for example if $p=4$, then part of the curve will protrude outside the polygon $\boldsymbol{Q}_0$, $\boldsymbol{Q}_{0e}$, $\boldsymbol{Q}_{1e}$, $\boldsymbol{Q}_1$. Condition (5.8) is the condition for the curve to remain inside the polygon regardless of the shape of the polygon. Therefore, the

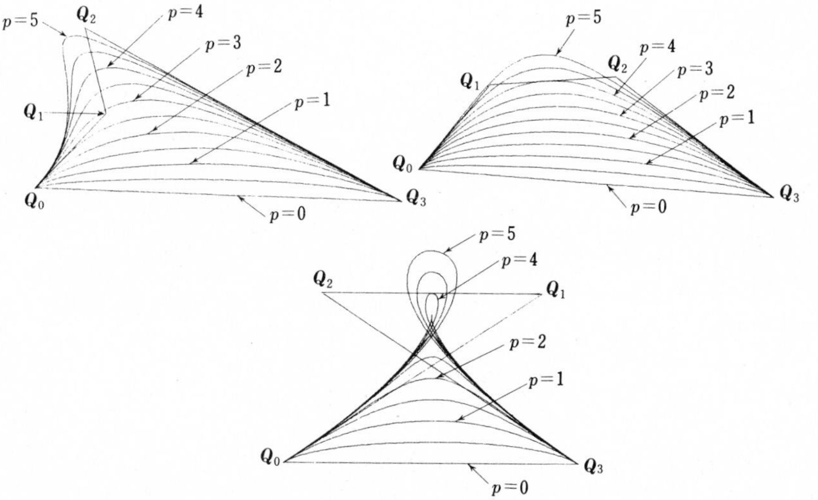

Fig. 5.2. Variation of curve shape with respect to the value of p.

closest the curve can come to the polygon and still be totally contained within the polygon is the shape it assumes when $p=3$. In this case, the functions $X_0(t)$, $X_1(t)$, $X_2(t)$ and $X_3(t)$ are:

$$\left.\begin{aligned} X_0(t) &= (1-t)^3 \\ X_1(t) &= 3(1-t)^2 t \\ X_2(t) &= 3(1-t)t^2 \\ X_3(t) &= t^3 \end{aligned}\right\}. \tag{5.9}$$

Let us now renumber the polygon vertices as follows:

$$\boldsymbol{Q}_0 \to \boldsymbol{Q}_0 \text{ (remains the same)}, \ \boldsymbol{Q}_{0e} \to \boldsymbol{Q}_1, \ \boldsymbol{Q}_{1e} \to \boldsymbol{Q}_2, \ \boldsymbol{Q}_1 \to \boldsymbol{Q}_3.$$

Then the curve segment described by Eq. (5.4) becomes:

$$\boldsymbol{P}(t) = [X_0(t) \ X_1(t) \ X_2(t) \ X_3(t)] \begin{bmatrix} \boldsymbol{Q}_0 \\ \boldsymbol{Q}_1 \\ \boldsymbol{Q}_2 \\ \boldsymbol{Q}_3 \end{bmatrix}$$

$$= \sum_{i=0}^{3} X_i(t) \boldsymbol{Q}_i \tag{5.10}$$

where:

$$\boldsymbol{P}(t) = [t^3 \ t^2 \ t \ 1] \begin{bmatrix} -1 & 3 & -3 & 1 \\ 3 & -6 & 3 & 0 \\ -3 & 3 & 0 & 0 \\ 1 & 0 & 0 & 0 \end{bmatrix} \begin{bmatrix} \boldsymbol{Q}_0 \\ \boldsymbol{Q}_1 \\ \boldsymbol{Q}_2 \\ \boldsymbol{Q}_3 \end{bmatrix} \tag{5.11}$$

$$\equiv [t^3 \ t^2 \ t \ 1] \boldsymbol{M}_B \begin{bmatrix} \boldsymbol{Q}_0 \\ \boldsymbol{Q}_1 \\ \boldsymbol{Q}_2 \\ \boldsymbol{Q}_3 \end{bmatrix} \tag{5.12}$$

where:

$$\boldsymbol{M}_B = \begin{bmatrix} -1 & 3 & -3 & 1 \\ 3 & -6 & 3 & 0 \\ -3 & 3 & 0 & 0 \\ 1 & 0 & 0 & 0 \end{bmatrix}. \tag{5.13}$$

Graphs of $X_0(t)$, $X_1(t)$, $X_2(t)$ and $X_3(t)$ are shown in Fig. 5.3. The curves described by Eqs. (5.10), (5.11) and (5.12) coincide with Bézier curve segments when the latter are of degree 3. The position vectors $\boldsymbol{Q}_0, \ldots, \boldsymbol{Q}_3$ which define

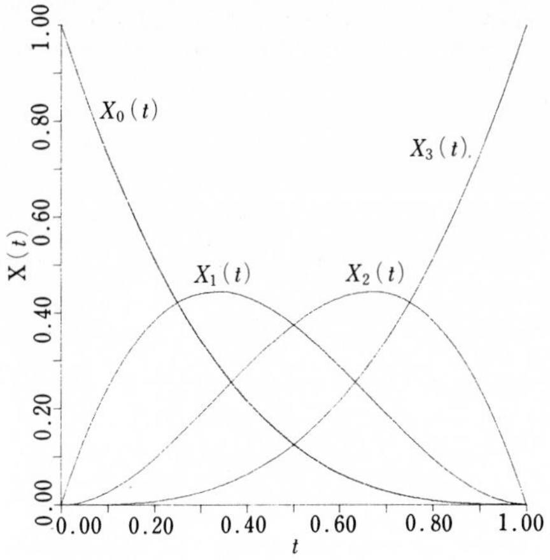

Fig. 5.3. Graphs of $X_0(t)$, $X_1(t)$, $X_2(t)$ and $X_3(t)$ (Cubic Bernstein basis)

a Bézier curve segment are called the curve defining vectors. The polygon formed by joining the points defined by the curve defining vectors is called by such names as the curve defining polygon, characteristic polygon or Bézier polygon.

5.1.2 Cubic Bézier Curve Segments

As can be seen from the derivation in the preceding section, the tangent vectors at the starting and end points of a cubic Bézier curve segment are, respectively:

$$\left.\begin{aligned} \dot{\boldsymbol{P}}(0)&=3(\boldsymbol{Q}_1-\boldsymbol{Q}_0)\\ \dot{\boldsymbol{P}}(1)&=3(\boldsymbol{Q}_3-\boldsymbol{Q}_2)\end{aligned}\right\}. \tag{5.14}$$

The slope of the curve at its starting and end points coincides with the directions of sides $\overrightarrow{\boldsymbol{Q}_0\boldsymbol{Q}_1}$ and $\overrightarrow{\boldsymbol{Q}_2\boldsymbol{Q}_3}$ of the curve defining polygon.

Next, let us find the curvature at the start and end points of the curve segment. The second derivative vectors at the two points are:

$$\left.\begin{aligned} \ddot{\boldsymbol{P}}(0)&=6\boldsymbol{Q}_0-12\boldsymbol{Q}_1+6\boldsymbol{Q}_2=6(\boldsymbol{Q}_2-\boldsymbol{Q}_1)-6(\boldsymbol{Q}_1-\boldsymbol{Q}_0)\\ \ddot{\boldsymbol{P}}(1)&=6\boldsymbol{Q}_1-12\boldsymbol{Q}_2+6\boldsymbol{Q}_3=6(\boldsymbol{Q}_3-\boldsymbol{Q}_2)-6(\boldsymbol{Q}_2-\boldsymbol{Q}_1)\end{aligned}\right\}. \tag{5.15}$$

The curvature κ_0 at the starting point of a curve segment is, from equation (1.26):

$$\kappa_0=\frac{|\dot{\boldsymbol{P}}(0)\times\ddot{\boldsymbol{P}}(0)|}{|\dot{\boldsymbol{P}}(0)|^3}=\frac{2}{3}\,\frac{|(\boldsymbol{Q}_1-\boldsymbol{Q}_0)\times(\boldsymbol{Q}_2-\boldsymbol{Q}_1)|}{|\boldsymbol{Q}_1-\boldsymbol{Q}_0|^3}. \tag{5.16}$$

Similarly, the curvature κ_1 at the end point is:

$$\kappa_1 = \frac{|\dot{\boldsymbol{P}}(1) \times \ddot{\boldsymbol{P}}(1)|}{|\dot{\boldsymbol{P}}(1)|^3} = \frac{2}{3} \frac{|(\boldsymbol{Q}_2 - \boldsymbol{Q}_1) \times (\boldsymbol{Q}_3 - \boldsymbol{Q}_2)|}{|\boldsymbol{Q}_3 - \boldsymbol{Q}_2|^3}. \tag{5.17}$$

As can be seen from Eq. (1.24), the location of the center of curvature depends on the direction of the 2nd derivative vector $\ddot{\boldsymbol{P}}$ with respect to the tangent vector $\dot{\boldsymbol{P}}$. At the start point of a Bézier curve segment, the center of curvature is in the same half space as $\boldsymbol{Q}_2$ with respect to the straight line $\boldsymbol{Q}_0\boldsymbol{Q}_1$. In Fig. 5.4(a) $\boldsymbol{Q}_2$ lies on the same side as $\boldsymbol{Q}_3$ does with respect to the straight line $\boldsymbol{Q}_0\boldsymbol{Q}_1$, so the center of curvature lies on the $\boldsymbol{Q}_3$ side. Similarly, at the end point of the curve the center of curvature lies on the $\boldsymbol{Q}_1$ side with respect to the straight line $\boldsymbol{Q}_2\boldsymbol{Q}_3$. In Fig. 5.4(a), $\boldsymbol{Q}_1$ is on the same side as $\boldsymbol{Q}_0$, so the center of curvature is on the $\boldsymbol{Q}_0$ side. In such a case, if the intersection of lines $\overline{\boldsymbol{Q}_0\boldsymbol{Q}_1}$ and $\overline{\boldsymbol{Q}_2\boldsymbol{Q}_3}$ is called $\boldsymbol{S}$, $\boldsymbol{S}$ is on the outside of both $\overline{\boldsymbol{Q}_0\boldsymbol{Q}_1}$ and $\overline{\boldsymbol{Q}_2\boldsymbol{Q}_3}$. In Fig. 5.4(b), $\boldsymbol{Q}_2$ is on the opposite side of $\overline{\boldsymbol{Q}_0\boldsymbol{Q}_1}$ as $\boldsymbol{Q}_3$, while $\boldsymbol{Q}_1$ is on the same side of $\overline{\boldsymbol{Q}_2\boldsymbol{Q}_3}$ as $\boldsymbol{Q}_0$. In such a case there is 1 inflection point on the curve. In this case, point $\boldsymbol{S}$ is on the outside of line $\overline{\boldsymbol{Q}_0\boldsymbol{Q}_1}$ and inside of line segment $\overline{\boldsymbol{Q}_2\boldsymbol{Q}_3}$.

Next, consider the polygon shape in Fig. 5.4(c). Points $\boldsymbol{Q}_1$ and $\boldsymbol{Q}_2$ are on extensions of $\overline{\boldsymbol{Q}_0\boldsymbol{S}}$ and $\overline{\boldsymbol{Q}_3\boldsymbol{S}}$, not far from $\boldsymbol{S}$. In this case, the center of curvature at the starting point is on the opposite side of line $\overline{\boldsymbol{Q}_0\boldsymbol{Q}_1}$ as $\boldsymbol{Q}_3$. The center of curvature at the end point is on the opposite side of line $\overline{\boldsymbol{Q}_2\boldsymbol{Q}_3}$ as $\boldsymbol{Q}_0$, and the curve segment has two inflection points. If $\boldsymbol{Q}_1$ and $\boldsymbol{Q}_2$ are removed farther from point $\boldsymbol{S}$ than in Fig. 5.4(c), the situation in Fig. 5.4(d) is obtained, with the curve having a cusp. If they are removed still farther, the curve has a loop, as in Fig. 5.4(e). This is easy to understand if we compare with the case of a Ferguson curve (refer to Sect. 3.2.1). Since the tangent vectors at the ends of a Bézier curve segment are given by Eqs. (5.14), as $\overline{\boldsymbol{Q}_1 - \boldsymbol{Q}_0}$ and $\overline{\boldsymbol{Q}_3 - \boldsymbol{Q}_2}$ become larger, when a certain relationship holds a cusp develops in the curve. If the magnitude of the tangent vectors becomes still larger, the curve develops a loop.

Let us find the condition for a cubic Bézier curve segment to have a cusp, following the same method as in Sect. 3.2.1. Letting:

$$\begin{aligned} &\boldsymbol{Q}_1 - \boldsymbol{Q}_0 = m(\boldsymbol{S} - \boldsymbol{Q}_0) \\ &\boldsymbol{Q}_3 - \boldsymbol{Q}_2 = n(\boldsymbol{Q}_3 - \boldsymbol{S}) \end{aligned}$$

the derivative of the Bézier curve segment is, from Eq. (5.12):

$$\dot{\boldsymbol{P}}(t) = [3t^2 \;\; 2t \;\; 1 \;\; 0]\, \boldsymbol{M}_B \begin{bmatrix} \boldsymbol{Q}_0 \\ m(\boldsymbol{S} - \boldsymbol{Q}_0) + \boldsymbol{Q}_0 \\ n(\boldsymbol{S} - \boldsymbol{Q}_3) + \boldsymbol{Q}_3 \\ \boldsymbol{Q}_3 \end{bmatrix}. \tag{5.18}$$

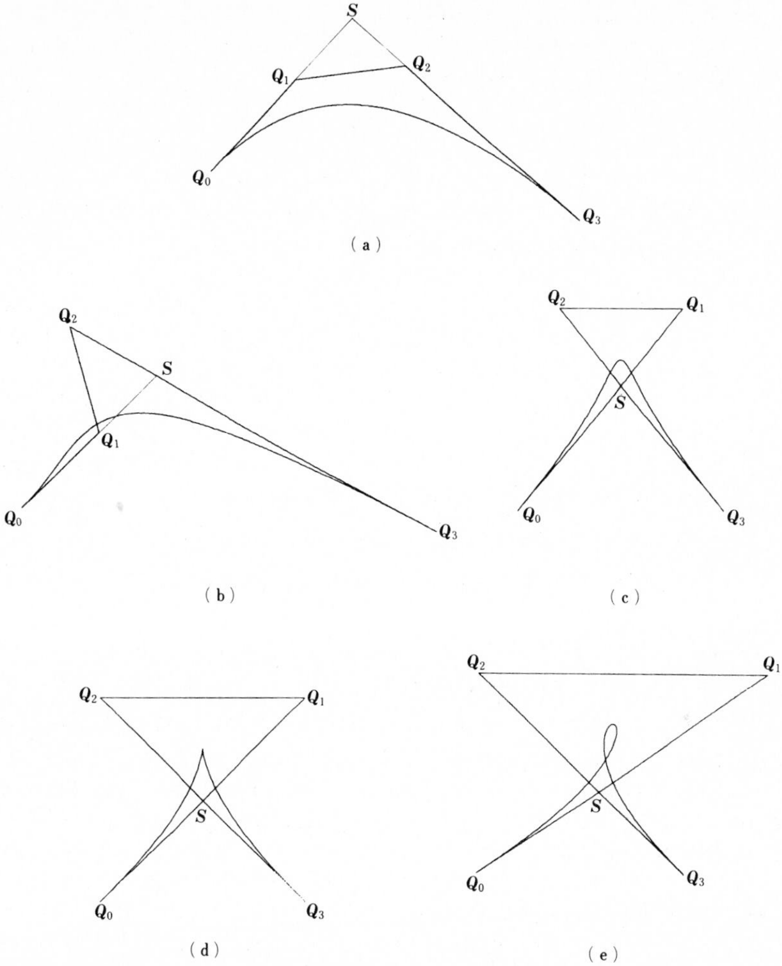

Fig. 5.4. Cubic Bézier curve segments

Since the shape of a Bézier curve segment is invariant under coordinate transformation, let us take the origin at $\boldsymbol{Q}_0$ and the positive x-axis in the $\boldsymbol{Q}_0 \rightarrow \boldsymbol{Q}_3$ direction. From the above equation we have:

$$\boldsymbol{Q}_0 = [0\ 0], \quad \boldsymbol{S} = [S_x\ S_y], \quad \boldsymbol{Q}_3 = [C\ 0].$$

Then (5.18) becomes:

$$\dot{\boldsymbol{P}}(t) = 3\,[t^2\ \ t\ \ 1] \begin{bmatrix} -1 & 3 & -3 & 1 \\ 2 & -4 & 2 & 0 \\ -1 & 1 & 0 & 0 \end{bmatrix} \begin{bmatrix} 0 & 0 \\ mS_x & mS_y \\ nS_x + (1-n)C & nS_y \\ C & 0 \end{bmatrix} =$$

$$= 3\,[t^2 \;\; t \;\; 1] \begin{bmatrix} 3(m-n)S_x+(3n-2)C & 3(m-n)S_y \\ 2(n-2m)S_x+2(1-n)C & 2(n-2m)S_y \\ mS_x & mS_y \end{bmatrix}. \tag{5.19}$$

To find the cusp, set $\dot{\boldsymbol{P}}(t)=\boldsymbol{0}$. Then:

$$\left.\begin{aligned} &[3(m-n)t^2+2(n-2m)t+m]\,S_x+[(3n-2)t^2+2(1-n)t]\,C=0 \\ &[3(m-n)t^2+2(n-2m)t+m]\,S_y=0 \end{aligned}\right\}. \tag{5.20}$$

Solving these equations gives:

$$\left(m-\frac{4}{3}\right)\left(n-\frac{4}{3}\right)=\frac{4}{9}. \tag{5.21}$$

When m and n have values that satisfy Eq. (5.21), the cubic Bézier curve segment has a cusp and the parameter has the value:

$$t=\frac{2(n-1)}{3n-2}. \tag{5.22}$$

If we take $m=n$, when $m=n=2$ the cusp occurs at $t=\frac{1}{2}$ (Fig. 5.4(d)).

A loop occurs in the curve when:

$$\left(m-\frac{4}{3}\right)\left(n-\frac{4}{3}\right)>\frac{4}{9}. \tag{5.23}$$

Representation of a Straight Line

Let us modify a cubic Bézier curve segment as follows:

$$\begin{aligned} \boldsymbol{P}(t) &= (1-t)^3\boldsymbol{Q}_0+3t(1-t)^2\boldsymbol{Q}_1+3t^2(1-t)\boldsymbol{Q}_2+t^3\boldsymbol{Q}_3 \\ &= (\boldsymbol{Q}_3-\boldsymbol{Q}_0+3\boldsymbol{Q}_1-3\boldsymbol{Q}_2)t^3+3(\boldsymbol{Q}_0-2\boldsymbol{Q}_1+\boldsymbol{Q}_2)t^2+3(\boldsymbol{Q}_1-\boldsymbol{Q}_0)t+\boldsymbol{Q}_0. \end{aligned} \tag{5.24}$$

The condition for the t^3 and t^2 terms to vanish is:

$$\boldsymbol{Q}_3-\boldsymbol{Q}_0+3\boldsymbol{Q}_1-3\boldsymbol{Q}_2=\boldsymbol{0} \tag{5.25}$$

$$\boldsymbol{Q}_0-2\boldsymbol{Q}_1+\boldsymbol{Q}_2=\boldsymbol{0}. \tag{5.26}$$

Using Eq. (5.26), (5.25) becomes:

$$\boldsymbol{Q}_1-2\boldsymbol{Q}_2+\boldsymbol{Q}_3=\boldsymbol{0}. \tag{5.27}$$

Equations (5.26) and (5.27) are also the conditions for the 2nd derivative vector to be $\boldsymbol{0}$ at the start and end points of the curve (refer to Eq. (5.15)). From Eqs. (5.26) and (5.27) we have:

$$Q_1 = \frac{2Q_0 + Q_3}{3}$$

$$Q_2 = \frac{Q_0 + 2Q_3}{3}.$$

Substituting these relations into Eq. (5.24), the Bézier curve segment becomes:

$$P(t) = (1-t)\,Q_0 + t\,Q_3$$

so that the curve degenerates to the line segment joining Q_0 and Q_3 (Fig. 5.5).

Approximate Representation of a Circle

Consider an approximate representation of the first quadrant of a circle of unit radius by a cubic Bézier curve segment (Fig. 5.6).

If we specify the condition that at $t=\frac{1}{2}$ the curve passes through the point $(1/\sqrt{2},\, 1/\sqrt{2})$:

$$P\left(\frac{1}{2}\right) = \frac{1}{8}\,Q_0 + \frac{3}{8}\,Q_1 + \frac{3}{8}\,Q_2 + \frac{1}{8}\,Q_3$$

$$= \frac{1}{8}\,[1\;0] + \frac{3}{8}\,[1\;k] + \frac{3}{8}\,[k\;1] + \frac{1}{8}\,[0\;1] = \left[\frac{1}{\sqrt{2}}\;\frac{1}{\sqrt{2}}\right]$$

then, solving the above equation, we have:

$$k = \frac{4}{3}\,(\sqrt{2}-1). \tag{5.28}$$

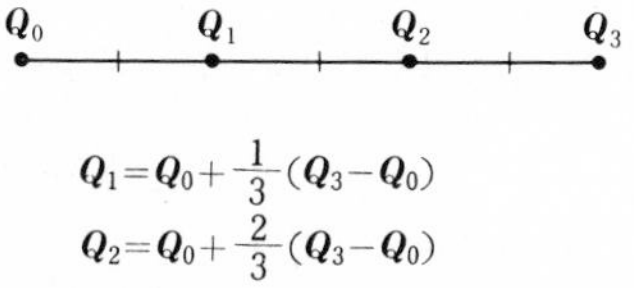

Fig. 5.5. Expression of a line segment as Bézier curve segment

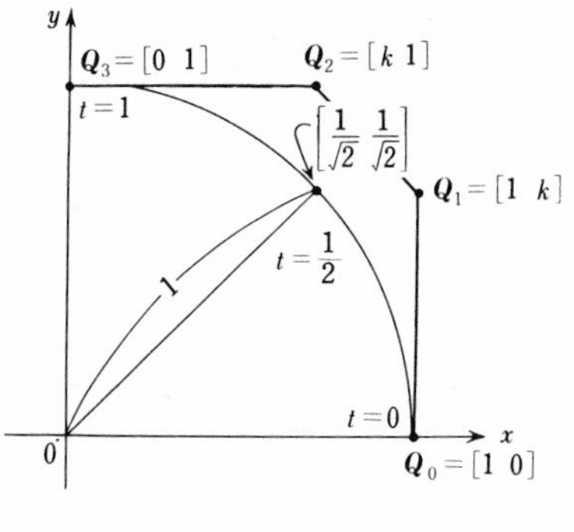

Fig. 5.6. Approximate expression of a $\frac{1}{4}$-circle by a Bézier curve segment

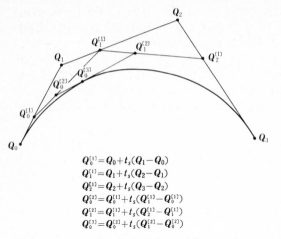

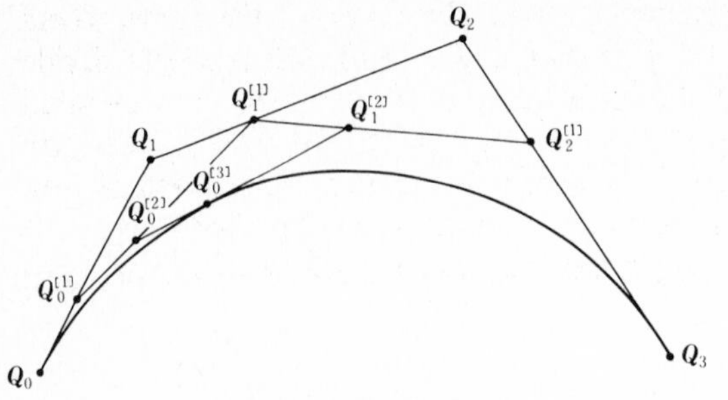

$$
\begin{aligned}
Q_0^{[1]} &= Q_0 + t_s(Q_1 - Q_0) \\
Q_1^{[1]} &= Q_1 + t_s(Q_2 - Q_1) \\
Q_2^{[1]} &= Q_2 + t_s(Q_3 - Q_2) \\
Q_0^{[2]} &= Q_0^{[1]} + t_s(Q_1^{[1]} - Q_0^{[1]}) \\
Q_1^{[2]} &= Q_1^{[1]} + t_s(Q_2^{[1]} - Q_1^{[1]}) \\
Q_0^{[3]} &= Q_0^{[2]} + t_s(Q_1^{[2]} - Q_0^{[2]})
\end{aligned}
$$

Fig. 5.7. Graphical method of determining a point on a curve corresponding to $t = t_s$

In this case, the maximum deviation from a true circle is +0.027% (refer to Table 3.2).

Determination of a Point on a Curve

Let the points on the 3 sides of the curve defining polygon which divide those sides in the ratio $t_s : 1 - t_s$ be $\boldsymbol{Q}_0^{[1]}$, $\boldsymbol{Q}_1^{[1]}$ and $\boldsymbol{Q}_2^{[1]}$ (Fig. 5.7):

$$
\left.\begin{aligned}
\boldsymbol{Q}_0^{[1]} &= (1-t_s)\boldsymbol{Q}_0 + t_s\boldsymbol{Q}_1 \\
\boldsymbol{Q}_1^{[1]} &= (1-t_s)\boldsymbol{Q}_1 + t_s\boldsymbol{Q}_2 \\
\boldsymbol{Q}_2^{[1]} &= (1-t_s)\boldsymbol{Q}_2 + t_s\boldsymbol{Q}_3
\end{aligned}\right\} . \qquad (5.29)
$$

Next, let $\boldsymbol{Q}_0^{[2]}$ and $\boldsymbol{Q}_1^{[2]}$ be 2 points on sides of the polygon $\boldsymbol{Q}_0^{[1]}$, $\boldsymbol{Q}_1^{[1]}$, $\boldsymbol{Q}_2^{[1]}$ that divide those sides in the ratio $t_s : 1 - t_s$:

$$
\left.\begin{aligned}
\boldsymbol{Q}_0^{[2]} &= (1-t_s)\boldsymbol{Q}_0^{[1]} + t_s\boldsymbol{Q}_1^{[1]} = (1-t_s)^2\boldsymbol{Q}_0 + 2(1-t_s)t_s\boldsymbol{Q}_1 + t_s^2\boldsymbol{Q}_2 \\
\boldsymbol{Q}_1^{[2]} &= (1-t_s)\boldsymbol{Q}_1^{[1]} + t_s\boldsymbol{Q}_2^{[1]} = (1-t_s)^2\boldsymbol{Q}_1 + 2(1-t_s)t_s\boldsymbol{Q}_2 + t_s^2\boldsymbol{Q}_3
\end{aligned}\right\} . \qquad (5.30)
$$

Finally, let $\boldsymbol{Q}_0^{[3]}$ be the point on the line segment $\boldsymbol{Q}_0^{[2]}$, $\boldsymbol{Q}_1^{[2]}$ that divide that line segment in the ratio $t_s : 1 - t_s$:

$$
\boldsymbol{Q}_0^{[3]} = (1-t_s)\boldsymbol{Q}_0^{[2]} + t_s\boldsymbol{Q}_1^{[2]} . \qquad (5.31)
$$

Substituting relations (5.30) into (5.31) gives:

$$
\begin{aligned}
\boldsymbol{Q}_0^{[3]} &= (1-t_s)^3\boldsymbol{Q}_0 + 3(1-t_s)^2 t_s\boldsymbol{Q}_1 + 3(1-t_s)t_s^2\boldsymbol{Q}_2 + t_s^3\boldsymbol{Q}_3 \\
&\equiv \boldsymbol{P}(t_s) \qquad (5.32)
\end{aligned}
$$

which is the point on the curve at which $t=t_s$. The point on the curve where $t=t_s$ can be found graphically by performing the following process: divide each side of the given polygon in the ratio of $t_s:1-t_s$. Connect the generated points and make a new polygon. Repeat the above process for the new polygon until the polygon becomes a line segment. Divide the line segment in the ratio $t_s:1-t_s$. Then the generated point is what we want to find.

Let us look at the vector $Q_1^{[2]}-Q_0^{[2]}$. From Eq. (5.30), we have:

$$Q_1^{[2]}-Q_0^{[2]}=[t_s^2+2t_s-1 \;\; 3t_s^2-4t_s+1 \;\; -3t_s^2+2t_s \;\; t_s^2]\begin{bmatrix} Q_0 \\ Q_1 \\ Q_2 \\ Q_3 \end{bmatrix}$$

$$=\frac{1}{3}[3t_s^2 \;\; 2t_s \;\; 1 \;\; 0]\begin{bmatrix} 1 & 3 & -3 & 1 \\ 3 & -6 & 3 & 0 \\ -3 & 3 & 0 & 0 \\ 1 & 0 & 0 & 0 \end{bmatrix}\begin{bmatrix} Q_0 \\ Q_1 \\ Q_2 \\ Q_3 \end{bmatrix}$$

$$\equiv\frac{1}{3}\dot{P}(t_s). \tag{5.33}$$

That is, among the polygons that are generated in sequence to find a point on the curve, the last line segment coincides with the tangent to the curve at the point $P(t_s)$ being found.

Partitioning

Consider the problem of dividing a cubic Bézier curve segment $P(t)$ at point $t=t_s$ into two Bézier curve segments $P_1(u)$ $(0\leqq u\leqq 1)$ and $P_2(u)$ $(0\leqq u\leqq 1)$ (Fig. 5.8).

$$P(t)=(1-t)^3Q_0+3(1-t)^2tQ_1+3(1-t)t^2Q_2+t^3Q_3$$

$$=[t^3 \;\; t^2 \;\; t \;\; 1]\,M_B\begin{bmatrix} Q_0 \\ Q_1 \\ Q_2 \\ Q_3 \end{bmatrix} \quad (0\leqq t\leqq 1).$$

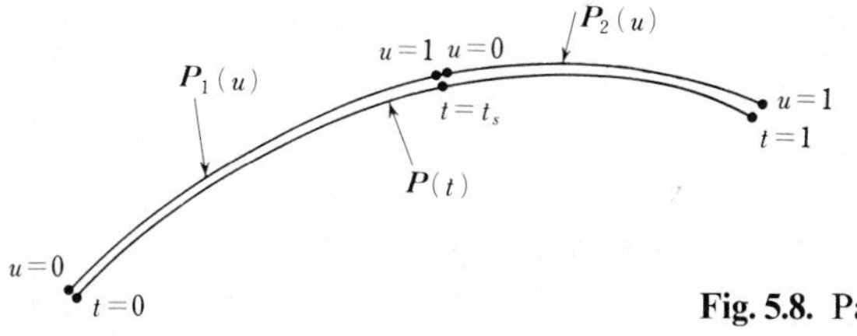

Fig. 5.8. Partitioning of a Bézier curve segment

Making the parameter transformation $u=t/t_s$ in the above equations, the part of the curve segment $\boldsymbol{P}(t)$ in the range $0 \leqq t \leqq t_s$ is now expressed as $\boldsymbol{P}_1(u)$ in the parameter range $0 \leqq u \leqq 1$:

$$\boldsymbol{P}_1(u)=[t_s^3 u^3 \;\; t_s^2 u^2 \;\; t_s u \;\; 1]\, \boldsymbol{M}_B \begin{bmatrix} \boldsymbol{Q}_0 \\ \boldsymbol{Q}_1 \\ \boldsymbol{Q}_2 \\ \boldsymbol{Q}_3 \end{bmatrix}$$

$$=[u^3 \;\; u^2 \;\; u \;\; 1] \begin{bmatrix} t_s^3 & 0 & 0 & 0 \\ 0 & t_s^2 & 0 & 0 \\ 0 & 0 & t_s & 0 \\ 0 & 0 & 0 & 1 \end{bmatrix} \boldsymbol{M}_B \begin{bmatrix} \boldsymbol{Q}_0 \\ \boldsymbol{Q}_1 \\ \boldsymbol{Q}_2 \\ \boldsymbol{Q}_3 \end{bmatrix}. \tag{5.34}$$

Since:

$$\begin{bmatrix} t_s^3 & 0 & 0 & 0 \\ 0 & t_s^2 & 0 & 0 \\ 0 & 0 & t_s & 0 \\ 0 & 0 & 0 & 1 \end{bmatrix} \boldsymbol{M}_B = \boldsymbol{M}_B \begin{bmatrix} 1 & 0 & 0 & 0 \\ 1-t_s & t_s & 0 & 0 \\ (1-t_s)^2 & 2(1-t_s)t_s & t_s^2 & 0 \\ (1-t_s)^3 & 3(1-t_s)^2 t_s & 3(1-t_s)t_s^2 & t_s^3 \end{bmatrix}$$

then (modifying so that the first matrix on the right side becomes $\boldsymbol{M}_B$):

$$\boldsymbol{P}_1(u)=[u^3 \;\; u^2 \;\; u \;\; 1]\, \boldsymbol{M}_B \begin{bmatrix} 1 & 0 & 0 & 0 \\ 1-t_s & t_s & 0 & 0 \\ (1-t_s)^2 & 2(1-t_s)t_s & t_s^2 & 0 \\ (1-t_s)^3 & 3(1-t_s)^2 t_s & 3(1-t_s)t_s^2 & t_s^3 \end{bmatrix} \begin{bmatrix} \boldsymbol{Q}_0 \\ \boldsymbol{Q}_1 \\ \boldsymbol{Q}_2 \\ \boldsymbol{Q}_3 \end{bmatrix}$$

$$=[u^3 \;\; u^2 \;\; u \;\; 1]\, \boldsymbol{M}_B \begin{bmatrix} \boldsymbol{Q}_0 \\ (1-t_s)\boldsymbol{Q}_0+t_s\boldsymbol{Q}_1 \\ (1-t_s)^2\boldsymbol{Q}_0+2(1-t_s)t_s\boldsymbol{Q}_1+t_s^2\boldsymbol{Q}_2 \\ (1-t_s)^3\boldsymbol{Q}_0+3(1-t_s)^2 t_s\boldsymbol{Q}_1+3(1-t_s)t_s^2\boldsymbol{Q}_2+t_s^3\boldsymbol{Q}_3 \end{bmatrix}$$

$$\equiv [u^3 \;\; u^2 \;\; u \;\; 1]\, \boldsymbol{M}_B \begin{bmatrix} \boldsymbol{Q}_0 \\ \boldsymbol{Q}_0^{[1]} \\ \boldsymbol{Q}_0^{[2]} \\ \boldsymbol{Q}_0^{[3]} \end{bmatrix} \tag{5.35}$$

where:

$$\left.\begin{aligned} &\boldsymbol{Q}_0^{[1]}=(1-t_s)\boldsymbol{Q}_0+t_s\boldsymbol{Q}_1, \;\; \boldsymbol{Q}_0^{[2]}=(1-t_s)^2\boldsymbol{Q}_0+2(1-t_s)t_s\boldsymbol{Q}_1+t_s^2\boldsymbol{Q}_2 \\ &\boldsymbol{Q}_0^{[3]}=(1-t_s)^3\boldsymbol{Q}_0+3(1-t_s)^2 t_s\boldsymbol{Q}_1+3(1-t_s)t_s^2\boldsymbol{Q}_2+t_s^3\boldsymbol{Q}_3 \end{aligned}\right\}. \tag{5.36}$$

From these equations, the vertex vectors of the curve defining polygon which expresses the part of the curve in the range $0 \leqq t \leqq t_s$ coincide with the vectors $\boldsymbol{Q}_0$, $\boldsymbol{Q}_0^{[1]}$, $\boldsymbol{Q}_0^{[2]}$ and $\boldsymbol{Q}_0^{[3]}$ found in the process of graphically determining the point $t = t_s$ on the curve (refer to Eqs. (5.29), (5.30) and (5.32)).

Next, find an expression $\boldsymbol{P}_2(u)$ which expresses the part of the curve $\boldsymbol{P}(t)$ in the range $t_s \leqq t \leqq 1$ in the parameter range $0 \leqq u \leqq 1$. To do this we perform the parameter transformation:

$$u = \frac{t - t_s}{1 - t_s}.$$

Substituting this into the formula for $\boldsymbol{P}(t)$ gives:

$$\boldsymbol{P}_2(u) = [\{(1-t_s)u + t_s\}^3 \;\; \{(1-t_s)u + t_s\}^2 \;\; (1-t_s)u + t_s \;\; 1]\, \boldsymbol{M}_B \begin{bmatrix} \boldsymbol{Q}_0 \\ \boldsymbol{Q}_1 \\ \boldsymbol{Q}_2 \\ \boldsymbol{Q}_3 \end{bmatrix}$$

$$= [u^3 \;\; u^2 \;\; u \;\; 1] \begin{bmatrix} (1-t_s)^3 & 0 & 0 & 0 \\ 3(1-t_s)^2 t_s & (1-t_s)^2 & 0 & 0 \\ 3(1-t_s)t_s^2 & 2(1-t_s)t_s & 1-t_s & 0 \\ t_s^3 & t_s^2 & t_s & 1 \end{bmatrix} \boldsymbol{M}_B \begin{bmatrix} \boldsymbol{Q}_0 \\ \boldsymbol{Q}_1 \\ \boldsymbol{Q}_2 \\ \boldsymbol{Q}_3 \end{bmatrix} \tag{5.37}$$

Since:

$$\begin{bmatrix} (1-t_s)^3 & 0 & 0 & 0 \\ 3(1-t_s)^2 t_s & (1-t_s)^2 & 0 & 0 \\ 3(1-t_s)t_s^2 & 2(1-t_s)t_s & 1-t_s & 0 \\ t_s^3 & t_s^2 & t_s & 1 \end{bmatrix} \boldsymbol{M}_B$$

$$= \boldsymbol{M}_B \begin{bmatrix} (1-t_s)^3 & 3(1-t_s)^2 t_s & 3(1-t_s)t_s^2 & t_s^3 \\ 0 & (1-t_s)^2 & 2(1-t_s)t_s & t_s^2 \\ 0 & 0 & 1-t_s & t_s \\ 0 & 0 & 0 & 1 \end{bmatrix}$$

then (modifying so that the first matrix on the right side becomes $\boldsymbol{M}_B$):

$$\boldsymbol{P}_2(u) = [u^3 \;\; u^2 \;\; u \;\; 1]\, \boldsymbol{M}_B \begin{bmatrix} (1-t_s)^3 & 3(1-t_s)^2 t_s & 3(1-t_s)t_s^2 & t_s^3 \\ 0 & (1-t_s)^2 & 2(1-t_s)t_s & t_s^2 \\ 0 & 0 & 1-t_s & t_s \\ 0 & 0 & 0 & 1 \end{bmatrix} \begin{bmatrix} \boldsymbol{Q}_0 \\ \boldsymbol{Q}_1 \\ \boldsymbol{Q}_2 \\ \boldsymbol{Q}_3 \end{bmatrix} =$$

$$= [u^3 \; u^2 \; u \; 1] \, \boldsymbol{M}_B \begin{bmatrix} (1-t_s)^3 \boldsymbol{Q}_0 + 3(1-t_s)^2 t_s \boldsymbol{Q}_1 + 3(1-t_s) t_s^2 \boldsymbol{Q}_2 + t_s^3 \boldsymbol{Q}_3 \\ (1-t_s)^2 \boldsymbol{Q}_1 + 2(1-t_s) t_s \boldsymbol{Q}_2 + t_s^2 \boldsymbol{Q}_3 \\ (1-t_s) \boldsymbol{Q}_2 + t_s \boldsymbol{Q}_3 \\ \boldsymbol{Q}_3 \end{bmatrix}$$

$$\equiv [u^3 \; u^2 \; u \; 1] \, \boldsymbol{M}_B \begin{bmatrix} \boldsymbol{Q}_0^{[3]} \\ \boldsymbol{Q}_1^{[2]} \\ \boldsymbol{Q}_2^{[1]} \\ \boldsymbol{Q}_3 \end{bmatrix} \tag{5.38}$$

where:

$$\left. \begin{array}{l} \boldsymbol{Q}_0^{[3]} = (1-t_s)^3 \boldsymbol{Q}_0 + 3(1-t_s)^2 t_s \boldsymbol{Q}_1 + 3(1-t_s) t_s^2 \boldsymbol{Q}_2 + t_s^3 \boldsymbol{Q}_3 \\ \boldsymbol{Q}_1^{[2]} = (1-t_s)^2 \boldsymbol{Q}_1 + 2(1-t_s) t_s \boldsymbol{Q}_2 + t_s^2 \boldsymbol{Q}_3, \;\; \boldsymbol{Q}_2^{[1]} = (1-t_s) \boldsymbol{Q}_2 + t_s \boldsymbol{Q}_3 \end{array} \right\} . \tag{5.39}$$

From these equations, we find that the vertex vectors of the curve defining polygon which expresses the part of the cubic Bézier curve segment $\boldsymbol{P}(t)$ in the range $t_s \leqq t \leqq 1$ coincide with the vectors $\boldsymbol{Q}_0^{[3]}$, $\boldsymbol{Q}_1^{[2]}$, $\boldsymbol{Q}_2^{[1]}$ and $\boldsymbol{Q}_3$ found in the process of graphically determining the point $t = t_s$ on the curve (refer to Eqs. (5.29), (5.30) and (5.32)).

5.1.3 Bézier Curve Segments

In Sect. 5.1.1, we derived a curve formula defined only by position vectors in order to solve some problems in the controllability of a Hermite-interpolated curve. Looking at the blending function $X_0(t)$, $X_1(t)$, $X_2(t)$, $X_3(t)$ of this curve formula, we see that these are just the terms of the binomial expansion $[(1-t)+t]^3$. That is to say, the newly derived curve is expressed as:

$$\boldsymbol{P}(t) = \sum_{i=0}^{3} X_i(t) \boldsymbol{Q}_i$$
$$= \sum_{i=0}^{3} \binom{3}{i} (1-t)^{3-i} t^i \boldsymbol{Q}_i .$$

If we replace 3 by n, we obtain the more general curve formula:

$$\boldsymbol{P}(t) = \sum_{i=0}^{n} \binom{n}{i} (1-t)^{n-i} t^i \boldsymbol{Q}_i$$
$$\equiv \sum_{i=0}^{n} B_{i,n}(t) \boldsymbol{Q}_i \tag{5.40}$$

where:

$$B_{i,n}(t) = \binom{n}{i} (1-t)^{n-i} t^i . \tag{5.41}$$

The $B_{i,n}(t)$ expresses the terms of the expansion $[(1-t)+t]^n$. This generalized curve formula (5.40) was proposed by P. Bézier, so it is called a Bézier curve segment. $B_{i,n}(t)$ is a Bernstein basis function. The Bézier curve segment for the case $n=8$ is shown in Fig. 5.9. Figures 5.10 through 5.20 show other examples of Bézier curve segments.

In a Bézier curve, note that if we take $n=2$ we obtain a parabola:

$$\boldsymbol{P}(t)=(1-t)^2\boldsymbol{Q}_0+2(1-t)t\boldsymbol{Q}_1+t^2\boldsymbol{Q}_2.$$

Refer to Sect. 7.3 for more details on parabolas.

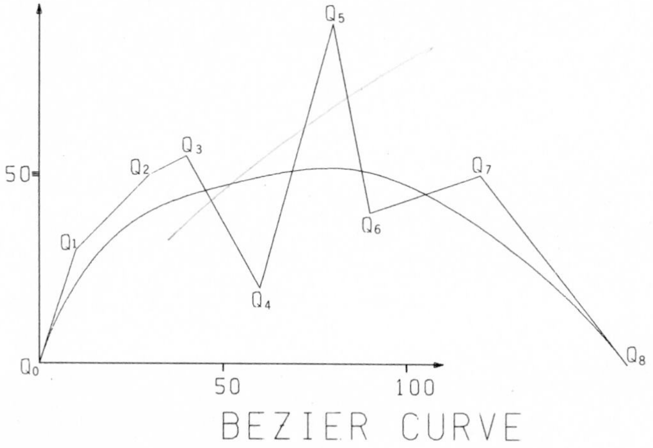

Fig. 5.9. A Bézier curve segment (1)

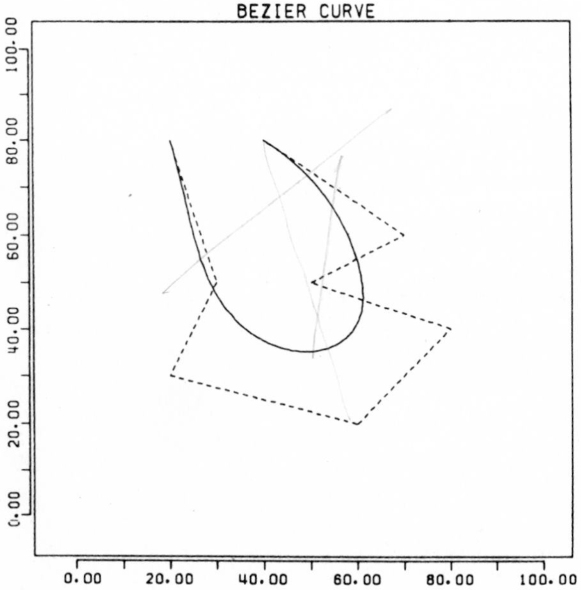

Fig. 5.10. A Bézier curve segment (2)

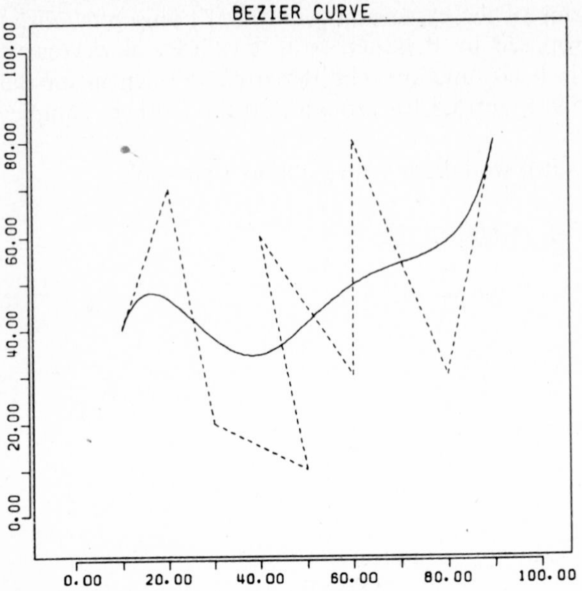

Fig. 5.11. A Bézier curve segment (3)

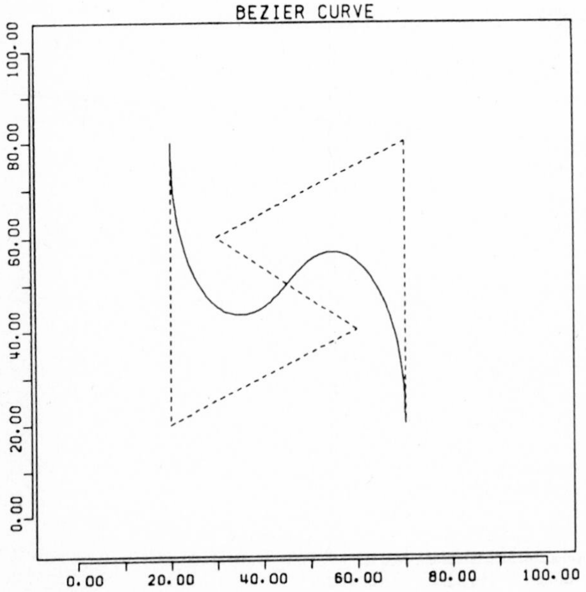

Fig. 5.12. A Bézier curve segment (4)

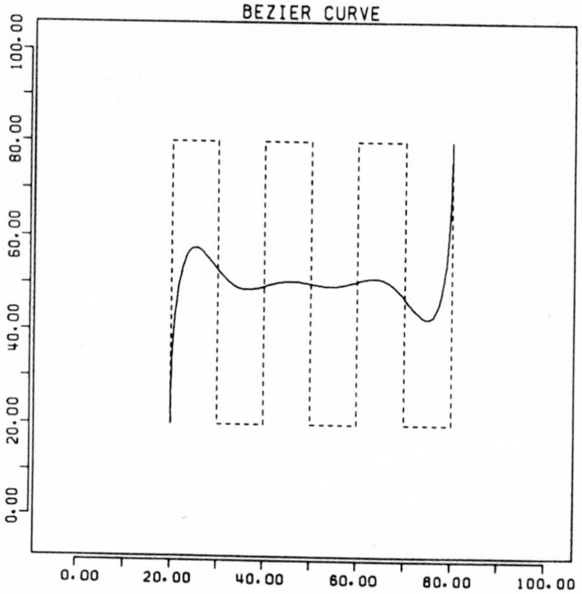

Fig. 5.13. A Bézier curve segment (5)

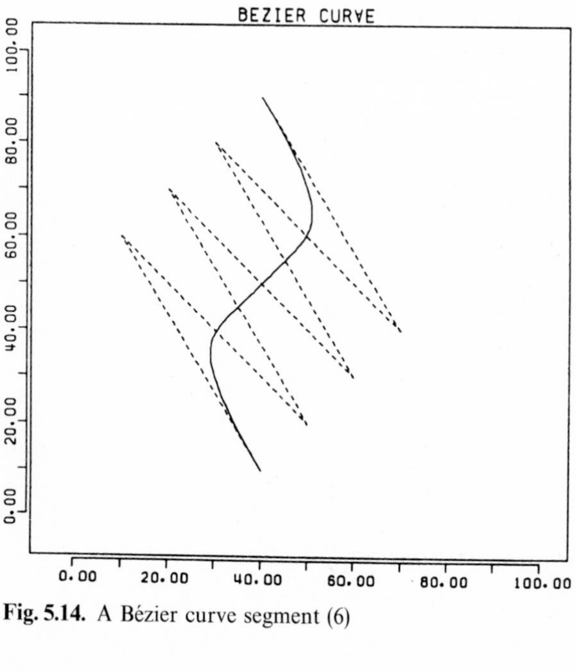

Fig. 5.14. A Bézier curve segment (6)

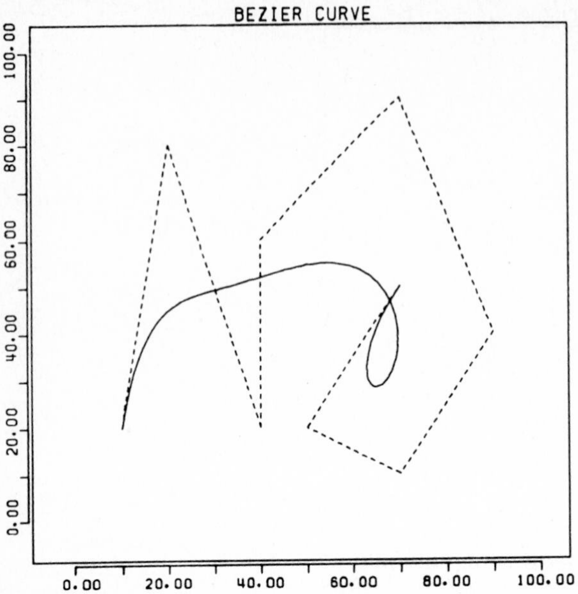

Fig. 5.15. A Bézier curve segment (7)

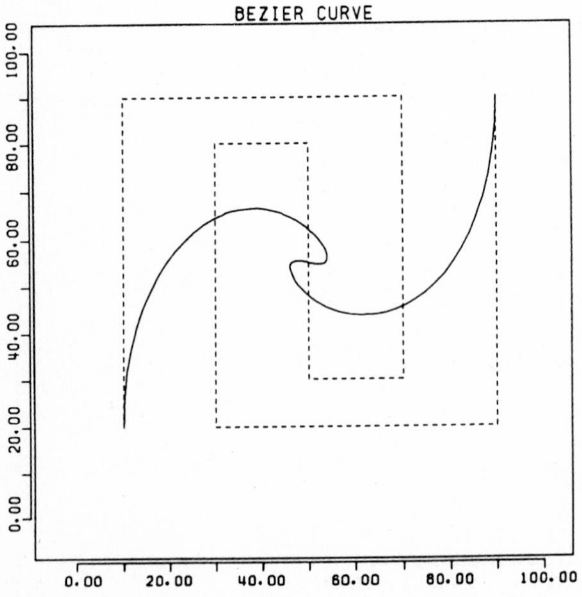

Fig. 5.16. A Bézier curve segment (8)

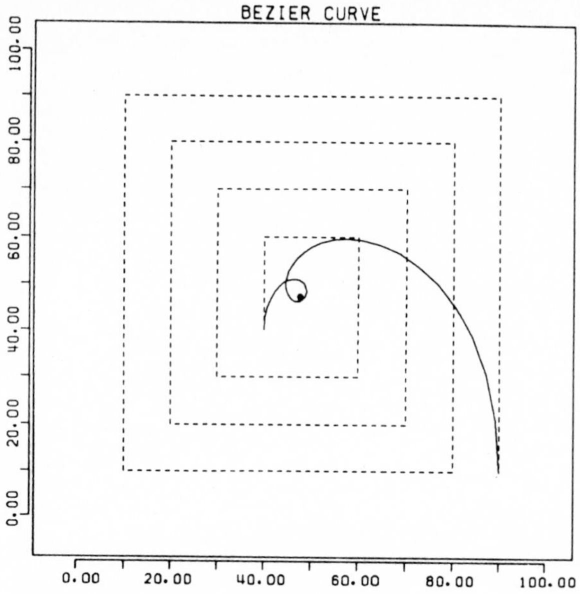

Fig. 5.17. A Bézier curve segment (9)

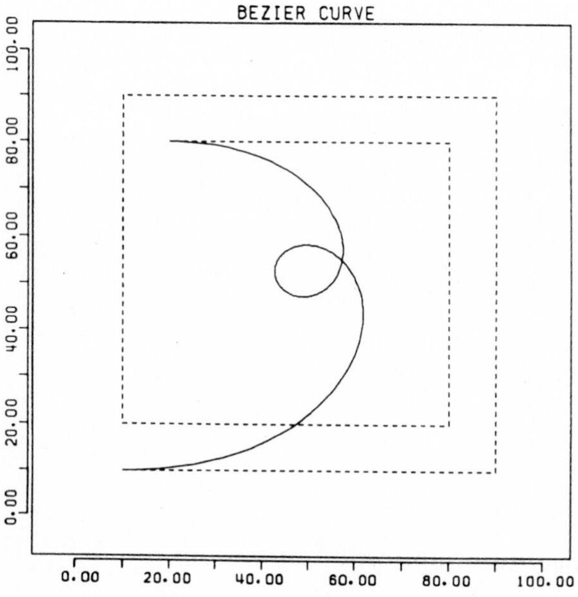

Fig. 5.18. A Bézier curve segment (10)

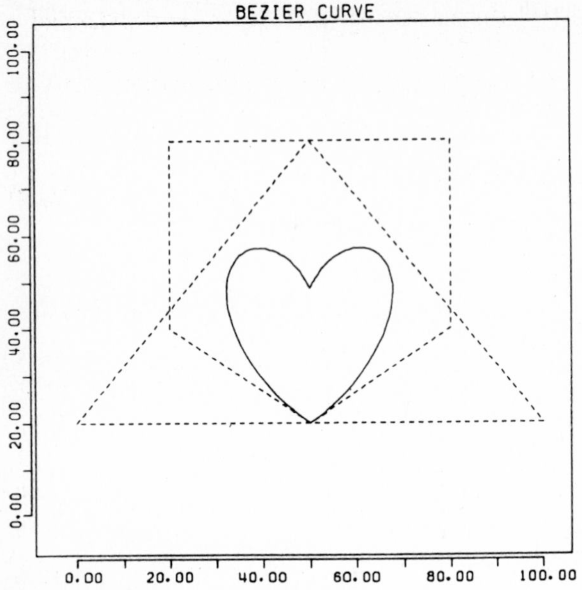

Fig. 5.19. A Bézier curve segment (11)

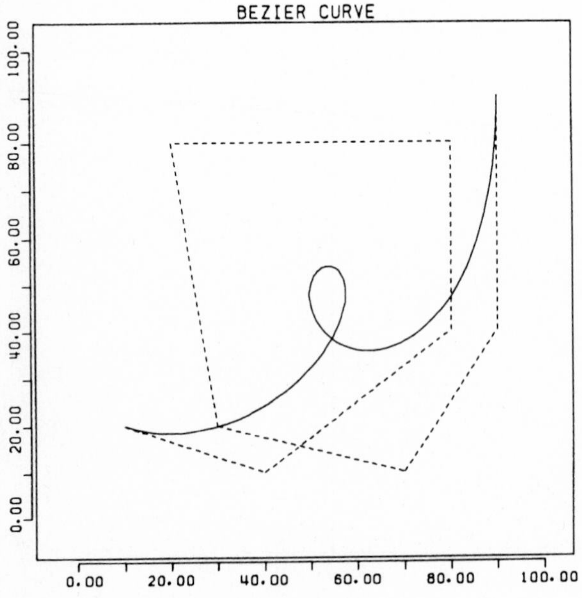

Fig. 5.20. A Bézier curve segment (12)

5.1.4 Properties of the Bernstein Basis Function and Bernstein Polynomial

A Bézier curve is based on a Bernstein polynomial approximation. An n-th order Bernstein polynomial is given by:

$$B_n(x;f)=\sum_{i=0}^{n} B_{i,n}(x)f\left(\frac{i}{n}\right) \tag{5.42}$$

where $f(x)$ $(x\in[0,1])$ is an arbitrary function.

Following are some important properties of Bernstein polynomials and Bernstein basis functions.

① For $x\in[0,1]$:

$$\left.\begin{array}{ll} B_{i,n}(x)\geqq 0 & (i=0,\ 1,\ \ldots,\ n) \\ B_{i,n}(x)<1 & (i=1,\ 2,\ \ldots,\ n-1) \\ B_{i,n}(x)\leqq 1 & (i=0,\ n) \end{array}\right\}. \tag{5.43}$$

Moreover:

$$B_{0,n}(0)=1,\ B_{0,n}(1)=0,\ B_{n,n}(0)=0,\ B_{n,n}(1)=1. \tag{5.44}$$

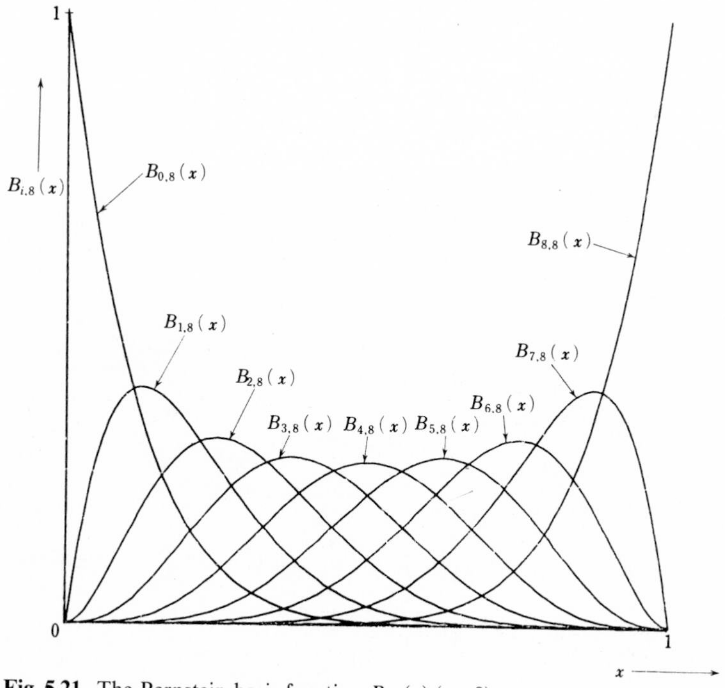

Fig. 5.21. The Bernstein basis function $B_{i,n}(x)$ $(n=8)$

From the above relations, in general the Bernstein polynomial approximation can be used to interpolate between $f(0)$ and $f(1)$. A graph of the Bernstein basis function $B_{i,n}(x)$ for the case $n=8$ is shown in Fig. 5.21.

② The derivatives of a Bernstein polynomial at $x=0$ and $x=1$ are given by the following formulas respectively:

at $x=0$:

$$\frac{d^r}{dx^r} B_n(x;f)\bigg|_{x=0} = \frac{n!}{(n-r)!} \sum_{j=0}^{r} (-1)^{r-j} \binom{r}{j} f\left(\frac{j}{n}\right) \tag{5.45}$$

at $x=1$:

$$\frac{d^r}{dx^r} B_n(x;f)\bigg|_{x=1} = \frac{n!}{(n-r)!} \sum_{j=0}^{r} (-1)^{j} \binom{r}{j} f\left(\frac{n-j}{n}\right). \tag{5.46}$$

From the above relations, the i-th derivative at one of the ends is determined by the value of $f(x)$ at the end and the i neighboring $f(x)$ values.

In particular, the 1st derivatives are given by:

$$\frac{d}{dx} B_n(x;f)\bigg|_{x=0} = n\left(f\left(\frac{1}{n}\right) - f(0)\right) \tag{5.47}$$

$$\frac{d}{dx} B_n(x;f)\bigg|_{x=1} = n\left(f(1) - f\left(\frac{n-1}{n}\right)\right). \tag{5.48}$$

From the above formulas, we see that the Bernstein polynomial functions at both ends are tangent to the straight lines joining the function values at the end points and at the neighboring interior points $f(1/n)$ and $f((n-1)/n)$.

③
$$\begin{aligned}\sum_{i=0}^{n} B_{i,n}(x) &= \sum_{i=0}^{n} \binom{n}{i} (1-x)^{n-i} x^i \\ &= [(1-x)+x]^n \\ &\equiv 1.\end{aligned} \tag{5.49}$$

That is, the Cauchy relation holds.

④ The maximum value of $B_{i,n}(x)$ occurs at $x=i/n$ $(i \neq 0, n)$, and its magnitude is:

$$B_{i,n}\left(\frac{i}{n}\right) = \binom{n}{i} \frac{i^i (n-i)^{n-i}}{n^n}. \tag{5.50}$$

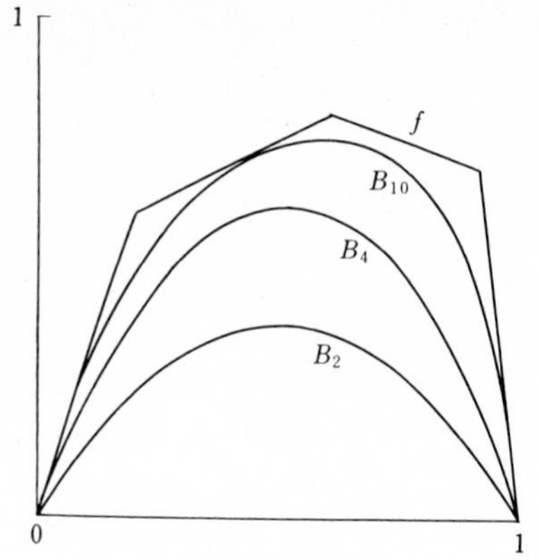

$$B_2(x\,;f)=\frac{3}{2}(x-x^2)$$

$$B_4(x\,;f)=\frac{5}{2}x-3x^2+\frac{3}{2}x^3-x^4$$

$$B_{10}(x\,;\,f)=3x-30x^3+105x^4-189x^5+210x^6-160x^7+90x^8-35x^9+6x^{10}$$

Fig. 5.22. Approximation of a function by Bernstein polynomials[19)]

⑤ **Theorem 5.1**[19,20)]. $f(x)$ is a bounded function on $x\in[0,1]$. At every point in $x\in[0,1]$ at which $f(x)$ is continuous, the following relation holds (refer to Fig. 5.22):

$$\lim_{n\to\infty} B_n(x;f)=f(x). \tag{5.51}$$

⑥ **Theorem 5.2**[19,20)]. Let $f^{(i)}(x)$ be between α_i and β_i:

$$\alpha_i \leqq f^{(i)}(x) \leqq \beta_i \quad x\in[0,1] \quad (i=0,1,\ldots,n). \tag{5.52}$$

Then the i-th derivative of its n-th order Bernstein polynomial satisfies the following inequalities:

$$\text{in the case } i=0: \qquad \alpha_0 \leqq B_n(x;f) \leqq \beta_0 \tag{5.53}$$

$$\text{in the case } 1\leqq i\leqq n: \quad \alpha_i \leqq \frac{n^i}{n(n-1)\ldots(n-i+1)} B_n^{(i)}(x;f) \leqq \beta_i. \tag{5.54}$$

What this theorem means is that for $i=0$, the Bernstein polynomial has values lying between the same maximum and minimum as those of $f(x)$ on $x\in[0,1]$. In general, if $f^{(i)}\geqq 0$ then $B_n^{(i)}(x;f)\geqq 0$. Consequently, for $i=1$, if $f(x)$ is monotonic $B_n(x;f)$ is also monotonic. For $i=2$, if $f(x)$ is convex (or concave) then $B_n(x;f)$ is also convex (or concave). That is, the Bernstein polynomial approximation expresses the overall form of $f(x)$ very well.

⑦ (Linearity). For arbitrary real numbers α and β, we have:

$$B_n(x;\alpha f+\beta g)=\alpha B_n(x;f)+\beta B_n(x;g). \tag{5.55}$$

⑧ (Bernstein approximation of linear functions)[21].

$$B_n(x;a+bx)=a+bx. \tag{5.56}$$

The Bernstein approximation of a linear function is the linear function itself.

⑨ (**Theorem 5.3:** Variation diminishing property)[21]. If f is a real-valued function defined in the interval $[0,1]$, we have:

$$\underset{0<x<1}{Z}(B_n(x;f)) \leqq v(f). \tag{5.57}$$

Here the left-hand side is the number of zeros of $B_n(x;f)$ in $x \in (0,1)$; the right-hand side is the number of sign changes of $f(x)$ in $x \in [0,1]$.

Proof: In the Bernstein polynomial $B_n(x;f)$, setting $z=x/(1-x)$ gives:

$$\frac{B_n(x;f)}{(1-x)^n}=\sum_{i=0}^{n} f\left(\frac{i}{n}\right)\binom{n}{i} z^i$$

so that we now have a polynomial in z in the range $0<z<\infty$ corresponding to $0<x<1$. Using Descartes' law of signs*) we obtain:

$$\begin{aligned}\underset{0<x<1}{Z}(B_n(x;f)) &= \underset{0<z<\infty}{Z}\left[\sum_{i=0}^{n} f\left(\frac{i}{n}\right)\binom{n}{i} z^i\right] \leqq v\left[f\left(\frac{i}{n}\right)\binom{n}{i}\right] \\ &= v\left[f\left(\frac{i}{n}\right)\right] \leqq v(f).\end{aligned} \tag{5.58}$$

Since we also have the relation:

$$v(B_n(x;f)) \leqq Z(B_n(x;f))$$

from this theorem the following inequality holds:

$$v(B_n(x;f)) \leqq v(f). \tag{5.59}$$

From properties ⑦ and ⑧ of a Bernstein polynomial, it follows that:

$$B_n(x;f(x)-a-bx)=B_n(x;f)-a-bx.$$

---

*) The number of positive roots of an n-th order real polynomial

$$f(x)=a_n x^n + a_{n-1} x^{n-1} + \ldots + a_0$$

is an even number less than the number of sign changes of its coefficients $a_n, a_{n-1}, \ldots, a_0$. "Even number" means 0 or a positive even number, and a root of multiplicity k is counted as k roots. This is called Descartes' law of signs.

Applying relation (5.59) to $f(x)-a-bx$, the following relation is obtained:

$$v[B_n(x;f)-a-bx] \leqq v[f(x)-a-bx]. \tag{5.60}$$

The geometrical significance of inequality (5.60) is that the number of intersections of $y=ax+b$ with $y=B_n(x;f)$ does not exceed the number of intersections of $y=ax+b$ with $y=f(x)$. This property is called the variation diminishing property of a Bernstein polynomial. That is to say, if a Bernstein approximation is applied to a certain function a smoothing effect is obtained. If the Bernstein approximation is applied repeatedly, the smoothing effect is repeated, until, in the limiting case of an infinite number of repetitions, the function reduces to the straight line joining $f(0)$ and $f(1)$.

5.1.5 Various Representations for Bézier Curve Segments

Let us try to express a Bézier curve segment by a formula other than (5.40).

① First, let us express it as a sum of powers of t.

$$\begin{aligned}
\boldsymbol{P}(t) &= \sum_{i=0}^{n} \binom{n}{i} (1-t)^{n-i} t^i \boldsymbol{Q}_i \\
&= \sum_{i=0}^{n} \boldsymbol{Q}_i \binom{n}{i} t^i \sum_{j=0}^{n-i} \binom{n-i}{j} (-1)^j t^j \\
&= \sum_{i=0}^{n} \boldsymbol{Q}_i \binom{n}{i} \sum_{j=0}^{n-i} \binom{n-i}{j} (-1)^j t^{i+j} \\
&= \sum_{i=0}^{n} \sum_{s=i}^{n} \boldsymbol{Q}_i \binom{n}{i} \binom{n-i}{s-i} (-1)^{s-i} t^s \\
&= \sum_{i=0}^{n} \sum_{s=0}^{n} \boldsymbol{Q}_i \binom{n}{s} \binom{s}{i} (-1)^{s-i} t^s \\
&= \sum_{s=0}^{n} t^s \sum_{i=0}^{s} (-1)^{s-i} \binom{n}{s} \binom{s}{i} \boldsymbol{Q}_i \\
&= \sum_{i=0}^{n} t^i \left[\sum_{j=0}^{i} (-1)^{i-j} \binom{n}{i} \binom{i}{j} \boldsymbol{Q}_j \right].
\end{aligned} \tag{5.61}$$

② Next, let us try to use an operator to express the Bézier curve segment. First, introduce the shift operator E to $\boldsymbol{Q}_0, \boldsymbol{Q}_1, \ldots, \boldsymbol{Q}_n$ of the curve defining polygon[22]:

$$\boldsymbol{Q}_i = E\boldsymbol{Q}_{i-1}. \tag{5.62}$$

From this we have:

$$\boldsymbol{Q}_i = E^i \boldsymbol{Q}_0 . \tag{5.63}$$

Substituting this relation into (5.40), we obtain:

$$\begin{aligned}\boldsymbol{P}(t) &= \sum_{i=0}^{n} \binom{n}{i} (1-t)^{n-i} t^i E^i \boldsymbol{Q}_0 = \left[\sum_{i=0}^{n} \binom{n}{i} (1-t)^{n-i} (tE)^i \right] \boldsymbol{Q}_0 \\ &= (1-t+tE)^n \boldsymbol{Q}_0 . \end{aligned} \tag{5.64}$$

Next, use the forward difference operator Δ to express the Bézier curve segment. From the definition of the forward difference operator, we have:

$$\Delta \boldsymbol{Q}_i = \boldsymbol{Q}_{i+1} - \boldsymbol{Q}_i .$$

From Eq. (5.62), we have:

$$\begin{aligned}&\Delta \boldsymbol{Q}_i = E\boldsymbol{Q}_i - \boldsymbol{Q}_i = (E-1)\boldsymbol{Q}_i \\ &\therefore \quad \Delta = E - 1 . \end{aligned} \tag{5.65}$$

From this equation, $E = \Delta + 1$. Substituting this into (5.64) gives:

$$\begin{aligned}\boldsymbol{P}(t) &= [1 - t + t(\Delta + 1)]^n \boldsymbol{Q}_0 = (1 + t\Delta)^n \boldsymbol{Q}_0 \\ &= \sum_{i=0}^{n} \binom{n}{i} (t\Delta)^i \boldsymbol{Q}_0 \\ &= \sum_{i=0}^{n} \binom{n}{i} t^i \Delta^i \boldsymbol{Q}_0 . \end{aligned} \tag{5.66}$$

This finite difference expression (5.66) of a Bézier curve segment can also be found immediately from (5.61) as follows. When a sequence of vectors $\boldsymbol{Q}_0$, $\boldsymbol{Q}_1$, $\boldsymbol{Q}_2$, ... is given, the differences of different orders at $\boldsymbol{Q}_k$ are:

$$\begin{aligned}\Delta \boldsymbol{Q}_k &= \boldsymbol{Q}_{k+1} - \boldsymbol{Q}_k \\ \Delta^2 \boldsymbol{Q}_k &= \Delta \boldsymbol{Q}_{k+1} - \Delta \boldsymbol{Q}_k \\ &= \boldsymbol{Q}_{k+2} - \boldsymbol{Q}_{k+1} - (\boldsymbol{Q}_{k+1} - \boldsymbol{Q}_k) \\ &= \boldsymbol{Q}_{k+2} - 2\boldsymbol{Q}_{k+1} + \boldsymbol{Q}_k \\ \Delta^3 \boldsymbol{Q}_k &= \Delta^2 \boldsymbol{Q}_{k+1} - \Delta^2 \boldsymbol{Q}_k \\ &= \boldsymbol{Q}_{k+3} - 2\boldsymbol{Q}_{k+2} + \boldsymbol{Q}_{k+1} - (\boldsymbol{Q}_{k+2} - 2\boldsymbol{Q}_{k+1} + \boldsymbol{Q}_k) \\ &= \boldsymbol{Q}_{k+3} - 3\boldsymbol{Q}_{k+2} + 3\boldsymbol{Q}_{k+1} - \boldsymbol{Q}_k \\ &= \sum_{j=0}^{3} (-1)^{3-j} \binom{3}{j} \boldsymbol{Q}_{k+j} \\ &\cdots \\ &\cdots \end{aligned}$$

Therefore, in general the i-th order finite difference is:

$$\Delta^i \boldsymbol{Q}_k = \sum_{j=0}^{i} (-1)^{i-j} \binom{i}{j} \boldsymbol{Q}_{k+j}. \tag{5.67}$$

In particular, if $k=0$ we have:

$$\Delta^i \boldsymbol{Q}_0 = \sum_{j=0}^{i} (-1)^{i-j} \binom{i}{j} \boldsymbol{Q}_j. \tag{5.68}$$

Then, from Eq. (5.61):

$$\begin{aligned}
\boldsymbol{P}(t) &= \sum_{i=0}^{n} t^i \sum_{j=0}^{i} (-1)^{i-j} \binom{n}{i} \binom{i}{j} \boldsymbol{Q}_j \\
&= \sum_{i=0}^{n} \binom{n}{i} t^i \left[\sum_{j=0}^{i} (-1)^{i-j} \binom{i}{j} \boldsymbol{Q}_j \right] \\
&= \sum_{i=0}^{n} \binom{n}{i} t^i \Delta^i \boldsymbol{Q}_0 .
\end{aligned}$$

③ Next, consider expressing a Bézier curve segment as a product of power-of-t basis vectors[23]:

$$\begin{aligned}
\boldsymbol{P}(t) &= [B_{0,n}(t) \;\; B_{1,n}(t) \;\; \ldots \;\; B_{n,n}(t)] \begin{bmatrix} \boldsymbol{Q}_0 \\ \boldsymbol{Q}_1 \\ \vdots \\ \boldsymbol{Q}_n \end{bmatrix} \\
&= [t^n \;\; t^{n-1} \;\; \ldots \;\; t \;\; 1] \, \beta \begin{bmatrix} \boldsymbol{Q}_0 \\ \boldsymbol{Q}_1 \\ \vdots \\ \boldsymbol{Q}_n \end{bmatrix}
\end{aligned} \tag{5.69}$$

and find the $(n+1) \times (n+1)$ matrix β for this expression to hold. β is the matrix that transforms power basis vectors to Bernstein basis vectors. Using Eq. (5.61), we obtain:

$$\boldsymbol{P}(t) = \sum_{i=0}^{n} t^i \sum_{j=0}^{i} (-1)^{i-j} \binom{n}{i} \binom{i}{j} \boldsymbol{Q}_j =$$

$$
=[t^n \; t^{n-1} \; \ldots \; t \; 1]\begin{bmatrix}
\sum_{j=0}^{n}(-1)^{n-j}\binom{n}{n}\binom{n}{j}\boldsymbol{Q}_j \\
\sum_{j=0}^{n-1}(-1)^{n-1-j}\binom{n}{n-1}\binom{n-1}{j}\boldsymbol{Q}_j \\
\vdots \\
\sum_{j=0}^{1}(-1)^{1-j}\binom{n}{1}\binom{1}{j}\boldsymbol{Q}_j \\
\sum_{j=0}^{0}(-1)^{0-j}\binom{n}{0}\binom{0}{j}\boldsymbol{Q}_j
\end{bmatrix}
$$

$$
=[t^n \; t^{n-1} \; \ldots \; t \; 1]
$$

$$
\times\begin{bmatrix}
(-1)^n\binom{n}{n}\binom{n}{0} & (-1)^{n-1}\binom{n}{n}\binom{n}{1} \ldots & (-1)^0\binom{n}{n}\binom{n}{n} \\
(-1)^{n-1}\binom{n}{n-1}\binom{n-1}{0} & (-1)^{n-2}\binom{n}{n-1}\binom{n-1}{1} \ldots & 0 \\
\vdots & \vdots & \vdots \\
(-1)^1\binom{n}{1}\binom{1}{0} & (-1)^0\binom{n}{1}\binom{1}{1} \ldots & \ldots \\
(-1)^0\binom{n}{0}\binom{0}{0} & 0 \ldots & 0
\end{bmatrix}\begin{bmatrix}\boldsymbol{Q}_0 \\ \boldsymbol{Q}_1 \\ \vdots \\ \boldsymbol{Q}_n\end{bmatrix}. \tag{5.70}
$$

The matrix β is:

$$
\beta=[b_{ij}]_{i,j=0}^{n} \tag{5.71}
$$

where:

$$
b_{i,j}=\begin{cases}(-1)^{n-i-j}\binom{n}{n-i}\binom{n-i}{j} & (0\leqq i+j\leqq n) \\ 0 & (\text{otherwise})\end{cases} \tag{5.72}
$$

④ The following simple relationship exists among Bernstein basis functions:

$$
\begin{aligned}
&tB_{i-1,r-1}(t)+(1-t)\,B_{i,r-1}(t) \\
&\quad=t\binom{r-1}{i-1}(1-t)^{r-i}t^{i-1}+(1-t)\binom{r-1}{i}(1-t)^{r-1-i}t^i \\
&\quad=\left[\binom{r-1}{i-1}+\binom{r-1}{i}\right](1-t)^{r-i}t^i=
\end{aligned}
$$

$$= \left[\frac{(r-1)!}{(i-1)!(r-i)!} + \frac{(r-1)!}{i!(r-i-1)!} \right] (1-t)^{r-i} t^i$$

$$= \frac{r!}{i!(r-i)!} (1-t)^{r-i} t^i$$

$$\equiv B_{i,r}(t)$$

which implies:

$$B_{i,r}(t) = t\, B_{i-1,r-1}(t) + (1-t)\, B_{i,r-1}(t) \tag{5.73}$$

$$= B_{1,1}(t)\, B_{i-1,r-1}(t) + B_{0,1}(t)\, B_{i,r-1}(t) \tag{5.74}$$

$$(1 \leqq r \leqq n;\ 0 \leqq i \leqq r).$$

We also have:

$$B_{-1,r-1}(t) = B_{r,r-1}(t) = 0. \tag{5.75}$$

This relation is shown in Fig. 5.23. As is clear from the figure, the n-th degree Bernstein basis function $B_{i,n}(t)$ can be expressed in terms of the $(n-1)$-st degree Bernstein basis function $B_{i-1,n-1}(t)$ and $B_{i,n-1}(t)$. Similarly, the $(n-1)$-st degree Bernstein basis functions can be expressed in terms of $(n-2)$-nd degree Bernstein basis functions. In general, an n-th degree Bernstein basis function can be expressed in terms of $(n-k)$-th degree Bernstein basis functions. If this relation among Bernstein basis functions is applied to Bézier curves, an n-th degree Bézier curve segment can be expressed in terms of $(n-k)$-th degree Bernstein basis functions:

$\left[(1-i/m) B_i^m + \frac{i+1}{m} B_{i+1}^m = B_i^{m-1} \right]$

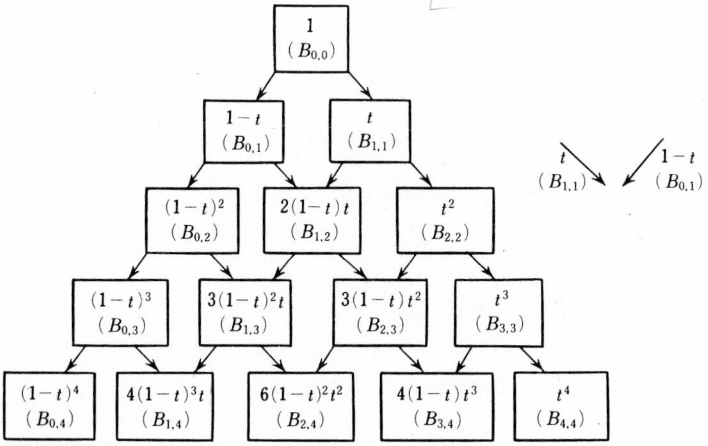

Fig. 5.23. Relation among Bernstein basis functions

$$\boldsymbol{P}(t)=\sum_{i=0}^{n-k} B_{i,n-k}(t)\,\boldsymbol{Q}_i^{[k]}(t) \tag{5.76}$$

where:

$$\boldsymbol{Q}_i^{[k]}(t)=\sum_{j=0}^{k} B_{j,k}(t)\,\boldsymbol{Q}_{i+j}. \tag{5.77}$$

As an example, consider such an expression of a Bézier curve segment for the case $n=4$.

$$\begin{aligned}
\boldsymbol{P}(t) &= B_{0,4}\boldsymbol{Q}_0+B_{1,4}\boldsymbol{Q}_1+B_{2,4}\boldsymbol{Q}_2+B_{3,4}\boldsymbol{Q}_3+B_{4,4}\boldsymbol{Q}_4\\
&= B_{0,1}B_{0,3}\boldsymbol{Q}_0+(B_{1,1}B_{0,3}+B_{0,1}B_{1,3})\boldsymbol{Q}_1+(B_{1,1}B_{1,3}+B_{0,1}B_{2,3})\boldsymbol{Q}_2\\
&\quad +(B_{1,1}B_{2,3}+B_{0,1}B_{3,3})\boldsymbol{Q}_3+B_{1,1}B_{3,3}\boldsymbol{Q}_4\\
&= B_{0,3}(B_{0,1}\boldsymbol{Q}_0+B_{1,1}\boldsymbol{Q}_1)+B_{1,3}(B_{0,1}\boldsymbol{Q}_1+B_{1,1}\boldsymbol{Q}_2)+B_{2,3}(B_{0,1}\boldsymbol{Q}_2\\
&\quad +B_{1,1}\boldsymbol{Q}_3)+B_{3,3}(B_{0,1}\boldsymbol{Q}_3+B_{1,1}\boldsymbol{Q}_4)\\
&= \sum_{i=0}^{4-1} B_{i,4-1}\left(\sum_{j=0}^{1} B_{j,1}\boldsymbol{Q}_{i+j}\right)\\
&= B_{0,1}B_{0,2}(B_{0,1}\boldsymbol{Q}_0+B_{1,1}\boldsymbol{Q}_1)+(B_{1,1}B_{0,2}+B_{0,1}B_{1,2})(B_{0,1}\boldsymbol{Q}_1+B_{1,1}\boldsymbol{Q}_2)\\
&\quad +(B_{1,1}B_{1,2}+B_{0,1}B_{2,2})(B_{0,1}\boldsymbol{Q}_2+B_{1,1}\boldsymbol{Q}_3)\\
&\quad +B_{1,1}B_{2,2}(B_{0,1}\boldsymbol{Q}_3+B_{1,1}\boldsymbol{Q}_4)\\
&= B_{0,2}(B_{0,1}^2\boldsymbol{Q}_0+B_{0,1}B_{1,1}\boldsymbol{Q}_1+B_{0,1}B_{1,1}\boldsymbol{Q}_1+B_{1,1}^2\boldsymbol{Q}_2)\\
&\quad +B_{1,2}(B_{0,1}^2\boldsymbol{Q}_1+B_{0,1}B_{1,1}\boldsymbol{Q}_2+B_{0,1}B_{1,1}\boldsymbol{Q}_2+B_{1,1}^2\boldsymbol{Q}_3)\\
&\quad +B_{2,2}(B_{0,1}^2\boldsymbol{Q}_2+B_{0,1}B_{1,1}\boldsymbol{Q}_3+B_{0,1}B_{1,1}\boldsymbol{Q}_3+B_{1,1}^2\boldsymbol{Q}_4)\\
&= B_{0,2}(B_{0,2}\boldsymbol{Q}_0+B_{1,2}\boldsymbol{Q}_1+B_{2,2}\boldsymbol{Q}_2)\\
&\quad +B_{1,2}(B_{0,2}\boldsymbol{Q}_1+B_{1,2}\boldsymbol{Q}_2+B_{2,2}\boldsymbol{Q}_3)+B_{2,2}(B_{0,2}\boldsymbol{Q}_2+B_{1,2}\boldsymbol{Q}_3+B_{2,2}\boldsymbol{Q}_4)\\
&= \sum_{i=0}^{n-k} B_{i,n-k}(t)\left(\sum_{j=0}^{k} B_{j,k}(t)\boldsymbol{Q}_{i+j}\right) \quad (n=4,\ k=2)
\end{aligned}$$

5.1.6 Derivative Vectors of Bézier Curve Segments

The r-th derivative vector of a Bézier curve segment can be expressed in terms of finite differences as:

$$\frac{d^r}{dt^r}\boldsymbol{P}(t)=\frac{n!}{(n-r)!}\sum_{i=0}^{n-r} B_{i,n-r}(t)\Delta^r\boldsymbol{Q}_i. \tag{5.78}$$

The derivative vectors at the ends $t=0$ and $t=1$ are:

$$\text{at } t=0: \quad \frac{d^r}{dt^r}\boldsymbol{P}(0)=\frac{n!}{(n-r)!}\,\Delta^r\boldsymbol{Q}_0 \tag{5.79}$$

$$\text{at } t=1: \qquad \frac{d^r}{dt^r}\boldsymbol{P}(1)=\frac{n!}{(n-r)!}\,\Delta^r\boldsymbol{Q}_{n-r}. \tag{5.80}$$

From Eq. (5.79), the r-th derivative vector at $t=0$ is determined in terms of the $(r+1)$ position vectors $\boldsymbol{Q}_0, \boldsymbol{Q}_1, \ldots, \boldsymbol{Q}_r$. Similarly, from Eq. (5.80) the r-th derivative vector at $t=1$ is determined in terms of the $(r+1)$ position vectors $\boldsymbol{Q}_{n-r}, \boldsymbol{Q}_{n-r+1}, \ldots, \boldsymbol{Q}_n$.

In particular, in the case $r=1$, that is, the case of the tangent vectors, we have:

$$\text{at } t=0: \qquad \dot{\boldsymbol{P}}(0)=n(\boldsymbol{Q}_1-\boldsymbol{Q}_0) \tag{5.81}$$

$$\text{at } t=1: \qquad \dot{\boldsymbol{P}}(1)=n(\boldsymbol{Q}_n-\boldsymbol{Q}_{n-1}). \tag{5.82}$$

The slope of the curve at the starting point is in the direction of side $\overrightarrow{\boldsymbol{Q}_0\boldsymbol{Q}_1}$; the slope at the end point is in the direction of side $\overrightarrow{\boldsymbol{Q}_{n-1}\boldsymbol{Q}_n}$.

In the case $r=2$, the equations become as follows.

$$\text{at } t=0: \qquad \ddot{\boldsymbol{P}}(0)=n(n-1)\,(\boldsymbol{Q}_2-2\boldsymbol{Q}_1+\boldsymbol{Q}_0) \tag{5.83}$$

$$\text{at } t=1: \qquad \ddot{\boldsymbol{P}}(1)=n(n-1)\,(\boldsymbol{Q}_n-2\boldsymbol{Q}_{n-1}+\boldsymbol{Q}_{n-2}). \tag{5.84}$$

5.1.7 Determination of a Point on a Curve Segment by Linear Operations

Applying relation (5.73) to Eq. (5.77) and rearranging, we obtain:

$$\begin{aligned}
\boldsymbol{Q}_i^{[k]}(t) &= \sum_{j=0}^{k} B_{j,k}(t)\,\boldsymbol{Q}_{i+j} \\
&= \sum_{j=0}^{k} \{t\,B_{j-1,k-1}(t)+(1-t)\,B_{j,k-1}(t)\}\,\boldsymbol{Q}_{i+j} \\
&= t\sum_{j=0}^{k} B_{j-1,k-1}(t)\,\boldsymbol{Q}_{i+j}+(1-t)\sum_{j=0}^{k} B_{j,k-1}(t)\,\boldsymbol{Q}_{i+j} \\
&= t\sum_{j=0}^{k-1} B_{j,k-1}(t)\,\boldsymbol{Q}_{i+j+1}+(1-t)\sum_{j=0}^{k-1} B_{j,k-1}(t)\,\boldsymbol{Q}_{i+j} \\
&\qquad\qquad (\because\ B_{-1,k-1}(t)=B_{k,k-1}(t)=0) \\
&= t\,\boldsymbol{Q}_{i+1}^{[k-1]}(t)+(1-t)\,\boldsymbol{Q}_i^{[k-1]}(t).
\end{aligned}$$

This implies that the following relation holds:

$$\boldsymbol{Q}_i^{[k]}(t)=(1-t)\,\boldsymbol{Q}_i^{[k-1]}(t)+t\,\boldsymbol{Q}_{i+1}^{[k-1]}(t). \tag{5.85}$$

Setting $k=n$ and $i=0$ in Eq. (5.77) gives:

$$Q_0^{[n]}(t)=\sum_{j=0}^{n} B_{j,n}(t)\,Q_j \equiv P(t). \tag{5.86}$$

Also, setting $k=0$ in Eq. (5.77) gives:

$$Q_i^{[0]}(t)=\sum_{j=0}^{0} B_{j,0}(t)\,Q_{i+j}=Q_i \quad (i=0,1,\ldots,n). \tag{5.87}$$

Using the above relations, a point on the curve $P(t)$ can be found by repeating simple calculations. This algorithm is as follows.

Algorithm

We are to find a point $P(t_s)$ on a curve where the parameter takes the value $t=t_s$ (refer to Fig. 5.24).

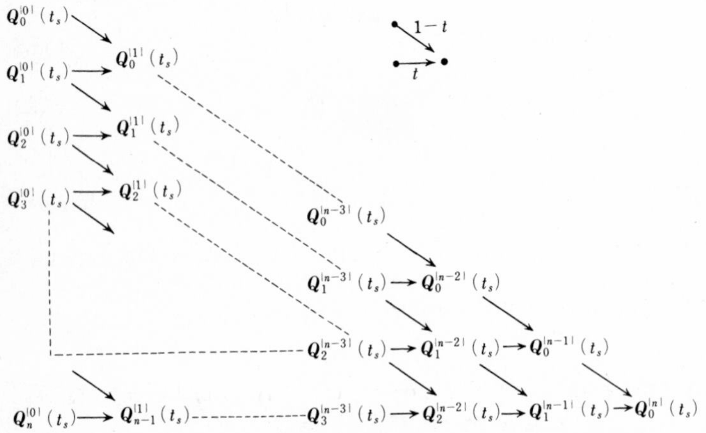

Fig. 5.24. Determination of a point on a Bézier curve segment by linear calculations

Step 1:

§7. de Casteljau.

$$Q_i^{[0]}(t_s)=Q_i \quad (i=0,1,\ldots,n). \tag{5.88}$$

Step 2:

Repeatedly apply:

$$Q_i^{[k]}(t_s)=(1-t_s)\,Q_i^{[k-1]}(t_s)+t_s\,Q_{i+1}^{[k-1]}(t_s) \tag{5.89}$$

for $k=1,2,\ldots,n$, to find:

$$Q_0^{[n]}(t_s)=P(t_s). \tag{5.90}$$

This algorithm can be easily carried out graphically. First draw the curve defining polygon with vertices $\boldsymbol{Q}_i^{[0]}(t_s) = \boldsymbol{Q}_i$ $(i = 0, 1, \ldots, n)$. Next, find the points which divide the polygon sides in the ratio $t_s : 1 - t_s$; call those points $\boldsymbol{Q}_i^{[1]}(t_s)$ $(i = 0, 1, \ldots, n-1)$. These new points form a new polygon. Again find the points $\boldsymbol{Q}_i^{[2]}(t_s)$ $(i = 0, \ldots, n-2)$ which divide the sides of this new polygon in the ratio $t_s : 1 - t_s$. Again a new polygon is formed. Finally, when the new polygon reduces to the line segment joining $\boldsymbol{Q}_0^{[n-1]}(t_s)$ and $\boldsymbol{Q}_1^{[n-1]}(t_s)$, find the point $\boldsymbol{Q}_0^{[n]}(t_s)$ that divides this line segment in the ratio $t_s : 1 - t_s$; it is the point $\boldsymbol{P}(t_s)$ (refer to Fig. 5.25).

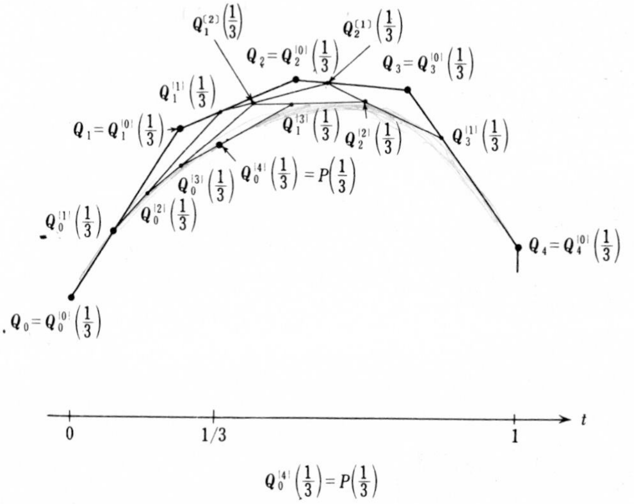

Fig. 5.25. Determination of the point $\boldsymbol{P}(\frac{1}{3})$ (Example 5.1)

Let us take a closer look at the vector $\boldsymbol{Q}_1^{[n-1]}(t_s) - \boldsymbol{Q}_0^{[n-1]}(t_s)$. Setting $k = n-1$ and $i = 0, 1$ in Eq. (5.77) we obtain the following.

$$
\begin{aligned}
\boldsymbol{Q}_1^{[n-1]}(t_s) - \boldsymbol{Q}_0^{[n-1]}(t_s) &= \sum_{j=0}^{n-1} B_{j,n-1}(t_s)\boldsymbol{Q}_{j+1} - \sum_{j=0}^{n-1} B_{j,n-1}(t_s)\boldsymbol{Q}_j \\
&= \sum_{j=0}^{n-1} B_{j,n-1}(t_s)\,(\boldsymbol{Q}_{j+1} - \boldsymbol{Q}_j) \\
&= \frac{1}{n}\left[\frac{n!}{(n-1)!}\sum_{j=0}^{n-1} B_{j,n-1}(t_s)\Delta\boldsymbol{Q}_j\right] \\
&\equiv \frac{1}{n}\,\dot{\boldsymbol{P}}(t_s).
\end{aligned}
\tag{5.91}
$$

That is, among the polygons which are generated in sequence in order to find a point on a curve, the final line segment coincides with the tangent at the point $\boldsymbol{P}(t_s)$ that is being found.

Example 5.1. On the Bézier curve $\boldsymbol{P}(t)=\sum_{i=0}^{4} B_{i,4}(t)\boldsymbol{Q}_i$, find the point $\boldsymbol{P}\left(\frac{1}{3}\right)$ using the above algorithm.

Solution.

Taking $k=1$, $\boldsymbol{Q}_0^{[1]}\left(\frac{1}{3}\right)$, $\boldsymbol{Q}_1^{[1]}\left(\frac{1}{3}\right)$, $\boldsymbol{Q}_2^{[1]}\left(\frac{1}{3}\right)$, $\boldsymbol{Q}_3^{[1]}\left(\frac{1}{3}\right)$ become:

$$\boldsymbol{Q}_0^{[1]}\left(\frac{1}{3}\right)=\frac{2}{3}\,\boldsymbol{Q}_0^{[0]}\left(\frac{1}{3}\right)+\frac{1}{3}\,\boldsymbol{Q}_1^{[0]}\left(\frac{1}{3}\right)$$

$$=\frac{2}{3}\,\boldsymbol{Q}_0+\frac{1}{3}\,\boldsymbol{Q}_1$$

$$\boldsymbol{Q}_1^{[1]}\left(\frac{1}{3}\right)=\frac{2}{3}\,\boldsymbol{Q}_1^{[0]}\left(\frac{1}{3}\right)+\frac{1}{3}\,\boldsymbol{Q}_2^{[0]}\left(\frac{1}{3}\right)$$

$$=\frac{2}{3}\,\boldsymbol{Q}_1+\frac{1}{3}\,\boldsymbol{Q}_2$$

$$\boldsymbol{Q}_2^{[1]}\left(\frac{1}{3}\right)=\frac{2}{3}\,\boldsymbol{Q}_2^{[0]}\left(\frac{1}{3}\right)+\frac{1}{3}\,\boldsymbol{Q}_3^{[0]}\left(\frac{1}{3}\right)$$

$$=\frac{2}{3}\,\boldsymbol{Q}_2+\frac{1}{3}\,\boldsymbol{Q}_3$$

$$\boldsymbol{Q}_3^{[1]}\left(\frac{1}{3}\right)=\frac{2}{3}\,\boldsymbol{Q}_3^{[0]}\left(\frac{1}{3}\right)+\frac{1}{3}\,\boldsymbol{Q}_4^{[0]}\left(\frac{1}{3}\right)$$

$$=\frac{2}{3}\,\boldsymbol{Q}_3+\frac{1}{3}\,\boldsymbol{Q}_4$$

Taking $k=2$, $\boldsymbol{Q}_0^{[2]}\left(\frac{1}{3}\right)$, $\boldsymbol{Q}_1^{[2]}\left(\frac{1}{3}\right)$, $\boldsymbol{Q}_2^{[2]}\left(\frac{1}{3}\right)$ become:

$$\boldsymbol{Q}_0^{[2]}\left(\frac{1}{3}\right)=\frac{2}{3}\,\boldsymbol{Q}_0^{[1]}\left(\frac{1}{3}\right)+\frac{1}{3}\,\boldsymbol{Q}_1^{[1]}\left(\frac{1}{3}\right)$$

$$=\frac{2}{3}\left(\frac{2}{3}\,\boldsymbol{Q}_0+\frac{1}{3}\,\boldsymbol{Q}_1\right)+\frac{1}{3}\left(\frac{2}{3}\,\boldsymbol{Q}_1+\frac{1}{3}\,\boldsymbol{Q}_2\right)$$

$$=\frac{4}{9}\,\boldsymbol{Q}_0+\frac{4}{9}\,\boldsymbol{Q}_1+\frac{1}{9}\,\boldsymbol{Q}_2$$

$$\boldsymbol{Q}_1^{[2]}\left(\frac{1}{3}\right)=\frac{2}{3}\,\boldsymbol{Q}_1^{[1]}\left(\frac{1}{3}\right)+\frac{1}{3}\,\boldsymbol{Q}_2^{[1]}\left(\frac{1}{3}\right)$$

$$=\frac{2}{3}\left(\frac{2}{3}\,\boldsymbol{Q}_1+\frac{1}{3}\,\boldsymbol{Q}_2\right)+\frac{1}{3}\left(\frac{2}{3}\,\boldsymbol{Q}_2+\frac{1}{3}\,\boldsymbol{Q}_3\right)=$$

$$= \frac{4}{9} \boldsymbol{Q}_1 + \frac{4}{9} \boldsymbol{Q}_2 + \frac{1}{9} \boldsymbol{Q}_3$$

$$\boldsymbol{Q}_2^{[2]}\left(\frac{1}{3}\right) = \frac{2}{3} \boldsymbol{Q}_2^{[1]}\left(\frac{1}{3}\right) + \frac{1}{3} \boldsymbol{Q}_3^{[1]}\left(\frac{1}{3}\right)$$

$$= \frac{2}{3}\left(\frac{2}{3} \boldsymbol{Q}_2 + \frac{1}{3} \boldsymbol{Q}_3\right) + \frac{1}{3}\left(\frac{2}{3} \boldsymbol{Q}_3 + \frac{1}{3} \boldsymbol{Q}_4\right)$$

$$= \frac{4}{9} \boldsymbol{Q}_2 + \frac{4}{9} \boldsymbol{Q}_3 + \frac{1}{9} \boldsymbol{Q}_4$$

Taking $k=3$, $\boldsymbol{Q}_0^{[3]}\left(\frac{1}{3}\right)$, $\boldsymbol{Q}_1^{[3]}\left(\frac{1}{3}\right)$ become:

$$\boldsymbol{Q}_0^{[3]}\left(\frac{1}{3}\right) = \frac{2}{3} \boldsymbol{Q}_0^{[2]}\left(\frac{1}{3}\right) + \frac{1}{3} \boldsymbol{Q}_1^{[2]}\left(\frac{1}{3}\right)$$

$$= \frac{2}{3}\left(\frac{4}{9} \boldsymbol{Q}_0 + \frac{4}{9} \boldsymbol{Q}_1 + \frac{1}{9} \boldsymbol{Q}_2\right) + \frac{1}{3}\left(\frac{4}{9} \boldsymbol{Q}_1 + \frac{4}{9} \boldsymbol{Q}_2 + \frac{1}{9} \boldsymbol{Q}_3\right)$$

$$= \frac{8}{27} \boldsymbol{Q}_0 + \frac{12}{27} \boldsymbol{Q}_1 + \frac{6}{27} \boldsymbol{Q}_2 + \frac{1}{27} \boldsymbol{Q}_3$$

$$\boldsymbol{Q}_1^{[3]}\left(\frac{1}{3}\right) = \frac{2}{3} \boldsymbol{Q}_1^{[2]}\left(\frac{1}{3}\right) + \frac{1}{3} \boldsymbol{Q}_2^{[2]}\left(\frac{1}{3}\right)$$

$$= \frac{2}{3}\left(\frac{4}{9} \boldsymbol{Q}_1 + \frac{4}{9} \boldsymbol{Q}_2 + \frac{1}{9} \boldsymbol{Q}_3\right) + \frac{1}{3}\left(\frac{4}{9} \boldsymbol{Q}_2 + \frac{4}{9} \boldsymbol{Q}_3 + \frac{1}{9} \boldsymbol{Q}_4\right)$$

$$= \frac{8}{27} \boldsymbol{Q}_1 + \frac{12}{27} \boldsymbol{Q}_2 + \frac{6}{27} \boldsymbol{Q}_3 + \frac{1}{27} \boldsymbol{Q}_4$$

Finally, taking $k=4$, $\boldsymbol{Q}_0^{[4]}\left(\frac{1}{3}\right)\left(=\boldsymbol{P}\left(\frac{1}{3}\right)\right)$ becomes:

$$\boldsymbol{Q}_0^{[4]}\left(\frac{1}{3}\right) = \frac{2}{3} \boldsymbol{Q}_0^{[3]}\left(\frac{1}{3}\right) + \frac{1}{3} \boldsymbol{Q}_1^{[3]}\left(\frac{1}{3}\right)$$

$$= \frac{2}{3}\left(\frac{8}{27} \boldsymbol{Q}_0 + \frac{12}{27} \boldsymbol{Q}_1 + \frac{6}{27} \boldsymbol{Q}_2 + \frac{1}{27} \boldsymbol{Q}_3\right)$$

$$+ \frac{1}{3}\left(\frac{8}{27} \boldsymbol{Q}_1 + \frac{12}{27} \boldsymbol{Q}_2 + \frac{6}{27} \boldsymbol{Q}_3 + \frac{1}{27} \boldsymbol{Q}_4\right)$$

$$= \frac{1}{81}(16 \boldsymbol{Q}_0 + 32 \boldsymbol{Q}_1 + 24 \boldsymbol{Q}_2 + 8 \boldsymbol{Q}_3 + \boldsymbol{Q}_4).$$

As a trial, take $n=4$ and $t=\frac{1}{3}$ in the Bézier curve formula, giving:

$$\boldsymbol{P}\left(\frac{1}{3}\right)=\sum_{i=0}^{4}\binom{4}{i}\left(1-\frac{1}{3}\right)^{4-i}\left(\frac{1}{3}\right)^{i}\boldsymbol{Q}_i$$

$$=\frac{1}{81}(16\boldsymbol{Q}_0+32\boldsymbol{Q}_1+24\boldsymbol{Q}_2+8\boldsymbol{Q}_3+\boldsymbol{Q}_4).$$

This confirms that $\boldsymbol{Q}_0^{[4]}\left(\frac{1}{3}\right)=\boldsymbol{P}\left(\frac{1}{3}\right)$. The graphical solution of this example is shown in Fig. 5.25.

5.1.8 Increase of the Degree of a Bézier Curve Segment

Consider the problem of formally increasing the degree of a Bézier curve segment without changing its shape. As an example suppose that we want to express a cubic Bézier curve segment by a quartic formula.

Let $\boldsymbol{Q}_0^3$, $\boldsymbol{Q}_1^3$, $\boldsymbol{Q}_2^3$ and $\boldsymbol{Q}_3^3$ be the curve defining vectors of the cubic Bézier curve segment $\boldsymbol{P}_3(t)$, and $\boldsymbol{Q}_0^4$, $\boldsymbol{Q}_1^4$, $\boldsymbol{Q}_2^4$, $\boldsymbol{Q}_3^4$ and $\boldsymbol{Q}_4^4$ the curve defining vectors of the quartic Bézier curve segment $\boldsymbol{P}_4(t)$.

Transform the cubic Bézier curve segment as follows.

$$\boldsymbol{P}_3(t)=[B_{0,3}(t)\ B_{1,3}(t)\ B_{2,3}(t)\ B_{3,3}(t)]\begin{bmatrix}\boldsymbol{Q}_0^3\\ \boldsymbol{Q}_1^3\\ \boldsymbol{Q}_2^3\\ \boldsymbol{Q}_3^3\end{bmatrix}$$

$$=[t^3\ t^2\ t\ 1]\begin{bmatrix}-1 & 3 & -3 & 1\\ 3 & -6 & 3 & 0\\ -3 & 3 & 0 & 0\\ 1 & 0 & 0 & 0\end{bmatrix}\begin{bmatrix}\boldsymbol{Q}_0^3\\ \boldsymbol{Q}_1^3\\ \boldsymbol{Q}_2^3\\ \boldsymbol{Q}_3^3\end{bmatrix}$$

$$=[t^4\ t^3\ t^2\ t\ 1]\begin{bmatrix}0 & 0 & 0 & 0\\ -1 & 3 & -3 & 1\\ 3 & -6 & 3 & 0\\ -3 & 3 & 0 & 0\\ 1 & 0 & 0 & 0\end{bmatrix}\begin{bmatrix}\boldsymbol{Q}_0^3\\ \boldsymbol{Q}_1^3\\ \boldsymbol{Q}_2^3\\ \boldsymbol{Q}_3^3\end{bmatrix}$$

$$\equiv[t^4\ t^3\ t^2\ t\ 1]\,\boldsymbol{M}_3\begin{bmatrix}\boldsymbol{Q}_0^3\\ \boldsymbol{Q}_1^3\\ \boldsymbol{Q}_2^3\\ \boldsymbol{Q}_3^3\end{bmatrix}.$$

The quartic Bézier curve segment $\boldsymbol{P}_4(t)$ becomes:

$$\boldsymbol{P}_4(t)=[B_{0,4}(t)\ B_{1,4}(t)\ B_{2,4}(t)\ B_{3,4}(t)\ B_{4,4}(t)]\begin{bmatrix}\boldsymbol{Q}_0^4\\ \boldsymbol{Q}_1^4\\ \boldsymbol{Q}_2^4\\ \boldsymbol{Q}_3^4\\ \boldsymbol{Q}_4^4\end{bmatrix}$$

$$=[t^4\ t^3\ t^2\ t\ 1]\begin{bmatrix}1 & -4 & 6 & -4 & 1\\ -4 & 12 & -12 & 4 & 0\\ 6 & -12 & 6 & 0 & 0\\ -4 & 4 & 0 & 0 & 0\\ 1 & 0 & 0 & 0 & 0\end{bmatrix}\begin{bmatrix}\boldsymbol{Q}_0^4\\ \boldsymbol{Q}_1^4\\ \boldsymbol{Q}_2^4\\ \boldsymbol{Q}_3^4\\ \boldsymbol{Q}_4^4\end{bmatrix}$$

$$\equiv[t^4\ t^3\ t^2\ t\ 1]\ \boldsymbol{M}_4\begin{bmatrix}\boldsymbol{Q}_0^4\\ \boldsymbol{Q}_1^4\\ \boldsymbol{Q}_2^4\\ \boldsymbol{Q}_3^4\\ \boldsymbol{Q}_4^4\end{bmatrix}.$$

The condition for $\boldsymbol{P}_3(t)$ and $\boldsymbol{P}_4(t)$ to be identically equal is:

$$\boldsymbol{M}_4\begin{bmatrix}\boldsymbol{Q}_0^4\\ \boldsymbol{Q}_1^4\\ \boldsymbol{Q}_2^4\\ \boldsymbol{Q}_3^4\\ \boldsymbol{Q}_4^4\end{bmatrix}=\boldsymbol{M}_3\begin{bmatrix}\boldsymbol{Q}_0^3\\ \boldsymbol{Q}_1^3\\ \boldsymbol{Q}_2^3\\ \boldsymbol{Q}_3^3\end{bmatrix}. \tag{5.92}$$

Multiplying both sides of the above equation by $\boldsymbol{M}_4^{-1}$ from the left gives:

$$\begin{bmatrix}\boldsymbol{Q}_0^4\\ \boldsymbol{Q}_1^4\\ \boldsymbol{Q}_2^4\\ \boldsymbol{Q}_3^4\\ \boldsymbol{Q}_4^4\end{bmatrix}=\boldsymbol{M}_4^{-1}\boldsymbol{M}_3\begin{bmatrix}\boldsymbol{Q}_0^3\\ \boldsymbol{Q}_1^3\\ \boldsymbol{Q}_2^3\\ \boldsymbol{Q}_3^3\end{bmatrix}=$$

$$= \begin{bmatrix} 0 & 0 & 0 & 0 & 1 \\ 0 & 0 & 0 & \frac{1}{4} & 1 \\ 0 & 0 & \frac{1}{6} & \frac{1}{2} & 1 \\ 0 & \frac{1}{4} & \frac{1}{2} & \frac{3}{4} & 1 \\ 1 & 1 & 1 & 1 & 1 \end{bmatrix} \begin{bmatrix} 0 & 0 & 0 & 0 \\ -1 & 3 & -3 & 1 \\ 3 & -6 & 3 & 0 \\ -3 & 3 & 0 & 0 \\ 1 & 0 & 0 & 0 \end{bmatrix} \begin{bmatrix} \boldsymbol{Q}_0^3 \\ \boldsymbol{Q}_1^3 \\ \boldsymbol{Q}_2^3 \\ \boldsymbol{Q}_3^3 \end{bmatrix}$$

$$= \begin{bmatrix} 1 & 0 & 0 & 0 \\ \frac{1}{4} & \frac{3}{4} & 0 & 0 \\ 0 & \frac{1}{2} & \frac{1}{2} & 0 \\ 0 & 0 & \frac{3}{4} & \frac{1}{4} \\ 0 & 0 & 0 & 1 \end{bmatrix} \begin{bmatrix} \boldsymbol{Q}_0^3 \\ \boldsymbol{Q}_1^3 \\ \boldsymbol{Q}_2^3 \\ \boldsymbol{Q}_3^3 \end{bmatrix}. \tag{5.93}$$

Expanding this matrix expression, we obtain:

$$\left.\begin{aligned} \boldsymbol{Q}_0^4 &= \boldsymbol{Q}_0^3 && = \frac{1}{4}(0 \cdot \boldsymbol{Q}_{-1}^3 + 4 \cdot \boldsymbol{Q}_0^3) \\ \boldsymbol{Q}_1^4 &= \frac{1}{4}\boldsymbol{Q}_0^3 + \frac{3}{4}\boldsymbol{Q}_1^3 && = \frac{1}{4}(1 \cdot \boldsymbol{Q}_0^3 + 3 \cdot \boldsymbol{Q}_1^3) \\ \boldsymbol{Q}_2^4 &= \frac{1}{2}\boldsymbol{Q}_1^3 + \frac{1}{2}\boldsymbol{Q}_2^3 && = \frac{1}{4}(2 \cdot \boldsymbol{Q}_1^3 + 2 \cdot \boldsymbol{Q}_2^3) \\ \boldsymbol{Q}_3^4 &= \frac{3}{4}\boldsymbol{Q}_2^3 + \frac{1}{4}\boldsymbol{Q}_3^3 && = \frac{1}{4}(3 \cdot \boldsymbol{Q}_2^3 + 1 \cdot \boldsymbol{Q}_3^3) \\ \boldsymbol{Q}_4^4 &= \boldsymbol{Q}_3^3 && = \frac{1}{4}(4 \cdot \boldsymbol{Q}_3^3 + 0 \cdot \boldsymbol{Q}_4^3) \end{aligned}\right\}. \tag{5.94}$$

This relation is shown in Fig. 5.26.

In general, when an n-th degree Bézier curve segment is rewritten in an $(n+1)$-st degree format, the expressions of the curve defining vectors can be predicted from the right-hand side of Eq. (5.94). We will derive these relations

$$\boldsymbol{P}_n(t) = \sum_{i=0}^{n} \binom{n}{i} (1-t)^{n-i} t^i \boldsymbol{Q}_i^n =$$

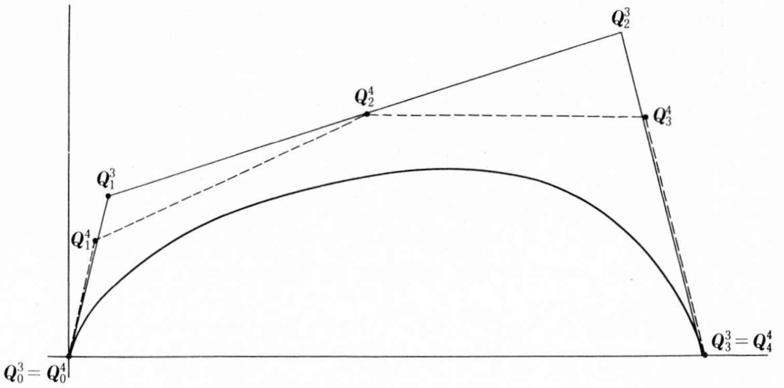

Fig. 5.26. Formal increase of the degree of a cubic Bézier curve segment (to quartic)

$$= \left[\sum_{i=0}^{n} \binom{n}{i} (1-t)^{n-i} t^i \boldsymbol{Q}_i^n\right] [(1-t)+t]$$

$$= \sum_{i=0}^{n} \binom{n}{i} (1-t)^{n-i+1} t^i \boldsymbol{Q}_i^n + \sum_{i=0}^{n} \binom{n}{i} (1-t)^{n-i} t^{i+1} \boldsymbol{Q}_i^n . \tag{5.95}$$

The first term of this expression can be expanded as follows:

$$\sum_{i=0}^{n} \binom{n}{i} (1-t)^{n-i+1} t^i \boldsymbol{Q}_i^n$$

$$= \binom{n}{0} (1-t)^{n+1} t^0 \boldsymbol{Q}_0^n + \ldots + \binom{n}{n} (1-t)^1 t^n \boldsymbol{Q}_n^n$$

$$= \binom{n}{0} (1-t)^{n+1} t^0 \boldsymbol{Q}_0^n + \ldots$$

$$+ \binom{n}{n} (1-t)^1 t^n \boldsymbol{Q}_n^n + \binom{n}{n+1} (1-t)^0 t^{n+1} \boldsymbol{Q}_{n+1}^n$$

$$= \sum_{i=0}^{n+1} \binom{n}{i} (1-t)^{n+1-i} t^i \boldsymbol{Q}_i^n .$$

Similarly, the second term can be expanded as:

$$\sum_{i=0}^{n} \binom{n}{i} (1-t)^{n-i} t^{i+1} \boldsymbol{Q}_i^n$$

$$= \binom{n}{0} (1-t)^n t^1 \boldsymbol{Q}_0^n + \ldots + \binom{n}{n} (1-t)^0 t^{n+1} \boldsymbol{Q}_n^n =$$

$$=\binom{n}{-1}(1-t)^{n+1}t^0 Q^n_{-1}+\binom{n}{0}(1-t)^n t^1 Q^n_0+\ldots$$

$$+\binom{n}{n}(1-t)^0 t^{n+1} Q^n_n$$

$$=\sum_{i=0}^{n+1}\binom{n}{i-1}(1-t)^{n+1-i}t^i Q^n_{i-1}.$$

Therefore, Eq. (5.95) becomes:

$$P_n(t)=\sum_{i=0}^{n+1}\binom{n}{i}(1-t)^{n+1-i}t^i Q^n_i+\sum_{i=0}^{n+1}\binom{n}{i-1}(1-t)^{n+1-i}t^i Q^n_{i-1} \tag{5.96}$$

$$P_{n+1}(t)=\sum_{i=0}^{n+1}\binom{n+1}{i}(1-t)^{n+1-i}t^i \; Q^{n+1}_i. \tag{5.97}$$

For all t in the range $0 \leqq t \leqq 1$ we must have:

$$P_n(t) \equiv P_{n+1}(t)$$

which gives:

$$\sum_{i=0}^{n+1}\left[\binom{n}{i}Q^n_i+\binom{n}{i-1}Q^n_{i-1}-\binom{n+1}{i}Q^{n+1}_i\right](1-t)^{n+1-i}t^i=0$$

$$\therefore \binom{n}{i}Q^n_i+\binom{n}{i-1}Q^n_{i-1}-\binom{n+1}{i}Q^{n+1}_i=0.$$

This equation can be rewritten as:

$$Q^{n+1}_i=\frac{1}{n+1}\left[i\,Q^n_{i-1}+(n+1-i)\,Q^n_i\right] \tag{5.98}$$

$$(i=0,\ 1,\ \ldots,\ n+1).$$

Here Q^n_{-1} and Q^n_{n+1} are undefined, but their coefficients are 0, so they are irrelevant in calculating Q^{n+1}_i.

Next, in the case of changing from an expression in terms of Q^n_i to one in terms of Q^{n+k}_i, let us focus on the expression for Q^{n+k}_1. Using Eq. (5.98) gives:

$$Q^{n+1}_1=\frac{1}{n+1}(Q^n_0+nQ^n_1)$$

$$Q^{n+2}_1=\frac{1}{n+2}[Q^{n+1}_0+(n+1)\,Q^{n+1}_1]=$$

$$= \frac{1}{n+2}\left[Q_0^n + (n+1)\frac{1}{n+1}\,(Q_0^n + nQ_1^n)\right]$$

$$= \frac{1}{n+2}\,(2Q_0^n + nQ_1^n).$$

Continuing this same procedure gives:

$$Q_1^{n+k} = \frac{1}{n+k}\,(kQ_0^n + nQ_1^n). \tag{5.99}$$

Of course we have:

$$Q_0^{n+k} = Q_0^n. \tag{5.100}$$

5.1.9 Partitioning of a Bézier Curve Segment

By a method to that used to divide a cubic Bézier curve segment (refer to Sect. 5.1.2), an n-th degree Bézier curve segment $P(t)$ can be divided at point $t = t_s$ into two Bézier curve segments $P_1(u)$ $(0 \leq u \leq 1)$ and $P_2(u)$ $(0 \leq u \leq 1)$. Equation (5.69) is used for $P(t)$. Performing the parameter transformation $u = t/t_s$ in Eq. (5.69) gives:

$$P_1(u) = [t_s^n u^n \;\; t_s^{n-1} u^{n-1} \;\; \ldots \;\; t_s u \;\; 1]\,\beta \begin{bmatrix} Q_0 \\ Q_1 \\ \vdots \\ Q_n \end{bmatrix}$$

$$= [u^n \;\; u^{n-1} \;\; \ldots \;\; u \;\; 1] \begin{bmatrix} t_s^n & 0 & \ldots & 0 \\ 0 & t_s^{n-1} & \ldots & \vdots \\ \vdots & \vdots & t_s & 0 \\ 0 & \ldots & 0 & 1 \end{bmatrix} \beta \begin{bmatrix} Q_0 \\ Q_1 \\ \vdots \\ Q_n \end{bmatrix}$$

Since:

$$\begin{bmatrix} t_s^n & \ldots & \ldots & 0 \\ 0 & t_s^{n-1} & \vdots & \vdots \\ \vdots & \vdots \ddots & t_s & 0 \\ 0 & \ldots & \ldots & 1 \end{bmatrix} \beta$$

$$= \beta \begin{bmatrix} 1 & 0 & 0 & \ldots & 0 \\ 1 - t_s & t_s & 0 & \ldots & 0 \\ (1-t_s)^2 & 2(1-t_s)t_s & t_s^2 & \ddots \;\; 0 & 0 \\ \vdots & \vdots & \vdots & \vdots & \vdots \\ (1-t_s)^n & \binom{n}{1}(1-t_s)^{n-1}t_s & \binom{n}{2}(1-t_s)^{n-2}t_s^2 & \ldots & t_s^n \end{bmatrix}$$

we obtain *):

$$P_1(u)=[u^n\ u^{n-1}\ \dots\ u\ 1]\,\beta$$

$$\times\begin{bmatrix} 1 & 0 & 0 & 0 & 0 \\ 1-t_s & t_s & 0 & 0 & 0 \\ (1-t_s)^2 & 2(1-t_s)t_s & t_s^2 & 0 & 0 \\ \vdots & \vdots & \vdots & \vdots & \vdots \\ (1-t_s)^n & \binom{n}{1}(1-t_s)^{n-1}t_s & \binom{n}{2}(1-t_s)^{n-2}t_s^2 & \dots & t_s^n \end{bmatrix}\begin{bmatrix} Q_0 \\ Q_1 \\ Q_2 \\ \vdots \\ Q_n \end{bmatrix}$$

$$=[u^n\ u^{n-1}\ \dots\ u\ 1]\,\beta$$

$$\times\begin{bmatrix} Q_0 \\ (1-t_s)Q_0+t_sQ_1 \\ (1-t_s)^2Q_0+2(1-t_s)t_sQ_1+t_s^2Q_2 \\ \dots \\ \dots \\ (1-t_s)^nQ_0+\binom{n}{1}(1-t_s)^{n-1}t_sQ_1+\binom{n}{2}(1-t_s)^{n-2}t_s^2Q_2+\dots+t_s^nQ_n \end{bmatrix}$$

$$\equiv[u^n\ u^{n-1}\ \dots\ u\ 1]\,\beta\begin{bmatrix} Q_0^{[0]} \\ Q_0^{[1]} \\ Q_0^{[2]} \\ \vdots \\ Q_0^{[n]} \end{bmatrix} \tag{5.101}$$

where:

$$\begin{aligned} Q_0^{[k]} &= (1-t_s)^kQ_0+\binom{k}{1}(1-t_s)^{k-1}t_sQ_1+\binom{k}{2}(1-t_s)^{k-2}t_s^2Q_2+\dots+t_s^kQ_k \\ &=\sum_{j=0}^{k}\binom{k}{j}(1-t_s)^{k-j}t_s^jQ_j \\ &\equiv\sum_{j=0}^{k}B_{j,k}(t_s)Q_j \quad (O\leq k\leq n). \end{aligned} \tag{5.102}$$

*) This could perhaps be anticipated from the $n=3$ case discussed in Sect. 5.1.2.

The vertex vectors for the curve defining polygon in the range $0 \leqq t \leqq t_s$ agree with the vectors $\boldsymbol{Q}_0^{[0]}, \boldsymbol{Q}_0^{[1]}, \boldsymbol{Q}_0^{[2]}, \ldots, \boldsymbol{Q}_0^{[n]}$ found in the process of graphically determining the point $t = t_s$ on the curve.

Next, to find $\boldsymbol{P}_2(u)$, perform the parameter transformation $u = (t - t_s)/(1 - t_s)$ in equation (5.69) to give:

$$\boldsymbol{P}_2(u) = [\{(1-t_s)u + t_s\}^n \ \{(1-t_s)u + t_s\}^{n-1} \ \ldots \ (1-t_s)u + t_s \ \ 1] \ \beta \begin{bmatrix} \boldsymbol{Q}_0 \\ \boldsymbol{Q}_1 \\ \vdots \\ \boldsymbol{Q}_n \end{bmatrix}$$

$$= [u^n \ u^{n-1} \ \ldots \ u \ \ 1] \begin{bmatrix} (1-t_s)^n & 0 & \ldots & 0 & 0 \\ \binom{n}{1}(1-t_s)^{n-1} t_s & (1-t_s)^{n-1} & \ldots & 0 & 0 \\ \binom{n}{2}(1-t_s)^{n-2} t_s^2 & \vdots & & \vdots & \vdots \\ \vdots & \vdots & & 1-t_s & 0 \\ t_s^n & t_s^{n-1} & \ldots & t_s & 1 \end{bmatrix} \beta \begin{bmatrix} \boldsymbol{Q}_0 \\ \boldsymbol{Q}_1 \\ \vdots \\ \boldsymbol{Q}_n \end{bmatrix}.$$

Since we also have:

$$\begin{bmatrix} (1-t_s)^n & 0 & \ldots & 0 & 0 \\ \binom{n}{1}(1-t_s)^{n-1} t_s & (1-t_s)^{n-1} & \ldots & 0 & 0 \\ \binom{n}{2}(1-t_s)^{n-2} t_s^2 & \vdots & & \vdots & \vdots \\ \vdots & \vdots & & 1-t_s & 0 \\ t_s^n & t_s^{n-1} & \ldots & t_s & 1 \end{bmatrix} \beta$$

$$= \beta \begin{bmatrix} (1-t_s)^n & \binom{n}{1}(1-t_s)^{n-1} t_s & \binom{n}{2}(1-t_s)^{n-2} t_s^2 & \ldots & t_s^n \\ 0 & (1-t_s)^{n-1} & \ldots & \ldots & t_s^{n-1} \\ \vdots & \vdots & & & \vdots \\ 0 & \ldots & 0 & 1-t_s & t_s \\ 0 & 0 & \ldots & 0 & 1 \end{bmatrix}$$

then*):

$$\boldsymbol{P}_2(u)=[u^n \ u^{n-1} \dots u \ 1]\,\beta$$

$$\times\begin{bmatrix} (1-t_s)^n & \binom{n}{1}(1-t_s)^{n-1}t_s & \binom{n}{2}(1-t_s)^{n-2}t_s^2 & \dots & t_s^n \\ 0 & (1-t_s)^{n-1} & \dots & \dots & t_s^{n-1} \\ \vdots & \vdots & & & \vdots \\ 0 & \dots & 0 & 1-t_s & t_s \\ 0 & 0 & \dots & 0 & 1 \end{bmatrix}\begin{bmatrix} \boldsymbol{Q}_0 \\ \boldsymbol{Q}_1 \\ \vdots \\ \boldsymbol{Q}_{n-1} \\ \boldsymbol{Q}_n \end{bmatrix}$$

$$=[u^n \ u^{n-1} \ \dots \ u \ 1]\,\beta$$

$$\times\begin{bmatrix} (1-t_s)^n\boldsymbol{Q}_0+\binom{n}{1}(1-t_s)^{n-1}t_s\boldsymbol{Q}_1+\binom{n}{2}(1-t_s)^{n-2}t_s^2\boldsymbol{Q}_2+\dots+t_s^n\boldsymbol{Q}_n \\ (1-t_s)^{n-1}\boldsymbol{Q}_1+\binom{n-1}{1}(1-t_s)^{n-2}t_s\boldsymbol{Q}_2+\dots+t_s^{n-1}\boldsymbol{Q}_n \\ \dots \\ \dots \\ (1-t_s)\boldsymbol{Q}_{n-1}+t_s\boldsymbol{Q}_n \\ \boldsymbol{Q}_n \end{bmatrix}$$

$$=[u^n \ u^{n-1} \ \dots \ u \ 1]\,\beta\begin{bmatrix} \boldsymbol{Q}_0^{[n]} \\ \boldsymbol{Q}_1^{[n-1]} \\ \vdots \\ \boldsymbol{Q}_{n-1}^{[1]} \\ \boldsymbol{Q}_n^{[0]} \end{bmatrix} \tag{5.103}$$

$$\begin{aligned}\boldsymbol{Q}_k^{[n-k]}&=(1-t_s)^{n-k}\boldsymbol{Q}_k+\binom{n-k}{1}(1-t_s)^{n-k-1}t_s\boldsymbol{Q}_{k+1}+\dots+t_s^{n-k}\boldsymbol{Q}_n\\ &=\sum_{j=0}^{n-k}\binom{n-k}{j}(1-t_s)^{n-k-j}t_s^j\boldsymbol{Q}_{k+j} \qquad (5.104)\\ &\equiv\sum_{j=0}^{n-k}B_{j,n-k}(t_s)\,\boldsymbol{Q}_{k+j} \quad (0\leqq k\leqq n).\end{aligned}$$

The vertex vectors for the curve defining polygon in the range $t_s\leqq t\leqq 1$ agree with the vectors $\boldsymbol{Q}_0^{[n]}$, $\boldsymbol{Q}_1^{[n-1]}$, ..., $\boldsymbol{Q}_n^{[0]}$ found in the process of graphically determining the point corresponding to $t=t_s$ on the curve.

---

*) This can be anticipated from the case $n=3$ in Sect. 5.1.2.

5.1.10 Connection of Bézier Curve Segments

To the Bézier curve segment:

$$P_{\mathrm{I}}(t)=\sum_{i=0}^{m}\binom{m}{i}(1-t)^{m-i}t^{i}Q_{\mathrm{I},i}\quad(0\leqq t\leqq 1)$$

we wish to connect a second Bézier curve segment:

$$P_{\mathrm{II}}(t)=\sum_{j=0}^{n}\binom{n}{j}(1-t)^{n-j}t^{j}Q_{\mathrm{II},j}\quad(0\leqq t\leqq 1)$$

with continuity up to the curvature vector (refer to Fig. 5.27).

First, at the connection point, from the condition of continuity of position we have, from (1.49):

$$Q_{\mathrm{II},0}=Q_{\mathrm{I},m}. \tag{5.105}$$

The condition of continuity of slope at the connection points is expressed by Eq. (1.50). Using (5.81) and (5.82) gives:

$$\frac{n}{\alpha_2}(Q_{\mathrm{II},1}-Q_{\mathrm{II},0})=\frac{m}{\alpha_1}(Q_{\mathrm{I},m}-Q_{\mathrm{I},m-1})=t. \tag{5.106}$$

Here α_1, α_2 are the magnitudes of the tangent vectors $\dot{P}_{\mathrm{I}}(1)$ and $\dot{P}_{\mathrm{II}}(0)$, respectively. Equation (5.106) requires that the 3 points $Q_{\mathrm{I},m-1}$, $Q_{\mathrm{I},m}=Q_{\mathrm{II},0}$ and $Q_{\mathrm{II},1}$ are colinear.

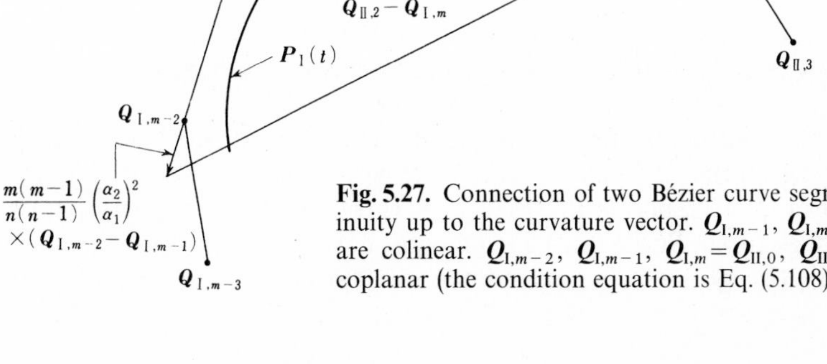

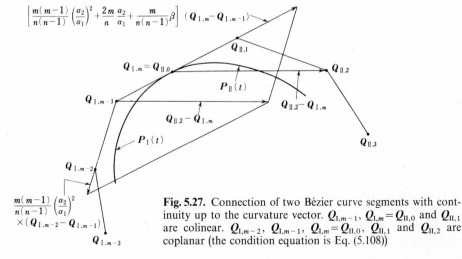

Fig. 5.27. Connection of two Bézier curve segments with continuity up to the curvature vector. $Q_{\mathrm{I},m-1}$, $Q_{\mathrm{I},m}=Q_{\mathrm{II},0}$ and $Q_{\mathrm{II},1}$ are colinear. $Q_{\mathrm{I},m-2}$, $Q_{\mathrm{I},m-1}$, $Q_{\mathrm{I},m}=Q_{\mathrm{II},0}$, $Q_{\mathrm{II},1}$ and $Q_{\mathrm{II},2}$ are coplanar (the condition equation is Eq. (5.108))

The condition of continuity of curvature vectors at the connection point is expressed by Eq. (1.52). Using Eqs. (5.83) and (5.84), the 2nd derivative vectors at the connection point are, respectively:

$$\ddot{\boldsymbol{P}}_{\mathrm{I}}(1) = m(m-1)\,(\boldsymbol{Q}_{\mathrm{I},m} - 2\,\boldsymbol{Q}_{\mathrm{I},m-1} + \boldsymbol{Q}_{\mathrm{I},m-2})$$
$$\ddot{\boldsymbol{P}}_{\mathrm{II}}(0) = n(n-1)\,(\boldsymbol{Q}_{\mathrm{II},2} - 2\,\boldsymbol{Q}_{\mathrm{II},1} + \boldsymbol{Q}_{\mathrm{II},0}).$$

Substituting these into Eq. (1.52) gives:

$$\begin{aligned} &n(n-1)\,(\boldsymbol{Q}_{\mathrm{II},2} - 2\,\boldsymbol{Q}_{\mathrm{II},1} + \boldsymbol{Q}_{\mathrm{II},0}) \\ &= m(m-1)\left(\frac{\alpha_2}{\alpha_1}\right)^2 (\boldsymbol{Q}_{\mathrm{I},m} - 2\,\boldsymbol{Q}_{\mathrm{I},m-1} + \boldsymbol{Q}_{\mathrm{I},m-2}) + m\beta(\boldsymbol{Q}_{\mathrm{I},m} - \boldsymbol{Q}_{\mathrm{I},m-1}). \end{aligned}$$

Using Eqs. (5.105) and (5.106) to eliminate $\boldsymbol{Q}_{\mathrm{II},0}$ and $\boldsymbol{Q}_{\mathrm{II},1}$ gives:

$$\begin{aligned} \boldsymbol{Q}_{\mathrm{II},2} = {} & \frac{m(m-1)}{n(n-1)}\left(\frac{\alpha_2}{\alpha_1}\right)^2 \boldsymbol{Q}_{\mathrm{I},m-2} \\ & - \left[\frac{2m(m-1)}{n(n-1)}\left(\frac{\alpha_2}{\alpha_1}\right)^2 + \frac{2m}{n}\,\frac{a_2}{\alpha_1} + \frac{m}{n(n-1)}\,\beta\right] \boldsymbol{Q}_{\mathrm{I},m-1} \\ & + \left[\frac{m(m-1)}{n(n-1)}\left(\frac{\alpha_2}{\alpha_1}\right)^2 + \frac{2m}{n}\,\frac{a_2}{\alpha_1} + 1 + \frac{m}{n(n-1)}\,\beta\right] \boldsymbol{Q}_{\mathrm{I},m}. \end{aligned} \tag{5.107}$$

This determines the $\boldsymbol{Q}_{\mathrm{II},2}$ that will make the curvature vectors continuous at the connection point. This equation can be changed into the form[7]:

$$\begin{aligned} \boldsymbol{Q}_{\mathrm{II},2} - \boldsymbol{Q}_{\mathrm{I},m} = {} & \frac{m(m-1)}{n(n-1)}\left(\frac{\alpha_2}{\alpha_1}\right)^2 (\boldsymbol{Q}_{\mathrm{I},m-2} - \boldsymbol{Q}_{\mathrm{I},m-1}) \\ & + \left[\frac{m(m-1)}{n(n-1)}\left(\frac{\alpha_2}{\alpha_1}\right)^2 + \frac{2m}{n}\,\frac{a_2}{\alpha_1} + \frac{m}{n(n-1)}\,\beta\right] (\boldsymbol{Q}_{\mathrm{I},m} - \boldsymbol{Q}_{\mathrm{I},m-1}). \end{aligned} \tag{5.108}$$

This equation shows that $\boldsymbol{Q}_{\mathrm{II},2}$ lies in the same plane as $\boldsymbol{Q}_{\mathrm{I},m-2}$, $\boldsymbol{Q}_{\mathrm{I},m-1}$, $\boldsymbol{Q}_{\mathrm{I},m} = \boldsymbol{Q}_{\mathrm{II},0}$ and $\boldsymbol{Q}_{\mathrm{II},1}$. The relation among the curve defining polygon vectors is shown in Fig. 5.27.

5.1.11 Creation of a Spline Curve with Cubic Bézier Curve Segments

Consider the problem of connecting cubic Bézier curve segments smoothly, with continuity up to the curvature vector, to generate a spline curve (Fig. 5.28).

As shown in Fig. 5.28, points $\boldsymbol{P}_0$, $\boldsymbol{P}_1$, ..., $\boldsymbol{P}_n$ through which the curve passes are given, and n cubic Bézier curve segments, defined by the $3n+1$ curve defining vectors $\boldsymbol{Q}_0$, $\boldsymbol{Q}_1$, ..., $\boldsymbol{Q}_{3n}$ are connected. The global parameter is t. t

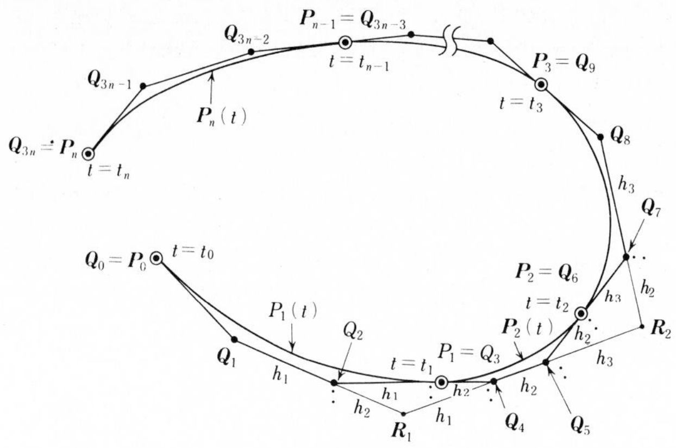

Fig. 5.28. Generation of a spline curve by connection of Bézier curve segments

assumes the knot values $t_0, t_1, \ldots, t_n$ at the ends of the segments (the points $\boldsymbol{P}_0$, $\boldsymbol{P}_1, \ldots, \boldsymbol{P}_n$ through which the curve passes). The i-th Bézier curve segment is given by the following formula.

$$\boldsymbol{P}_i(t)=\sum_{j=0}^{3} B_{j,3}(u)\boldsymbol{Q}_{3(i-1)+j} \tag{5.109}$$

where the parameter u is:

$$0 \leqq u = \frac{t-t_{i-1}}{t_i-t_{i-1}} \leqq 1. \tag{5.110}$$

From here on we set $t_i - t_{i-1} = h_i$.

The condition for two curve segments $\boldsymbol{P}_i(t)$ and $\boldsymbol{P}_{i+1}(t)$ to have continuous slopes and curvature vectors can be expressed by the following two equations, with $m=n=3$, $\alpha_2/\alpha_1=(t_{i+1}-t_i)/(t_i-t_{i-1})\equiv h_{i+1}/h_i$, $\boldsymbol{Q}_{\mathrm{I},m}=\boldsymbol{Q}_{3i}$, $\boldsymbol{Q}_{\mathrm{I},m-1}=\boldsymbol{Q}_{3i-1}$, $\boldsymbol{Q}_{\mathrm{I},m-2}=\boldsymbol{Q}_{3i-2}$, $\boldsymbol{Q}_{\mathrm{II},0}=\boldsymbol{Q}_{3i}$, $\boldsymbol{Q}_{\mathrm{II},1}=\boldsymbol{Q}_{3i+1}$, $\boldsymbol{Q}_{\mathrm{II},2}=\boldsymbol{Q}_{3i+2}$ and, for simplicity, $\beta=0$[24] in Eqs. (1.51) and (1.52):

$$h_{i+1}\boldsymbol{Q}_{3i-1}+h_i\boldsymbol{Q}_{3i+1}=(h_i+h_{i+1})\boldsymbol{Q}_{3i} \tag{5.111}$$

$$[(h_i+h_{i+1})\boldsymbol{Q}_{3i+1}-h_i\boldsymbol{Q}_{3i+2}]h_i=[(h_i+h_{i+1})\boldsymbol{Q}_{3i-1}-h_{i+1}\boldsymbol{Q}_{3i-2}]h_{i+1} \tag{5.112}$$

where:

$$1 \leqq i \leqq n-1.$$

The connection conditions (5.111) and (5.112) can be rewritten as follows:

$$h_i(\boldsymbol{Q}_{3i+1}-\boldsymbol{Q}_{3i})=h_{i+1}(\boldsymbol{Q}_{3i}-\boldsymbol{Q}_{3i-1}) \tag{5.113}$$

$$Q_{3i-1}+\frac{h_{i+1}}{h_i}(Q_{3i-1}-Q_{3i-2})=Q_{3i+1}+\frac{h_i}{h_{i+1}}(Q_{3i+1}-Q_{3i+2}). \quad (5.114)$$

From these equations, the connection conditions can be shown graphically as in Fig. 5.29.

Using Eqs. (5.113) and (5.114), a spline curve can be generated by the Bézier curves as follows.

If we take the knot values of parameter t to be equal to the cumulative chord lengths at the predetermined points through which the curve must pass, h_i is the chord length of $\overline{P_{i-1}P_i}$:

$$h_i=|P_{i-1}-P_i|.$$

First, set $Q_0=P_0$ and $Q_3=P_1$. Choose Q_1 and Q_2 to have reasonable values; then the first segment $P_1(t)$, with end points P_0 and P_1, is determined. Next, on the extension of the line segment $\overline{Q_1Q_2}$, find point R_1 such that $\overline{Q_1Q_2}:\overline{Q_2R_1}=h_1:h_2$. Similarly, on the extension of line segment $\overline{Q_2Q_3}$, find point Q_4 such that $\overline{Q_2Q_3}:\overline{Q_3Q_4}=h_1:h_2$. In addition, find point Q_5 on the extension of line segment $\overline{R_1Q_4}$ such that a similar ratio holds. Then Q_3, Q_4, Q_5 and Q_6 $(=P_2)$ are the curve defining vectors for the second curve segment; these two segments are connected with continuity up to the curvature vector. Generate the complete spline curve by repeating similar operations until the end at $Q_{3n}=P_n$.

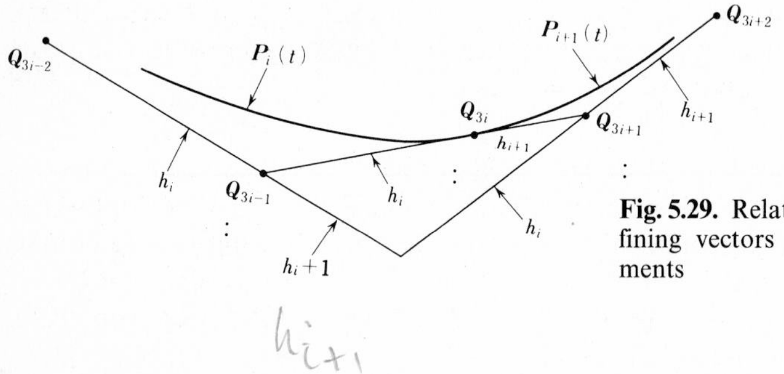

Fig. 5.29. Relation between curve defining vectors of adjacent curve segments

5.2 Surfaces

5.2.1 Bézier Surface Patches

By using Bernstein basis functions, a cartesian product surface (tensor product surface) can be expressed as follows.

$$P(u,w)=\sum_{i=0}^{m}\sum_{j=0}^{n}B_{i,m}(u)B_{j,n}(w)Q_{ij} \quad (5.115)$$

where:

$$B_{i,m}(u) = \binom{m}{i} (1-u)^{m-i} u^i .$$

Normally, by Bézier surface patch, we mean a cartesian product surface as expressed by Eq. (5.115). Equation (5.115) can be expressed in matrix form as:

$$P(u,w) = [B_{0,m}(u)\ B_{1,m}(u)\ \ldots\ B_{m,m}(u)] \times \begin{bmatrix} Q_{00} & Q_{01} & \cdots & Q_{0n} \\ Q_{10} & Q_{11} & \cdots & Q_{1n} \\ \vdots & \cdots & \cdots & \vdots \\ \vdots & \cdots & \cdots & \vdots \\ Q_{m0} & Q_{m1} & \cdots & Q_{mn} \end{bmatrix} \begin{bmatrix} B_{0,n}(w) \\ B_{1,n}(w) \\ \vdots \\ \vdots \\ B_{n,n}(w) \end{bmatrix} . \tag{5.116}$$

Figure 5.30 shows the relation between the surface defining vectors Q_{ij} and the surface (this figure shows the case $m=5$, $n=4$). The shape defined by the Q_{ij} is called a surface defining net, Bézier net or characteristic net.

Figures 5.31 to 5.33 show examples of this kind of Bézier surface patch.

The 4 boundary curves of such a surface patch are the Bézier curve segments expressed by the 0-th row, m-th row, 0-th column and n-th column of the center matrix. In Fig. 5.30, for example the boundary curve $P(0,w)$ is the Bézier curve segment having Q_{00}, Q_{01}, Q_{02}, Q_{03} and Q_{04} as curve defining vectors. At the 4 corner points of the surface, there must be agreement with the corresponding surface defining vectors:

$$P(0,0) = Q_{00}, \quad P(0,1) = Q_{0n}, \quad P(1,0) = Q_{m0}, \quad P(1,1) = Q_{mn}.$$

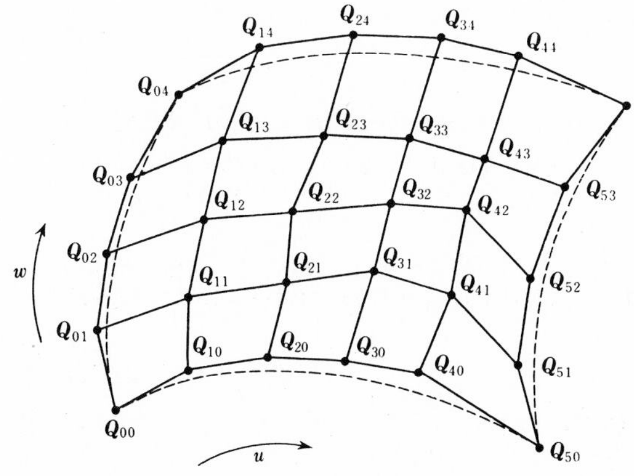

Fig. 5.30. Cartesian product Bézier surface patch ($m=5$, $n=4$)

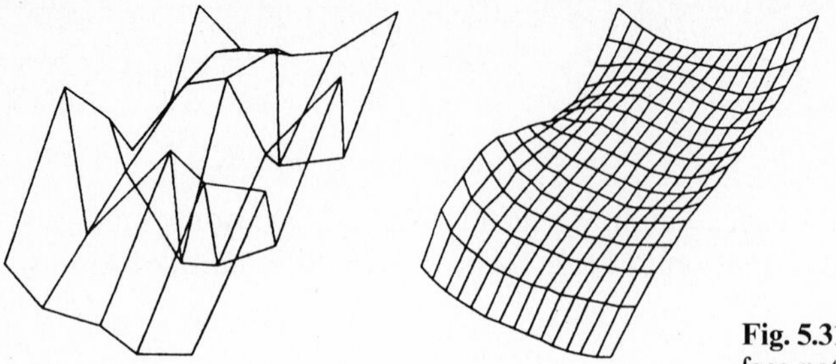

Fig. 5.31. Examples of Bézier surface patches (1)

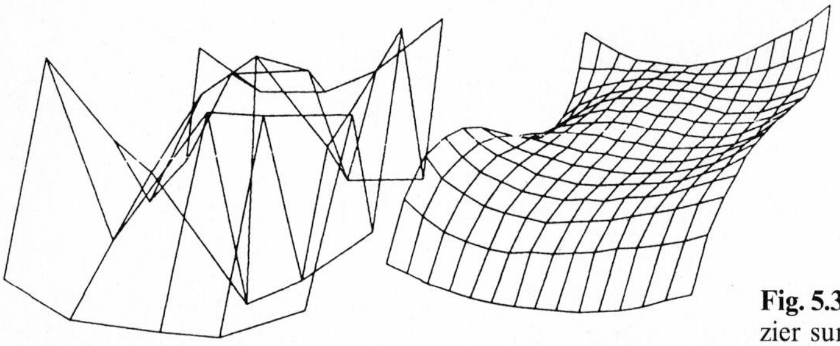

Fig. 5.32. Examples of Bézier surface patches (2)

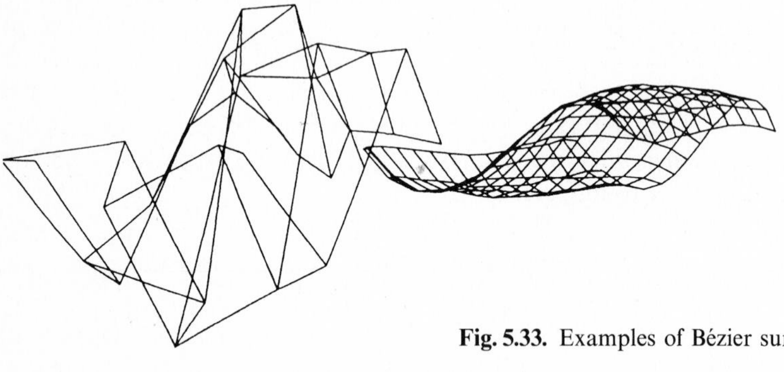

Fig. 5.33. Examples of Bézier surface patches (3)

5.2.2 The Relation Between a Bi-cubic Bézier Surface Patch and a Bi-cubic Coons Surface Patch

Just as we could express a Ferguson curve segment as a cubic Bézier curve segment, the relation between the surface defining vectors of a bi-cubic Coons surface and the surface defining vectors of a bi-cubic Bézier surface patch can be derived as follows.

The bi-cubic Coons surface patch can be expressed by the following equation.

$$\boldsymbol{P}_C(u, w) = \boldsymbol{U}\boldsymbol{M}_C\boldsymbol{B}_C\boldsymbol{M}_C^T\boldsymbol{W}^T . \qquad (5.117)^{*)}$$

*) For definitions of $\boldsymbol{M}_C$ and $\boldsymbol{B}_C$, refer to (3.10) and (3.103), respectively.

The bi-cubic Bézier surface patch can be expressed as follows:

$$\boldsymbol{P}_B(u, w) = \boldsymbol{U}\boldsymbol{M}_B\boldsymbol{B}_B\boldsymbol{M}_B^T\boldsymbol{W}^T. \tag{5.118}$$

$\boldsymbol{M}_B$ is given by Eq. (5.13). $\boldsymbol{B}_B$ is:

$$\boldsymbol{B}_B = \begin{bmatrix} \boldsymbol{Q}_{00} & \boldsymbol{Q}_{01} & \boldsymbol{Q}_{02} & \boldsymbol{Q}_{03} \\ \boldsymbol{Q}_{10} & \boldsymbol{Q}_{11} & \boldsymbol{Q}_{12} & \boldsymbol{Q}_{13} \\ \boldsymbol{Q}_{20} & \boldsymbol{Q}_{21} & \boldsymbol{Q}_{22} & \boldsymbol{Q}_{23} \\ \boldsymbol{Q}_{30} & \boldsymbol{Q}_{31} & \boldsymbol{Q}_{32} & \boldsymbol{Q}_{33} \end{bmatrix}. \tag{5.119}$$

Setting $\boldsymbol{P}_C(u, w) \equiv \boldsymbol{P}_B(u, w)$:

$$\boldsymbol{M}_C\boldsymbol{B}_C\boldsymbol{M}_C^T = \boldsymbol{M}_B\boldsymbol{B}_B\boldsymbol{M}_B^T. \tag{5.120}$$

This implies:

$$\begin{aligned} \boldsymbol{B}_B &= (\boldsymbol{M}_B^{-1}\boldsymbol{M}_C)\,\boldsymbol{B}_C\,\boldsymbol{M}_C^T(\boldsymbol{M}_B^T)^{-1} \\ &= (\boldsymbol{M}_B^{-1}\boldsymbol{M}_C)\,\boldsymbol{B}_C\,\boldsymbol{M}_C^T(\boldsymbol{M}_B^{-1})^T \\ &= (\boldsymbol{M}_B^{-1}\boldsymbol{M}_C)\,\boldsymbol{B}_C(\boldsymbol{M}_B^{-1}\boldsymbol{M}_C)^T. \end{aligned} \tag{5.121}$$

We also have:

$$\boldsymbol{M}_B^{-1} = \begin{bmatrix} 0 & 0 & 0 & 1 \\ 0 & 0 & \frac{1}{3} & 1 \\ 0 & \frac{1}{3} & \frac{2}{3} & 1 \\ 1 & 1 & 1 & 1 \end{bmatrix} \tag{5.122}$$

$$\boldsymbol{M}_B^{-1}\boldsymbol{M}_C = \begin{bmatrix} 1 & 0 & 0 & 0 \\ 1 & 0 & \frac{1}{3} & 0 \\ 0 & 1 & 0 & -\frac{1}{3} \\ 0 & 1 & 0 & 0 \end{bmatrix}. \tag{5.123}$$

So that:

$$\boldsymbol{B}_B = \begin{bmatrix} 1 & 0 & 0 & 0 \\ 1 & 0 & \frac{1}{3} & 0 \\ 0 & 1 & 0 & -\frac{1}{3} \\ 0 & 1 & 0 & 0 \end{bmatrix} \begin{bmatrix} \boldsymbol{Q}(0,0) & \boldsymbol{Q}(0,1) & \boldsymbol{Q}_w(0,0) & \boldsymbol{Q}_w(0,1) \\ \boldsymbol{Q}(1,0) & \boldsymbol{Q}(1,1) & \boldsymbol{Q}_w(1,0) & \boldsymbol{Q}_w(1,1) \\ \boldsymbol{Q}_u(0,0) & \boldsymbol{Q}_u(0,1) & \boldsymbol{Q}_{uw}(0,0) & \boldsymbol{Q}_{uw}(0,1) \\ \boldsymbol{Q}_u(1,0) & \boldsymbol{Q}_u(1,1) & \boldsymbol{Q}_{uw}(1,0) & \boldsymbol{Q}_{uw}(1,1) \end{bmatrix} \times$$

$$\times\begin{bmatrix} 1 & 1 & 0 & 0 \\ 0 & 0 & 1 & 1 \\ 0 & \frac{1}{3} & 0 & 0 \\ 0 & 0 & -\frac{1}{3} & 0 \end{bmatrix} \equiv \begin{bmatrix} \boldsymbol{Q}_{00} & \boldsymbol{Q}_{01} & \boldsymbol{Q}_{02} & \boldsymbol{Q}_{03} \\ \boldsymbol{Q}_{10} & \boldsymbol{Q}_{11} & \boldsymbol{Q}_{12} & \boldsymbol{Q}_{13} \\ \boldsymbol{Q}_{20} & \boldsymbol{Q}_{21} & \boldsymbol{Q}_{22} & \boldsymbol{Q}_{23} \\ \boldsymbol{Q}_{30} & \boldsymbol{Q}_{31} & \boldsymbol{Q}_{32} & \boldsymbol{Q}_{33} \end{bmatrix}. \tag{5.124}$$

Therefore:

$$\left.\begin{aligned}
\boldsymbol{Q}_{00} &= \boldsymbol{Q}(0,0) \\
\boldsymbol{Q}_{01} &= \boldsymbol{Q}(0,0) + \frac{1}{3}\,\boldsymbol{Q}_w(0,0) \\
\boldsymbol{Q}_{02} &= \boldsymbol{Q}(0,1) - \frac{1}{3}\,\boldsymbol{Q}_w(0,1) \\
\boldsymbol{Q}_{03} &= \boldsymbol{Q}(0,1) \\
\boldsymbol{Q}_{10} &= \boldsymbol{Q}(0,0) + \frac{1}{3}\,\boldsymbol{Q}_u(0,0) \\
\boldsymbol{Q}_{11} &= \boldsymbol{Q}(0,0) + \frac{1}{3}\,\boldsymbol{Q}_u(0,0) + \frac{1}{3}\,\boldsymbol{Q}_w(0,0) + \frac{1}{9}\,\boldsymbol{Q}_{uw}(0,0) \\
\boldsymbol{Q}_{12} &= \boldsymbol{Q}(0,1) + \frac{1}{3}\,\boldsymbol{Q}_u(0,1) - \frac{1}{3}\,\boldsymbol{Q}_w(0,1) + \frac{1}{9}\,\boldsymbol{Q}_{uw}(0,1) \\
\boldsymbol{Q}_{13} &= \boldsymbol{Q}(0,1) + \frac{1}{3}\,\boldsymbol{Q}_u(0,1) \\
\boldsymbol{Q}_{20} &= \boldsymbol{Q}(1,0) - \frac{1}{3}\,\boldsymbol{Q}_u(1,0) \\
\boldsymbol{Q}_{21} &= \boldsymbol{Q}(1,0) - \frac{1}{3}\,\boldsymbol{Q}_u(1,0) + \frac{1}{3}\,\boldsymbol{Q}_w(1,0) - \frac{1}{9}\,\boldsymbol{Q}_{uw}(1,0) \\
\boldsymbol{Q}_{22} &= \boldsymbol{Q}(1,1) - \frac{1}{3}\,\boldsymbol{Q}_u(1,1) - \frac{1}{3}\,\boldsymbol{Q}_w(1,1) + \frac{1}{9}\,\boldsymbol{Q}_{uw}(1,1) \\
\boldsymbol{Q}_{23} &= \boldsymbol{Q}(1,1) - \frac{1}{3}\,\boldsymbol{Q}_u(1,1) \\
\boldsymbol{Q}_{30} &= \boldsymbol{Q}(1,0) \\
\boldsymbol{Q}_{31} &= \boldsymbol{Q}(1,0) + \frac{1}{3}\,\boldsymbol{Q}_w(1,0) \\
\boldsymbol{Q}_{32} &= \boldsymbol{Q}(1,1) - \frac{1}{3}\,\boldsymbol{Q}_w(1,1) \\
\boldsymbol{Q}_{33} &= \boldsymbol{Q}(1,1)
\end{aligned}\right\} \tag{5.125}$$

Or, conversely, the surface defining vectors of a bi-cubic Coons surface patch can be expressed in terms of the defining vectors of a Bézier surface patch. From Eq. (5.120):

$$\boldsymbol{B}_C = (\boldsymbol{M}_C^{-1}\boldsymbol{M}_B)\boldsymbol{B}_B(\boldsymbol{M}_C^{-1}\boldsymbol{M}_B)^T.$$

Using $\boldsymbol{M}_C^{-1}$ of Eq. (3.14), we have:

$$\boldsymbol{M}_C^{-1}\boldsymbol{M}_B = \begin{bmatrix} 1 & 0 & 0 & 0 \\ 0 & 0 & 0 & 1 \\ -3 & 3 & 0 & 0 \\ 0 & 0 & -3 & 3 \end{bmatrix}.$$

Therefore:

$$\boldsymbol{B}_C \equiv \begin{bmatrix} \boldsymbol{Q}(0,0) & \boldsymbol{Q}(0,1) & \boldsymbol{Q}_w(0,0) & \boldsymbol{Q}_w(0,1) \\ \boldsymbol{Q}(1,0) & \boldsymbol{Q}(1,1) & \boldsymbol{Q}_w(1,0) & \boldsymbol{Q}_w(1,1) \\ \boldsymbol{Q}_u(0,0) & \boldsymbol{Q}_u(0,1) & \boldsymbol{Q}_{uw}(0,0) & \boldsymbol{Q}_{uw}(0,1) \\ \boldsymbol{Q}_u(1,0) & \boldsymbol{Q}_u(1,1) & \boldsymbol{Q}_{uw}(1,0) & \boldsymbol{Q}_{uw}(1,1) \end{bmatrix}$$

$$= \begin{bmatrix} 1 & 0 & 0 & 0 \\ 0 & 0 & 0 & 1 \\ -3 & 3 & 0 & 0 \\ 0 & 0 & -3 & 3 \end{bmatrix} \begin{bmatrix} \boldsymbol{Q}_{00} & \boldsymbol{Q}_{01} & \boldsymbol{Q}_{02} & \boldsymbol{Q}_{03} \\ \boldsymbol{Q}_{10} & \boldsymbol{Q}_{11} & \boldsymbol{Q}_{12} & \boldsymbol{Q}_{13} \\ \boldsymbol{Q}_{20} & \boldsymbol{Q}_{21} & \boldsymbol{Q}_{22} & \boldsymbol{Q}_{23} \\ \boldsymbol{Q}_{30} & \boldsymbol{Q}_{31} & \boldsymbol{Q}_{32} & \boldsymbol{Q}_{33} \end{bmatrix} \begin{bmatrix} 1 & 0 & -3 & 0 \\ 0 & 0 & 3 & 0 \\ 0 & 0 & 0 & -3 \\ 0 & 1 & 0 & 3 \end{bmatrix}$$

$$= \begin{bmatrix} \boldsymbol{Q}_{00} & \boldsymbol{Q}_{03} & 3(\boldsymbol{Q}_{01}-\boldsymbol{Q}_{00}) & 3(\boldsymbol{Q}_{03}-\boldsymbol{Q}_{02}) \\ \boldsymbol{Q}_{30} & \boldsymbol{Q}_{33} & 3(\boldsymbol{Q}_{31}-\boldsymbol{Q}_{30}) & 3(\boldsymbol{Q}_{33}-\boldsymbol{Q}_{32}) \\ 3(\boldsymbol{Q}_{10}-\boldsymbol{Q}_{00}) & 3(\boldsymbol{Q}_{13}-\boldsymbol{Q}_{03}) & 9(\boldsymbol{Q}_{00}-\boldsymbol{Q}_{01}-\boldsymbol{Q}_{10}+\boldsymbol{Q}_{11}) & 9(\boldsymbol{Q}_{02}-\boldsymbol{Q}_{03}-\boldsymbol{Q}_{12}+\boldsymbol{Q}_{13}) \\ 3(\boldsymbol{Q}_{30}-\boldsymbol{Q}_{20}) & 3(\boldsymbol{Q}_{33}-\boldsymbol{Q}_{23}) & 9(\boldsymbol{Q}_{20}-\boldsymbol{Q}_{21}-\boldsymbol{Q}_{30}+\boldsymbol{Q}_{31}) & 9(\boldsymbol{Q}_{22}-\boldsymbol{Q}_{23}-\boldsymbol{Q}_{32}+\boldsymbol{Q}_{33}) \end{bmatrix} \tag{5.126}$$

From Eq. (5.126), we see that among the surface defining vectors of a bi-cubic Coons surface patch, the tangent vectors are related only to the vertex vectors adjacent to the 4 corners of the Bézier surface; while the internal vertex vectors $\boldsymbol{Q}_{11}$, $\boldsymbol{Q}_{12}$, $\boldsymbol{Q}_{21}$ and $\boldsymbol{Q}_{22}$ affect the twist vector.

5.2.3 Connection of Bézier Surface Patches[24)]

Let:

$$\boldsymbol{P}_{\mathrm{I}}(u,w) = [u^3 \;\; u^2 \;\; u \;\; 1]\boldsymbol{M}_B\boldsymbol{B}_{B,\mathrm{I}}\boldsymbol{M}_B^T[w^3 \;\; w^2 \;\; w \;\; 1]^T \quad (0 \leqq u,\; w \leqq 1)$$

$$\boldsymbol{P}_{\mathrm{II}}(u,w) = [u^3 \;\; u^2 \;\; u \;\; 1]\boldsymbol{M}_B\boldsymbol{B}_{B,\mathrm{II}}\boldsymbol{M}_B^T[w^3 \;\; w^2 \;\; w \;\; 1]^T \quad (0 \leqq u,\; w \leqq 1)$$

be two bi-cubic Bézier surface patches. Suppose that we are to connect the $u=0$ boundary curve of $_{\mathrm{II}}(u, w)$ to the $u=1$ boundary curve of $\boldsymbol{P}_{\mathrm{I}}(u, w)$ (Fig. 5.34).

The condition for positions to be connected is:

$$[0\ 0\ 0\ 1]\boldsymbol{M}_B\boldsymbol{B}_{B,\mathrm{II}}\boldsymbol{M}_B^T[w^3\ \ w^2\ \ w\ \ 1]^T$$
$$=[1\ 1\ 1\ 1]\boldsymbol{M}_B\boldsymbol{B}_{B,\mathrm{I}}\boldsymbol{M}_B^T[w^3\ \ w^2\ \ w\ \ 1]^T \quad (0\leqq w\leqq 1).$$

Therefore:

$$[0\ 0\ 0\ 1]\boldsymbol{M}_B\boldsymbol{B}_{B,\mathrm{II}}=[1\ 1\ 1\ 1]\boldsymbol{M}_B\boldsymbol{B}_{B,\mathrm{I}}.$$

Expanding this gives:

$$[0\ 0\ 0\ 1]\begin{bmatrix}-1 & 3 & -3 & 1\\ 3 & -6 & 3 & 0\\ -3 & 3 & 0 & 0\\ 1 & 0 & 0 & 0\end{bmatrix}\begin{bmatrix}\boldsymbol{Q}_{\mathrm{II},00} & \boldsymbol{Q}_{\mathrm{II},01} & \boldsymbol{Q}_{\mathrm{II},02} & \boldsymbol{Q}_{\mathrm{II},03}\\ \boldsymbol{Q}_{\mathrm{II},10} & \boldsymbol{Q}_{\mathrm{II},11} & \boldsymbol{Q}_{\mathrm{II},12} & \boldsymbol{Q}_{\mathrm{II},13}\\ \boldsymbol{Q}_{\mathrm{II},20} & \boldsymbol{Q}_{\mathrm{II},21} & \boldsymbol{Q}_{\mathrm{II},22} & \boldsymbol{Q}_{\mathrm{II},23}\\ \boldsymbol{Q}_{\mathrm{II},30} & \boldsymbol{Q}_{\mathrm{II},31} & \boldsymbol{Q}_{\mathrm{II},32} & \boldsymbol{Q}_{\mathrm{II},33}\end{bmatrix}$$
$$=[1\ 1\ 1\ 1]\begin{bmatrix}-1 & 3 & -3 & 1\\ 3 & -6 & 3 & 0\\ -3 & 3 & 0 & 0\\ 1 & 0 & 0 & 0\end{bmatrix}\begin{bmatrix}\boldsymbol{Q}_{\mathrm{I},00} & \boldsymbol{Q}_{\mathrm{I},01} & \boldsymbol{Q}_{\mathrm{I},02} & \boldsymbol{Q}_{\mathrm{I},03}\\ \boldsymbol{Q}_{\mathrm{I},10} & \boldsymbol{Q}_{\mathrm{I},11} & \boldsymbol{Q}_{\mathrm{I},12} & \boldsymbol{Q}_{\mathrm{I},13}\\ \boldsymbol{Q}_{\mathrm{I},20} & \boldsymbol{Q}_{\mathrm{I},21} & \boldsymbol{Q}_{\mathrm{I},22} & \boldsymbol{Q}_{\mathrm{I},23}\\ \boldsymbol{Q}_{\mathrm{I},30} & \boldsymbol{Q}_{\mathrm{I},31} & \boldsymbol{Q}_{\mathrm{I},32} & \boldsymbol{Q}_{\mathrm{I},33}\end{bmatrix}$$

which implies:

$$\boldsymbol{Q}_{\mathrm{II},0i}=\boldsymbol{Q}_{\mathrm{I},3i} \quad (i=0,\ 1,\ 2,\ 3). \tag{5.127}$$

This relation expresses the fact that in order for the two surface patches to have the same boundary curve, they must have the same surface defining vectors corresponding to that boundary curve.

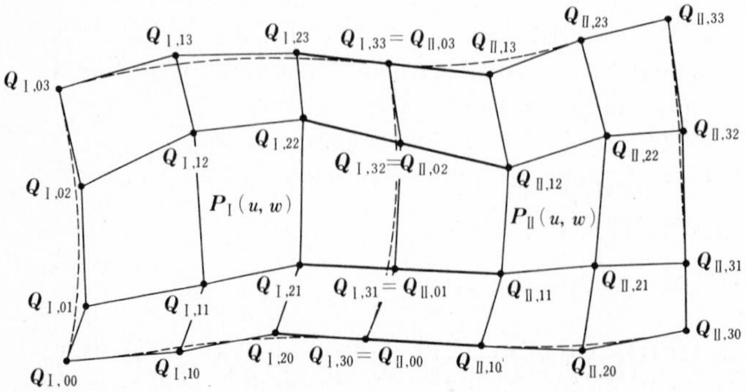

Fig. 5.34. Connection of cubic Bézier surface patches (simple method)

Next, let us find the condition that the slopes be continuous in the direction across the boundary curve. In general, this relation is given by Eq. (1.122), but first let us consider the special case given in Eq. (1.123).

The partial derivatives of each surface with respect to u are:

$$\boldsymbol{P}_{\mathrm{I},u}(u,w)=[3u^2\ 2u\ 1\ 0]\boldsymbol{M}_B\boldsymbol{B}_{B,\mathrm{I}}\boldsymbol{M}_B^T[w^3\ w^2\ w\ 1]^T \tag{5.128}$$

$$\boldsymbol{P}_{\mathrm{II},u}(u,w)=[3u^2\ 2u\ 1\ 0]\boldsymbol{M}_B\boldsymbol{B}_{B,\mathrm{II}}\boldsymbol{M}_B^T[w^3\ w^2\ w\ 1]^T. \tag{5.129}$$

We find $\boldsymbol{P}_{\mathrm{I},u}(1,w)$ and $\boldsymbol{P}_{\mathrm{II},u}(0,w)$ from these relations, then substitute in Eq. (1.123) to obtain:

$$\begin{aligned}&[0\ 0\ 1\ 0]\boldsymbol{M}_B\boldsymbol{B}_{B,\mathrm{II}}\boldsymbol{M}_B^T[w^3\ w^2\ w\ 1]^T\\ &=\mu(w)[3\ 2\ 1\ 0]\boldsymbol{M}_B\boldsymbol{B}_{B,\mathrm{I}}\boldsymbol{M}_B^T[w^3\ w^2\ w\ 1]^T.\end{aligned} \tag{5.130}$$

Since the surface on the left-hand side of this equation is cubic in w, take $\mu(w)=\mu$. This can be simplified to:

$$[0\ 0\ 1\ 0]\boldsymbol{M}_B\boldsymbol{B}_{B,\mathrm{II}}=\mu[3\ 2\ 1\ 0]\boldsymbol{M}_B\boldsymbol{B}_{B,\mathrm{I}}$$

which becomes:

$$\begin{aligned}&[0\ 0\ 1\ 0]\begin{bmatrix}-1&3&-3&1\\3&-6&3&0\\-3&3&0&0\\1&0&0&0\end{bmatrix}\begin{bmatrix}\boldsymbol{Q}_{\mathrm{II},00}&\boldsymbol{Q}_{\mathrm{II},01}&\boldsymbol{Q}_{\mathrm{II},02}&\boldsymbol{Q}_{\mathrm{II},03}\\\boldsymbol{Q}_{\mathrm{II},10}&\boldsymbol{Q}_{\mathrm{II},11}&\boldsymbol{Q}_{\mathrm{II},12}&\boldsymbol{Q}_{\mathrm{II},13}\\\boldsymbol{Q}_{\mathrm{II},20}&\boldsymbol{Q}_{\mathrm{II},21}&\boldsymbol{Q}_{\mathrm{II},22}&\boldsymbol{Q}_{\mathrm{II},23}\\\boldsymbol{Q}_{\mathrm{II},30}&\boldsymbol{Q}_{\mathrm{II},31}&\boldsymbol{Q}_{\mathrm{II},32}&\boldsymbol{Q}_{\mathrm{II},33}\end{bmatrix}\\ &=\mu[3\ 2\ 1\ 0]\begin{bmatrix}-1&3&-3&1\\3&-6&3&0\\-3&3&0&0\\1&0&0&0\end{bmatrix}\begin{bmatrix}\boldsymbol{Q}_{\mathrm{I},00}&\boldsymbol{Q}_{\mathrm{I},01}&\boldsymbol{Q}_{\mathrm{I},02}&\boldsymbol{Q}_{\mathrm{I},03}\\\boldsymbol{Q}_{\mathrm{I},10}&\boldsymbol{Q}_{\mathrm{I},11}&\boldsymbol{Q}_{\mathrm{I},12}&\boldsymbol{Q}_{\mathrm{I},13}\\\boldsymbol{Q}_{\mathrm{I},20}&\boldsymbol{Q}_{\mathrm{I},21}&\boldsymbol{Q}_{\mathrm{I},22}&\boldsymbol{Q}_{\mathrm{I},23}\\\boldsymbol{Q}_{\mathrm{I},30}&\boldsymbol{Q}_{\mathrm{I},31}&\boldsymbol{Q}_{\mathrm{I},32}&\boldsymbol{Q}_{\mathrm{I},33}\end{bmatrix}.\end{aligned}$$

This can be expanded to:

$$\boldsymbol{Q}_{\mathrm{II},1i}-\boldsymbol{Q}_{\mathrm{II},0i}=\mu(\boldsymbol{Q}_{\mathrm{I},3i}-\boldsymbol{Q}_{\mathrm{I},2i})\quad(i=0,\ 1,\ 2,\ 3). \tag{5.131}$$

This means that $\boldsymbol{Q}_{\mathrm{I},2i}$, $\boldsymbol{Q}_{\mathrm{I},3i}=\boldsymbol{Q}_{\mathrm{II},0i}$, $\boldsymbol{Q}_{\mathrm{II},1i}$ must be colinear (Fig. 5.34).

Next, let us look for general connection conditions based on Eq. (1.122). For convenience let us repeat Eq. (1.122) here.

$$\boldsymbol{P}_{\mathrm{II},u}(0,w)=\mu(w)\boldsymbol{P}_{\mathrm{I},u}(1,w)+\gamma(w)\boldsymbol{P}_{\mathrm{I},w}(1,w). \tag{5.132}$$

This connection condition expresses the fact that $\boldsymbol{P}_{\mathrm{II},u}(0,w)$ on the No. 2 surface patch must lie on the tangent plane to a point on the boundary curve of No. 1 surface patch. Equation (5.132) becomes the following:

$$[0\ 0\ 1\ 0]\,\boldsymbol{M}_B\boldsymbol{B}_{B,\mathrm{II}}\boldsymbol{M}_B^T\begin{bmatrix}w^3\\w^2\\w\\1\end{bmatrix}=\mu(w)\,[3\ 2\ 1\ 0]\,\boldsymbol{M}_B\boldsymbol{B}_{B,\mathrm{I}}\boldsymbol{M}_B^T\begin{bmatrix}w^3\\w^2\\w\\1\end{bmatrix}$$

$$+\gamma(w)\,[1\ 1\ 1\ 1]\,\boldsymbol{M}_B\boldsymbol{B}_{B,\mathrm{I}}\boldsymbol{M}_B^T\begin{bmatrix}3w^3\\2w\\1\\0\end{bmatrix}. \tag{5.133}$$

If we assume that both of the surface patches are bi-cubic, then, for the same reason as before, $\mu(w)$ is an arbitrary positive scalar μ. Also, $\gamma(w)$ can be expressed as the linear form $\gamma(w)=\gamma_0+\gamma_1 w$. In this case, the left-hand side of Eq. (5.133) becomes:

$$\text{left-hand side}=[0\ 0\ 1\ 0]\,\boldsymbol{M}_B\boldsymbol{B}_{B,\mathrm{II}}\boldsymbol{M}_B^T\,[w^3\ w^2\ w\ 1]^T$$

$$=[0\ 0\ 1\ 0]\begin{bmatrix}-1&3&-3&1\\3&-6&3&0\\-3&3&0&0\\1&0&0&0\end{bmatrix}\begin{bmatrix}\boldsymbol{Q}_{\mathrm{II},00}&\boldsymbol{Q}_{\mathrm{II},01}&\boldsymbol{Q}_{\mathrm{II},02}&\boldsymbol{Q}_{\mathrm{II},03}\\\boldsymbol{Q}_{\mathrm{II},10}&\boldsymbol{Q}_{\mathrm{II},11}&\boldsymbol{Q}_{\mathrm{II},12}&\boldsymbol{Q}_{\mathrm{II},13}\\\boldsymbol{Q}_{\mathrm{II},20}&\boldsymbol{Q}_{\mathrm{II},21}&\boldsymbol{Q}_{\mathrm{II},22}&\boldsymbol{Q}_{\mathrm{II},23}\\\boldsymbol{Q}_{\mathrm{II},30}&\boldsymbol{Q}_{\mathrm{II},31}&\boldsymbol{Q}_{\mathrm{II},32}&\boldsymbol{Q}_{\mathrm{II},33}\end{bmatrix}$$

$$\times\begin{bmatrix}-1&3&-3&1\\3&-6&3&0\\-3&3&0&0\\1&0&0&0\end{bmatrix}\begin{bmatrix}w^3\\w^2\\w\\1\end{bmatrix}$$

$$=3\,[w^3\ w^2\ w\ 1]$$

$$\times\begin{bmatrix}\boldsymbol{Q}_{\mathrm{II},00}-3\boldsymbol{Q}_{\mathrm{II},01}+3\boldsymbol{Q}_{\mathrm{II},02}-\boldsymbol{Q}_{\mathrm{II},03}-\boldsymbol{Q}_{\mathrm{II},10}+3\boldsymbol{Q}_{\mathrm{II},11}-3\boldsymbol{Q}_{\mathrm{II},12}+\boldsymbol{Q}_{\mathrm{II},13}\\-3\boldsymbol{Q}_{\mathrm{II},00}+6\boldsymbol{Q}_{\mathrm{II},01}-3\boldsymbol{Q}_{\mathrm{II},02}+3\boldsymbol{Q}_{\mathrm{II},10}-6\boldsymbol{Q}_{\mathrm{II},11}+3\boldsymbol{Q}_{\mathrm{II},12}\\3\boldsymbol{Q}_{\mathrm{II},00}-3\boldsymbol{Q}_{\mathrm{II},01}-3\boldsymbol{Q}_{\mathrm{II},10}+3\boldsymbol{Q}_{\mathrm{II},11}\\-\boldsymbol{Q}_{\mathrm{II},00}+\boldsymbol{Q}_{\mathrm{II},10}\end{bmatrix}$$

The 1-st term on the right-hand side of Eq. (5.133) is:

$$\text{1-st term on the right-hand side}=\mu(w)\,[3\ 2\ 1\ 0]\,\boldsymbol{M}_B\boldsymbol{B}_{B,\mathrm{I}}\boldsymbol{M}_B^T\,[w^3\ w^2\ w\ 1]^T$$

$$=\mu\,[3\ 2\ 1\ 0]\begin{bmatrix}-1&3&-3&1\\3&-6&3&0\\-3&3&0&0\\1&0&0&0\end{bmatrix}\begin{bmatrix}\boldsymbol{Q}_{\mathrm{I},00}&\boldsymbol{Q}_{\mathrm{I},01}&\boldsymbol{Q}_{\mathrm{I},02}&\boldsymbol{Q}_{\mathrm{I},03}\\\boldsymbol{Q}_{\mathrm{I},10}&\boldsymbol{Q}_{\mathrm{I},11}&\boldsymbol{Q}_{\mathrm{I},12}&\boldsymbol{Q}_{\mathrm{I},13}\\\boldsymbol{Q}_{\mathrm{I},20}&\boldsymbol{Q}_{\mathrm{I},21}&\boldsymbol{Q}_{\mathrm{I},22}&\boldsymbol{Q}_{\mathrm{I},23}\\\boldsymbol{Q}_{\mathrm{I},30}&\boldsymbol{Q}_{\mathrm{I},31}&\boldsymbol{Q}_{\mathrm{I},32}&\boldsymbol{Q}_{\mathrm{I},33}\end{bmatrix}\times$$

$$\times\begin{bmatrix}-1 & 3 & -3 & 1\\ 3 & -6 & 3 & 0\\ -3 & 3 & 0 & 0\\ 1 & 0 & 0 & 0\end{bmatrix}\begin{bmatrix}w^3\\ w^2\\ w\\ 1\end{bmatrix}$$

$$=3\,[w^3\ \ w^2\ \ w\ \ 1]$$

$$\times\begin{bmatrix}\mu(\boldsymbol{Q}_{\mathrm{I},20}-3\boldsymbol{Q}_{\mathrm{I},21}+3\boldsymbol{Q}_{\mathrm{I},22}-\boldsymbol{Q}_{\mathrm{I},23}-\boldsymbol{Q}_{\mathrm{I},30}+3\boldsymbol{Q}_{\mathrm{I},31}-3\boldsymbol{Q}_{\mathrm{I},32}+\boldsymbol{Q}_{\mathrm{I},33})\\ \mu(-3\boldsymbol{Q}_{\mathrm{I},20}+6\boldsymbol{Q}_{\mathrm{I},21}-3\boldsymbol{Q}_{\mathrm{I},22}+3\boldsymbol{Q}_{\mathrm{I},30}-6\boldsymbol{Q}_{\mathrm{I},31}+3\boldsymbol{Q}_{\mathrm{I},32})\\ \mu(3\boldsymbol{Q}_{\mathrm{I},20}-3\boldsymbol{Q}_{\mathrm{I},21}-3\boldsymbol{Q}_{\mathrm{I},30}+3\boldsymbol{Q}_{\mathrm{I},31})\\ \mu(-\boldsymbol{Q}_{\mathrm{I},20}+\boldsymbol{Q}_{\mathrm{I},30})\end{bmatrix}.$$

The 2-nd term on the right-hand side of Eq. (5.133) is:

2-nd term on right-hand side $=\gamma(w)\ [1\ \ 1\ \ 1\ \ 1]\,\boldsymbol{M}_B\boldsymbol{B}_{B,\mathrm{I}}\boldsymbol{M}_B^T\,[3w^2\ \ 2w\ \ 1\ \ 0]^T$

$$=(\gamma_0+\gamma_1 w)\,[1\ \ 1\ \ 1\ \ 1]\begin{bmatrix}-1 & 3 & -3 & 1\\ 3 & -6 & 3 & 0\\ -3 & 3 & 0 & 0\\ 1 & 0 & 0 & 0\end{bmatrix}\begin{bmatrix}\boldsymbol{Q}_{\mathrm{I},00} & \boldsymbol{Q}_{\mathrm{I},01} & \boldsymbol{Q}_{\mathrm{I},02} & \boldsymbol{Q}_{\mathrm{I},03}\\ \boldsymbol{Q}_{\mathrm{I},10} & \boldsymbol{Q}_{\mathrm{I},11} & \boldsymbol{Q}_{\mathrm{I},12} & \boldsymbol{Q}_{\mathrm{I},13}\\ \boldsymbol{Q}_{\mathrm{I},20} & \boldsymbol{Q}_{\mathrm{I},21} & \boldsymbol{Q}_{\mathrm{I},22} & \boldsymbol{Q}_{\mathrm{I},23}\\ \boldsymbol{Q}_{\mathrm{I},30} & \boldsymbol{Q}_{\mathrm{I},31} & \boldsymbol{Q}_{\mathrm{I},32} & \boldsymbol{Q}_{\mathrm{I},33}\end{bmatrix}$$

$$\times\begin{bmatrix}-1 & 3 & -3 & 1\\ 3 & -6 & 3 & 0\\ -3 & 3 & 0 & 0\\ 1 & 0 & 0 & 0\end{bmatrix}\begin{bmatrix}0 & 3 & 0 & 0\\ 0 & 0 & 2 & 0\\ 0 & 0 & 0 & 1\\ 0 & 0 & 0 & 0\end{bmatrix}\begin{bmatrix}w^3\\ w^2\\ w\\ 1\end{bmatrix}$$

$$=3\,[w^3\ \ w^2\ \ w\ \ 1]$$

$$\times\begin{bmatrix}\gamma_1(-\boldsymbol{Q}_{\mathrm{I},30}+3\boldsymbol{Q}_{\mathrm{I},31}-3\boldsymbol{Q}_{\mathrm{I},32}+\boldsymbol{Q}_{\mathrm{I},33})\\ \gamma_0(-\boldsymbol{Q}_{\mathrm{I},30}+3\boldsymbol{Q}_{\mathrm{I},31}-3\boldsymbol{Q}_{\mathrm{I},32}+\boldsymbol{Q}_{\mathrm{I},33})+\gamma_1(2\boldsymbol{Q}_{\mathrm{I},30}-4\boldsymbol{Q}_{\mathrm{I},31}+2\boldsymbol{Q}_{\mathrm{I},32})\\ \gamma_0(2\boldsymbol{Q}_{\mathrm{I},30}-4\boldsymbol{Q}_{\mathrm{I},31}+2\boldsymbol{Q}_{\mathrm{I},32})+\gamma_1(-\boldsymbol{Q}_{\mathrm{I},30}+\boldsymbol{Q}_{\mathrm{I},31})\\ \gamma_0(-\boldsymbol{Q}_{\mathrm{I},30}+\boldsymbol{Q}_{\mathrm{I},31})\end{bmatrix}.$$

Comparing the constant terms on both sides of Eq. (5.133) we obtain:

$$\boldsymbol{Q}_{\mathrm{II},10}-\boldsymbol{Q}_{\mathrm{II},00}=\mu(\boldsymbol{Q}_{\mathrm{I},30}-\boldsymbol{Q}_{\mathrm{I},20})+\gamma_0(\boldsymbol{Q}_{\mathrm{I},31}-\boldsymbol{Q}_{\mathrm{I},30}). \tag{5.134}$$

Next, comparing the coefficients of the w terms:

$$\begin{aligned}3\boldsymbol{Q}_{\mathrm{II},00}-3\boldsymbol{Q}_{\mathrm{II},01}-3\boldsymbol{Q}_{\mathrm{II},10}+3\boldsymbol{Q}_{\mathrm{II},11}=\ &\mu\,(3\boldsymbol{Q}_{\mathrm{I},20}-3\boldsymbol{Q}_{\mathrm{I},21}-3\boldsymbol{Q}_{\mathrm{I},30}+3\boldsymbol{Q}_{\mathrm{I},31})\\ &+\gamma_0(2\boldsymbol{Q}_{\mathrm{I},30}-4\boldsymbol{Q}_{\mathrm{I},31}+2\boldsymbol{Q}_{\mathrm{I},32})\\ &+\gamma_1(-\boldsymbol{Q}_{\mathrm{I},30}+\boldsymbol{Q}_{\mathrm{I},31}).\end{aligned}$$

Rearranging terms and using Eq. (5.134):

$$Q_{\text{II},11}-Q_{\text{II},01}=\mu(Q_{\text{I},31}-Q_{\text{I},21})+\frac{\gamma_0}{3}(2Q_{\text{I},32}-Q_{\text{I},30}-Q_{\text{I},31})$$

$$+\frac{\gamma_1}{3}(Q_{\text{I},31}-Q_{\text{I},30}). \tag{5.135}$$

Similarly, when the coefficients of the w^2 and w^3 terms are compared the following 2 equations are obtained:

$$Q_{\text{II},12}-Q_{\text{II},02}=\mu(Q_{\text{I},32}-Q_{\text{I},22})+\frac{\gamma_0}{3}(Q_{\text{I},32}+Q_{\text{I},33}-2Q_{\text{I},31})$$

$$+\frac{2\gamma_1}{3}(Q_{\text{I},32}-Q_{\text{I},31}). \tag{5.136}$$

$$Q_{\text{II},13}-Q_{\text{II},03}=\mu(Q_{\text{I},33}-Q_{\text{I},23})+(\gamma_0+\gamma_1)(Q_{\text{I},33}-Q_{\text{I},32}). \tag{5.137}$$

In the above 4 equations, (5.134), (5.135), (5.136) and (5.137), setting $\gamma_0=\gamma_1=0$ we can confirm that these reduce to Eq. (5.131).

The geometrical significance of Eq. (5.134) is that $Q_{\text{II},10}$ of the No. 2 surface patch lies on the plane formed by the vectors $Q_{\text{I},30}-Q_{\text{I},20}$ and $Q_{\text{I},31}-Q_{\text{I},30}$ of the No. 1 surface patch (refer to Fig. 5.35). Similarly, the significance of Eq. (5.137) is that $Q_{\text{II},13}$ of the No. 2 surface patch lies on the plane formed by $Q_{\text{I},33}-Q_{\text{I},23}$ and $Q_{\text{I},33}-Q_{\text{I},32}$.

The geometrical interpretation of Eq. (5.135) and (5.136) is not as simple as the above 2 cases.

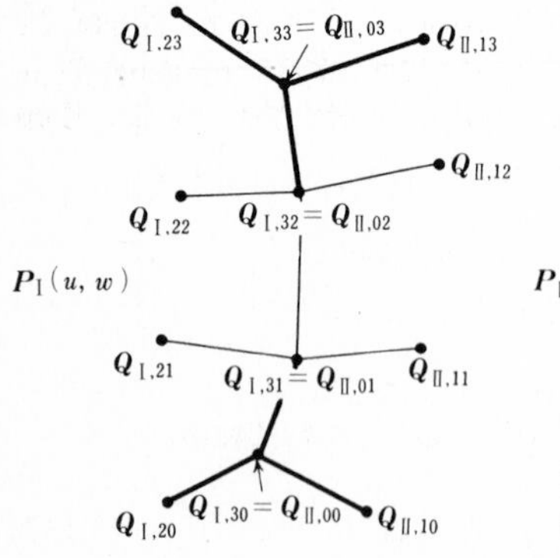

Fig. 5.35. Connection of cubic Bézier surface patches (general method)

5.2.4 Triangular Patches Formed by Degeneration

When a triangular surface patch is to be expressed by an oridnary quadrangular Bézier surface patch, let us consider the conditions for the unit normal vector at the degenerate point to be uniquely determined (refer to Sect. 1.3.8). Assume that we are dealing with a bi-cubic Bézier surface patch. Then we have:

$$P(u,w) = UM_B B_B M_B^T W^T$$

$P_u(0,w)$ in Eq. (1.128) can be found to be:

$$\begin{aligned} P_u(0,w) &= [0\ 0\ 1\ 0]\, M_B B_B M_B^T\, [w^3\ w^2\ w\ 1]^T \\ &= 3\,B_{0,3}(w)\,(Q_{10}-Q_{00}) + 3\,B_{1,3}(w)\,(Q_{11}-Q_{01}) \\ &\quad + 3\,B_{2,3}(w)\,(Q_{12}-Q_{02}) + 3\,B_{3,3}(w)\,(Q_{13}-Q_{03}). \end{aligned}$$

In Fig. 5.36, at the degenerate point D we have $Q_{00}=Q_{01}=Q_{02}=Q_{03}$, so:

$$\begin{aligned} P_u(0,w) &= 3\,B_{0,3}(w)\,(Q_{10}-Q_{00}) + 3\,B_{1,3}(w)\,(Q_{11}-Q_{00}) \\ &\quad + 3\,B_{2,3}(w)\,(Q_{12}-Q_{00}) + 3\,B_{3,3}(w)\,(Q_{13}-Q_{00}). \end{aligned} \tag{5.138}$$

We also have, for $P_{uw}(0,w)$:

$$\begin{aligned} P_{uw}(0,w) &= [0\ 0\ 1\ 0]\, M_B B_B M_B^T\, [3w^2\ 2w\ 1\ 0]^T \\ &= 3\,\dot{B}_{0,3}(w)\,(Q_{10}-Q_{00}) + 3\,\dot{B}_{1,3}(w)\,(Q_{11}-Q_{00}) \\ &\quad + 3\,\dot{B}_{2,3}(w)\,(Q_{12}-Q_{00}) + 3\,\dot{B}_{3,3}(w)\,(Q_{13}-Q_{00}). \end{aligned} \tag{5.139}$$

As can be seen from Eqs. (5.138) and (5.139), $P_u(0,w)$ and $P_{uw}(0,w)$ are both expressed as linear combinations of the four vectors $Q_{10}-Q_{00}$, $Q_{11}-Q_{00}$, $Q_{12}-Q_{00}$ and $Q_{13}-Q_{00}$. From the properties of a Bézier surface, the direction of the tangent of curve $P(u,0)$ at point D coincides with the direction of $Q_{10}-Q_{00}$; also, the direction of the tangent to curve $P(u,1)$ at point D coincides with the direction of $Q_{13}-Q_{00}$, so the tangent plane at point D is formed by these 2 vectors. Therefore, if Q_{11} and Q_{12} lie on this plane, then

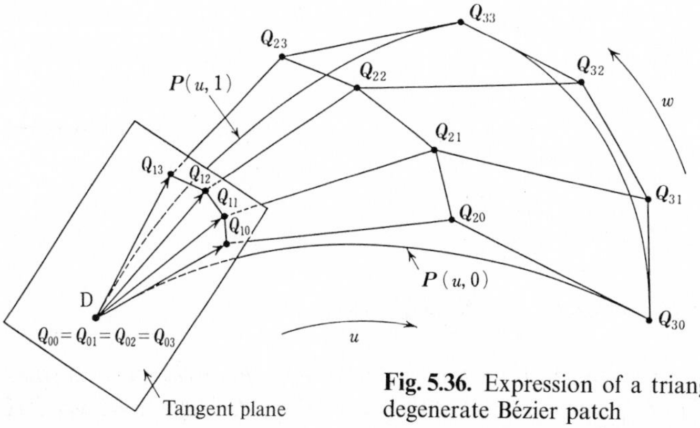

Fig. 5.36. Expression of a triangular surface patch as a degenerate Bézier patch

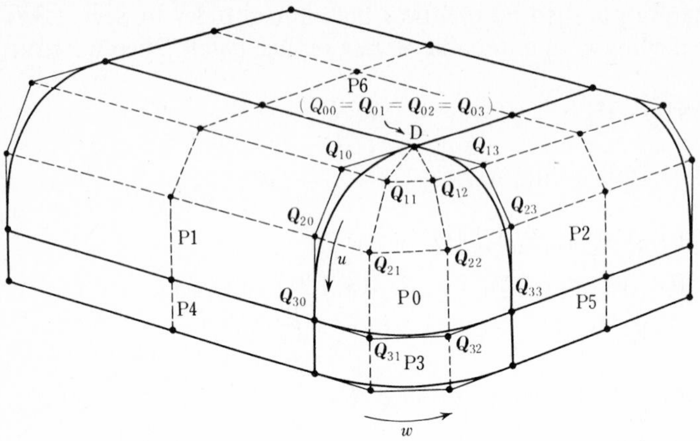

Fig. 5.37. Application of a degenerate Bézier surface patch (rounded convex corner). D, degenerate point of patch; •, surface defining point $\boldsymbol{Q}_{ij}$

$\boldsymbol{P}_u(0, w)$ and $\boldsymbol{P}_{uw}(0, w)$ are vectors on the tangent plane regardless of the value of w, so the unit normal vector $\boldsymbol{e}$ is uniquely determined as:

$$\boldsymbol{e} = \frac{\boldsymbol{P}_u(0, w) \times \boldsymbol{P}_{uw}(0, w)}{|\boldsymbol{P}_u(0, w) \times \boldsymbol{P}_{uw}(0, w)|}.$$

If three edges of a cuboid are rounded, the triangular part that is produced at a corner can be expressed as a quadrangular Bézier surface patch, as shown in Fig. 5.37. Point D is the degenerate point of the patch, where the four surface defining vectors of the patch P0, $\boldsymbol{Q}_{00}$, $\boldsymbol{Q}_{01}$, $\boldsymbol{Q}_{02}$ and $\boldsymbol{Q}_{03}$, coincide. Since $\boldsymbol{Q}_{10}$, $\boldsymbol{Q}_{11}$, $\boldsymbol{Q}_{12}$ and $\boldsymbol{Q}_{13}$ are on an extension of patch P6 (plane), the normal vector at point D can be found uniquely by the method described above. P0 share the same boundary curve defining vectors with surface patches P1, P2, P3, P4 and P5, and, moreover, the three surface defining vectors in the direction across the boundary curve are colinear, so the patches are connected with continuity up to the slope.

5.2.5 Triangular Patches

Parametric curves can be expressed as the sum of products of a number of basis functions and the same number of vectors. In the case of a Bézier curve, the basis functions (Bernstein basis functions) are the terms of the binomial expansion:

$$[(1-t)+t]^n.$$

A Bézier surface can be expressed as a sum of products of basis functions, which

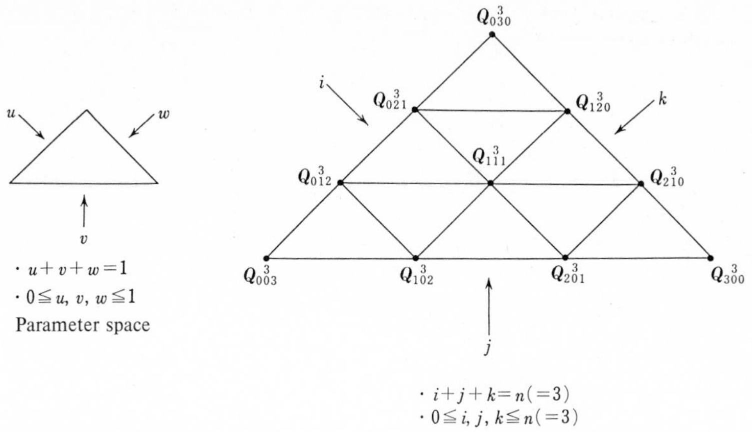

Fig. 5.38. Triangular patch expressed by the Bézier surface formula

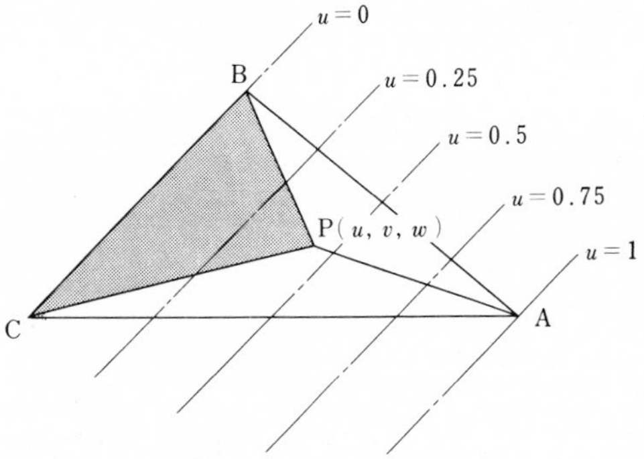

Fig. 5.39. Triangular area coordinates

are expressed as the terms of products of two binomial expansions, and position vectors which are given as a lattice.

$$[(1-u)+u]^m [(1-w)+w]^n. \tag{5.140}$$

A triangular patch can be defined by position vectors which are given in triangular form (Fig. 5.38). Let u, v and w be the area coordinates of a triangle. Specifically, in the triangle ABC (Fig. 5.39), one point inside the triangle is expressed uniquely by the three coordinates u, v and w:

$$u = (\text{area of } \triangle\text{PBC})/(\text{area of } \triangle\text{ABC}),$$

$$v = (\text{area of } \triangle\text{PCA})/(\text{area of } \triangle\text{ABC}),$$

$$w = (\text{area of } \triangle\text{PAB})/(\text{area of } \triangle\text{ABC}).$$

u, v and w are not mutually independent.

$$u+v+w=1. \tag{5.141}$$

The parameter u is 0 when point P is on side BC, and 1 when point P coincides with vertex A. Similar relations hold for the other parameters v and w:

$$0 \leqq u,\ v,\ w \leqq 1. \tag{5.142}$$

Also, u has same values when point P is on a line parallel to side BC. Similar relations hold for v and w.

As shown in Fig. 5.38, a sequence of points given in a triangular lattice can be expressed in the notation $\boldsymbol{Q}^n_{ijk}$. Here we have the relations:

$$0 \leqq i,\ j,\ k \leqq n$$
$$i+j+k=n.$$

Since relation (5.141) holds among area coordinates, we can conceive of a surface formula of the following form using the coefficients of the terms in the expansion:

$$[u+(v+w)]^n$$

corresponding to Eq. (5.140):

$$\begin{aligned}\boldsymbol{P}(u,v,w) &= \sum_{i=0}^{n} \binom{n}{i} u^i \sum_{j=0}^{n-i} \binom{n-i}{j} v^j w^{n-i-j} \boldsymbol{Q}^n_{i,j,n-i-j} \\ &= \sum_{i=0}^{n} \sum_{j=0}^{n-i} \frac{n!}{i!\,j!\,k!} u^i v^j w^k \boldsymbol{Q}^n_{ijk} \quad (i+j+k=n).\end{aligned} \tag{5.143}$$

For example, taking $n=2$ this becomes:

$$\begin{aligned}\boldsymbol{P}(u,v,w) &= \sum_{i=0}^{2} \sum_{j=0}^{2-i} \frac{2!}{i!\,j!\,k!} u^i v^j w^k \boldsymbol{Q}^2_{ijk} \\ &= u^2\boldsymbol{Q}^2_{200} + v^2\boldsymbol{Q}^2_{020} + w^2\boldsymbol{Q}^2_{002} + 2uv\boldsymbol{Q}^2_{110} + 2vw\boldsymbol{Q}^2_{011} + 2uw\boldsymbol{Q}^2_{101}.\end{aligned}$$

Similarly, taking $n=3$ gives:

$$\begin{aligned}\boldsymbol{P}(u,v,w) &= \sum_{i=0}^{3} \sum_{j=0}^{3-i} \frac{3!}{i!\,j!\,k!} u^i v^j w^k \boldsymbol{Q}^3_{ijk} \\ &= u^3\boldsymbol{Q}^3_{300} + v^3\boldsymbol{Q}^3_{030} + w^3\boldsymbol{Q}^3_{003} + 3u^2v\boldsymbol{Q}^3_{210} + 3u^2w\boldsymbol{Q}^3_{201} \\ &\quad + 3v^2u\boldsymbol{Q}^3_{120} + 3v^2w\boldsymbol{Q}^3_{021} + 3w^2u\boldsymbol{Q}^3_{102} + 3w^2v\boldsymbol{Q}^3_{012} + 6uvw\boldsymbol{Q}^3_{111}.\end{aligned}$$

Now let us consider formula (5.143) with one of the three parameters, for example w, taken to be 0:

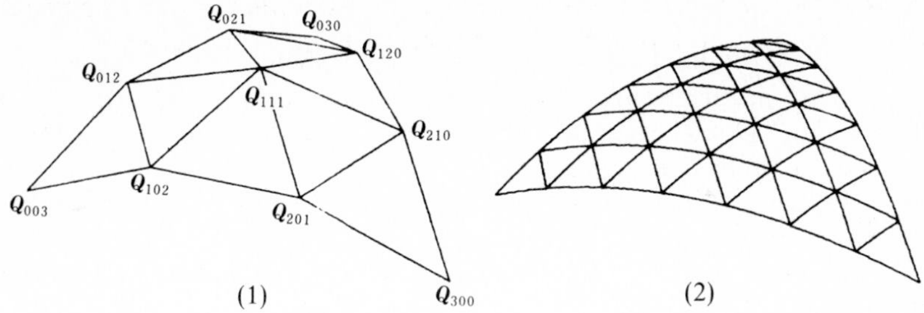

Fig. 5.40. Example of producing a triangular patch. (1) surface defining net; (2) surface

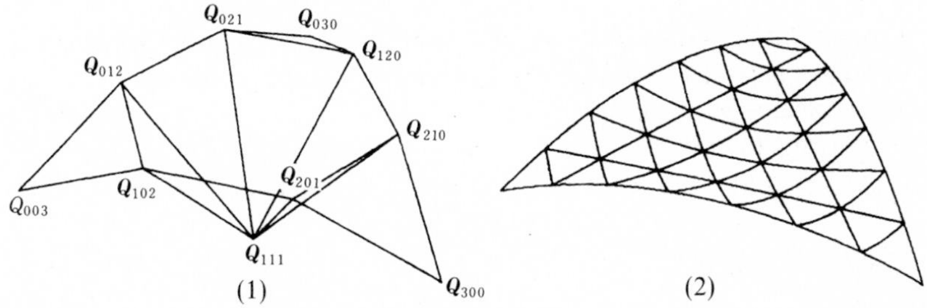

Fig. 5.41. Shape control of the triangular patch in Fig. 5.40. (1) surface defining net; (2) surface

$$\begin{aligned}\boldsymbol{P}(u,v,0) &= \sum_{i=0}^{n} \sum_{j=0}^{n-i} \frac{n!}{i!\,j!\,k!} u^i v^j \boldsymbol{Q}^n_{ij0} \quad (i+j=n;\ k=0;\ u+v=1)\\ &= \sum_{i=0}^{n} \frac{n!}{i!\,(n-i)!} u^i (1-u)^{n-i} \boldsymbol{Q}^n_{i,n-i,0}\,.\end{aligned}$$

This has the form of an n-th degree Bézier curve segment. Since the same is true of the other boundaries, the boundary curves of a triangular patch are n-th degree Bézier curve segments, with the position vectors corresponding to boundary curve segments as curve defining vectors. The triangular patch produced is shown in Fig. 5.40.

5.2.6 Some Considerations on Bézier Curves and Surfaces

Bézier curves and surfaces are defined only by position vectors. They do not require analytical data which is difficult to understand intuitively, such as tangent vectors and twist vectors, as do Hermite interpolation curves and surfaces. What are needed are geometrical data which are easily understood intuitively, such as polygons and polyhedrons. Moreover, Bézier curves and surfaces have the following superior geometrical properties.

① *(Convex Hull Property)*

Relation (5.43) and (5.49) hold for the coefficient function $B_{i,n}$ of a Bézier curve, so the Bézier curve segments are convex combinations of the curve defining vectors $\boldsymbol{Q}_0, \boldsymbol{Q}_1, \ldots, \boldsymbol{Q}_n$. That is to say, a Bézier curve segment is enclosed within the convex hull determined by these points. Consequently, if $\boldsymbol{Q}_0 = \boldsymbol{Q}_1 = \ldots \boldsymbol{Q}_n$, the curve degenerates to the single point $\boldsymbol{Q}_0$; in addition, if all of the curve defining vectors are placed to be colinear, it is the line that joins $\boldsymbol{Q}_0$ to $\boldsymbol{Q}_n$.

② *(Variation Diminishing Property)*

The number of intersections of an arbitrary straight line and a Bézier curve is never more than the number of intersections of that straight line with the curve defining polygon. Consequently, a Bézier curve takes a shape that is a smoothed version of the curve defining polygon shape, and does not contain any variations that are not in the shape of the curve defining polygon.

In summary, Bézier curves and surfaces preserve the principal features of the shapes of their respective curve defining polygons and nets (refer to Theorem 5.2), and do not include any important shape variations that are not in the curve defining polygons and nets, so shape control is easy for humans. In addition, the degree of a curve can be increased and the curve can be divided into two curve segments while preserving the shape, which increases flexibility in shape control.

One problem with Bézier curves is connection. It is theoretically possible to connect curve segments and surface patches smoothly, but when it comes to shape control by humans, a program that will preserve those conditions with flexibility and generality becomes complicated. In addition, the resemblance of Bézier curves and surfaces to their respective curve defining polygons and nets is not comparatively good (refer to the discussion of *B*-spline curves and surfaces in Chap. 6).

References (Chap. 5)

18) Yamaguchi, F.: "A Taxonomical Study of Computer Aided Geometric Design Systems", CS 653 Note 3, Computer Science, Univ. of Utah, 1979.
19) Davis, P. J.: "Interpolation & Approximation", Dover Publications, 1963.
20) Riesenfeld, R. F.: "Application of *B*-Spline Approximations to Geometric Problems of Computer-aided Design", *Univ. Utah Comput. Sci. Dept.* UTECCSc-73-126, March 1973, University Microfilms OP 65903.
21) Schoenberg, I. J.: "On Spline Functions", with Supplement by T.N.E. Greville, Inequalities (O. Shisha, editor), Academic Press, 1967.
22) Hosaka, M., and F. Kimura: "3-dimensional free shape design control theory and its methodology", *Joho Shori (Information Processing),* Vol. 21, No. 5, May 1980 (in Japanese).
23) Cohen, E., and R. Riesenfeld: "General Matrix Representations for Bézier and *B*-Spline Curves", unpublished, Dec. 1977.
24) Böhm, W.: "Cubic *B*-Spline Curves and Surfaces in Computer Aided Geometric Design", Computing, 19, 1977.

6. The B-Spline Approximation

6.1 Uniform Cubic B-Spline Curves

6.1.1 Derivation of the Curve Formula[26)]

If $(n+1)$ ordered position vectors $\boldsymbol{Q}_0, \boldsymbol{Q}_1, \ldots, \boldsymbol{Q}_{n-1}, \boldsymbol{Q}_n$ are given (Fig. 6.1), consider the $(n-2)$ linear combinations:

$$\boldsymbol{P}_i(t)=X_0(t)\boldsymbol{Q}_{i-1}+X_1(t)\boldsymbol{Q}_i+X_2(t)\boldsymbol{Q}_{i+1}+X_3(t)\boldsymbol{Q}_{i+2}$$
$$(i=1, 2, \ldots, n-2) \qquad (6.1)^{*)}$$

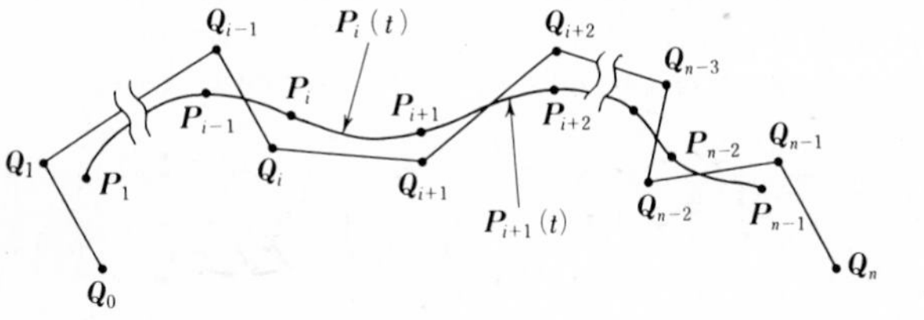

Fig. 6.1. Derivation of a B-spline curve (case of $M=4$)

each formed from four successive points. $X_0(t)$, $X_1(t)$, $X_2(t)$ and $X_3(t)$ are polynomials in the parameter $t\,(0\leqq t\leqq 1)$. $\boldsymbol{P}_i(t)$ is a curve segment expressed in terms of the varying parameter. The condition for two neighboring curve segments $\boldsymbol{P}_i(t)$ and $\boldsymbol{P}_{i+1}(t)$ to be continuous at the point corresponding to $t=1$ for the first segment and $t=0$ for the second, that is, for $\boldsymbol{P}_i(1)=\boldsymbol{P}_{i+1}(0)$ to hold for all $\boldsymbol{Q}_j$ $(j=i-1, i, \ldots, i+3)$, is:

$$\left.\begin{aligned}
X_0(1)&=X_3(0)=0\\
X_1(1)&=X_0(0)\\
X_2(1)&=X_1(0)\\
X_3(1)&=X_2(0)
\end{aligned}\right\} \qquad (6.2)$$

Similarly, the conditions that the 1-st derivative vector and the 2-nd derivative vector be continuous, that is, that $\dot{\boldsymbol{P}}_i(1)=\dot{\boldsymbol{P}}_{i+1}(0)$ and $\ddot{\boldsymbol{P}}_i(1)=\ddot{\boldsymbol{P}}_{i+1}(0)$, for all $\boldsymbol{Q}_j$ $(j=i-1, i, \ldots, i+3)$ become:

---

*) These functions $X_i(t)$ are different from the $X_i(t)$ in Sect. 5.1.1. In this book $X_i(t)$ is used for an unknown function.

$$\left.\begin{aligned}\dot{X}_0(1)&=\dot{X}_3(0)=0\\ \dot{X}_1(1)&=\dot{X}_0(0)\\ \dot{X}_2(1)&=\dot{X}_1(0)\\ \dot{X}_3(1)&=\dot{X}_2(0)\end{aligned}\right\} \tag{6.3}$$

$$\left.\begin{aligned}\ddot{X}_0(1)&=\ddot{X}_3(0)=0\\ \ddot{X}_1(1)&=\ddot{X}_0(0)\\ \ddot{X}_2(1)&=\ddot{X}_1(0)\\ \ddot{X}_3(1)&=\ddot{X}_2(0)\end{aligned}\right\} \tag{6.4}$$

In addition, we add the condition that shape be invariant under coordinate transformation (Cauchy's condition) (refer to Sect. 1.1.3):

$$X_0(t)+X_1(t)+X_2(t)+X_3(t)\equiv 1. \tag{6.5}$$

Assuming that the functions $X_0(t)$, $X_1(t)$, $X_2(t)$ and $X_3(t)$ are cubic polynomials, there are a total of 16 unknowns. There are also 16 condition Eqs. (6.2) to (6.5), so the forms of functions $X_0(t)$, ..., $X_3(t)$ can be determined.

Considering the symmetry of Eqs. (6.2) to (6.5), the functions $X_0(t)$, ..., $X_3(t)$ can be written as follows.

$$X_0(t)=X_3(1-t) \tag{6.6}$$

$$X_1(t)=X_2(1-t) \tag{6.7}$$

$$X_2(t)=a_{20}t^3+a_{21}t^2+a_{22}t+a_{23} \tag{6.8}$$

$$X_3(t)=a_{30}t^3+a_{31}t^2+a_{32}t+a_{33}. \tag{6.9}$$

From the first equations of sets (6.2) to (6.4), we immediately find that:

$$a_{31}=a_{32}=a_{33}=0. \tag{6.10}$$

From the fourth equations of each set (6.2) to (6.4) we have:

$$a_{21}=3a_{30},\ a_{22}=3a_{30},\ a_{23}=a_{30}. \tag{6.11}$$

From Eqs. (6.6), (6.7) and (6.5) we have:

$$\left.\begin{aligned}a_{20}&=-3a_{30}\\ a_{30}&=1/6\end{aligned}\right\}. \tag{6.12}$$

From these results we have:

$$\left.\begin{aligned}
X_0(t)&=\frac{1}{6}(1-t)^3\\
X_1(t)&=\frac{1}{2}t^3-t^2+\frac{2}{3}\\
X_2(t)&=-\frac{1}{2}t^3+\frac{1}{2}t^2+\frac{1}{2}t+\frac{1}{6}\\
X_3(t)&=\frac{1}{6}t^3
\end{aligned}\right\} \tag{6.13}$$

Using these functions, the curve segment formula (6.1) can be expressed as:

$$\begin{aligned}
\boldsymbol{P}_i(t)&=X_0(t)\boldsymbol{Q}_{i-1}+X_1(t)\boldsymbol{Q}_i+X_2(t)\boldsymbol{Q}_{i+1}+X_3(t)\boldsymbol{Q}_{i+2}\\
&=[t^3\ t^2\ t\ 1]\begin{bmatrix} -\frac{1}{6} & \frac{1}{2} & -\frac{1}{2} & \frac{1}{6}\\ \frac{1}{2} & -1 & \frac{1}{2} & 0\\ -\frac{1}{2} & 0 & \frac{1}{2} & 0\\ \frac{1}{6} & \frac{2}{3} & \frac{1}{6} & 0 \end{bmatrix}\begin{bmatrix}\boldsymbol{Q}_{i-1}\\ \boldsymbol{Q}_i\\ \boldsymbol{Q}_{i+1}\\ \boldsymbol{Q}_{i+2}\end{bmatrix}\\
&=[t^3\ t^2\ t\ 1]\,\boldsymbol{M}_R\begin{bmatrix}\boldsymbol{Q}_{i-1}\\ \boldsymbol{Q}_i\\ \boldsymbol{Q}_{i+1}\\ \boldsymbol{Q}_{i+2}\end{bmatrix}
\end{aligned} \tag{6.14}$$

where:

$$\boldsymbol{M}_R=\begin{bmatrix} -\frac{1}{6} & \frac{1}{2} & -\frac{1}{2} & \frac{1}{6}\\ \frac{1}{2} & -1 & \frac{1}{2} & 0\\ -\frac{1}{2} & 0 & \frac{1}{2} & 0\\ \frac{1}{6} & \frac{2}{3} & \frac{1}{6} & 0 \end{bmatrix} \tag{6.15}$$

Functions (6.13) make the slopes and curvature vectors continuous at the connection point between curve segments expressed in the form (6.1). In fact, as will be discussed more fully later, these functions agree with the so-called uniform B-spline functions of order $M=4$ and degree 3 in the case when the knot vector is specified as:

$$T=[t_0\ t_1\ t_2\ t_3\ t_4\ t_5\ t_6\ t_7]=[-3\ -2\ -1\ 0\ 1\ 2\ 3\ 4].$$

In this book, these functions will be expressed by the notation $N_{i,4}(t)$:

$$\left.\begin{array}{ll} X_0(t)\equiv N_{0,4}(t), & X_1(t)\equiv N_{1,4}(t) \\ X_2(t)\equiv N_{2,4}(t), & X_3(t)\equiv N_{3,4}(t) \end{array}\right\}. \tag{6.16}$$

As can be seen from the above derivation, the continuity of the segments at the connection points is assured for all $\boldsymbol{Q}_j (j=0, 1, \ldots, n)$, so the difficult connection problems encountered with Ferguson curves and Bézier curves are absent. An example of a curve is shown in Fig. 6.2.

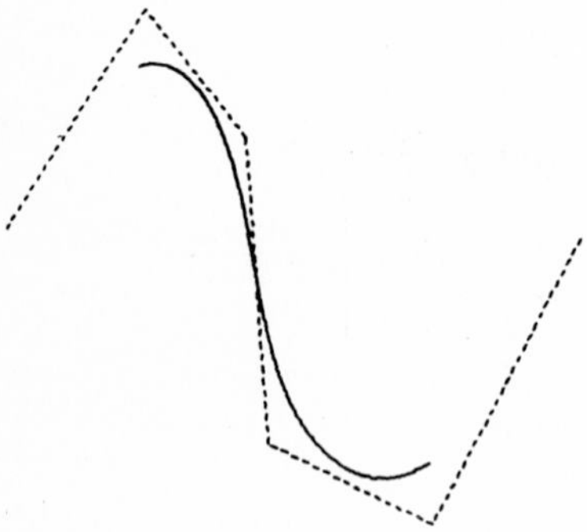

Fig. 6.2. Example of a B-spline curve ($M=4$)

Similarly, a curve can be formed from uniform B-spline functions of order 3 degree 2 by assuming quadratic functions and forming linear combinations of 3 points; more generally, a curve can be formed from uniform B-spline functions of order $n+1$, degree n as linear combinations of $(n+1)$ points. For example, the uniform B-spline functions of order 3, degree 2 $Y_0(t)$, $Y_1(t)$, $Y_2(t)$ when the knot vector is specified as

$$\mathrm{T}=[t_0\ t_1\ t_2\ t_3\ t_4\ t_5]=[-2\ -1\ 0\ 1\ 2\ 3]$$

are given by the following formulas:

$$\left.\begin{array}{l} Y_0(t)=\dfrac{1}{2}(1-t)^2\equiv N_{0,3}(t) \\ Y_1(t)=-t^2+t+\dfrac{1}{2}\equiv N_{1,3}(t) \\ Y_2(t)=\dfrac{1}{2}t^2\equiv N_{2,3}(t) \end{array}\right\} \tag{6.17}$$

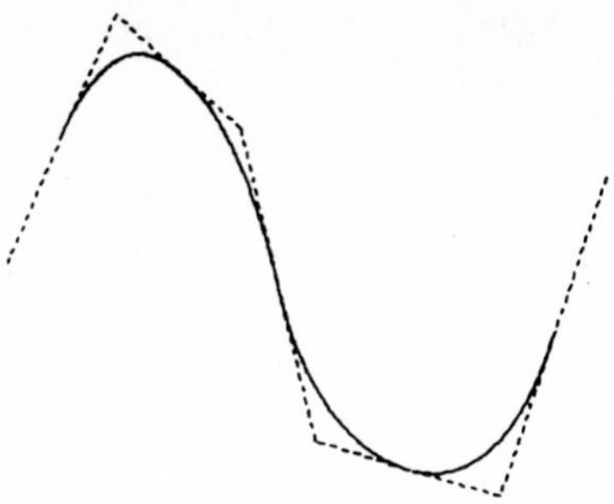

Fig. 6.3. Example of a *B*-spline curve ($M=3$)

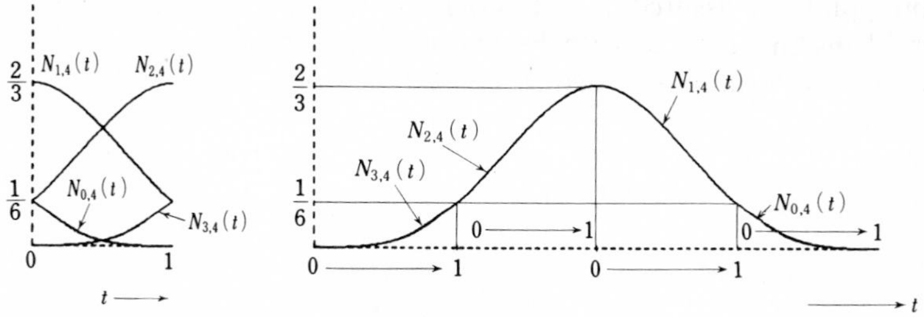

Fig. 6.4. Uniform *B*-spline basis functions ($M=4$)

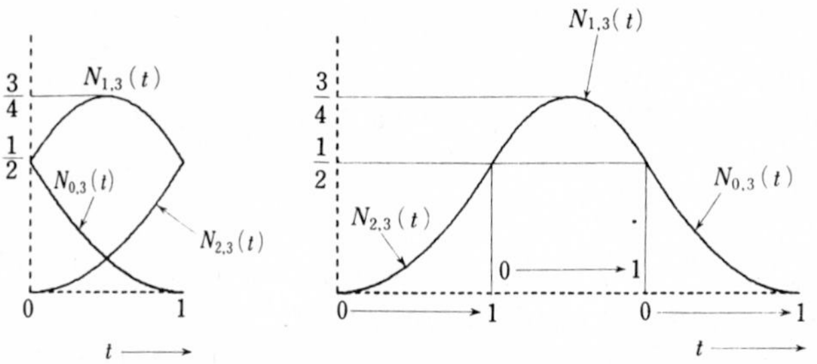

Fig. 6.5. Uniform *B*-spline basis functions ($M=3$)

The curve segment formed from these *B*-spline functions becomes:

$$\begin{aligned} \boldsymbol{P}_i(t) &= Y_0(t)\boldsymbol{Q}_{i-1} + Y_1(t)\boldsymbol{Q}_i + Y_2(t)\boldsymbol{Q}_{i+1} \\ &= \frac{1}{2}[t^2 \ \ t \ \ 1]\begin{bmatrix} 1 & -2 & 1 \\ -2 & 2 & 0 \\ 1 & 1 & 0 \end{bmatrix}\begin{bmatrix} \boldsymbol{Q}_{i-1} \\ \boldsymbol{Q}_i \\ \boldsymbol{Q}_{i+1} \end{bmatrix} \quad (i=1, 2, \ldots, n-1). \end{aligned} \tag{6.18}$$

An example of such a curve is shown in Fig. 6.3.

Graphs of $N_{i,3}(t)$ and $N_{i,4}(t)$ in the range $0 \leqq t \leqq 1$ are shown in Figs. 6.4 and 6.5 respectively.

In order to express the *B*-spline curve segment of Eq. (6.14) by a Ferguson curve segment, we take:

$$[t^3\ t^2\ t\ 1]\, \boldsymbol{M}_R \begin{bmatrix} \boldsymbol{Q}_{i-1} \\ \boldsymbol{Q}_i \\ \boldsymbol{Q}_{i+1} \\ \boldsymbol{Q}_{i+2} \end{bmatrix} = [t^3\ t^2\ t\ 1]\, \boldsymbol{M}_C \begin{bmatrix} \boldsymbol{Q}_0 \\ \boldsymbol{Q}_1 \\ \dot{\boldsymbol{Q}}_0 \\ \dot{\boldsymbol{Q}}_1 \end{bmatrix}$$

which gives:

$$\begin{bmatrix} \boldsymbol{Q}_0 \\ \boldsymbol{Q}_1 \\ \dot{\boldsymbol{Q}}_0 \\ \dot{\boldsymbol{Q}}_1 \end{bmatrix} = \boldsymbol{M}_C^{-1} \boldsymbol{M}_R \begin{bmatrix} \boldsymbol{Q}_{i-1} \\ \boldsymbol{Q}_i \\ \boldsymbol{Q}_{i+1} \\ \boldsymbol{Q}_{i+2} \end{bmatrix}$$

$$= \begin{bmatrix} \frac{1}{6} & \frac{2}{3} & \frac{1}{6} & 0 \\ 0 & \frac{1}{6} & \frac{2}{3} & \frac{1}{6} \\ -\frac{1}{2} & 0 & \frac{1}{2} & 0 \\ 0 & -\frac{1}{2} & 0 & \frac{1}{2} \end{bmatrix} \begin{bmatrix} \boldsymbol{Q}_{i-1} \\ \boldsymbol{Q}_i \\ \boldsymbol{Q}_{i+1} \\ \boldsymbol{Q}_{i+2} \end{bmatrix}$$

$$= \begin{bmatrix} \frac{1}{6}\boldsymbol{Q}_{i-1} + \frac{2}{3}\boldsymbol{Q}_i + \frac{1}{6}\boldsymbol{Q}_{i+1} \\ \frac{1}{6}\boldsymbol{Q}_i + \frac{2}{3}\boldsymbol{Q}_{i+1} + \frac{1}{6}\boldsymbol{Q}_{i+2} \\ \frac{1}{2}\mathrm{Q}_{i+1} - \frac{1}{2}\boldsymbol{Q}_{i-1} \\ \frac{1}{2}\boldsymbol{Q}_{i+2} - \frac{1}{2}\boldsymbol{Q}_i \end{bmatrix} \tag{6.19}$$

Alternatively, to express a *B*-spline curve segment by a Bézier curve segment we take:

$$[t^3\ t^2\ t\ 1]\, \boldsymbol{M}_R \begin{bmatrix} \boldsymbol{Q}_{i-1} \\ \boldsymbol{Q}_i \\ \boldsymbol{Q}_{i+1} \\ \boldsymbol{Q}_{i+2} \end{bmatrix} = [t^3\ t^2\ t\ 1]\, \boldsymbol{M}_B \begin{bmatrix} \boldsymbol{Q}_0 \\ \boldsymbol{Q}_1 \\ \boldsymbol{Q}_2 \\ \boldsymbol{Q}_3 \end{bmatrix}.$$

Then the Bézier curve vectors $\boldsymbol{Q}_0, \ldots, \boldsymbol{Q}_3$ are expressed in terms of the *B*-spline curve vectors $\boldsymbol{Q}_{i-1}$, $\boldsymbol{Q}_i$, $\boldsymbol{Q}_{i+1}$, $\boldsymbol{Q}_{i+2}$ as:

$$\begin{bmatrix} \boldsymbol{Q}_0 \\ \boldsymbol{Q}_1 \\ \boldsymbol{Q}_2 \\ \boldsymbol{Q}_3 \end{bmatrix} = \boldsymbol{M}_B^{-1}\boldsymbol{M}_R \begin{bmatrix} \boldsymbol{Q}_{i-1} \\ \boldsymbol{Q}_i \\ \boldsymbol{Q}_{i+1} \\ \boldsymbol{Q}_{i+2} \end{bmatrix}$$

$$= \begin{bmatrix} \frac{1}{6} & \frac{2}{3} & \frac{1}{6} & 0 \\ 0 & \frac{2}{3} & \frac{1}{3} & 0 \\ 0 & \frac{1}{3} & \frac{2}{3} & 0 \\ 0 & \frac{1}{6} & \frac{2}{3} & \frac{1}{6} \end{bmatrix} \begin{bmatrix} \boldsymbol{Q}_{i-1} \\ \boldsymbol{Q}_i \\ \boldsymbol{Q}_{i+1} \\ \boldsymbol{Q}_{i+2} \end{bmatrix}$$

$$= \begin{bmatrix} \frac{1}{6}\boldsymbol{Q}_{i-1} + \frac{2}{3}\boldsymbol{Q}_i + \frac{1}{6}\boldsymbol{Q}_{i+1} \\ \frac{2}{3}\boldsymbol{Q}_i + \frac{1}{3}\boldsymbol{Q}_{i+1} \\ \frac{1}{3}\boldsymbol{Q}_i + \frac{2}{3}\boldsymbol{Q}_{i+1} \\ \frac{1}{6}\boldsymbol{Q}_i + \frac{2}{3}\boldsymbol{Q}_{i+1} + \frac{1}{6}\boldsymbol{Q}_{i+2} \end{bmatrix} \tag{6.20}$$

From this we obtain:

$$\boldsymbol{Q}_0 = \frac{1}{6}\boldsymbol{Q}_{i-1} + \frac{2}{3}\boldsymbol{Q}_i + \frac{1}{6}\boldsymbol{Q}_{i+1}$$

$$\boldsymbol{Q}_1 = \frac{2}{3}\boldsymbol{Q}_i + \frac{1}{3}\boldsymbol{Q}_{i+1}$$

$$= \boldsymbol{Q}_0 + \frac{1}{6}(\boldsymbol{Q}_{i+1} - \boldsymbol{Q}_{i-1})$$

$$\boldsymbol{Q}_2 = \frac{1}{3}\boldsymbol{Q}_i + \frac{2}{3}\boldsymbol{Q}_{i+1}$$

$$= \boldsymbol{Q}_3 - \frac{1}{6}(\boldsymbol{Q}_{i+2} - \boldsymbol{Q}_i)$$

$$\boldsymbol{Q}_3 = \frac{1}{6}\boldsymbol{Q}_i + \frac{2}{3}\boldsymbol{Q}_{i+1} + \frac{1}{6}\boldsymbol{Q}_{i+2}.$$

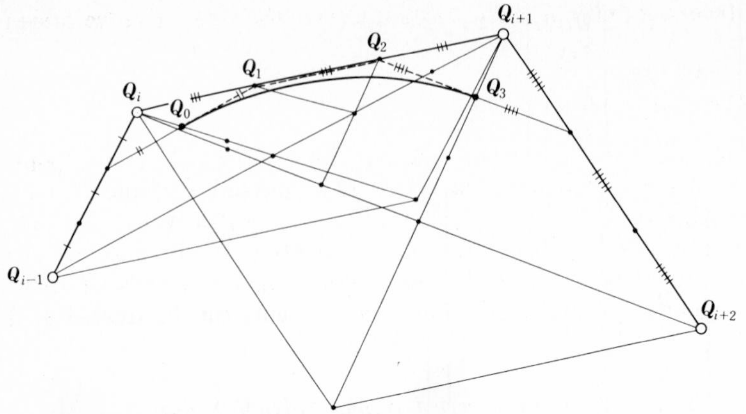

Fig. 6.6. Expression of a uniform cubic *B*-spline curve segment in terms of a Bézier curve segments. Q_{i-1}, Q_i, Q_{i+1}, Q_{i+2}: Curve defining vectors of the *B*-spline curve segment. Q_0, Q_1, Q_2, Q_3: Curve defining vectors of the Bézier curve segment.

The geometric relation between the *B*-spline curve defining vectors Q_{i-1}, Q_i, Q_{i+1}, Q_{i+2} and the Bézier curve defining vectors is shown in Fig. 6.6. To express a curve defined by given cubic uniform *B*-spline curve defining vectors in terms of a Bézier curve formula, first trisect the 3 sides $\overline{Q_{i-1}Q_i}$, $\overline{Q_iQ_{i+1}}$, $\overline{Q_{i+1}Q_{i+2}}$. In this case, on side $\overline{Q_iQ_{i+1}}$ the trisecting point closer to Q_i is Q_1 and the trisecting point closer to Q_{i+1} is Q_2. Next, join the trisecting point on side $\overline{Q_{i-1}Q_i}$ that is closer to Q_i to Q_1 by a straight line. The midpoint of that line segment becomes the starting point of the curve, Q_0. Similarly, join the trisecting point on side $\overline{Q_{i+1}Q_{i+2}}$ closer to Q_{i+1} to Q_2 by a straight line; the midpoint of that line segment becomes the end point of the curve, Q_3.

6.1.2 Properties of Curves[26)]

The uniform cubic *B*-spline curve formula (6.14) is relatively simple and has superior properties, so it is often used in practical applications.

In formula (6.14), the Q_j $(j=0, 1, \ldots, n)$ are the vectors that define the curve, so they are called curve defining vectors. As in the case of a Bézier curve segment, the vectors Q_j define the vertices of the polygon that determines the characteristics of the curve shape, that is, the characteristic polygon. The points P_i shown in Fig. 6.1 $(i=1, 2, \ldots, n-1)$ correspond to parameter values $t=0$ or $t=1$. These are the position vectors corresponding to knots on the curve, as will be discussed below. In this book the P_i will be called knot point vectors.

Normally, when the curve defining vectors Q_j are given $(j=0, 1, \ldots, n)$ and the knot point vectors P_i are found from them, the transformation is called a normal transformation; conversely, when the P_i are given and the Q_i are found from them, it is called an inverse transformation.

As shown below, a uniform cubic B-spline curve has superior curve design properties.

① C^2 *Class Continuity*

As is clear from the derivation of B-spline functions in Sect. 6.1.1, at a point where uniform cubic B-spline curve segments are connected, $\dot{\boldsymbol{P}}(t)$ and $\ddot{\boldsymbol{P}}(t)$ are continuous regardless of the values of the vectors $\boldsymbol{Q}_j$; in other words, not only the slopes but also the curvature vectors are continuous. Because of this property, there are no difficult problems in connecting the ends of curve segments, and a curve consisting of connected segments can be treated as a single unit.

② *Faithfulness with Respect to the Curve Defining Polygon Shape*

B-spline curves, like Bézier curve segments, have the variation diminishing property.

Also, the B-spline function $N_{i,4}(t)$ is always 0 or positive in $0 \leqq t \leqq 1$, and, in addition, the Cauchy relation holds:

$$\sum_{i=0}^{3} N_{i,4}(t) \equiv 1 . \tag{6.21}$$

Therefore, the curve segment is expressed as a convex combination of local sequences of points $\boldsymbol{Q}_{i-1}$, $\boldsymbol{Q}_i$, $\boldsymbol{Q}_{i+1}$, $\boldsymbol{Q}_{i+2}$. This convex hull is shown by slanted lines in Figs. 6.7 and 6.8. In contrast to a Bézier curve, of which the convex hull is a global region made by all points $\boldsymbol{Q}_j$ ($j=0, 1, \ldots, n$) (the region enclosed by dotted lines in Fig. 6.8), the convex hull of a cubic B-spline curve is a local region formed by 4 points. Compared to a Bézier curve, a B-spline curve has a shape that is more faithful to the shape of the curve defining polygon.

In addition, as mentioned in Sect. 6.1.1, each segment of a B-spline curve can be expressed as a Bézier curve defining polygon $\boldsymbol{Q}_0$, $\boldsymbol{Q}_1$, $\boldsymbol{Q}_2$, $\boldsymbol{Q}_3$. Consequently, the region in which a B-spline curve segment exists, is inside the convex hull $\boldsymbol{Q}_0$, $\boldsymbol{Q}_1$, $\boldsymbol{Q}_2$, $\boldsymbol{Q}_3$. This convex hull is shaded in Figs. 6.7 and 6.8.

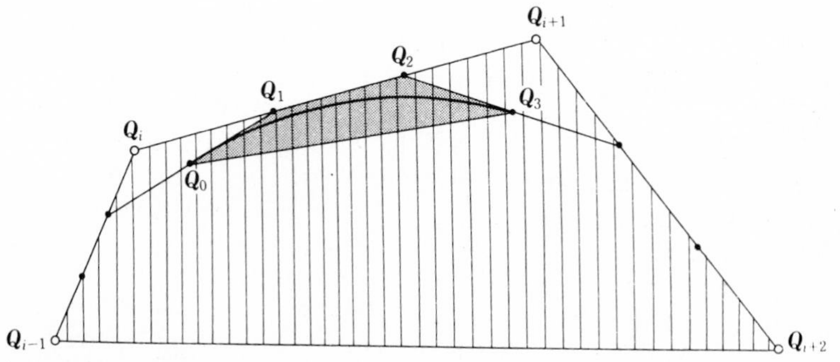

Fig. 6.7. Shrinking of a convex hull. $\boldsymbol{Q}_{i-1}$, $\boldsymbol{Q}_i$, $\boldsymbol{Q}_{i+1}$, $\boldsymbol{Q}_{i+2}$: Curve defining vectors of the B-spline curve segment. $\boldsymbol{Q}_0$, $\boldsymbol{Q}_1$, $\boldsymbol{Q}_2$, $\boldsymbol{Q}_3$: Curve defining vectors of the Bézier curve segment.

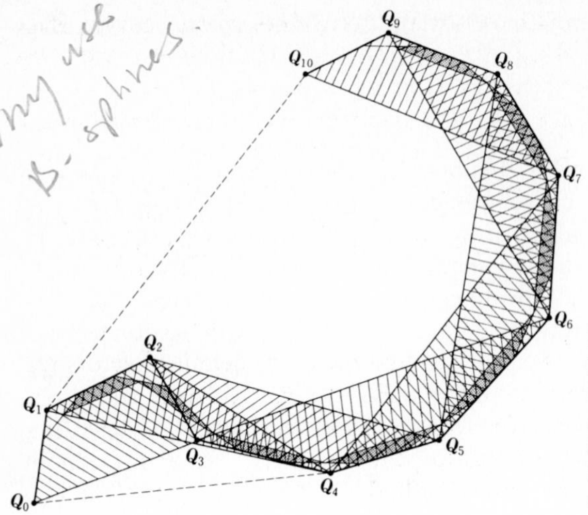

Fig. 6.8. Relation between convex hulls of Bézier curve segment and *B*-spline curve. *Slanted line area:* *B*-spline curve convex hull; *dotted lines:* Bézier curve segment convex hull; *shaded area:* shrunken *B*-spline curve convex hull

③ *Local Uniqueness*

Since a cubic *B*-spline curve is locally defined by only 4 adjacent vertex vectors of the curve defining polygon, the change of shape produced by varying one of the $\boldsymbol{Q}_j$ is locally confined. This is an important property in increasing the efficiency of curve design work. An interpolating spline curve does not have this property; if one part of it is changed, the effect extends over the whole curve.

④ *Geometrical Relations Between Derivative Vectors at Ends of Segments*

Position vectors, tangent vectors and 2-nd derivative vectors can be easily determined graphically at the ends of a uniform cubic *B*-spline curve segment (Fig. 6.9). That is to say, when a parallelogram is formed by the 3 points $\boldsymbol{Q}_{i-1}$, $\boldsymbol{Q}_i$ and $\boldsymbol{Q}_{i+1}$, the starting point $\boldsymbol{P}_i(0)$ of the curve segment is on the diagonal joining $\boldsymbol{Q}_i$ to the fourth corner, 1/6 of the way from $\boldsymbol{Q}_i$ to the fourth corner. The tangent vector $\dot{\boldsymbol{P}}_i(0)$ at the starting point is half-way from $\boldsymbol{Q}_{i-1}$ to $\boldsymbol{Q}_{i+1}$. The 2-nd derivative vector $\ddot{\boldsymbol{P}}_i(0)$ at the starting point is a vector along the diagonal from $\boldsymbol{Q}_i$ to the fourth corner, having the length of that diagonal as its magnitude.

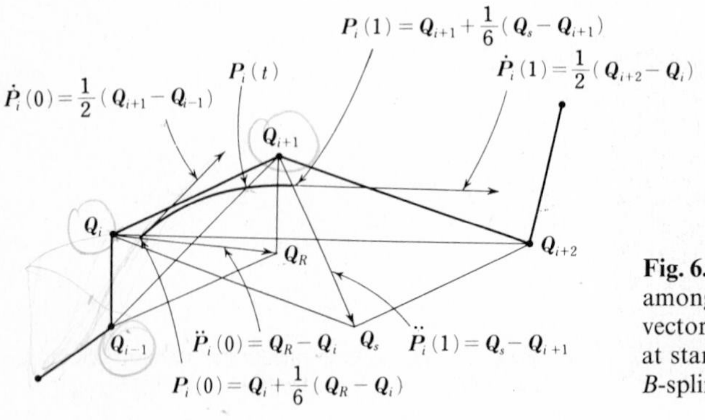

Fig. 6.9. Geometrical relation among position vectors, tangent vectors and 2-nd derivative vectors at starting and end points of *B*-spline curve segment

Similar relations exist among the position vector $\boldsymbol{P}_i(1)$, tangent vector $\dot{\boldsymbol{P}}_i(1)$ and 2-nd derivative vector $\ddot{\boldsymbol{P}}_i(1)$ at the end point of the segment, with the parallelogram being formed by the 3 points $\boldsymbol{Q}_i$, $\boldsymbol{Q}_{i+1}$ and $\boldsymbol{Q}_{i+2}$.

The parallelogram used to determine the position vector, tangent vector and 2-nd derivative vector at the end point of one curve segment is the same as the parallelogram used to determine the position vector, tangent vector and 2-nd derivative vector at the starting point of the next curve segment. This fact helps to demonstrate the continuity between segments at the connecting points.

⑤ *Ease of Inverse Transformation*

In shape design, it is sometimes necessary to find the curve defining vectors $\boldsymbol{Q}_0$, $\boldsymbol{Q}_1$, ..., $\boldsymbol{Q}_n$ given the knot point vectors $\boldsymbol{P}_1$, $\boldsymbol{P}_2$, ..., $\boldsymbol{P}_{n-1}$ (inverse transformation). As will be discussed later (refer to Sect. 6.1.4), in the case of a uniform cubic B-spline curve, inverse transformation can be carried out at high speed, by simple computations, using the same method for either an open curve or a closed curve.

⑥ *Expression of Straight Lines, Circles, Cusps*

In the case of a cubic B-spline curve, if the curve defining vectors $\boldsymbol{Q}_{i-1}$, $\boldsymbol{Q}_i$, $\boldsymbol{Q}_{i+1}$ and $\boldsymbol{Q}_{i+2}$ lie colinear, the curve segment defined by those four position vectors degenerates to a line segment. This line segment connects continuously to the neighboring curve segments with continuity up to the curvature vector. This feature is extremely useful in practical applications (Fig. 6.10).

Formula (6.14) is not a rational expression in t, so it cannot rigorously express a circle, but by making the curve defining polygon equilateral a circle can be expressed approximately.

When the curve defining polygon is an equilateral m-sided polygon, let us find the curve defining vectors $\boldsymbol{Q}_i = [Q_{ix}, Q_{iy}]$ which will approximately generate a circle of unit radius (Fig. 6.11). In Fig. 6.11, the radius R of the circle that passes through each vertex $\boldsymbol{Q}_i$ can be found by the geometrical construction discussed in ④:

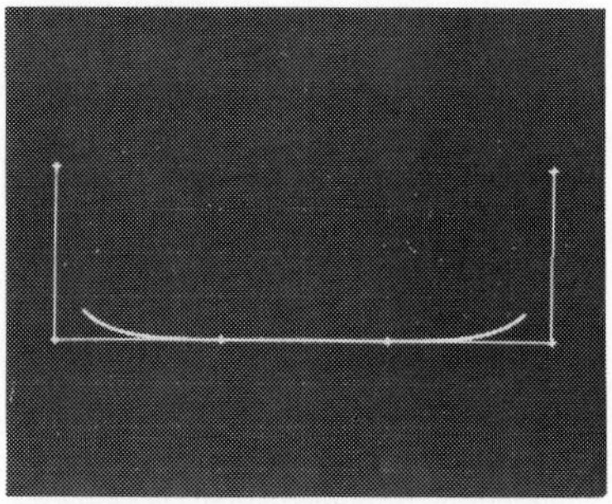

Fig. 6.10. B-spline curves connected to a straight line with class C^2 continuity

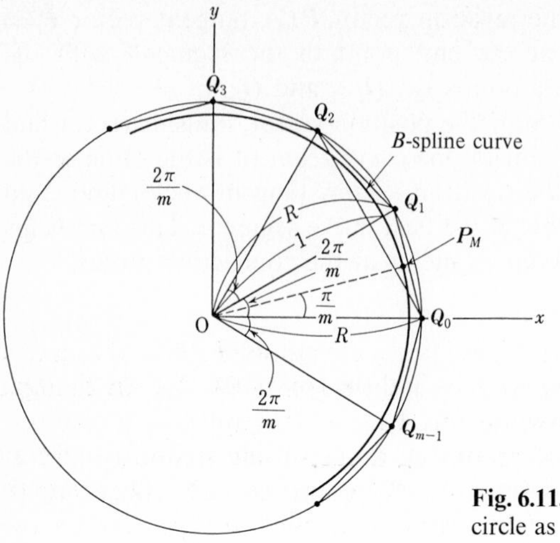

Fig. 6.11. Method of generating an approximate circle as a B-spline curve ($M=4$)

$$3(R-1)=R\left(1-\cos\frac{2\pi}{m}\right)$$

$$\therefore\ R=\frac{3}{2+\cos\dfrac{2\pi}{m}}. \tag{6.22}$$

Consequently, the vertex vectors $\boldsymbol{Q}_i$ that will generate a unit circle are:

$$\boldsymbol{Q}_i=\left[R\cos\frac{2\pi}{m}i \quad R\sin\frac{2\pi}{m}i\right]$$

$$=\left[\frac{3\cos\dfrac{2\pi}{m}i}{2+\cos\dfrac{2\pi}{m}} \quad \frac{3\sin\dfrac{2\pi}{m}i}{2+\cos\dfrac{2\pi}{m}}\right] \quad (i=0,1,\dots,m-1). \tag{6.23}$$

Next, let us find the deviation from a true circle. In Fig. 6.11, the midpoint $\boldsymbol{P}_M$ of the curve segment generated by the 4 points $\boldsymbol{Q}_{m-1}$, $\boldsymbol{Q}_0$, $\boldsymbol{Q}_1$ and $\boldsymbol{Q}_2$ is found by setting $t=\frac{1}{2}$ in formula (6.14):

$$\boldsymbol{P}_M=\frac{1}{48}\boldsymbol{Q}_{m-1}+\frac{23}{48}\boldsymbol{Q}_0+\frac{23}{48}\boldsymbol{Q}_1+\frac{1}{48}\boldsymbol{Q}_2$$

$$=\left[\frac{\left(1+\cos\dfrac{2\pi}{m}\right)\left(11+\cos\dfrac{2\pi}{m}\right)}{8\left(2+\cos\dfrac{2\pi}{m}\right)} \quad \frac{\sin\dfrac{2\pi}{m}\left(11+\cos\dfrac{2\pi}{m}\right)}{8\left(2+\cos\dfrac{2\pi}{m}\right)}\right] \tag{6.24}$$

so that the deviation $E(\%)$ from a true circle is:

$$E = (1 - \sqrt{P_M^2}) \times 100$$

$$= \frac{\left(1 - \cos\frac{\pi}{m}\right)^2 \left(2 - \cos\frac{\pi}{m}\right)}{2\left(2 + \cos\frac{2\pi}{m}\right)} \times 100. \qquad (6.25)$$

From formula (6.25) we find that a square gives a deviation of 2.8% from a true circle; an equilateral 12-sided polygon gives a deviation of 0.02%. An example of generation of a semicircle is shown in Fig. 6.12.

Let us consider the problem of producing a cusp using uniform cubic B-spline curves. In Fig. 6.13, the sequence of vertices $\boldsymbol{Q}_0$, $\boldsymbol{Q}_1$, $\boldsymbol{Q}_2$, $\boldsymbol{Q}_3$, $\boldsymbol{Q}_4$ determines a curve $\boldsymbol{P}_1\boldsymbol{P}_2\boldsymbol{P}_3$. Next, if two vertices are made to coincide so that the vertex sequence becomes $\boldsymbol{Q}_0$, $\boldsymbol{Q}_1$, $\boldsymbol{Q}_2$, $\boldsymbol{Q}_2$, $\boldsymbol{Q}_3$, $\boldsymbol{Q}_4$, the curve becomes $\boldsymbol{P}_1\boldsymbol{P}_4\boldsymbol{P}_5\boldsymbol{P}_3$, increasing the curvature in the vicinity of $\boldsymbol{Q}_2$. If three vertices are overlapped at point $\boldsymbol{Q}_2$, the vertex sequence that defines the curve becomes $\boldsymbol{Q}_0$, $\boldsymbol{Q}_1$, $\boldsymbol{Q}_2$, $\boldsymbol{Q}_2$, $\boldsymbol{Q}_2$, $\boldsymbol{Q}_3$, $\boldsymbol{Q}_4$. The curve becomes $\boldsymbol{P}_1\boldsymbol{P}_4\boldsymbol{Q}_2\boldsymbol{P}_5\boldsymbol{P}_3$, and the slope becomes discontinuous at point $\boldsymbol{Q}_2$, making that point a cusp. In this case there are straight line segments, $\boldsymbol{P}_4\boldsymbol{Q}_2$ and $\boldsymbol{Q}_2\boldsymbol{P}_5$, on both sides of the cusp. By suitably controlling the arrangement of the 3 points centered on $\boldsymbol{Q}_2$, a sharp cusp can be produced.

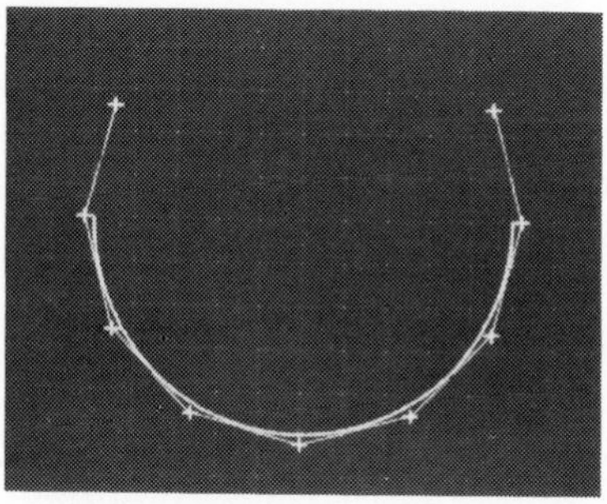

Fig. 6.12. Example of generation of a circular arc (half circle) as a B-spline curve ($M=4$)

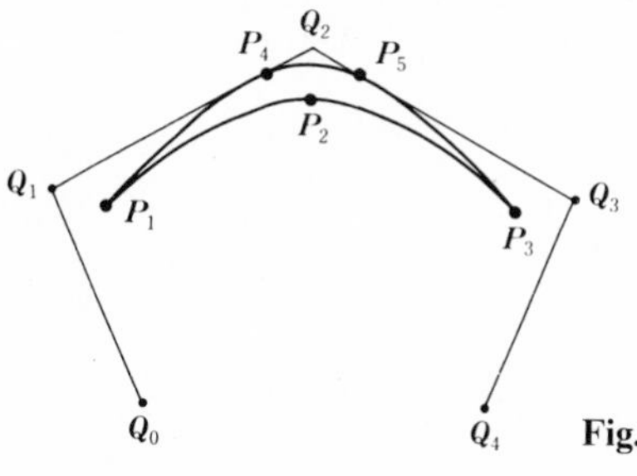

Fig. 6.13. Formation of a cusp by B-spline curves ($M=4$)

6.1.3 Determination of a Point on a Curve by Finite Difference Operations[27)]

The finite difference matrix needed to calculate a point on a uniform cubic *B*-spline curve is found as follows (refer to Sect. 1.2.4).

In formula (6.14), abbreviating $N_{i,4}(t) \equiv N_i(t)$ we obtain:

$$\boldsymbol{P}_i(t) = N_0(t)\boldsymbol{Q}_{i-1} + N_1(t)\boldsymbol{Q}_i + N_2(t)\boldsymbol{Q}_{i+1} + N_3(t)\boldsymbol{Q}_{i+2}. \tag{6.26}$$

The 1-st, 2-nd and 3-rd finite differences of $\boldsymbol{P}_i(t)$ at $t=0$ are $\Delta\boldsymbol{P}_i(0)$, $\Delta^2\boldsymbol{P}_i(0)$ and $\Delta^3\boldsymbol{P}_i(0)$, and the finite differences of $N_i(t)$ are $\Delta N_i(0)$, $\Delta^2 N_i(0)$ and $\Delta^3 N_i(0)$. Then we have:

$$\begin{aligned}
\boldsymbol{P}_i(0) &= N_0(0)\boldsymbol{Q}_{i-1} + N_1(0)\boldsymbol{Q}_i + N_2(0)\boldsymbol{Q}_{i+1} + N_3(0)\boldsymbol{Q}_{i+2}\\
\Delta\boldsymbol{P}_i(0) &= \Delta N_0(0)\boldsymbol{Q}_{i-1} + \Delta N_1(0)\boldsymbol{Q}_i + \Delta N_2(0)\boldsymbol{Q}_{i+1} + \Delta N_3(0)\boldsymbol{Q}_{i+2}\\
\Delta^2\boldsymbol{P}_i(0) &= \Delta^2 N_0(0)\boldsymbol{Q}_{i-1} + \Delta^2 N_1(0)\boldsymbol{Q}_i + \Delta^2 N_2(0)\boldsymbol{Q}_{i+1} + \Delta^2 N_3(0)\boldsymbol{Q}_{i+2}\\
\Delta^3\boldsymbol{P}_i(0) &= \Delta^3 N_0(0)\boldsymbol{Q}_{i-1} + \Delta^3 N_1(0)\boldsymbol{Q}_i + \Delta^3 N_2(0)\boldsymbol{Q}_{i+1} + \Delta^3 N_3(0)\boldsymbol{Q}_{i+2}
\end{aligned} \tag{6.27}$$

where:

$$N_0(0) = \frac{1}{6}, \quad N_1(0) = \frac{2}{3}, \quad N_2(0) = \frac{1}{6}, \quad N_3(0) = 0,$$

$$\Delta N_0(0) = -\frac{1}{6}\delta^3 + \frac{1}{2}\delta^2 - \frac{1}{2}\delta, \quad \Delta N_1(0) = \frac{1}{2}\delta^3 - \delta^2,$$

$$\Delta N_2(0) = -\frac{1}{2}\delta^3 + \frac{1}{2}\delta^2 + \frac{1}{2}\delta, \quad \Delta N_3(0) = \frac{1}{6}\delta^3,$$

$$\Delta^2 N_0(0) = -\delta^3 + \delta^2, \qquad \Delta^2 N_1(0) = 3\delta^3 - 2\delta^2$$

$$\Delta^2 N_2(0) = -3\delta^3 + \delta^2, \qquad \Delta^2 N_3(0) = \delta^3,$$

$$\Delta^3 N_0(0) = -\delta^3, \; \Delta^3 N_1(0) = 3\delta^3, \; \Delta^3 N_2(0) = -3\delta^3, \; \Delta^3 N_0(0) = \delta^3$$

Then we can find the finite difference matrix to be:

$$\begin{bmatrix} \boldsymbol{P}_i(0) \\ \frac{1}{\delta}\Delta\boldsymbol{P}_i(0) \\ \frac{1}{\delta^2}\Delta^2\boldsymbol{P}_i(0) \\ \frac{1}{\delta^3}\Delta^3\boldsymbol{P}_i(0) \end{bmatrix} = \begin{bmatrix} N_0(0) & N_1(0) & N_2(0) & N_3(0) \\ \frac{1}{\delta}\Delta N_0(0) & \frac{1}{\delta}\Delta N_1(0) & \frac{1}{\delta}\Delta N_2(0) & \frac{1}{\delta}\Delta N_3(0) \\ \frac{1}{\delta^2}\Delta^2 N_0(0) & \frac{1}{\delta^2}\Delta^2 N_1(0) & \frac{1}{\delta^2}\Delta^2 N_2(0) & \frac{1}{\delta^2}\Delta^2 N_3(0) \\ \frac{1}{\delta^3}\Delta^3 N_0(0) & \frac{1}{\delta^3}\Delta^3 N_1(0) & \frac{1}{\delta^3}\Delta^3 N_2(0) & \frac{1}{\delta^3}\Delta^3 N_3(0) \end{bmatrix} \begin{bmatrix} \boldsymbol{Q}_{i-1} \\ \boldsymbol{Q}_i \\ \boldsymbol{Q}_{i+1} \\ \boldsymbol{Q}_{i+2} \end{bmatrix} =$$

$$
=\begin{bmatrix}
\frac{1}{6} & \frac{2}{3} & \frac{1}{6} & 0 \\
-\frac{1}{6}\delta^2+\frac{1}{2}\delta-\frac{1}{2} & \frac{1}{2}\delta^2-\delta & -\frac{1}{2}\delta^2+\frac{1}{2}\delta+\frac{1}{2} & \frac{1}{6}\delta^2 \\
-\delta+1 & 3\delta-2 & -3\delta+1 & \delta \\
-1 & 3 & -3 & 1
\end{bmatrix}
\begin{bmatrix} \boldsymbol{Q}_{i-1} \\ \boldsymbol{Q}_i \\ \boldsymbol{Q}_{i+1} \\ \boldsymbol{Q}_{i+2} \end{bmatrix}. \tag{6.28}
$$

Using the matrix in Eq. (6.28), a point on the curve can be found at high-speed without multiplications.

In Eq. (6.28), if a certain $\boldsymbol{Q}_{i+j}(j=1,0,1,2)$ varies, the finite difference matrix for $\boldsymbol{Q}_{i+j}$ has to be recalculated, but it is wasteful to repeat the full calculation of (6.28). It is more efficient to calculate the change due to only the change $\blacktriangle\boldsymbol{Q}_{i+j}$ of $\boldsymbol{Q}_{i+j}$:

$$
\begin{bmatrix} \boldsymbol{P}_i(0) \\ \frac{1}{\delta}\Delta\boldsymbol{P}_i(0) \\ \frac{1}{\delta^2}\Delta^2\boldsymbol{P}_i(0) \\ \frac{1}{\delta^3}\Delta^3\boldsymbol{P}_i(0) \end{bmatrix}
\longleftarrow
\begin{bmatrix} \boldsymbol{P}_i(0) \\ \frac{1}{\delta}\Delta\boldsymbol{P}_i(0) \\ \frac{1}{\delta^2}\Delta^2\boldsymbol{P}_i(0) \\ \frac{1}{\delta^3}\Delta^3\boldsymbol{P}_i(0) \end{bmatrix}
+
\begin{bmatrix} N_{j+1}(0) \\ \frac{1}{\delta}\Delta N_{j+1}(0) \\ \frac{1}{\delta^2}\Delta^2 N_{j+1}(0) \\ \frac{1}{\delta^3}\Delta^3 N_{j+1}(0) \end{bmatrix}
\blacktriangle\boldsymbol{Q}_{i+j}. \tag{6.29}
$$

$(j=-1, 0, 1, 2)$

6.1.4 Inverse Transformation of a Curve[27]

Consider the inverse transformation problem, of finding the curve defining vectors $\boldsymbol{Q}_j$ $(j=0, 1, \ldots, n)$ given the knot point vectors $\boldsymbol{P}_i$ $(i=1, 2, \ldots, n-1)$ of a

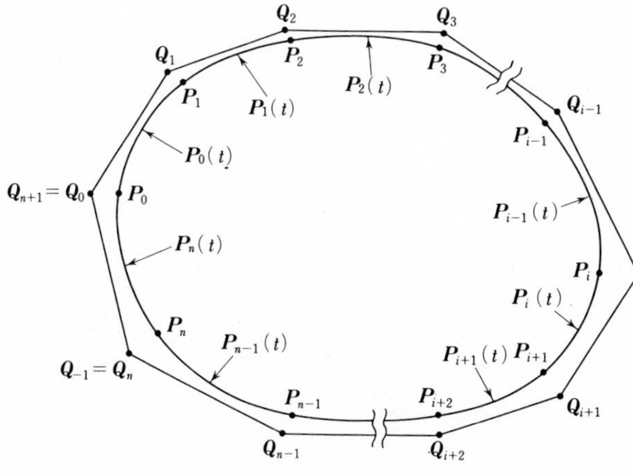

Fig. 6.14. Inverse transformation of a curve ($M=4$) (case of a closed curve)

uniform cubic B-spline curve (Fig. 6.1). In the case of a closed curve, the problem is one of finding $\boldsymbol{Q}_j$ $(j=-1, 0, \ldots, n, n+1)$ given $\boldsymbol{P}_i (i=0, 1, \ldots, n)$ (Fig. 6.14).

For both open and closed curves, the following system of simultaneous equation holds:

$$\frac{1}{6}\boldsymbol{Q}_{i-1}+\frac{2}{3}\boldsymbol{Q}_i+\frac{1}{6}\boldsymbol{Q}_{i+1}=\boldsymbol{P}_i \quad \begin{pmatrix} i=1,2,\ldots,n-1 & \text{(open curve)} \\ i=0,1,\ldots,n & \text{(closed curve)} \end{pmatrix}. \tag{6.30}$$

Since there are two fewer equations than unknowns, the following conditions are applied for open curves and closed curves, respectively:

① for open curve generation:

$$\boldsymbol{Q}_0=\boldsymbol{Q}_1, \quad \boldsymbol{Q}_n=\boldsymbol{Q}_{n-1} \tag{6.31}$$

② for closed curve generation:

$$\boldsymbol{Q}_{-1}=\boldsymbol{Q}_n, \quad \boldsymbol{Q}_{n+1}=\boldsymbol{Q}_0. \tag{6.32}$$

Conditions (6.31) are that the curvature is 0 at both ends of the open curve. This condition simplifies the inverse transformation algorithm to the same form as that for a closed curve. Conditions (6.32) are those for joining the starting and end points with continuity up to the curvature vector to form a closed curve.

By adding conditions (6.31) and (6.32) for their respective cases, the number of equations is made equal to the number of unknowns, and the simultaneous Eqs. (6.30) can be solved. Note that equations satisfy the convergence condition of the Gauss-Seidel method and that there is a special relationship among the coefficients. Denote the approximation to $\boldsymbol{Q}_i$ obtained on the k-th iteration by $\boldsymbol{Q}_i^k$. After $\boldsymbol{Q}_0^k, \ldots, \boldsymbol{Q}_i^k$ have been found on the k-th iteration, the following equation holds:

$$\frac{1}{6}\boldsymbol{Q}_{i-1}^k+\frac{2}{3}\boldsymbol{Q}_i^k+\frac{1}{6}\boldsymbol{Q}_{i+1}^{k-1}=\boldsymbol{P}_i. \tag{6.33}$$

This equation can be changed into the following form.

$$\boldsymbol{Q}_i^k=\boldsymbol{P}_i+\frac{1}{2}\left\{\boldsymbol{P}_i-\frac{1}{2}(\boldsymbol{Q}_{i-1}^k+\boldsymbol{Q}_{i+1}^{k-1})\right\}. \tag{6.34}$$

Letting δ_i^k be the difference between the k-th and $(k-1)$-st iterations of $\boldsymbol{Q}_i$, we obtain:

$$\begin{aligned}\delta_i^k &= \boldsymbol{Q}_i^k-\boldsymbol{Q}_i^{k-1} \\ &= \boldsymbol{P}_i-\boldsymbol{Q}_i^{k-1}+\frac{1}{2}\left\{\boldsymbol{P}_i-\frac{1}{2}\,\boldsymbol{Q}_{i-1}^k+\boldsymbol{Q}_{i+1}^{k-1})\right\}.\end{aligned} \tag{6.35}$$

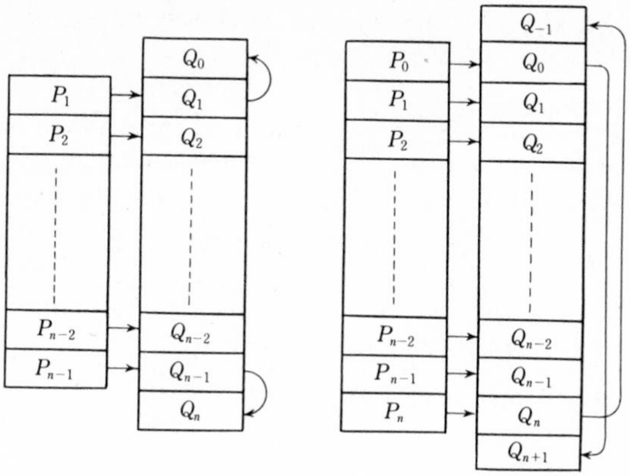

Fig. 6.15. P-table and Q-table in an inverse transformation algorithm. (a) open curve; (b) closed curve

For initial values, in the case of an open curve for $i=1, 2, \ldots, n-1$ (in the case of a closed curve for $i=0, 1, \ldots, n$) $\boldsymbol{Q}_i^0=\boldsymbol{P}_i$, $\boldsymbol{Q}_0^0=\boldsymbol{P}_1$ and $\boldsymbol{Q}_n^0=\boldsymbol{P}_{n-1}$ (in the case of a closed curve, $\boldsymbol{Q}_{-1}^0=\boldsymbol{P}_n$, $\boldsymbol{Q}_{n+1}^0=\boldsymbol{P}_0$). In addition, in Eqs. (6.34) and (6.35), in the case $i=1$, $\boldsymbol{Q}_0^{k-1}$ is used in place of $\boldsymbol{Q}_0^k$ (in the case of a closed curve, for $i=0$ $\boldsymbol{Q}_{-1}^{k-1}$ is used in place of $\boldsymbol{Q}_{-1}^k$). By making such changes that do not affect the essential content of the algorithm, the inverse transformation can be carried out very simply and at high speed. This algorithm is given below. δ_s in the algorithm is the allowable error, which is set initially. Symbols outside parentheses give processing for generating an open curve; those inside parentheses, for a closed curve.

Inverse transformation algorithm (refer to Fig. 6.15)

Perform the following processing with respect to each of the x, y and z coordinates.

Step 1
For $i=1, 2, \ldots, n-1$ $(i=0, 1, \ldots, n)$

$$P_i \rightarrow Q_i$$

then $Q_1 \rightarrow Q_0$, $Q_{n-1} \rightarrow Q_n$ $(Q_n \rightarrow Q_{-1},\ Q_0 \rightarrow Q_{n+1})$.

Step 2
For $i=1, 2, \ldots, n-1$ $(i=0, 1, \ldots, n)$

$$\delta_i = P_i - Q_i + \frac{1}{2}\left\{P_i - \frac{1}{2}(Q_{i-1}+Q_{i+1})\right\} \tag{6.36}$$

$$\delta_i + Q_i \rightarrow Q_i$$

then $Q_1 \rightarrow Q_0$, $Q_{n-1} \rightarrow Q_n$ $(Q_n \rightarrow Q_{-1},\ Q_0 \rightarrow Q_{n+1})$.

Step 3
If $\max\{\delta_i\} > \delta_s$, then return to Step 2.

If $\max\{\delta_i\} \leqq \delta_s$, the calculation is completed.

Equation (6.36) can be applied for generation of either an open or closed curve, and the algorithm is extremely simple. In addition, the multiplication by $\frac{1}{2}$ in Eq. (6.36) can be replaced by a shift operation, so the inverse transformation can be performed at very high speed.

As will be discussed below (refer to Sect. 6.2.3) this inverse transformation algorithm for curves can also be applied to inverse transformation of surfaces.

6.1.5 Change of Polygon Vertices[28)]

Let us try to change the polygon vertex vectors of a uniform cubic *B*-spline curve so that at the starting and end points it will appear to have properties similar to those of a Bézier curve (refer to Sect. 6.1.1).

In Fig. 6.16, the direction of the tangent vector at the end point $\boldsymbol{P}_1$ of the curve segment $\boldsymbol{P}_1(t)$ is the same as that of the vector from $\boldsymbol{Q}_0$ to $\boldsymbol{Q}_2$. Draw a straight line from point $\boldsymbol{P}_1$ parallel to line $\boldsymbol{Q}_0\boldsymbol{Q}_2$; the intersection of this line with side $\boldsymbol{Q}_1\boldsymbol{Q}_2$ of the polygon is $\boldsymbol{Q}_{1,c}$. As was mentioned in Sect. 6.1.2, point $\boldsymbol{P}_1$ is on the line joining the midpoint of line segment $\boldsymbol{Q}_0\boldsymbol{Q}_2$ to $\boldsymbol{Q}_1$, and is 1/3 of the

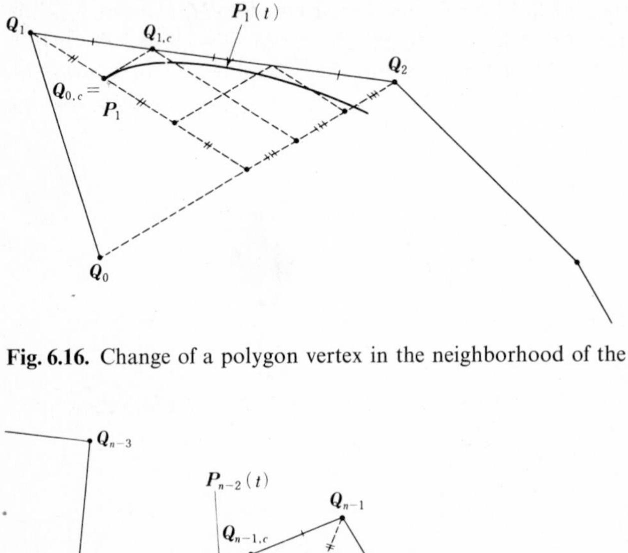

Fig. 6.16. Change of a polygon vertex in the neighborhood of the starting point of a curve

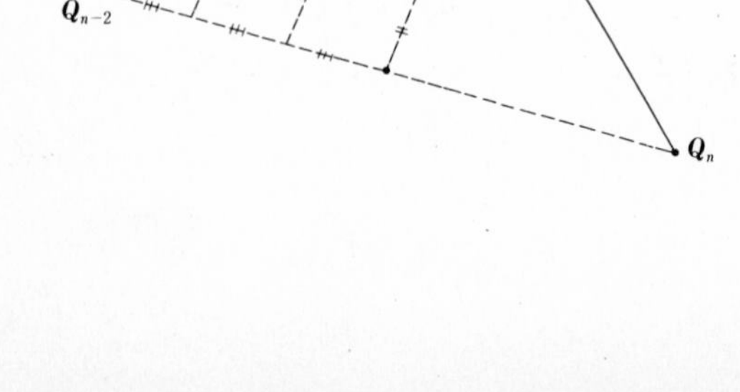

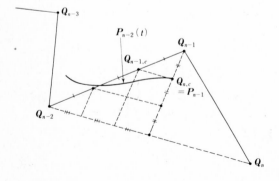

Fig. 6.17. Change of a polygon vertex in the neighborhood of the end point of a curve

way from Q_1. Consequently, when $Q_{0,c}$, $Q_{1,c}$ and Q_2 are given, the true curve defining vectors Q_0 and Q_1 are found as:

$$Q_0 = Q_2 + 6(Q_{0,c} - Q_{1,c}) \tag{6.37}$$

$$Q_1 = Q_{1,c} + \frac{Q_{1,c} - Q_2}{2} = \frac{3Q_{1,c} - Q_2}{2}. \tag{6.38}$$

Similarly, for the end point of the curve (Fig. 6.17), if Q_{n-2}, $Q_{n-1,c}$ and $Q_{n,c}$ are given, the true curve defining vectors Q_{n-1} and Q_n are:

$$Q_{n-1} = Q_{n-1,c} + \frac{Q_{n-1,c} - Q_{n-2}}{2} = \frac{3Q_{n-1,c} - Q_{n-2}}{2} \tag{6.39}$$

$$Q_n = Q_{n-2} + 6(Q_{n,c} - Q_{n-1,c}). \tag{6.40}$$

The polygon $Q_{0,c}$, $Q_{1,c}$, Q_2, ..., Q_{n-2}, $Q_{n-1,c}$, $Q_{n,c}$ and the curve are displayed on the screen. If any of the polygon vertex vectors $Q_{0,c}$, $Q_{1,c}$, $Q_{n-1,c}$, $Q_{n,c}$ that is displayed should vary, Q_0, Q_1 or Q_{n-1}, Q_n can be calculated from Eqs. (6.37), (6.38), (6.39) and (6.40). If the curve segment that is affected is recalculated, then it appears to resemble a Bézier curve, and the *B*-spline curve shape can be controlled.

6.2 Uniform Bi-Cubic *B*-Spline Surfaces

6.2.1 Surface Patch Formulas

A bi-cubic surface patch can be defined using the uniform *B*-spline functions (6.16) (Fig. 6.18):

$$P_{i,j}(u,w) = [N_{0,4}(u)\ N_{1,4}(u)\ N_{2,4}(u)\ N_{3,4}(u)]$$

$$\times \begin{bmatrix} Q_{i-1,j-1} & Q_{i-1,j} & Q_{i-1,j+1} & Q_{i-1,j+2} \\ Q_{i,j-1} & Q_{i,j} & Q_{i,j+1} & Q_{i,j+2} \\ Q_{i+1,j-1} & Q_{i+1,j} & Q_{i+1,j+1} & Q_{i+1,j+2} \\ Q_{i+2,j-1} & Q_{i+2,j} & Q_{i+2,j+1} & Q_{i+2,j+2} \end{bmatrix} \begin{bmatrix} N_{0,4}(w) \\ N_{1,4}(w) \\ N_{2,4}(w) \\ N_{3,4}(w) \end{bmatrix} \tag{6.41}$$

$$= U M_R B_R M_R^T W^T \tag{6.42}$$

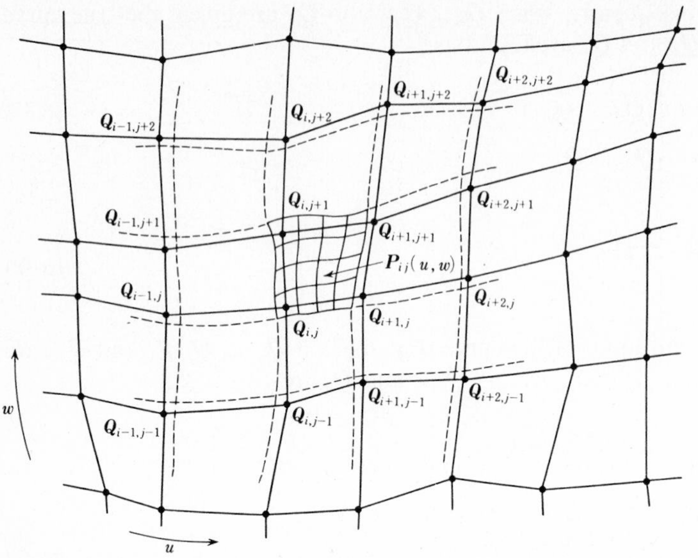

Fig. 6.18. Uniform bi-cubic *B*-spline surface patch and surface defining vectors

where:

$$B_R = \begin{bmatrix} Q_{i-1,j-1} & Q_{i-1,j} & Q_{i-1,j+1} & Q_{i-1,j+2} \\ Q_{i,j-1} & Q_{i,j} & Q_{i,j+1} & Q_{i,j+2} \\ Q_{i+1,j-1} & Q_{i+1,j} & Q_{i+1,j+1} & Q_{i+1,j+2} \\ Q_{i+2,j-1} & Q_{i+2,j} & Q_{i+2,j+1} & Q_{i+2,j+2} \end{bmatrix} \tag{6.43}$$

$$U = [u^3 \; u^2 \; u \; 1], \quad W = [w^3 \; w^2 \; w \; 1].$$

As shown in Fig. 6.18, one *B*-spline surface patch is defined by a net consisting of 16 vertices. In contrast to the case of a Bézier surface patch, the uniform *B*-spline surface patch of Eq. (6.41) generally does not pass through any of the surface defining vector (net vertices). The 4 corner points of the surface patch are in the vicinity of the 4 surface defining vectors $Q_{i,j}$, $Q_{i,j+1}$, $Q_{i+1,j}$ and $Q_{i+1,j+1}$.

Let us look at the relation between the *B*-spline surface of Eq. (6.41) and the Coons bi-cubic surface. Equating formulas (6.42) and (3.102) gives:

$$M_R B_R M_R^T = M_C B_C M_C^T.$$

Solving this equation for B_C gives:

$$B_C = (M_C^{-1} M_R) \, B_R (M_C^{-1} M_R)^T.$$

We also have:

$$M_C^{-1}M_R = \begin{bmatrix} \frac{1}{6} & \frac{2}{3} & \frac{1}{6} & 0 \\ 0 & \frac{1}{6} & \frac{2}{3} & \frac{1}{6} \\ -\frac{1}{2} & 0 & \frac{1}{2} & 0 \\ 0 & -\frac{1}{2} & 0 & \frac{1}{2} \end{bmatrix}$$

Therefore:

$$B_C \equiv \begin{bmatrix} Q(0,0) & Q(0,1) & Q_w(0,0) & Q_w(0,1) \\ Q(1,0) & Q(1,1) & Q_w(1,0) & Q_w(1,1) \\ Q_u(0,0) & Q_u(0,1) & Q_{uw}(0,0) & Q_{uw}(0,1) \\ Q_u(1,0) & Q_u(1,1) & Q_{uw}(1,0) & Q_{uw}(1,1) \end{bmatrix}$$

$$= \begin{bmatrix} \frac{1}{6} & \frac{2}{3} & \frac{1}{6} & 0 \\ 0 & \frac{1}{6} & \frac{2}{3} & \frac{1}{6} \\ -\frac{1}{2} & 0 & \frac{1}{2} & 0 \\ 0 & -\frac{1}{2} & 0 & \frac{1}{2} \end{bmatrix}$$

$$\times \begin{bmatrix} Q_{i-1,j-1} & Q_{i-1,j} & Q_{i-1,j+1} & Q_{i-1,j+2} \\ Q_{i,j-1} & Q_{i,j} & Q_{i,j+1} & Q_{i,j+2} \\ Q_{i+1,j-1} & Q_{i+1,j} & Q_{i+1,j+1} & Q_{i+1,j+2} \\ Q_{i+2,j-1} & Q_{i+2,j} & Q_{i+2,j+1} & Q_{i+2,j+2} \end{bmatrix} \begin{bmatrix} \frac{1}{6} & 0 & -\frac{1}{2} & 0 \\ \frac{2}{3} & \frac{1}{6} & 0 & -\frac{1}{2} \\ \frac{1}{6} & \frac{2}{3} & \frac{1}{2} & 0 \\ 0 & \frac{1}{6} & 0 & \frac{1}{2} \end{bmatrix}$$

This implies:

$$Q(0,0) = \frac{1}{6}\left(\frac{1}{6}Q_{i-1,j-1} + \frac{2}{3}Q_{i,j-1} + \frac{1}{6}Q_{i+1,j-1}\right) = \qquad (6.44)$$

$$+\frac{2}{3}\left(\frac{1}{6}\boldsymbol{Q}_{i-1,j}+\frac{2}{3}\boldsymbol{Q}_{i,j}+\frac{1}{6}\boldsymbol{Q}_{i+1,j}\right)$$

$$+\frac{1}{6}\left(\frac{1}{6}\boldsymbol{Q}_{i-1,j+1}+\frac{2}{3}\boldsymbol{Q}_{i,j+1}+\frac{1}{6}\boldsymbol{Q}_{i+1,j+1}\right)$$

$$\boldsymbol{Q}(0,1)=\frac{1}{6}\left(\frac{1}{6}\boldsymbol{Q}_{i-1,j}+\frac{2}{3}\boldsymbol{Q}_{i,j}+\frac{1}{6}\boldsymbol{Q}_{i+1,j}\right)$$

$$+\frac{2}{3}\left(\frac{1}{6}\boldsymbol{Q}_{i-1,j+1}+\frac{2}{3}\boldsymbol{Q}_{i,j+1}+\frac{1}{6}\boldsymbol{Q}_{i+1,j+1}\right)$$

$$+\frac{1}{6}\left(\frac{1}{6}\boldsymbol{Q}_{i-1,j+2}+\frac{2}{3}\boldsymbol{Q}_{i,j+2}+\frac{1}{6}\boldsymbol{Q}_{i+1,j+2}\right)$$

$$\boldsymbol{Q}(1,0)=\frac{1}{6}\left(\frac{1}{6}\boldsymbol{Q}_{i,j-1}+\frac{2}{3}\boldsymbol{Q}_{i+1,j-1}+\frac{1}{6}\boldsymbol{Q}_{i+2,j-1}\right)$$

$$+\frac{2}{3}\left(\frac{1}{6}\boldsymbol{Q}_{i,j}+\frac{2}{3}\boldsymbol{Q}_{i+1,j}+\frac{1}{6}\boldsymbol{Q}_{i+2,j}\right)$$

$$+\frac{1}{6}\left(\frac{1}{6}\boldsymbol{Q}_{i,j+1}+\frac{2}{3}\boldsymbol{Q}_{i+1,j+1}+\frac{1}{6}\boldsymbol{Q}_{i+2,j+1}\right)$$

$$\boldsymbol{Q}(1,1)=\frac{1}{6}\left(\frac{1}{6}\boldsymbol{Q}_{i,j}+\frac{2}{3}\boldsymbol{Q}_{i+1,j}+\frac{1}{6}\boldsymbol{Q}_{i+2,j}\right) \tag{6.44}$$

$$+\frac{2}{3}\left(\frac{1}{6}\boldsymbol{Q}_{i,j+1}+\frac{2}{3}\boldsymbol{Q}_{i+1,j+1}+\frac{1}{6}\boldsymbol{Q}_{i+2,j+1}\right)$$

$$+\frac{1}{6}\left(\frac{1}{6}\boldsymbol{Q}_{i,j+2}+\frac{2}{3}\boldsymbol{Q}_{i+1,j+2}+\frac{1}{6}\boldsymbol{Q}_{i+2,j+2}\right)$$

$$\boldsymbol{Q}_u(0,0)=\frac{1}{6}\left(\frac{1}{2}\boldsymbol{Q}_{i+1,j-1}-\frac{1}{2}\boldsymbol{Q}_{i-1,j-1}\right)$$

$$+\frac{2}{3}\left(\frac{1}{2}\boldsymbol{Q}_{i+1,j}-\frac{1}{2}\boldsymbol{Q}_{i-1,j}\right)+\frac{1}{6}\left(\frac{1}{2}\boldsymbol{Q}_{i+1,j+1}-\frac{1}{2}\boldsymbol{Q}_{i-1,j+1}\right)$$

$$\boldsymbol{Q}_u(0,1)=\frac{1}{6}\left(\frac{1}{2}\boldsymbol{Q}_{i+1,j}-\frac{1}{2}\boldsymbol{Q}_{i-1,j}\right)$$

$$+\frac{2}{3}\left(\frac{1}{2}\boldsymbol{Q}_{i+1,j+1}-\frac{1}{2}\boldsymbol{Q}_{i-1,j+1}\right)+\frac{1}{6}\left(\frac{1}{2}\boldsymbol{Q}_{i+1,j+2}-\frac{1}{2}\boldsymbol{Q}_{i-1,j+2}\right)$$

$$\boldsymbol{Q}_u(1,0)=\frac{1}{6}\left(\frac{1}{2}\boldsymbol{Q}_{i+2,j-1}-\frac{1}{2}\boldsymbol{Q}_{i,j-1}\right)+\frac{2}{3}\left(\frac{1}{2}\boldsymbol{Q}_{i+2,j}-\frac{1}{2}\boldsymbol{Q}_{i,j}\right)$$

$$+\frac{1}{6}\left(\frac{1}{2}\boldsymbol{Q}_{i+2,j+1}-\frac{1}{2}\boldsymbol{Q}_{i,j+1}\right)$$

$$
\left.
\begin{aligned}
\boldsymbol{Q}_u(1,1) &= \frac{1}{6}\left(\frac{1}{2}\boldsymbol{Q}_{i+2,j}-\frac{1}{2}\boldsymbol{Q}_{i,j}\right)+\frac{2}{3}\left(\frac{1}{2}\boldsymbol{Q}_{i+2,j+1}-\frac{1}{2}\boldsymbol{Q}_{i,j+1}\right)\\
&\quad+\frac{1}{6}\left(\frac{1}{2}\boldsymbol{Q}_{i+2,j+2}-\frac{1}{2}\boldsymbol{Q}_{i,j+2}\right)\\
\boldsymbol{Q}_w(0,0) &= \frac{1}{2}\left(\frac{1}{6}\boldsymbol{Q}_{i-1,j+1}+\frac{2}{3}\boldsymbol{Q}_{i,j+1}+\frac{1}{6}\boldsymbol{Q}_{i+1,j+1}\right)\\
&\quad-\frac{1}{2}\left(\frac{1}{6}\boldsymbol{Q}_{i-1,j-1}+\frac{2}{3}\boldsymbol{Q}_{i,j-1}+\frac{1}{6}\boldsymbol{Q}_{i+1,j-1}\right)\\
\boldsymbol{Q}_w(0,1) &= \frac{1}{2}\left(\frac{1}{6}\boldsymbol{Q}_{i-1,j+2}+\frac{2}{3}\boldsymbol{Q}_{i,j+2}+\frac{1}{6}\boldsymbol{Q}_{i+1,j+2}\right)\\
&\quad-\frac{1}{2}\left(\frac{1}{6}\boldsymbol{Q}_{i-1,j}+\frac{2}{3}\boldsymbol{Q}_{i,j}+\frac{1}{6}\boldsymbol{Q}_{i+1,j}\right)\\
\boldsymbol{Q}_w(1,0) &= \frac{1}{2}\left(\frac{1}{6}\boldsymbol{Q}_{i,j+1}+\frac{2}{3}\boldsymbol{Q}_{i+1,j+1}+\frac{1}{6}\boldsymbol{Q}_{i+2,j+1}\right)\\
&\quad-\frac{1}{2}\left(\frac{1}{6}\boldsymbol{Q}_{i,j-1}+\frac{2}{3}\boldsymbol{Q}_{i+1,j-1}+\frac{1}{6}\boldsymbol{Q}_{i+2,j-1}\right)\\
\boldsymbol{Q}_w(1,1) &= \frac{1}{2}\left(\frac{1}{6}\boldsymbol{Q}_{i,j+2}+\frac{2}{3}\boldsymbol{Q}_{i+1,j+2}+\frac{1}{6}\boldsymbol{Q}_{i+2,j+2}\right)\\
&\quad-\frac{1}{2}\left(\frac{1}{6}\boldsymbol{Q}_{i,j}+\frac{2}{3}\boldsymbol{Q}_{i+1,j}+\frac{1}{6}\boldsymbol{Q}_{i+2,j}\right)\\
\boldsymbol{Q}_{uw}(0,0) &= \frac{1}{2}\left(\frac{1}{2}\boldsymbol{Q}_{i+1,j+1}-\frac{1}{2}\boldsymbol{Q}_{i-1,j+1}\right)\\
&\quad-\frac{1}{2}\left(\frac{1}{2}\boldsymbol{Q}_{i+1,j-1}-\frac{1}{2}\boldsymbol{Q}_{i-1,j-1}\right)\\
\boldsymbol{Q}_{uw}(0,1) &= \frac{1}{2}\left(\frac{1}{2}\boldsymbol{Q}_{i+1,j+2}-\frac{1}{2}\boldsymbol{Q}_{i-1,j+2}\right)\\
&\quad-\frac{1}{2}\left(\frac{1}{2}\boldsymbol{Q}_{i+1,j}-\frac{1}{2}\boldsymbol{Q}_{i-1,j}\right)\\
\boldsymbol{Q}_{uw}(1,0) &= \frac{1}{2}\left(\frac{1}{2}\boldsymbol{Q}_{i+2,j+1}-\frac{1}{2}\boldsymbol{Q}_{i,j+1}\right)\\
&\quad-\frac{1}{2}\left(\frac{1}{2}\boldsymbol{Q}_{i+2,j-1}-\frac{1}{2}\boldsymbol{Q}_{i,j-1}\right)
\end{aligned}
\right| \qquad (6.44)
$$

$$\left.\begin{aligned} Q_{uw}(1,1) = \frac{1}{2}\left(\frac{1}{2}Q_{i+2,j+2} - \frac{1}{2}Q_{i,j+2}\right) \\ -\frac{1}{2}\left(\frac{1}{2}Q_{i+2,j} - \frac{1}{2}Q_{i,j}\right). \end{aligned}\right\} \tag{6.44}$$

Next, let us find the relation between the B-spline surface given by formula (6.41) and a Bézier bi-cubic surface patch (5.118). Equating formulas (6.42) and (5.118) gives:

$$M_R B_R M_R^T = M_B B_B M_B^T.$$

Solving this equation for B_B gives:

$$B_B = (M_B^{-1} M_R)\, B_R (M_B^{-1} M_R)^T.$$

We also have:

$$M_B^{-1} M_R = \begin{bmatrix} \frac{1}{6} & \frac{2}{3} & \frac{1}{6} & 0 \\ 0 & \frac{2}{3} & \frac{1}{3} & 0 \\ 0 & \frac{1}{3} & \frac{2}{3} & 0 \\ 0 & \frac{1}{6} & \frac{2}{3} & \frac{1}{6} \end{bmatrix}$$

Therefore:

$$B_B \equiv \begin{bmatrix} Q_{00} & Q_{01} & Q_{02} & Q_{03} \\ Q_{10} & Q_{11} & Q_{12} & Q_{13} \\ Q_{20} & Q_{21} & Q_{22} & Q_{23} \\ Q_{30} & Q_{31} & Q_{32} & Q_{33} \end{bmatrix}$$

$$= \begin{bmatrix} \frac{1}{6} & \frac{2}{3} & \frac{1}{6} & 0 \\ 0 & \frac{2}{3} & \frac{1}{3} & 0 \\ 0 & \frac{1}{3} & \frac{2}{3} & 0 \\ 0 & \frac{1}{6} & \frac{2}{3} & \frac{1}{6} \end{bmatrix} \times$$

$$\times\begin{bmatrix} \boldsymbol{Q}_{i-1,j-1} & \boldsymbol{Q}_{i-1,j} & \boldsymbol{Q}_{i-1,j+1} & \boldsymbol{Q}_{i-1,j+2} \\ \boldsymbol{Q}_{i,j-1} & \boldsymbol{Q}_{i,j} & \boldsymbol{Q}_{i,j+1} & \boldsymbol{Q}_{i,j+2} \\ \boldsymbol{Q}_{i+1,j-1} & \boldsymbol{Q}_{i+1,j} & \boldsymbol{Q}_{i+1,j+1} & \boldsymbol{Q}_{i+1,j+2} \\ \boldsymbol{Q}_{i+2,j-1} & \boldsymbol{Q}_{i+2,j} & \boldsymbol{Q}_{i+2,j+1} & \boldsymbol{Q}_{i+2,j+2} \end{bmatrix}\begin{bmatrix} \frac{1}{6} & 0 & 0 & 0 \\ \frac{2}{3} & \frac{2}{3} & \frac{1}{3} & \frac{1}{6} \\ \frac{1}{6} & \frac{1}{3} & \frac{2}{3} & \frac{2}{3} \\ 0 & 0 & 0 & \frac{1}{6} \end{bmatrix}$$

This implies:

$$\left.\begin{aligned}
\boldsymbol{Q}_{00} &= \frac{1}{6}\left(\frac{1}{6}\boldsymbol{Q}_{i-1,j-1}+\frac{2}{3}\boldsymbol{Q}_{i,j-1}+\frac{1}{6}\boldsymbol{Q}_{i+1,j-1}\right) \\
&\quad+\frac{2}{3}\left(\frac{1}{6}\boldsymbol{Q}_{i-1,j}+\frac{2}{3}\boldsymbol{Q}_{i,j}+\frac{1}{6}\boldsymbol{Q}_{i+1,j}\right) \\
&\quad+\frac{1}{6}\left(\frac{1}{6}\boldsymbol{Q}_{i-1,j+1}+\frac{2}{3}\boldsymbol{Q}_{i,j+1}+\frac{1}{6}\boldsymbol{Q}_{i+1,j+1}\right) \\
\boldsymbol{Q}_{01} &= \frac{2}{3}\left(\frac{1}{6}\boldsymbol{Q}_{i-1,j}+\frac{2}{3}\boldsymbol{Q}_{i,j}+\frac{1}{6}\boldsymbol{Q}_{i+1,j}\right) \\
&\quad+\frac{1}{3}\left(\frac{1}{6}\boldsymbol{Q}_{i-1,j+1}+\frac{2}{3}\boldsymbol{Q}_{i,j+1}+\frac{1}{6}\boldsymbol{Q}_{i+1,j+1}\right) \\
\boldsymbol{Q}_{02} &= \frac{1}{3}\left(\frac{1}{6}\boldsymbol{Q}_{i-1,j}+\frac{2}{3}\boldsymbol{Q}_{i,j}+\frac{1}{6}\boldsymbol{Q}_{i+1,j}\right) \\
&\quad+\frac{2}{3}\left(\frac{1}{6}\boldsymbol{Q}_{i-1,j+1}+\frac{2}{3}\boldsymbol{Q}_{i,j+1}+\frac{1}{6}\boldsymbol{Q}_{i+1,j+1}\right) \\
\boldsymbol{Q}_{03} &= \frac{1}{6}\left(\frac{1}{6}\boldsymbol{Q}_{i-1,j}+\frac{2}{3}\boldsymbol{Q}_{i,j}+\frac{1}{6}\boldsymbol{Q}_{i+1,j}\right) \\
&\quad+\frac{2}{3}\left(\frac{1}{6}\boldsymbol{Q}_{i-1,j+1}+\frac{2}{3}\boldsymbol{Q}_{i,j+1}+\frac{1}{6}\boldsymbol{Q}_{i+1,j+1}\right) \\
&\quad+\frac{1}{6}\left(\frac{1}{6}\boldsymbol{Q}_{i-1,j+2}+\frac{2}{3}\boldsymbol{Q}_{i,j+2}+\frac{1}{6}\boldsymbol{Q}_{i+1,j+2}\right) \\
\boldsymbol{Q}_{10} &= \frac{1}{6}\left(\frac{2}{3}\boldsymbol{Q}_{i,j-1}+\frac{1}{3}\boldsymbol{Q}_{i+1,j-1}\right)+\frac{2}{3}\left(\frac{2}{3}\boldsymbol{Q}_{i,j}+\frac{1}{3}\boldsymbol{Q}_{i+1,j}\right) \\
&\quad+\frac{1}{6}\left(\frac{2}{3}\boldsymbol{Q}_{i,j+1}+\frac{1}{3}\boldsymbol{Q}_{i+1,j+1}\right)
\end{aligned}\right\} \tag{6.45}$$

$$\left.\begin{aligned}
Q_{11} &= \frac{2}{3}\left(\frac{2}{3}Q_{i,j}+\frac{1}{3}Q_{i+1,j}\right)+\frac{1}{3}\left(\frac{2}{3}Q_{i,j+1}+\frac{1}{3}Q_{i+1,j+1}\right)\\
Q_{12} &= \frac{1}{3}\left(\frac{2}{3}Q_{i,j}+\frac{1}{3}Q_{i+1,j}\right)+\frac{2}{3}\left(\frac{2}{3}Q_{i,j+1}+\frac{1}{3}Q_{i+1,j+1}\right)\\
Q_{13} &= \frac{1}{6}\left(\frac{2}{3}Q_{i,j}+\frac{1}{3}Q_{i+1,j}\right)+\frac{2}{3}\left(\frac{2}{3}Q_{i,j+1}+\frac{1}{3}Q_{i+1,j+1}\right)\\
&\quad+\frac{1}{6}\left(\frac{2}{3}Q_{i,j+2}+\frac{1}{3}Q_{i+1,j+2}\right)\\
Q_{20} &= \frac{1}{6}\left(\frac{1}{3}Q_{i,j-1}+\frac{2}{3}Q_{i+1,j-1}\right)+\frac{2}{3}\left(\frac{1}{3}Q_{i,j}+\frac{2}{3}Q_{i+1,j}\right)\\
&\quad+\frac{1}{6}\left(\frac{1}{3}Q_{i,j+1}+\frac{2}{3}Q_{i+1,j+1}\right)\\
Q_{21} &= \frac{2}{3}\left(\frac{1}{3}Q_{i,j}+\frac{2}{3}Q_{i+1,j}\right)+\frac{1}{3}\left(\frac{1}{3}Q_{i,j+1}+\frac{2}{3}Q_{i+1,j+1}\right)\\
Q_{22} &= \frac{1}{3}\left(\frac{1}{3}Q_{i,j}+\frac{2}{3}Q_{i+1,j}\right)+\frac{2}{3}\left(\frac{1}{3}Q_{i,j+1}+\frac{2}{3}Q_{i+1,j+1}\right)\\
Q_{23} &= \frac{1}{6}\left(\frac{1}{3}Q_{i,j}+\frac{2}{3}Q_{i+1,j}\right)+\frac{2}{3}\left(\frac{1}{3}Q_{i,j+1}+\frac{2}{3}Q_{i+1,j+1}\right)\\
&\quad+\frac{1}{6}\left(\frac{1}{3}Q_{i,j+2}+\frac{2}{3}Q_{i+1,j+2}\right)\\
Q_{30} &= \frac{1}{6}\left(\frac{1}{6}Q_{i,j-1}+\frac{2}{3}Q_{i+1,j-1}+\frac{1}{6}Q_{i+2,j-1}\right)\\
&\quad+\frac{2}{3}\left(\frac{1}{6}Q_{i,j}+\frac{2}{3}Q_{i+1,j}+\frac{1}{6}Q_{i+2,j}\right)\\
&\quad+\frac{1}{6}\left(\frac{1}{6}Q_{i,j+1}+\frac{2}{3}Q_{i+1,j+1}+\frac{1}{6}Q_{i+2,j+1}\right)\\
Q_{31} &= \frac{2}{3}\left(\frac{1}{6}Q_{i,j}+\frac{2}{3}Q_{i+1,j}+\frac{1}{6}Q_{i+2,j}\right)\\
&\quad+\frac{1}{3}\left(\frac{1}{6}Q_{i,j+1}+\frac{2}{3}Q_{i+1,j+1}+\frac{1}{6}Q_{i+2,j+1}\right)\\
Q_{32} &= \frac{1}{3}\left(\frac{1}{6}Q_{i,j}+\frac{2}{3}Q_{i+1,j}+\frac{1}{6}Q_{i+2,j}\right)\\
&\quad+\frac{2}{3}\left(\frac{1}{6}Q_{i,j+1}+\frac{2}{3}Q_{i+1,j+1}+\frac{1}{6}Q_{i+2,j+1}\right)
\end{aligned}\right\}\qquad(6.45)$$

$$
\left.\begin{aligned}
\boldsymbol{Q}_{33} = & \frac{1}{6}\left(\frac{1}{6}\boldsymbol{Q}_{i,j}+\frac{2}{3}\boldsymbol{Q}_{i+1,j}+\frac{1}{6}\boldsymbol{Q}_{i+2,j}\right) \\
& +\frac{2}{3}\left(\frac{1}{6}\boldsymbol{Q}_{i,j+1}+\frac{2}{3}\boldsymbol{Q}_{i+1,j+1}+\frac{1}{6}\boldsymbol{Q}_{i+2,j+1}\right) \\
& +\frac{1}{6}\left(\frac{1}{6}\boldsymbol{Q}_{i,j+2}+\frac{2}{3}\boldsymbol{Q}_{i+1,j+2}+\frac{1}{6}\boldsymbol{Q}_{i+2,j+2}\right).
\end{aligned}\right\} \tag{6.45}
$$

6.2.2 Determination of a Point on a Surface by Finite Difference Operations

When a *B*-spline surface patch is given by formula (6.41), then, using the abbreviation $N_{i,4}(t) \equiv N_i(t)$, the finite difference matrices *A*, *B*, *C* and *D* given by Eqs. (1.113), (1.114), (1.115) and (1.116) are, respectively, as follows.

Finite difference matrix *A* is:

$$
\begin{bmatrix}
\boldsymbol{P}_{ij}(0,0) \\
\frac{1}{\delta}\Delta\boldsymbol{P}_{ij}(0,0) \\
\frac{1}{\delta^2}\Delta^2\boldsymbol{P}_{ij}(0,0) \\
\frac{1}{\delta^3}\Delta^3\boldsymbol{P}_{ij}(0,0)
\end{bmatrix}
=
\begin{bmatrix}
N_0(0) & N_1(0) & N_2(0) & N_3(0) \\
\frac{1}{\delta}\Delta N_0(0) & \frac{1}{\delta}\Delta N_1(0) & \frac{1}{\delta}\Delta N_2(0) & \frac{1}{\delta}\Delta N_3(0) \\
\frac{1}{\delta^2}\Delta^2 N_0(0) & \frac{1}{\delta^2}\Delta^2 N_1(0) & \frac{1}{\delta^2}\Delta^2 N_2(0) & \frac{1}{\delta^2}\Delta^2 N_3(0) \\
\frac{1}{\delta^3}\Delta^3 N_0(0) & \frac{1}{\delta^3}\Delta^3 N_1(0) & \frac{1}{\delta^3}\Delta^3 N_2(0) & \frac{1}{\delta^3}\Delta^3 N_3(0)
\end{bmatrix}
$$

$$
\times
\begin{bmatrix}
\boldsymbol{Q}_{i-1,j-1} & \boldsymbol{Q}_{i-1,j} & \boldsymbol{Q}_{i-1,j+1} & \boldsymbol{Q}_{i-1,j+2} \\
\boldsymbol{Q}_{i,j-1} & \boldsymbol{Q}_{i,j} & \boldsymbol{Q}_{i,j+1} & \boldsymbol{Q}_{i,j+2} \\
\boldsymbol{Q}_{i+1,j-1} & \boldsymbol{Q}_{i+1,j} & \boldsymbol{Q}_{i+1,j+1} & \boldsymbol{Q}_{i+1,j+2} \\
\boldsymbol{Q}_{i+2,j-1} & \boldsymbol{Q}_{i+2,j} & \boldsymbol{Q}_{i+2,j+1} & \boldsymbol{Q}_{i+2,j+2}
\end{bmatrix}
\begin{bmatrix}
N_0(0) \\ N_1(0) \\ N_2(0) \\ N_3(0)
\end{bmatrix}
\tag{6.46}
$$

Denoting differences in the *w*-direction by ∇, finite difference matrix *B* is:

$$
\begin{bmatrix}
\frac{1}{\delta}\nabla\boldsymbol{P}_{ij}(0,0) \\
\frac{1}{\delta^2}\nabla(\Delta\boldsymbol{P}_{ij}(0,0)) \\
\frac{1}{\delta^3}\nabla(\Delta^2\boldsymbol{P}_{ij}(0,0)) \\
\frac{1}{\delta^4}\nabla(\Delta^3\boldsymbol{P}_{ij}(0,0))
\end{bmatrix}
=
\begin{bmatrix}
N_0(0) & N_1(0) & N_2(0) & N_3(0) \\
\frac{1}{\delta}\Delta N_0(0) & \frac{1}{\delta}\Delta N_1(0) & \frac{1}{\delta}\Delta N_2(0) & \frac{1}{\delta}\Delta N_3(0) \\
\frac{1}{\delta^2}\Delta^2 N_0(0) & \frac{1}{\delta^2}\Delta^2 N_1(0) & \frac{1}{\delta^2}\Delta^2 N_2(0) & \frac{1}{\delta^2}\Delta^2 N_3(0) \\
\frac{1}{\delta^3}\Delta^3 N_0(0) & \frac{1}{\delta^3}\Delta^3 N_1(0) & \frac{1}{\delta^3}\Delta^3 N_2(0) & \frac{1}{\delta^3}\Delta^3 N_3(0)
\end{bmatrix}
\times
$$

$$\times \begin{bmatrix} Q_{i-1,j-1} & Q_{i-1,j} & Q_{i-1,j+1} & Q_{i-1,j+2} \\ Q_{i,j-1} & Q_{i,j} & Q_{i,j+1} & Q_{i,j+2} \\ Q_{i+1,j-1} & Q_{i+1,j} & Q_{i+1,j+1} & Q_{i+1,j+2} \\ Q_{i+2,j-1} & Q_{i+2,j} & Q_{i+2,j+1} & Q_{i+2,j+2} \end{bmatrix} \begin{bmatrix} \frac{1}{\delta}\nabla N_0(0) \\ \frac{1}{\delta}\nabla N_1(0) \\ \frac{1}{\delta}\nabla N_2(0) \\ \frac{1}{\delta}\nabla N_3(0) \end{bmatrix} \tag{6.47}$$

The finite difference matrix C is:

$$\begin{bmatrix} \frac{1}{\delta^2}\nabla^2 \boldsymbol{P}_{ij}(0,0) \\ \frac{1}{\delta^3}\nabla^2(\Delta \boldsymbol{P}_{ij}(0,0)) \\ \frac{1}{\delta^4}\nabla^2(\Delta^2 \boldsymbol{P}_{ij}(0,0)) \\ \frac{1}{\delta^5}\nabla^2(\Delta^3 \boldsymbol{P}_{ij}(0,0)) \end{bmatrix} = \begin{bmatrix} N_0(0) & N_1(0) & N_2(0) & N_3(0) \\ \frac{1}{\delta}\Delta N_0(0) & \frac{1}{\delta}\Delta N_1(0) & \frac{1}{\delta}\Delta N_2(0) & \frac{1}{\delta}\Delta N_3(0) \\ \frac{1}{\delta^2}\Delta^2 N_0(0) & \frac{1}{\delta^2}\Delta^2 N_1(0) & \frac{1}{\delta^2}\Delta^2 N_2(0) & \frac{1}{\delta^2}\Delta^2 N_3(0) \\ \frac{1}{\delta^3}\Delta^3 N_0(0) & \frac{1}{\delta^3}\Delta^3 N_1(0) & \frac{1}{\delta^3}\Delta^3 N_2(0) & \frac{1}{\delta^3}\Delta^3 N_3(0) \end{bmatrix}$$

$$\times \begin{bmatrix} Q_{i-1,j-1} & Q_{i-1,j} & Q_{i-1,j+1} & Q_{i-1,j+2} \\ Q_{i,j-1} & Q_{i,j} & Q_{i,j+1} & Q_{i,j+2} \\ Q_{i+1,j-1} & Q_{i+1,j} & Q_{i+1,j+1} & Q_{i+1,j+2} \\ Q_{i+2,j-1} & Q_{i+2,j} & Q_{i+2,j+1} & Q_{i+2,j+2} \end{bmatrix} \begin{bmatrix} \frac{1}{\delta^2}\nabla^2 N_0(0) \\ \frac{1}{\delta^2}\nabla^2 N_1(0) \\ \frac{1}{\delta^2}\nabla^2 N_2(0) \\ \frac{1}{\delta^2}\nabla^2 N_3(0) \end{bmatrix} \tag{6.48}$$

The finite difference matrix D is:

$$\begin{bmatrix} \frac{1}{\delta^3}\nabla^3 \boldsymbol{P}_{ij}(0,0) \\ \frac{1}{\delta^4}\nabla^3(\Delta \boldsymbol{P}_{ij}(0,0)) \\ \frac{1}{\delta^5}\nabla^3(\Delta^2 \boldsymbol{P}_{ij}(0,0)) \\ \frac{1}{\delta^6}\nabla^3(\Delta^3 \boldsymbol{P}_{ij}(0,0)) \end{bmatrix} = \begin{bmatrix} N_0(0) & N_1(0) & N_2(0) & N_3(0) \\ \frac{1}{\delta}\Delta N_0(0) & \frac{1}{\delta}\Delta N_1(0) & \frac{1}{\delta}\Delta N_2(0) & \frac{1}{\delta}\Delta N_3(0) \\ \frac{1}{\delta^2}\Delta^2 N_0(0) & \frac{1}{\delta^2}\Delta^2 N_1(0) & \frac{1}{\delta^2}\Delta^2 N_2(0) & \frac{1}{\delta^2}\Delta^2 N_3(0) \\ \frac{1}{\delta^3}\Delta^3 N_0(0) & \frac{1}{\delta^3}\Delta^3 N_1(0) & \frac{1}{\delta^3}\Delta^3 N_2(0) & \frac{1}{\delta^3}\Delta^3 N_3(0) \end{bmatrix} \times$$

$$\times \begin{bmatrix} Q_{i-1,j-1} & Q_{i-1,j} & Q_{i-1,j+1} & Q_{i-1,j+2} \\ Q_{i,j-1} & Q_{i,j} & Q_{i,j+1} & Q_{i,j+2} \\ Q_{i+1,j-1} & Q_{i+1,j} & Q_{i+1,j+1} & Q_{i+1,j+2} \\ Q_{i+2,j-1} & Q_{i+2,j} & Q_{i+2,j+1} & Q_{i+2,j+2} \end{bmatrix} \begin{bmatrix} \frac{1}{\delta^3} \nabla^3 N_0(0) \\ \frac{1}{\delta^3} \nabla^3 N_1(0) \\ \frac{1}{\delta^3} \nabla^3 N_2(0) \\ \frac{1}{\delta^3} \nabla^3 N_3(0) \end{bmatrix} \tag{6.49}$$

Equations (6.46), (6.47), (6.48) and (6.49) can be combined into one equation as follows:

$$\begin{bmatrix} \frac{1}{\delta^k} \nabla^k P_{ij}(0,0) \\ \frac{1}{\delta^{k+1}} \nabla^k(\Delta P_{ij}(0,0)) \\ \frac{1}{\delta^{k+2}} \nabla^k(\Delta^2 P_{ij}(0,0)) \\ \frac{1}{\delta^{k+3}} \nabla^k(\Delta^3 P_{ij}(0,0)) \end{bmatrix} = \begin{bmatrix} N_0(0) & N_1(0) & N_2(0) & N_3(0) \\ \frac{1}{\delta}\Delta N_0(0) & \frac{1}{\delta}\Delta N_1(0) & \frac{1}{\delta}\Delta N_2(0) & \frac{1}{\delta}\Delta N_3(0) \\ \frac{1}{\delta^2}\Delta^2 N_0(0) & \frac{1}{\delta^2}\Delta^2 N_1(0) & \frac{1}{\delta^2}\Delta^2 N_2(0) & \frac{1}{\delta^2}\Delta^2 N_3(0) \\ \frac{1}{\delta^3}\Delta^3 N_0(0) & \frac{1}{\delta^3}\Delta^3 N_1(0) & \frac{1}{\delta^3}\Delta^3 N_2(0) & \frac{1}{\delta^3}\Delta^3 N_3(0) \end{bmatrix}$$

$$\times \begin{bmatrix} Q_{i-1,j-1} & Q_{i-1,j} & Q_{i-1,j+1} & Q_{i-1,j+2} \\ Q_{i,j-1} & Q_{i,j} & Q_{i,j+1} & Q_{i,j+2} \\ Q_{i+1,j-1} & Q_{i+1,j} & Q_{i+1,j+1} & Q_{i+1,j+2} \\ Q_{i+2,j-1} & Q_{i+2,j} & Q_{i+2,j+1} & Q_{i+2,j+2} \end{bmatrix} \begin{bmatrix} \frac{1}{\delta^k} \nabla^k N_0(0) \\ \frac{1}{\delta^k} \nabla^k N_1(0) \\ \frac{1}{\delta^k} \nabla^k N_2(0) \\ \frac{1}{\delta^k} \nabla^k N_3(0) \end{bmatrix} \tag{6.50}$$

$(k=0,\ 1,\ 2,\ 3)$.

In surface shape control, if one of the $Q_{i,j}$ varies, it is necessary to recalculate all of the finite difference matrices *A*, *B*, *C* and *D*, but to do this using Eq. (6.50) is wasteful. To perform this calculation efficiently, it should be performed only for the change $\blacktriangle Q_{i,j}$ ($\blacktriangle$ denotes the change) and then add this change. The equation for this can be derived from (6.50):

$$\begin{bmatrix} \frac{1}{\delta^k} \nabla^k \boldsymbol{P}_{ij}(0,0) \\ \frac{1}{\delta^{k+1}} \nabla^k(\Delta \boldsymbol{P}_{ij}(0,0)) \\ \frac{1}{\delta^{k+2}} \nabla^k(\Delta^2 \boldsymbol{P}_{ij}(0,0)) \\ \frac{1}{\delta^{k+3}} \nabla^k(\Delta^3 \boldsymbol{P}_{ij}(0,0)) \end{bmatrix} \longleftarrow \begin{bmatrix} \frac{1}{\delta^k} \nabla^k \boldsymbol{P}_{ij}(0,0) \\ \frac{1}{\delta^{k+1}} \nabla^k(\Delta \boldsymbol{P}_{ij}(0,0)) \\ \frac{1}{\delta^{k+2}} \nabla^k(\Delta^2 \boldsymbol{P}_{ij}(0,0)) \\ \frac{1}{\delta^{k+3}} \nabla^k(\Delta^3 \boldsymbol{P}_{ij}(0,0)) \end{bmatrix}$$

$$+\begin{bmatrix} \frac{1}{\delta^k} N_{p+1}(0)\,\nabla^k N_{q+1}(0) \\ \frac{1}{\delta^{k+1}} \Delta N_{p+1}(0)\,\nabla^k N_{q+1}(0) \\ \frac{1}{\delta^{k+2}} \Delta^2 N_{p+1}(0)\,\nabla^k N_{q+1}(0) \\ \frac{1}{\delta^{k+3}} \Delta^3 N_{p+1}(0)\,\nabla^k N_{q+1}(0) \end{bmatrix} \blacktriangle \boldsymbol{Q}_{i+p,j+q} \tag{6.51}$$

$$(k=0,\ 1,\ 2,\ 3;\ \ p, q=-1,\ 0,\ 1,\ 2).$$

6.2.3 Inverse Transformation of a Surface

When a lattice of points $\boldsymbol{P}_{i,j}$ ($i=1, 2, \ldots, m-1$; $j=1, 2, \ldots, n-1$) is given the surface defining vectors of the uniform bi-cubic B-spline surface formula (6.41), with the lattice points as the 4 corners of surface patches, can be found very easily and at high speed using the curve inverse transformation algorithm discussed in Sect. 6.1.4 (Fig. 6.19).

Noting that the B-spline functions become:

$$N_{0,4}(0)=1/6,\quad N_{1,4}(0)=2/3,\quad N_{2,4}(0)=1/6,\quad N_{3,4}(0)=0$$
$$N_{0,4}(1)=0,\quad N_{1,4}(1)=1/6,\quad N_{2,4}(1)=2/3,\quad N_{3,4}(1)=1/6$$

we see that the following equation hold:

$$\left.\begin{aligned} \boldsymbol{P}_{i,j} &= \frac{1}{6}\boldsymbol{V}_{i-1,j} + \frac{2}{3}\boldsymbol{V}_{i,j} + \frac{1}{6}\boldsymbol{V}_{i+1,j} \quad (1 \leqq i \leqq m-1;\ 1 \leqq j \leqq n-1) \\ \boldsymbol{V}_{i,j} &= \frac{1}{6}\boldsymbol{Q}_{i,j-1} + \frac{2}{3}\boldsymbol{Q}_{i,j} + \frac{1}{6}\boldsymbol{Q}_{i,j+1} \quad (0 \leqq i \leqq m;\ 1 \leqq j \leqq n-1) \end{aligned}\right\} \tag{6.52}$$

or:

$$\left.\begin{aligned} \boldsymbol{P}_{ij} &= \frac{1}{6}\boldsymbol{U}_{i,j-1} + \frac{2}{3}\boldsymbol{U}_{ij} + \frac{1}{6}\boldsymbol{U}_{i,j+1} \quad (1 \leqq i \leqq m-1;\ 1 \leqq j \leqq n-1) \\ \boldsymbol{U}_{ij} &= \frac{1}{6}\boldsymbol{Q}_{i-1,j} + \frac{2}{3}\boldsymbol{Q}_{ij} + \frac{1}{6}\boldsymbol{Q}_{i+1,j} \quad (1 \leqq i \leqq m-1;\ 0 \leqq j \leqq n) \end{aligned}\right\} \tag{6.53}$$

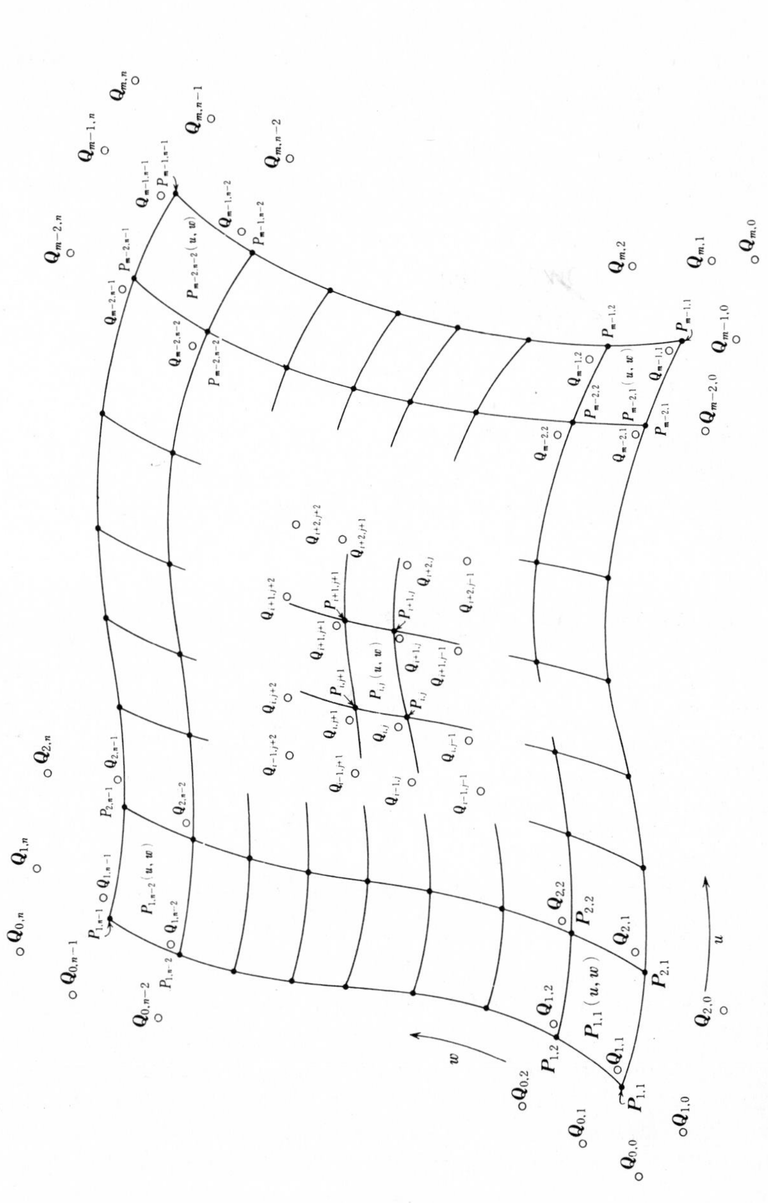

Fig. 6.19. Inverse transformation of a surface

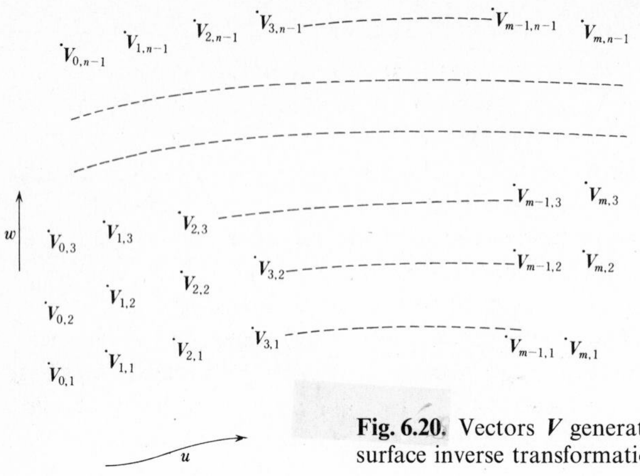

Fig. 6.20. Vectors V generated as intermediate steps in the surface inverse transformation algorithm

Regardless of whether Eqs. (6.52) or (6.53) are used, the same $Q_{i,j}$ can be determined. The following discussion will use Eqs. (6.52) to explain the inverse transformation procedure.

First, apply the inverse transformation algorithm for a curve to the sequence of $m-1$ points in the u-direction $P_{1,1}, P_{2,1}, \ldots, P_{m-2,1}, P_{m-1,1}$. Using relations (6.52) gives the sequence of points $V_{0,1}, V_{1,1}, \ldots, V_{m-1,1}, V_{m,1}$. Similarly, apply the curve inverse transformation algorithm to the 2-nd sequence of points in the u-direction $P_{1,2}, P_{2,2}, \ldots, P_{m-2,2}, P_{m-1,2}$, thus generating the sequence of points $V_{0,2}, V_{1,2}, \ldots, V_{m-1,2}, V_{m,2}$. Repeat this procedure for all $n-1$ sequences of points in the u-direction to give the lattice of sequences of points $V_{i,j}$ shown in Fig. 6.20. Next, focus attention on the $m+1$ sequences of points in the w-direction in Fig. 6.20. First, apply the curve inverse transformation algorithm to the first sequence of points $V_{0,1}, V_{0,2}, \ldots, V_{0,n-1}$. Then apply relations (6.52) to generate the sequence of points $Q_{0,0}, Q_{0,1}, Q_{0,2}, \ldots, Q_{0,n-1}, Q_{0,n}$. Similarly, apply the inverse transformation algorithm to the other m point sequences to generate all of the $Q_{i,j}$ ($i=0, 1, 2, \ldots, m$; $j=0, 1, \ldots, n$). These are the surface defining vectors in Eq. (6.41).

If the order of application of the curve inverse transformation algorithm is reversed, Eqs. (6.53) apply. First apply the inverse transformation algorithm to the sequence of points $P_{i,j}$ in the w-direction, then, with respect to the sequences of points $U_{i,j}$ that are thus generated in the u-direction, apply the inverse transformation algorithm again. The final result is the same with either method.

In the inverse transformation algorithm for a curve, we specified that the two vectors at the ends of the point sequence that is generated must be the same: $Q_0=Q_1$; $Q_{n-1}=Q_n$. For this reason, the surface defining vectors that are generated satisfy similar conditions:

$$\left.\begin{array}{lll} Q_{i,0}=Q_{i,1}, & Q_{i,n-1}=Q_{i,n} & (i=0, 1, \ldots, m) \\ Q_{0,j}=Q_{1,j}, & Q_{m-1,j}=Q_{m,j} & (j=0, 1, \ldots, n) \end{array}\right\}. \tag{6.54}$$

6.2.4 Surfaces of Revolution

In industry, surfaces of revolution are used frequently, and are very important. As long as rational polynomial expressions are not used, a circle (or surface of revolution) cannot be rigorously expressed (refer to Sect. 7.4), but, as we saw in Sect. 6.1.2, by taking the appropriate number of vertices a uniform cubic B-spline curve can be used to approximate a circle to sufficient accuracy. Similarly, uniform cubic B-spline curved surfaces can be used to generate approximate surfaces of revolution, as will now be explained.

In Fig. 6.21 are shown a curve $\boldsymbol{P}_1(u)$ in the plane $x=z$, and a circle (or circular arc) of unit radius with center at (0,1,0) in the plane $y=1$. The vector $\boldsymbol{P}(u,w)$ has components which are expressed as products of the corresponding coordinate components of $\boldsymbol{P}_1(u)$ and $\boldsymbol{P}_2(w)$. $\boldsymbol{P}(u,w)$ describes the curved surface (or part of it) formed by rotating the projection of curve $\boldsymbol{P}_1(u)$ onto the $x-y$ plane around the y axis:

$$\begin{aligned}\boldsymbol{P}(u,w) &= [P_x(u,w) \;\; P_y(u,w) \;\; P_z(u,w)]\\ &= [P_{1x}(u)P_{2x}(w) \;\; P_{1y}(u)P_{2y}(w) \;\; P_{1z}(u)P_{2z}(w)]\\ &= [P_{1x}(u)P_{2x}(w) \;\; P_{1y}(u) \;\; P_{1x}(u)P_{2z}(w)].\end{aligned} \tag{6.55}$$

Therefore:

$$\begin{aligned}P_x^2(u,w)+P_z^2(u,w) &= P_{1x}^2(u)(P_{2x}^2(w)+P_{2z}^2(w))\\ &= P_{1x}^2(u).\end{aligned} \tag{6.56}$$

From Eqs. (6.55) and (6.56), we see that the surface $\boldsymbol{P}(u,w)$ has a cross-section in the plane $y=P_{1y}(u)$ that is a circle or circular arc of radius P_{1x}.

Suppose that $\boldsymbol{P}_1(u)$ is a uniform cubic B-spline curve and that the circle $\boldsymbol{P}_2(w)$ is approximated by a uniform cubic B-spline curve:

$$\boldsymbol{P}_{1,i}(u)=N_{0,4}(u)\,\boldsymbol{V}_{i-1}+N_{1,4}(u)\,\boldsymbol{V}_i+N_{2,4}(u)\,\boldsymbol{V}_{i+1}+N_{3,4}(u)\,\boldsymbol{V}_{i+2} \tag{6.57}$$

$$\boldsymbol{P}_{2,j}(w)=N_{0,4}(w)\,\boldsymbol{W}_{j-1}+N_{1,4}(w)\,\boldsymbol{W}_j+N_{2,4}(w)\,\boldsymbol{W}_{j+1}+N_{3,4}(w)\,\boldsymbol{W}_{j+2} \tag{6.58}$$

$(i=1, 2, \ldots, m-2;\; j=1, 2, \ldots, n-2)$.

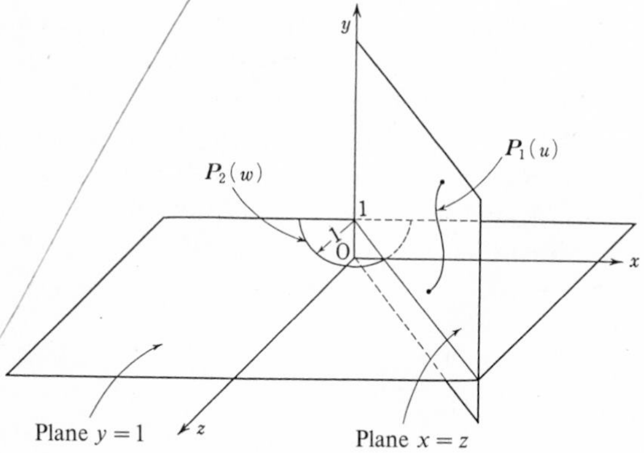

Fig. 6.21. Generation of a surface of revolution

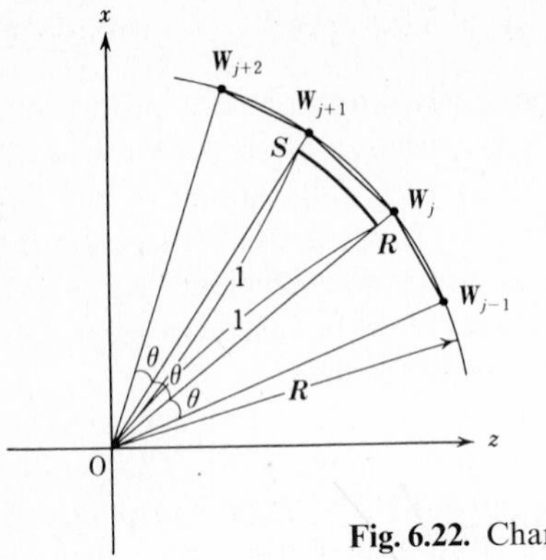

Fig. 6.22. Characteristic polygon of an approximate unit circular arc

$\boldsymbol{W}_{j-1}$, $\boldsymbol{W}_j$, $\boldsymbol{W}_{j+1}$ and $\boldsymbol{W}_{j+2}$ (Fig. 6.22) in the above formula (6.58) can be obtained immediately from Fig. 6.22 when the starting point $\boldsymbol{R}$ and end point $\boldsymbol{S}$ of the unit circular arc are given (when one circular arc is expressed by multiple curve segments, $\boldsymbol{R}$ and $\boldsymbol{S}$ are the starting and end points of one segment) (refer to Sect. 6.1.2):

$$\boldsymbol{W}_j = R\boldsymbol{R} = \frac{3}{2+\boldsymbol{R}\cdot\boldsymbol{S}}\boldsymbol{R} \tag{6.59}$$

$$\boldsymbol{W}_{j+1} = R\boldsymbol{S} = \frac{3}{2+\boldsymbol{R}\cdot\boldsymbol{S}}\boldsymbol{S}. \tag{6.60}$$

The rotation matrices $\boldsymbol{M}_0$ and $\boldsymbol{M}_1$ for clockwise and counterclockwise rotation, respectively, by an angle θ are:

$$\boldsymbol{M}_0 = \begin{bmatrix} \boldsymbol{R}\cdot\boldsymbol{S} & 0 & \sqrt{1-(\boldsymbol{R}\cdot\boldsymbol{S})^2} \\ 0 & 1 & 0 \\ -\sqrt{1-(\boldsymbol{R}\cdot\boldsymbol{S})^2} & 0 & \boldsymbol{R}\cdot\boldsymbol{S} \end{bmatrix}$$

$$\boldsymbol{M}_1 = \begin{bmatrix} \boldsymbol{R}\cdot\boldsymbol{S} & 0 & -\sqrt{1-(\boldsymbol{R}\cdot\boldsymbol{S})^2} \\ 0 & 1 & 0 \\ \sqrt{1-(\boldsymbol{R}\cdot\boldsymbol{S})^2} & 0 & \boldsymbol{R}\cdot\boldsymbol{S} \end{bmatrix}$$

so that $\boldsymbol{W}_{j-1}$ and $\boldsymbol{W}_{j+2}$ are given by:

$$\boldsymbol{W}_{j-1} = \boldsymbol{W}_j\boldsymbol{M}_0 \tag{6.61}$$

$$\boldsymbol{W}_{j+2} = \boldsymbol{W}_{j+1}\boldsymbol{M}_1. \tag{6.62}$$

The x component (for example) $\boldsymbol{P}_{ij,x}(u,w)$ of the surface of revolution $\boldsymbol{P}_{ij}(u,w)$ is:

$$P_{i,j,x}(u,w) = [N_{0,4}(u)\ N_{1,4}(u)\ N_{2,4}(u)\ N_{3,4}(u)] \times \begin{bmatrix} V_{i-1,x} \\ V_{i,x} \\ V_{i+1,x} \\ V_{i+2,x} \end{bmatrix} [W_{j-1,x}\ W_{j,x}\ W_{j+1,x}\ W_{j+2,x}] \begin{bmatrix} N_{0,4}(w) \\ N_{1,4}(w) \\ N_{2,4}(w) \\ N_{3,4}(w) \end{bmatrix} =$$

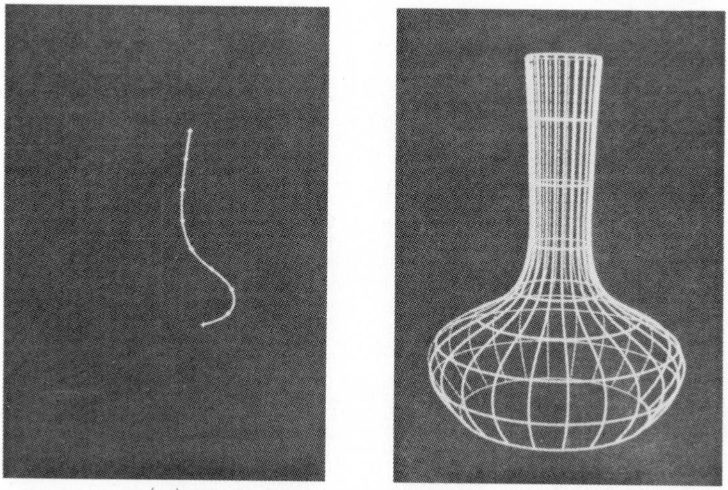
(a) (b)

Fig. 6.23. Example of a surface of revolution (1). (a) curve; (b) surface of revolution

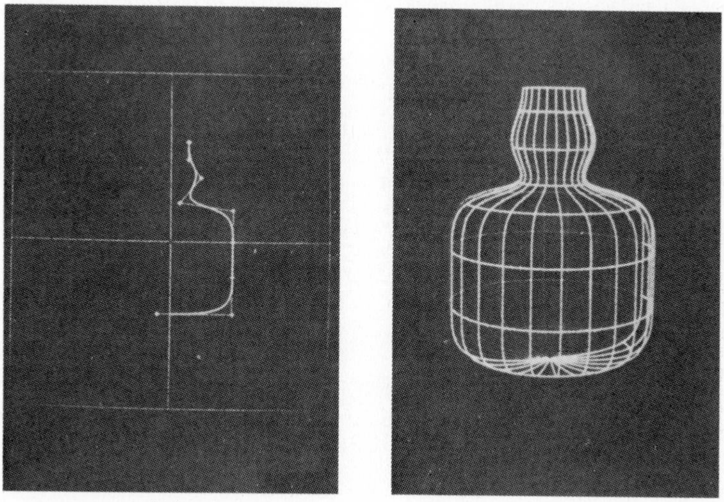
(a) (b)

Fig. 6.24. Example of a surface of revolution (2). (a) curve; (b) surface of revolution

$$
\begin{aligned}
&= [N_{0,4}(u) \;\; N_{1,4}(u) \;\; N_{2,4}(u) \;\; N_{3,4}(u)] \\
&\times \begin{bmatrix}
V_{i-1,x}W_{j-1,x} & V_{i-1,x}W_{j,x} & V_{i-1,x}W_{j+1,x} & V_{i-1,x}W_{j+2,x} \\
V_{i,x}W_{j-1,x} & V_{i,x}W_{j,x} & V_{i,x}W_{j+1,x} & V_{i,x}W_{j+2,x} \\
V_{i+1,x}W_{j-1,x} & V_{i+1,x}W_{j,x} & V_{i+1,x}W_{j+1,x} & V_{i+1,x}W_{j+2,x} \\
V_{i+2,x}W_{j-1,x} & V_{i+2,x}W_{j,x} & V_{i+2,x}W_{j+1,x} & V_{i+2,x}W_{j+2,x}
\end{bmatrix} \\
&\times \begin{bmatrix} N_{0,4}(w) \\ N_{1,4}(w) \\ N_{2,4}(w) \\ N_{3,4}(w) \end{bmatrix}
\end{aligned}
\tag{6.63}
$$

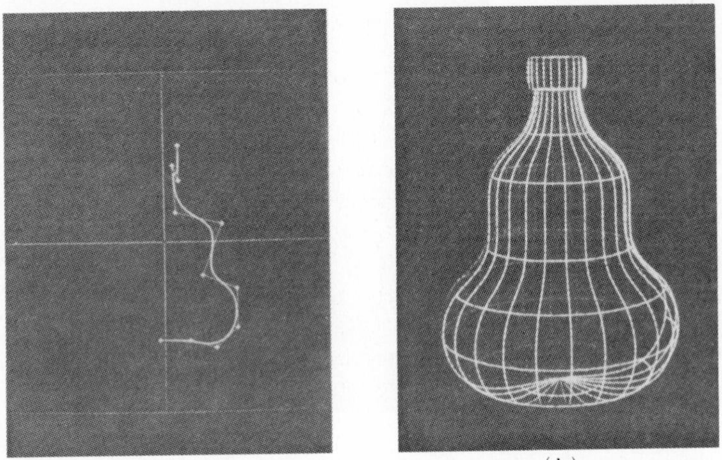

(a) (b)

Fig. 6.25. Example of a surface of revolution (3). (a) curve; (b) surface of revolution

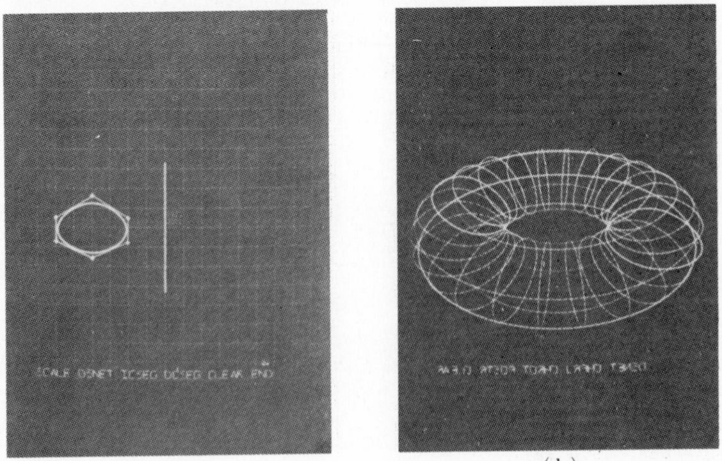

(a) (b)

Fig. 6.26. Example of a surface of revolution (4). (a) curve; (b) surface of revolution

Similarly, $P_{ij,y}(u,w)$ and $P_{ij,z}(u,w)$ can be found, so that $\boldsymbol{P}_{ij}(u,w)$ becomes:

$$\boldsymbol{P}_{ij}(u,w)=[N_{0,4}(u)\ N_{1,4}(u)\ N_{2,4}(u)\ N_{3,4}(u)]$$
$$\times\begin{bmatrix}\boldsymbol{Q}_{i-1,j-1} & \boldsymbol{Q}_{i-1,j} & \boldsymbol{Q}_{i-1,j+1} & \boldsymbol{Q}_{i-1,j+2}\\ \boldsymbol{Q}_{i,j-1} & \boldsymbol{Q}_{i,j} & \boldsymbol{Q}_{i,j+1} & \boldsymbol{Q}_{i,j+2}\\ \boldsymbol{Q}_{i+1,j-1} & \boldsymbol{Q}_{i+1,j} & \boldsymbol{Q}_{i+1,j+1} & \boldsymbol{Q}_{i+1,j+2}\\ \boldsymbol{Q}_{i+2,j-1} & \boldsymbol{Q}_{i+2,j} & \boldsymbol{Q}_{i+2,j+1} & \boldsymbol{Q}_{i+2,j+2}\end{bmatrix}\begin{bmatrix}N_{0,4}(w)\\ N_{1,4}(w)\\ N_{2,4}(w)\\ N_{3,4}(w)\end{bmatrix} \tag{6.64}$$

where:

$$\boldsymbol{Q}_{p,r}=[V_{p,x}W_{r,x}\ \ V_{p,y}\ \ V_{p,x}W_{r,z}] \tag{6.65}$$
$$(p=i-1,\ i,\ i+1,\ i+2;\ r=j-1,\ j,\ j+1,\ j+2).$$

Equation (6.64) has the same form as that of a general free form surface. That is to say, in the very general curved surface formula (6.64), when $\boldsymbol{Q}_{p,r}$ is a vector calculated from Eq. (6.65), this surface is a surface of revolution with the y axis as its rotation axis. This is very important, suggesting that it is possible to first design a surface of revolution and then vary its shape. Figures 6.23 to 6.26 show examples of surfaces of revolution that were generated according to the theory described above. Figures 6.27 to 6.29 show examples of shapes designed by first designing a plane curve, then rotating it to create a surface of revolution in accordance with the above theory, and finally applying control to the $\boldsymbol{Q}_{p,r}$ data. This design method can be applied widely to design of shapes which are surfaces of revolution or relatively close to surfaces of revolution, such as various kinds of containers.

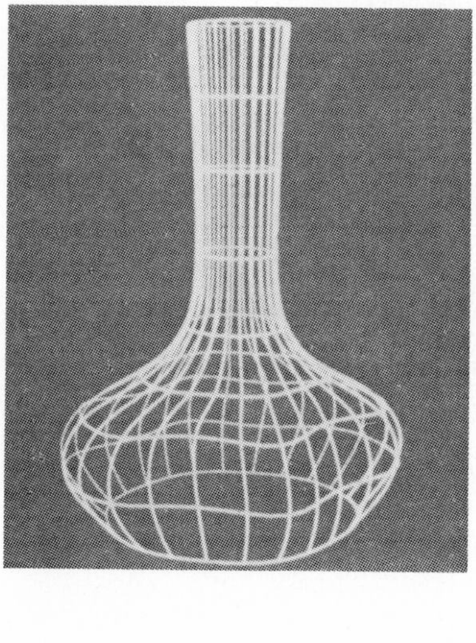

Fig. 6.27. Example of shape control on a surface of revolution (1)

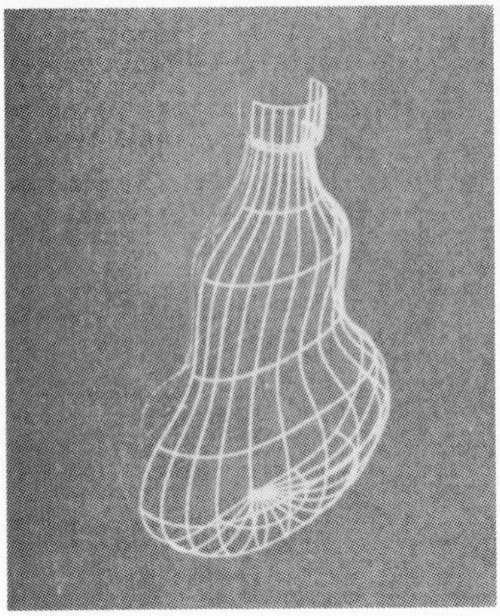

Fig. 6.28. Example of shape control on a surface of revolution (2)

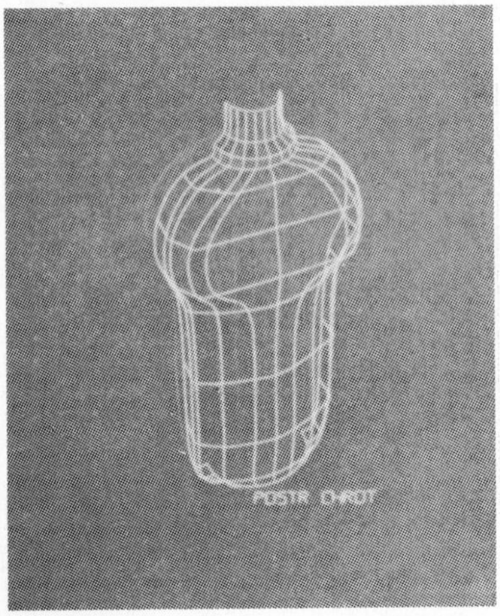

Fig. 6.29. Example of shape control on a surface of revolution (3)

6.3 *B*-Spline Functions and Their Properties (1)

Below, we give the definition of a *B*-spline function.

Definition 6.1[29]**.** In the following 2 variable functions of x and y:

$$M(x; y) = (y-x)_+^{M-1} = \begin{cases} (y-x)^{M-1} & (y \geqq x) \\ 0 & (y < x) \end{cases} \tag{6.66}$$

define the M-th order divided difference function $M_{j,M}(x)$ with respect to the variable y based on $x_0, x_1, \ldots, x_n$:

$$M_{j,M}(x) = M[x; x_j, x_{j+1}, \ldots, x_{j+M}] \quad (j = 0, 1, \ldots, n-M). \tag{6.67}$$

The functions $M_{j,M}(x)$ in Eq. (6.67) are called M-th order basis spline functions, or *B*-spline functions.

Following are some important properties of *B*-spline functions and *B*-spline function approximations.

① Let us show that the functions expressed by Eq. (6.67) are *C*-splines of degree $M-1$ having the knots $x_j, x_{j+1}, \ldots, x_{j+M}$[8]. Use the following identity.

$$(y-x)_+^{M-1} = (-1)^M (x-y)_+^{M-1} + (y-x)^{M-1}. \tag{6.68}$$

Take the M-th order divided difference, with respect to y, of both sides of this identity and write the result as follows:

$$L_y[(y-x)_+^{M-1}] = (-1)^M L_y[(x-y)_+^{M-1}] + L_y[(y-x)^{M-1}].$$

Since the 2-nd term on the right-hand side of this equation is an M-th order divided difference of a polynomial of degree $M-1$, we have:

$$L_y[(y-x)^{M-1}] = 0.$$

Therefore:

$$L_y[(y-x)_+^{M-1}] = (-1)^M L_y[(x-y)_+^{M-1}].$$

Here we introduce the following polynomial $W(x)$:

$$W(x) = (x-x_j)(x-x_{j+1}) \ldots (x-x_{j+M}). \tag{6.69}$$

Let $W_i(x)$ be this same product with the factor $(x-x_{j+i})$ omitted. Then, from Eq. (2.16), we have:

$$L_y[(y-x)_+^{M-1}] \equiv M_{j,M}(x)$$
$$= (-1)^M \left[\frac{(x-x_j)_+^{M-1}}{W_0(x_j)} + \frac{(x-x_{j+1})_+^{M-1}}{W_1(x_{j+1})} + \ldots + \frac{(x-x_{j+M})_+^{M-1}}{W_M(x_{j+M})} \right]$$

This equation shows that the *B*-spline function $M_{j,M}(x) \equiv L_y[(y-x)_+^{M-1}]$ is a spline function of degree $M-1$ having the knots $x_j, x_{j+1}, \ldots, x_{j+M}$ (refer to Eq. (4.5)). However, since $(y-x)_+^{M-1}$ is 0 at $y = x_j, x_{j+1}, \ldots, x_{j+M}$ when $x \geqq x_{j+M}$, the divided difference with respect to y knots are 0; in other words, for $x \geqq x_{j+M}$, $M_{j,M}(x) \equiv 0$. In addition, when $x \leqq x_j$, $(y-x)_+^{M-1} = (y-x)^{M-1}$ for

$y=x_j, x_{j+1}, \ldots, x_{j+M}$. Since this is a polynomial of degree $M-1$ in y, the M-th order divided difference for it is 0; in other words, for $x \leqq x_j$, $M_{j,M}(x) \equiv 0$.

From the above discussion, we see that a *B*-spline is a *C*-spline; in fact, a *B*-spline is a special case of a *C*-spline.

② "*B*" in *B*-spline function stands for basis. This refers to the important property that an arbitrary *C*-spline of degree $M-1$ having knots $x_0, x_1, \ldots, x_n$ can be uniquely expressed as a linear combination of $n-M+1$ *B*-splines[8]:

$$S(x) = \sum_{j=0}^{n-M} b_j M_{j,M}(x). \tag{6.70}$$

③ Consider the case in which $M-1$ knots each are added at both ends of the knots $x_0, x_1, \ldots, x_n$. These knots are called extended knots:

$$\begin{aligned} &[x_0^* \; x_1^* \; \ldots \; x_{M-2}^* \; x_{M-1}^* \; x_M^* \; \ldots \; x_{n+M-1}^* \; x_{n+M}^* \; \ldots \; x_{n+2M-2}^*] \\ &\quad = [x_{-(M-1)} \; x_{-(M-2)} \; \ldots \; x_{-1} \; x_0 \; x_1 \; \ldots \; x_n \; x_{n+1} \; \ldots \; x_{n+M-1}]. \end{aligned} \tag{6.71}$$

In extended knots, $x_0, x_1, \ldots, x_n$ are called interior knots, and $x_{-(M-1)}, \ldots, x_{-1}$ and $x_{n+1}, \ldots, x_{n+M-1}$ are called additional knots.

$n+M-1$ *B*-spline functions are determined by the extended knots. It can be shown that an arbitrary spline function $S(x)$ of degree $M-1$, having the interior knots as its knots, can be uniquely expressed as a linear combination of these $n+M-1$ *B*-spline functions[7]. Denoting constants by b_j, we have:

$$S(x) = \sum_{j=0}^{n+M-2} b_j M_{j,M}(x). \tag{6.72}$$

In the following discussion, except when it is specifically necessary to make the distinction, "*" will be abbreviated in the extended knots.

6.4 *B*-Spline Functions and Their Properties (2)

This book deals mainly with the use of parametric vector functions for describing curves and surfaces. Let us now define a *B*-spline function with knots expressed by the symbol t.

Definition 6.2. In the following 2 variable functions of t and u:

$$M(t;u) = (u-t)_+^{M-1} \tag{6.73}$$

$$= \begin{cases} (u-t)^{M-1} & (u \geqq t) \\ 0 & (u < t) \end{cases} \tag{6.74}$$

define an M-th order divided difference function $M_{j,M}(t)$ with respect to the variable u based on $t_0, t_1, \ldots, t_n$:

$$M_{j,M}(t) = M[t; t_j, t_{j+1}, \ldots, t_{j+M}]. \tag{6.75}$$

The function $M_{j,M}(t)$ in Eq. (6.75) is called a *B*-spline function of order M.

The *B*-spline functions used in practical applications are the normalized *B*-splines $N_{j,M}(t)$. $N_{j,M}(t)$ is given by the following formula:

$$N_{j,M}(t) = (t_{j+M} - t_j)\, M_{j,M}(t) \tag{6.76}$$

$$= M[t; t_{j+1}, \ldots, t_{j+M}] - M[t; t_j, \ldots, t_{j+M-1}] \tag{6.77}$$

$M_{j,M}(t)$ in (6.75) and $N_{j,M}(t)$ in (6.76) have no meaning at $t = t_j$ with $M = 1$. For $M = 1$ we define:

$$M_{j,1}(t) = \begin{cases} (t_{j+1} - t_j)^{-1} & (t_j \leqq t < t_{j+1})^{*)} \\ 0 & \text{(outside above range)} \end{cases}. \tag{6.78}$$

In this case:

$$N_{j,1}(t) = \begin{cases} 1 & (t_j \leqq t < t_{j+1})^{*)} \\ 0 & \text{(outside above range)} \end{cases}. \tag{6.79}$$

It is necessary to note that (6.78) and (6.79) imply that:

$$M_{j,1}(t) \equiv N_{j,1}(t) \equiv 0 \qquad (t_j = t_{j+1}). \tag{6.80}$$

There is also a relation corresponding to Eq. (6.72) for the normalized *B*-spline functions $N_{j,M}(t)$; it is clear that:

$$S(t) = \sum_{j=0}^{n+M-2} c_j N_{j,M}(t). \tag{6.81}$$

Let us now continue our discussion of the properties of *B*-spline functions.

④ The following relations hold for *B*-spline functions (refer to Appendix B.1).

$$\mathrm{M}_{j,M}(t) = \frac{t - t_j}{t_{j+M} - t_j} M_{j,M-1}(t) + \frac{t_{j+M} - t}{t_{j+M} - t_j} M_{j+1,M-1}(t) \tag{6.82}$$

$$N_{j,M}(t) = \frac{t - t_j}{t_{j+M-1} - t_j} N_{j,M-1}(t) + \frac{t_{j+M} - t}{t_{j+M} - t_{j+1}} N_{j+1,M-1}(t). \tag{6.83}$$

---

*) In (6.78) and (6.79), note carefully where the equal sign is in describing the domains of t.

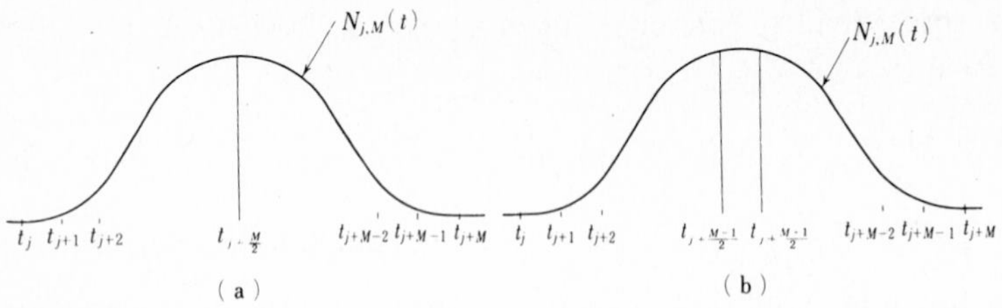

Fig. 6.30. Relation between maximum value and knots of a *B*-spline function. (a) M: even; (b) M: odd

Equation (6.82) means that for an arbitrary t, $M_{j,M}(t)$ is the average of $M_{j,M-1}(t)$ and $M_{j+1,M-1}(t)$, and that for $t_j<t<t_{j+M}$, $M_{j,M}(t)$ is a convex combination of $M_{j,M-1}(t)$ and $M_{j+1,M-1}(t)$. Consequently, from Eq. (6.78), $M_{j,1}(t)$ is positive for $t_j \leqq t < t_{j+1}$ and zero elsewhere. Accordingly, for $M>1$:

$$\left.\begin{aligned} &M_{j,M}(t)>0 \quad (t_j<t<t_{j+M}) \\ &M_{j,M}(t)=0 \quad (t \leqq t_j \quad \text{or} \quad t \geqq t_{j+M}) \end{aligned}\right\}. \tag{6.84}$$

Similarly:

$$\left.\begin{aligned} &N_{j,M}(t)>0 \quad (t_j<t<t_{j+M}) \\ &N_{j,M}(t)=0 \quad (t \leqq t_j \quad \text{or} \quad t \geqq t_{j+M}) \end{aligned}\right\}. \tag{6.85}$$

As shown in Fig. 6.30, the graph of the *B*-spline function $N_{j,M}(t)$ is, in general, bell-shaped in $t_j \leqq t \leqq t_{j+M}$. The maximum value of the function occurs at $t_{j+M/2}$ or in its vicinity when M is even, midway between $t_{j+(M-1)/2}$ and $t_{j+(M+1)/2}$ when M is odd.

⑤ As can be seen from Appendix B, in Eq. (6.81) the following relation holds:

$$\sum_{j=0}^{n+M-2} N_{j,M}(t) \equiv 1. \tag{6.86}$$

Therefore, $S(t)$ is a convex combination of c_j ($j=0, \ldots, n+M-2$).

6.5 Derivation of *B*-Spline Functions

Let us find the *B*-spline functions of orders $M=1, 2, 3, 4$ from the definition of a divided difference.

In general, the M-th order divided difference of a function f can be expressed as a linear combination of the $M+1$ function values $f_0, f_1, \ldots, f_M$ according to the following equation:

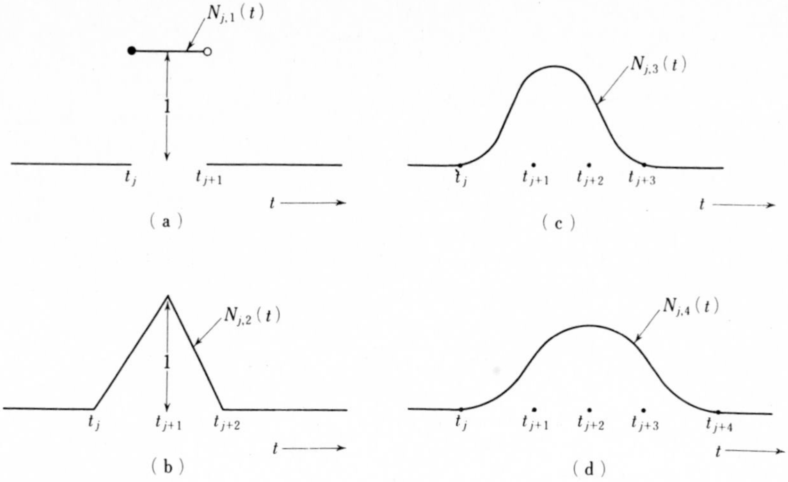

Fig. 6.31. *B*-spline functions for $M=1, 2, 3, 4$

$$f[t_0, t_1, \ldots, t_M] = \frac{f_0}{(t_0-t_1)\ldots(t_0-t_M)} + \frac{f_1}{(t_1-t_0)(t_1-t_2)\ldots(t_1-t_M)} + \ldots + \frac{f_M}{(t_M-t_0)\ldots(t_M-t_{M-1})}. \tag{6.87}$$

Case of $M=1$:
From Eq. (6.79) we have:

$$N_{j,1}(t) = \begin{cases} 1 & (t_j \leqq t < t_{j+1}) \\ 0 & \text{(outside above range)} \end{cases} \tag{6.88}$$

The graph of $N_{j,1}(t)$ is shown in Fig. 6.31(a).

Case of $M=2$:
Setting $M=2$ in Eq. (6.76) gives:

$$\begin{aligned} N_{j,2}(t) &= (t_{j+2}-t_j)\,M_{j,2}(t) \\ &= (t_{j+2}-t_j)\,M[t;\, t_j,\, t_{j+1},\, t_{j+2}] \\ &= (t_{j+2}-t_j)\left\{\frac{(t_j-t)_+}{(t_j-t_{j+1})(t_j-t_{j+2})} + \frac{(t_{j+1}-t)_+}{(t_{j+1}-t_j)(t_{j+1}-t_{j+2})}\right. \\ &\quad \left. + \frac{(t_{j+2}-t)_+}{(t_{j+2}-t_j)(t_{j+2}-t_{j+1})}\right\} = \end{aligned}$$

$$
= \begin{cases} 0 & (t \leqq t_j) \\ \dfrac{t - t_j}{t_{j+1} - t_j} & (t_j < t \leqq t_{j+1}) \\ \dfrac{t_{j+2} - t}{t_{j+2} - t_{j+1}} & (t_{j+1} < t < t_{j+2}) \\ 0 & (t \geqq t_{j+2}) \end{cases} \tag{6.89}
$$

The graph of $N_{j,2}(t)$ is shown in Fig. 6.31(b).

Case of $M=3$:
Setting $M=3$ in Eq. (6.76) gives:

$$
\begin{aligned}
N_{j,3}(t) &= (t_{j+3} - t_j)\, M_{j,3}(t) \\
&= (t_{j+3} - t_j) M\,[t;\, t_j,\, t_{j+1},\, t_{j+2},\, t_{j+3}] \\
&= (t_{j+3} - t_j) \left\{ \frac{(t_j - t)_+^2}{(t_j - t_{j+1})(t_j - t_{j+2})(t_j - t_{j+3})} \right. \\
&\quad + \frac{(t_{j+1} - t)_+^2}{(t_{j+1} - t_j)(t_{j+1} - t_{j+2})(t_{j+1} - t_{j+3})} \\
&\quad + \frac{(t_{j+2} - t)_+^2}{(t_{j+2} - t_j)(t_{j+2} - t_{j+1})(t_{j+2} - t_{j+3})} \\
&\quad \left. + \frac{(t_{j+3} - t)_+^2}{(t_{j+3} - t_j)(t_{j+3} - t_{j+1})(t_{j+3} - t_{j+2})} \right\}
\end{aligned}
$$

$$
= \begin{cases} 0 & (t \leqq t_j) \\ \dfrac{(t_j - t)^2}{(t_{j+1} - t_j)(t_{j+2} - t_j)} & (t_j < t \leqq t_{j+1}) \\ (t_{j+3} - t_j) \left\{ \dfrac{(t_{j+3} - t)^2}{(t_{j+3} - t_j)(t_{j+3} - t_{j+1})(t_{j+3} - t_{j+2})} \right. \\ \qquad \left. - \dfrac{(t_{j+2} - t)^2}{(t_{j+2} - t_j)(t_{j+2} - t_{j+1})(t_{j+3} - t_{j+2})} \right\} & (t_{j+1} < t \leqq t_{j+2}) \\ \dfrac{(t_{j+3} - t)^2}{(t_{j+3} - t_{j+1})(t_{j+3} - t_{j+2})} & (t_{j+2} < t < t_{j+3}) \\ 0 & (t \geqq t_{j+3}) \end{cases} \tag{6.90}
$$

The graph of $N_{j,3}(t)$ is shown in Fig. 6.31(c).

Case of $M=4$:
Setting $M=4$ in Eq. (6.76) gives:

$$
\begin{aligned}
N_{j,4}(t) &= (t_{j+4}-t_j)M_{j,4}(t)\\
&= (t_{j+4}-t_j)M\,[t;\, t_j,\, t_{j+1},\, t_{j+2},\, t_{j+3},\, t_{j+4}]\\
&= (t_{j+4}-t_j)\left\{\frac{(t_j-t)_+^3}{(t_j-t_{j+1})(t_j-t_{j+2})(t_j-t_{j+3})(t_j-t_{j+4})}\right.\\
&\quad+\frac{(t_{j+1}-t)_+^3}{(t_{j+1}-t_j)(t_{j+1}-t_{j+2})(t_{j+1}-t_{j+3})(t_{j+1}-t_{j+4})}\\
&\quad+\frac{(t_{j+2}-t)_+^3}{(t_{j+2}-t_j)(t_{j+2}-t_{j+1})(t_{j+2}-t_{j+3})(t_{j+2}-t_{j+4})}\\
&\quad+\frac{(t_{j+3}-t)_+^3}{(t_{j+3}-t_j)(t_{j+3}-t_{j+1})(t_{j+3}-t_{j+2})(t_{j+3}-t_{j+4})}\\
&\quad\left.+\frac{(t_{j+4}-t)_+^3}{(t_{j+4}-t_j)(t_{j+4}-t_{j+1})(t_{j+4}-t_{j+2})(t_{j+4}-t_{j+3})}\right\}
\end{aligned}
$$

$$
=\begin{cases}
0 & (t\leqq t_j)\\[4pt]
(t_{j+4}-t_j)\left\{\dfrac{(t_{j+1}-t)^3}{(t_{j+1}-t_j)(t_{j+1}-t_{j+2})(t_{j+1}-t_{j+3})(t_{j+1}-t_{j+4})}\right. & \\
\quad+\dfrac{(t_{j+2}-t)^3}{(t_{j+2}-t_j)(t_{j+2}-t_{j+1})(t_{j+2}-t_{j+3})(t_{j+2}-t_{j+4})} & \\
\quad+\dfrac{(t_{j+3}-t)^3}{(t_{j+3}-t_j)(t_{j+3}-t_{j+1})(t_{j+3}-t_{j+2})(t_{j+3}-t_{j+4})} & \\
\quad\left.+\dfrac{(t_{j+4}-t)^3}{(t_{j+4}-t_j)(t_{j+4}-t_{j+1})(t_{j+4}-t_{j+2})(t_{j+4}-t_{j+3})}\right\} & (t_j<t\leqq t_{j+1})\\[4pt]
(t_{j+4}-t_j)\left\{\dfrac{(t_{j+2}-t)^3}{(t_{j+2}-t_j)(t_{j+2}-t_{j+1})(t_{j+2}-t_{j+3})(t_{j+2}-t_{j+4})}\right. & \\
\quad+\dfrac{(t_{j+3}-t)^3}{(t_{j+3}-t_j)(t_{j+3}-t_{j+1})(t_{j+3}-t_{j+2})(t_{j+3}-t_{j+4})} & \\
\quad\left.+\dfrac{(t_{j+4}-t)^3}{(t_{j+4}-t_j)(t_{j+4}-t_{j+1})(t_{j+4}-t_{j+2})(t_{j+4}-t_{j+3})}\right\} & (t_{j+1}<t\leqq t_{j+2})\\[4pt]
(t_{j+4}-t_j)\left\{\dfrac{(t_{j+3}-t)^3}{(t_{j+3}-t_j)(t_{j+3}-t_{j+1})(t_{j+3}-t_{j+2})(t_{j+3}-t_{j+4})}\right. & \\
\quad\left.+\dfrac{(t_{j+4}-t)^3}{(t_{j+4}-t_j)(t_{j+4}-t_{j+1})(t_{j+4}-t_{j+2})(t_{j+4}-t_{j+3})}\right\} & (t_{j+2}<t\leqq t_{j+3})\\[4pt]
\dfrac{(t_{j+4}-t)^3}{(t_{j+4}-t_{j+1})(t_{j+4}-t_{j+2})(t_{j+4}-t_{j+3})} & (t_{j+3}<t<t_{j+4})\\[4pt]
0 & (t\geqq t_{j+4})
\end{cases}
\tag{6.91}
$$

The graph of $N_{j,4}(t)$ is shown in Fig. 6.31(d).

Next, for the cases $M=3$ and 4, let us calculate $N_{j,M}(t)$ with $t_j=j-(M-1)$ $(j=0, 1, \ldots, 2M-1)$ as the knot vector elements.

Case of $M=3$:
Since $T=[t_0\ t_1\ t_2\ t_3\ t_4\ t_5]=[-2\ -1\ 0\ 1\ 2\ 3]$, from Eq. (6.90) we have:

$$N_{0,3}(t)=\begin{cases} 0 & (t\leqq -2) \\ \frac{1}{2}(t+2)^2 & (-2<t\leqq -1) \\ -t^2-t+\frac{1}{2} & (-1<t\leqq 0) \\ \boxed{\frac{1}{2}(t-1)^2} & (0<t<1) \\ 0 & (t\geqq 1) \end{cases} \tag{6.92}$$

$$N_{1,3}(t)=\begin{cases} 0 & (t\leqq -1) \\ \frac{1}{2}(t+1)^2 & (-1<t\leqq 0) \\ \boxed{-\frac{3}{2}(1-t)^2+\frac{1}{2}(2-t)^2\left(=-t^2+t+\frac{1}{2}\right)} & (0<t\leqq 1) \\ \frac{1}{2}(t-2)^2 & (1<t<2) \\ 0 & (t\geqq 2) \end{cases} \tag{6.93}$$

$$N_{2,3}(t)=\begin{cases} 0 & (t\leqq 0) \\ \boxed{\frac{1}{2}t^2} & (0<t\leqq 1) \\ -t^2+3t-\frac{3}{2} & (1<t\leqq 2) \\ \frac{1}{2}(t-3)^2 & (2<t<3) \\ 0 & (t\geqq 3) \end{cases} \tag{6.94}$$

Case of $M=4$:

Since $T=[t_0\ t_1\ t_2\ t_3\ t_4\ t_5\ t_6\ t_7]=[-3\ -2\ -1\ 0\ 1\ 2\ 3\ 4]$, from Eq. (6.91) we have:

$$N_{0,4}(t)=\begin{cases} 0 & (t\leqq -3)\\ \frac{1}{6}(t+3)^3 & (-3<t\leqq -2)\\ -(t+1)^3+\frac{2}{3}t^3-\frac{1}{6}(t-1)^3 & (-2<t\leqq -1)\\ \frac{2}{3}t^3-\frac{1}{6}(t-1)^3 & (-1<t\leqq 0)\\ -\frac{1}{6}(t-1)^3 & (0<t<1)\\ 0 & (t\geqq 1)\end{cases} \tag{6.95}$$

$$N_{1,4}(t)=\begin{cases} 0 & (t\leqq -2)\\ \frac{1}{6}(t+2)^3 & (-2<t\leqq -1)\\ -t^3+\frac{2}{3}(t-1)^3-\frac{1}{6}(t-2)^3 & (-1<t\leqq 0)\\ \frac{2}{3}(t-1)^3-\frac{1}{6}(t-2)^3\left(=\frac{1}{2}t^3-t^2+\frac{2}{3}\right) & (0<t\leqq 1)\\ -\frac{1}{6}(t-2)^3 & (1<t<2)\\ 0 & (t\geqq 2)\end{cases} \tag{6.96}$$

$$N_{2,4}(t)=\begin{cases} 0 & (t\leqq -1)\\ \frac{1}{6}(t+1)^3 & (-1<t\leqq 0)\\ -(t-1)^3+\frac{2}{3}(t-2)^3-\frac{1}{6}(t-3)^3 \\ \left(=-\frac{1}{2}t^3+\frac{1}{2}t^2+\frac{1}{2}t+\frac{1}{6}\right) & (0<t\leqq 1)\\ \frac{2}{3}(t-2)^3-\frac{1}{6}(t-3)^3 & (1<t\leqq 2)\\ -\frac{1}{6}(t-3)^3 & (2<t<3)\\ 0 & (t\geqq 3)\end{cases} \tag{6.97}$$

$$N_{3,4}(t)=\begin{cases} 0 & (t \leqq 0) \\ \boxed{\dfrac{1}{6}\,t^3} & (0<t\leqq 1) \\ -(t-2)^3+\dfrac{2}{3}(t-3)^3-\dfrac{1}{6}(t-4)^3 & (1<t\leqq 2) \\ \dfrac{2}{3}(t-3)^3-\dfrac{1}{6}(t-4)^3 & (2<t\leqq 3) \\ -\dfrac{1}{6}(t-4)^3 & (3<t<4) \\ 0 & (t\geqq 4) \end{cases} \tag{6.98}$$

From the above equations, we see that for $0\leqq t\leqq 1$, $N_{i,3}(t)$ $(i=0,1,2)$ and $N_{i,4}(t)$ $(i=0,1,2,3)$ are the same as $Y_i(t)$ $(i=0,1,2)$ and $X_i(t)$ $(i=0,1,2,3)$, respectively, which were derived in Sect. 6.1.1. If the graph of $N_{0,M}(t)$ can be obtained, then it is also clear from the above equations that $N_{j,M}(t)$ can be obtained by horizontal translation (Fig. 6.32). In the case of $t_j=j-(M-1)$, we have:

$$N_{j,M}(t)=N_{0,M}(t-j). \tag{6.99}$$

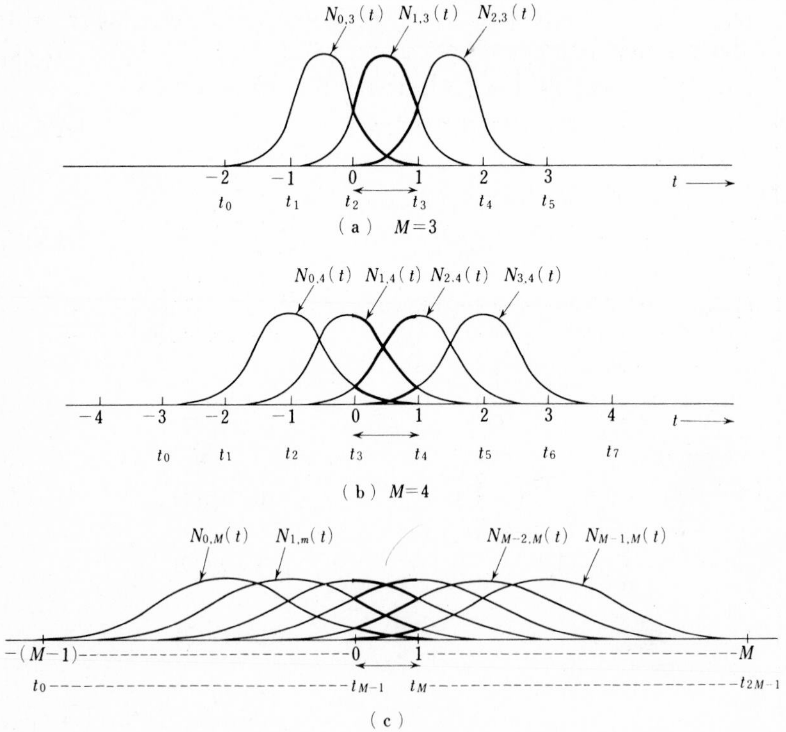

Fig. 6.32. How $N_{j,M}(t)$ $(j=0, 1, \ldots)$ are related to each other through horizontal translation. (a) $M=3$; (b) $M=4$; (c) general case

$N_{0,M}(t)$ can be expressed in the following form:

$$N_{0,M}(t)=\frac{1}{(M-1)!}\sum_{i=0}^{M}(-1)^i\binom{M}{i}(t+M-1-i)_+^{M-1}. \tag{6.100}$$

Therefore:

$$N_{j,M}(t)=\frac{1}{(M-1)!}\sum_{i=0}^{M}(-1)^i\binom{M}{i}(t+M-1-i-j)_+^{M-1} \tag{6.101}$$

$$(j=0,1,\ldots,M-1).$$

6.6 *B*-Spline Curve Type (1)

Using the uniform *B*-spline functions that were derived in Sect. 6.1.1, the following types of *B*-spline curves can be determined.

(A) Open Curve

A class C^{M-2} open curve (of degree $M-1$) consisting of $n-M+2$ curve segments $\boldsymbol{P}_i(t)$, with a curve defining polygon determined by the $n+1$ position vectors $\boldsymbol{Q}_0, \boldsymbol{Q}_1, \ldots, \boldsymbol{Q}_n$, can be expressed by the following functions, where M is the order of the *B*-spline functions (Fig. 6.33).

$$\boldsymbol{P}_i(t)=\sum_{j=i-1}^{i+M-2}N_{j-i+1,M}(t)\boldsymbol{Q}_j \quad (0\leqq t\leqq 1) \tag{6.102}$$

$$\boldsymbol{P}_i(0)\equiv\boldsymbol{P}_i, \qquad \boldsymbol{P}_{n-M+2}(1)\equiv\boldsymbol{P}_{n-M+3}$$

$$(i=1,2,\ldots,n-M+2)$$

$\boldsymbol{P}_1, \boldsymbol{P}_2, \ldots, \boldsymbol{P}_{n-M+3}$ are the ends of the curve segments.

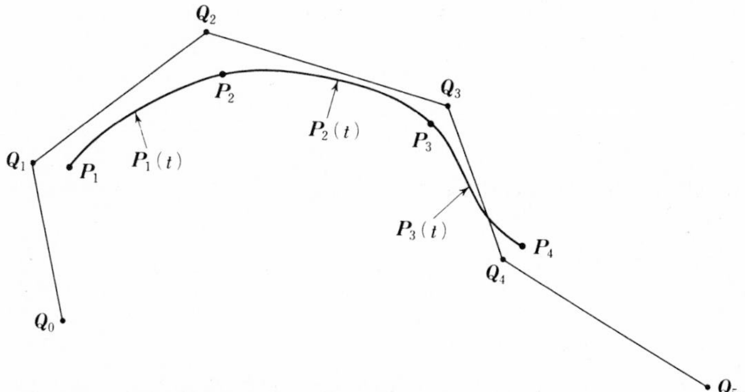

Fig. 6.33. *B*-spline curve form (1) (open curve)

(B) Closed Curve

A class C^{M-2} closed curve (of degree $M-1$) consisting of $n+1$ curve segments $\boldsymbol{P}_i(t)$ $(i=0, 1, \ldots, n)$, with a curve defining polygon determined by the $n+1$ position vectors $\boldsymbol{Q}_0, \boldsymbol{Q}_1, \ldots, \boldsymbol{Q}_n$, can be expressed by the following function (Fig. 6.34):

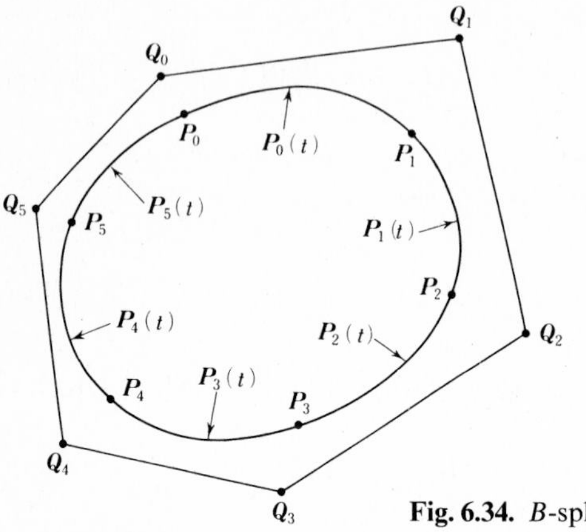

Fig. 6.34. *B*-spline curve form (2) (closed curve)

$$\boldsymbol{P}_i(t) = \sum_{j=i-1}^{i+M-2} N_{j-i+1,M}(t)\, \boldsymbol{Q}_{j \bmod (n+1)} \quad (0 \leqq t \leqq 1) \tag{6.103}$$

$$\boldsymbol{P}_i(0) = \boldsymbol{P}_i$$

$$(i=0, 1, \ldots, n).$$

For example, if the order of the *B*-spline function is $M=4$ and $n=5$, function (6.102) becomes:

$$\boldsymbol{P}_i(t) = N_{0,4}(t)\boldsymbol{Q}_{i-1} + N_{1,4}(t)\boldsymbol{Q}_i + N_{2,4}(t)\boldsymbol{Q}_{i+1} + N_{3,4}(t)\boldsymbol{Q}_{i+2}$$
$$(i=1, 2, 3)$$

and (6.103) becomes:

$$\boldsymbol{P}_i(t) = N_{0,4}(t)\boldsymbol{Q}_{(i-1)\bmod 6} + N_{1,4}(t)\boldsymbol{Q}_{i \bmod 6} + N_{2,4}(t)\boldsymbol{Q}_{(i+1)\bmod 6} + N_{3,4}(t)\boldsymbol{Q}_{(i+2)\bmod 6}$$
$$(i=0, 1, \ldots, 5).$$

The relations between these curves and their curve defining vectors are shown in Figs. 6.33 and 6.34.

6.7 *B*-Spline Curve Type (2)

In functions (6.102) and (6.103), each *B*-spline curve segment is defined within a parameter range of $0 \leqq t \leqq 1$. Noting that $N_{0,M}(t)$, $N_{1,M}(t)$, $N_{2,M}(t) \ldots$ are related by horizontal translation, the total curve can be defined as follows.

(A) Open Curve

A class C^{M-2} open curve (of degree $M-1$) consisting of $n-M+2$ segments, with a curve defining polygon determined by the $n+1$ position vectors $\boldsymbol{Q}_0$, $\boldsymbol{Q}_1$, $\boldsymbol{Q}_2$, ..., $\boldsymbol{Q}_n$, can be expressed by the following function:

$$\boldsymbol{P}(t) = \sum_{j=0}^{n} N_{j,M}(t)\,\boldsymbol{Q}_j \quad (0 \leqq t \leqq n-M+2) \tag{6.104}$$

where the knot vector T that determines the *B*-spline function is specified as (Fig. 6.35):

$$T = [t_0 \ t_1 \ \ldots \ t_{n+M}]^{*)} = [-(M-1) \ -(M-2) \ \ldots \ n+1]. \tag{6.105}$$

(B) Closed Curve

A class C^{M-2} closed curve (of degree $M-1$), with a curve defining polygon consisting of the $n+1$ position vectors $\boldsymbol{Q}_0$, $\boldsymbol{Q}_1$, ..., $\boldsymbol{Q}_n$, can be expressed by a single function as follows:

$$\boldsymbol{P}(t) = \sum_{j=0}^{n+M-1} N_{j,M}(t)\,\boldsymbol{Q}_{j \bmod (n+1)} \quad (0 \leqq t \leqq n+1) \tag{6.106}$$

where the knot vector T that determines the *B*-spline function is specified as (refer to Fig. 6.35):

$$T = [t_0 \ t_1 \ \ldots \ t_{n+2M-1}] = [-(M-1) \ -(M-2) \ \ldots \ n+M]. \tag{6.107}$$

B-spline curves defined in type (2) are exactly the same as those defined in type (1).

---

*) In function (6.104), it is necessary to determine $n+1$ *B*-spline functions. From Sect. 6.3 (3), the extended knots that include the interior knots t_0, t_1, ..., t_n define $n+M-1$ *B*-spline functions, so to determine $n+1$ *B*-spline functions we must specify the internal knots t_0, t_t, ..., t_{n-M+2}; therefore, the extended knots are:

$$[t_{-M+1} \ t_{-M+2} \ \ldots \ t_0 \ t_1 \ \ldots \ t_{n-M+2} \ t_{n-M+3} \ \ldots \ t_{n+1}] = [t_0^* \ t_1^* \ \ldots \ t_{M-1}^* \ t_M^* \ \ldots \ t_{n+1}^* \ t_{n+2}^* \ \ldots \ t_{n+M}^*]$$

			Additional knots	Interior knots	Additional knots	
Extended knots		Before changing subscripts	$t_{-(M-1)} \cdots\cdots t_{-1}$	$t_0 \quad t_1 \cdots\cdots t_{n-M+1} \quad t_{n-M+2}$	$t_{n-M+3} \cdots\cdots t_{n+1}$	$t_{n+2} \cdots\cdots t_{n+M}$
		After changing subscripts	$t_0^* \cdots\cdots t_{M-2}^*$	$t_{M-1}^* \quad t_M^* \cdots\cdots t_n^* \quad t_{n+1}^*$	$t_{n+2}^* \cdots\cdots t_{n+M}^*$	$t_{n+M+1}^* \cdots\cdots t_{n+2M-1}^*$
Knot values	Curve type (2)	Open curve	$-(M-1) \cdots\cdots -1$	$0 \quad 1 \cdots\cdots n-M+1 \quad n-M+2$	$n-M+3 \cdots\cdots n+1$	
		Closed curve	$-(M-1) \cdots\cdots -1$	$0 \quad 1 \cdots\cdots n-M+1 \quad n-M+2$	$n-M+3 \cdots\cdots n+1$	$n+2 \cdots\cdots n+M$
	Curve type (3)	Open curve	$a_0 \cdots\cdots a_0$	$a_0 \quad a_1 \cdots\cdots a_{n-M+1} \quad a_{n-M+2}$	$a_{n-M+2} \cdots\cdots a_{n-M+2}$	
		Closed curve	$a_0-(a_{n+1}-a_{n-M+2}) \cdots a_0-(a_{n+1}-a_n)$	$a_0 \quad a_1 \cdots\cdots a_{n-M+1} \quad a_{n-M+2}$	$a_{n-M+3} \cdots\cdots a_{n+1}$	$a_{n+1}+(a_1-a_0) \cdots\cdots a_{n+1}+(a_{M-1}-a_0)$

Fig. 6.35. Relation between (extended) knots of curve type (2) and curve type (3)

6.8 Recursive Calculation of B-Spline Functions

B-spline functions are easy to calculate recursively using Eq. (6.83) (for its derivation refer to Appendix B.1).

If order M (degree $M-1$) and knots $t_0, t_1, \dots$ are specified, the B-spline functions $N_{j,M}(t)$ are found from the following formula (refer to Sect. 6.4):

$$N_{j,1}(t)=\begin{cases} 1 & (t_j \leqq t < t_{j+1})^{*)} \\ 0 & \text{(outside above range).} \end{cases} \tag{6.108}$$

For the case $M>1$ we have:

$$N_{j,M}(t)=\frac{t-t_j}{t_{j+M-1}-t_j} N_{j,M-1}(t)+\frac{t_{j+M}-t}{t_{j+M}-t_{j+1}} N_{j+1,M-1}(t). \tag{6.109}$$

Two or more coincident knots are permitted, with the convention that $0/0=0$.

To find $N_{j,M}(t)$ from formula (6.109), it is generally necessary to know $N_{j,M-1}(t)$ and $N_{j+1,M-1}(t)$. Also, to find $N_{j,M-1}(t)$ and $N_{j+1,M-1}(t)$ it is necessary to know $N_{j,M-2}(t)$, $N_{j+1,M-2}(t)$ and $N_{j+2,M-2}(t)$. Continuing this to the last step, we see that we need to know $N_{j,1}(t)$, $N_{j+1,1}(t)$, ..., $N_{j+M-1,1}(t)$. Since these are known, $N_{j,M}(t)$ is determined.

Now we will use formulas (6.108) and (6.109) to find some B-spline functions.

Example 6.1. With the knot vector

$$T=[t_0\, t_1 \dots t_{2M-1}]=[-(M-1) \quad -(M-2) \dots M-1 \quad M],$$

calculate B-spline functions recursively for the cases $M=2$, 3, 4 using formulas (6.108) and (6.109).

Solution. Case of $M=2$: When $M=2$,

$$T=[t_0 \;\; t_1 \;\; t_2 \;\; t_3]=[-1 \;\; 0 \;\; 1 \;\; 2].$$

In formula (6.109), setting $j=0$, $M=2$ gives:

$$\begin{aligned} N_{0,2}(t)&=\frac{t-t_0}{t_1-t_0} N_{0,1}(t)+\frac{t_2-t}{t_2-t_1} N_{1,1}(t) \\ &=(t+1)\, N_{0,1}(t)+(1-t)\, N_{1,1}(t). \end{aligned}$$

---

*) Note that there is no equal sign on the t_{j+1} side. Therefore, when $t_j=t_{j+1}$, $N_{j,1}(t)\equiv 0$.

From (6.108) we have:

$$N_{0,2}(t)=\begin{cases}0 & (t<-1)\\ t+1 & (-1\leqq t<0)\\ 1-t & (0\leqq t\leqq 1)\\ 0 & (t>1)\end{cases}$$

Setting $j=1$, $M=2$ in formula (6.109) gives:

$$\begin{aligned}N_{1,2}(t)&=\frac{t-t_1}{t_2-t_1}N_{1,1}(t)+\frac{t_3-t}{t_3-t_2}N_{2,1}(t)\\ &=t\,N_{1,1}(t)+(2-t)\,N_{2,1}(t).\end{aligned}$$

Therefore:

$$N_{1,2}(t)=\begin{cases}0 & (t<0)\\ t & (0\leqq t<1)\\ 2-t & (1\leqq t\leqq 2)\\ 0 & (t>2)\end{cases}$$

Case of $M=3$: When $M=3$,

$$T=[t_0\ t_1\ t_2\ t_3\ t_4\ t_5]=[-2\ -1\ 0\ 1\ 2\ 3].$$

Setting $j=0$, $M=3$ in formula (6.109) gives:

$$\begin{aligned}N_{0,3}(t)&=\frac{t-t_0}{t_2-t_0}N_{0,2}(t)+\frac{t_3-t}{t_3-t_1}N_{1,2}(t)\\ &=\frac{1}{2}(t+2)\,N_{0,2}(t)+\frac{1}{2}(1-t)\,N_{1,2}(t).\end{aligned}$$

Similarly:

$$\begin{aligned}N_{1,3}(t)&=\frac{t-t_1}{t_3-t_1}N_{1,2}(t)+\frac{t_4-t}{t_4-t_2}N_{2,2}(t)\\ &=\frac{1}{2}(t+1)\,N_{1,2}(t)+\frac{1}{2}(2-t)\,N_{2,2}(t)\end{aligned}$$

$$\begin{aligned}N_{2,3}(t)&=\frac{t-t_2}{t_4-t_2}N_{2,2}(t)+\frac{t_5-t}{t_5-t_3}N_{3,2}(t)\\ &=\frac{t}{2}N_{2,2}(t)+\frac{1}{2}(3-t)\,N_{3,2}(t)\end{aligned}$$

$$\begin{aligned}
N_{0,2}(t) &= \frac{t-t_0}{t_1-t_0} N_{0,1}(t) + \frac{t_2-t}{t_2-t_1} N_{1,1}(t) \\
&= (t+2)\, N_{0,1}(t) - t\, N_{1,1}(t) \\
N_{1,2}(t) &= \frac{t-t_1}{t_2-t_1} N_{1,1}(t) + \frac{t_3-t}{t_3-t_2} N_{2,1}(t) \\
&= (t+1)\, N_{1,1}(t) + (1-t)\, N_{2,1}(t) \\
N_{2,2}(t) &= \frac{t-t_2}{t_3-t_2} N_{2,1}(t) + \frac{t_4-t}{t_4-t_3} N_{3,1}(t) \\
&= t\, N_{2,1}(t) + (2-t)\, N_{3,1}(t) \\
N_{3,2}(t) &= \frac{t-t_3}{t_4-t_3} N_{3,1}(t) + \frac{t_5-t}{t_5-t_4} N_{4,1}(t) \\
&= (t-1)\, N_{3,1}(t) + (3-t)\, N_{4,1}(t).
\end{aligned}$$

From the above equations we have:

$$N_{0,3}(t) = \frac{1}{2}(t+2)^2 N_{0,1}(t) + \left(-t^2 - t + \frac{1}{2}\right) N_{1,1}(t) + \frac{1}{2}(t-1)^2 N_{2,1}(t)$$

$$N_{1,3}(t) = \frac{1}{2}(t+1)^2 N_{1,1}(t) + \left(-t^2 + t + \frac{1}{2}\right) N_{2,1}(t) + \frac{1}{2}(t-2)^2 N_{3,1}(t)$$

$$N_{2,3}(t) = \frac{t^2}{2} N_{2,1}(t) + \left(-t^2 + 3t - \frac{3}{2}\right) N_{3,1}(t) + \frac{1}{2}(t-3)^2 N_{4,1}(t).$$

Therefore:

$$N_{0,3}(t) = \begin{cases} 0 & (t < -2) \\ \frac{1}{2}(t+2)^2 & (-2 \leqq t < -1) \\ -t^2 - t + \frac{1}{2} & (-1 \leqq t < 0) \\ \frac{1}{2}(t-1)^2 & (0 \leqq t \leqq 1) \\ 0 & (t > 1) \end{cases}$$

$$N_{1,3}(t)=\begin{cases}0 & (t<-1)\\ \frac{1}{2}(t+1)^2 & (-1\leqq t<0)\\ -t^2+t+\frac{1}{2} & (0\leqq t<1)\\ \frac{1}{2}(t-2)^2 & (1\leqq t\leqq 2)\\ 0 & (t>2)\end{cases}$$

$$N_{2,3}(t)=\begin{cases}0 & (t<0)\\ \frac{t^2}{2} & (0\leqq t<1)\\ -t^2+3t-\frac{3}{2} & (1\leqq t<2)\\ \frac{1}{2}(t-3)^2 & (2\leqq t\leqq 3)\\ 0 & (t>3)\end{cases}$$

Case of $M=4$: When $M=4$, the knot vector is:

$$T=[t_0\ t_1\ t_2\ t_3\ t_4\ t_5\ t_6\ t_7]=[-3\ -2\ -1\ 0\ 1\ 2\ 3\ 4].$$

Therefore:

$$N_{0,4}(t)=\frac{1}{3}(t+3)\,N_{0,3}(t)+\frac{1}{3}(1-t)\,N_{1,3}(t)$$

$$N_{1,4}(t)=\frac{1}{3}(t+2)\,N_{1,3}(t)+\frac{1}{3}(2-t)\,N_{2,3}(t)$$

$$N_{2,4}(t)=\frac{1}{3}(t+1)\,N_{2,3}(t)+\frac{1}{3}(3-t)\,N_{3,3}(t)$$

$$N_{3,4}(t)=\frac{1}{3}t\,N_{3,3}(t)+\frac{1}{3}(4-t)\,N_{4,3}(t)$$

Next, find $N_{0,3}(t)$, $N_{1,3}(t)$, $N_{2,3}(t)$, $N_{3,3}(t)$ and $N_{4,3}(t)$:

$$N_{0,3}(t)=\frac{1}{2}(t+3)\,N_{0,2}(t)+\frac{1}{2}(0-t)\,N_{1,2}(t)$$

$$N_{1,3}(t)=\frac{1}{2}(t+2)\,N_{1,2}(t)+\frac{1}{2}(1-t)\,N_{2,2}(t)$$

$$N_{2,3}(t)=\frac{1}{2}(t+1)\,N_{2,2}(t)+\frac{1}{2}(2-t)\,N_{3,2}(t)$$

$$N_{3,3}(t)=\frac{1}{2}(t+0)\,N_{3,2}(t)+\frac{1}{2}(3-t)\,N_{4,2}(t)$$

$$N_{4,3}(t)=\frac{1}{2}(t-1)\,N_{4,2}(t)+\frac{1}{2}(4-t)\,N_{5,2}(t)$$

Next, find $N_{0,2}(t)$, $N_{1,2}(t)$, $N_{2,2}(t)$, $N_{3,2}(t)$, $N_{4,2}(t)$, $N_{5,2}(t)$:

$$N_{0,2}(t)=(t+3)\,N_{0,1}(t)+(-1-t)\,N_{1,1}(t)$$
$$N_{1,2}(t)=(t+2)\,N_{1,1}(t)+(0-t)\,N_{2,1}(t)$$
$$N_{2,2}(t)=(t+1)\,N_{2,1}(t)+(1-t)\,N_{3,1}(t)$$
$$N_{3,2}(t)=t\,N_{3,1}(t)+(2-t)\,N_{4,1}(t)$$
$$N_{4,2}(t)=(t-1)\,N_{4,1}(t)+(3-t)\,N_{5,1}(t)$$
$$N_{5,2}(t)=(t-2)\,N_{5,1}(t)+(4-t)\,N_{6,1}(t)$$

From the above equations we have:

$$N_{0,4}(t)=\frac{1}{6}(t+3)^3\,N_{0,1}(t)+\left\{-(t+1)^3+\frac{2}{3}t^3-\frac{1}{6}(t-1)^3\right\}N_{1,1}(t)$$
$$+\left\{\frac{2}{3}t^3-\frac{1}{6}(t-1)^3\right\}N_{2,1}(t)-\frac{1}{6}(t-1)^3\,N_{3,1}(t)$$

$$N_{1,4}(t)=\frac{1}{6}(t+2)^3\,N_{1,1}(t)+\left\{-t^3+\frac{2}{3}(t-1)^3-\frac{1}{6}(t-2)^3\right\}N_{2,1}(t)$$
$$+\left\{\frac{2}{3}(t-1)^3-\frac{1}{6}(t-2)^3\right\}N_{3,1}(t)-\frac{1}{6}(t-2)^3\,N_{4,1}(t)$$

$$N_{2,4}(t)=\frac{1}{6}(t+1)^3\,N_{2,1}(t)+\left\{-(t-1)^3+\frac{2}{3}(t-2)^3-\frac{1}{6}(t-3)^3\right\}N_{3,1}(t)$$
$$+\left\{\frac{2}{3}(t-2)^3-\frac{1}{6}(t-3)^3\right\}N_{4,1}(t)-\frac{1}{6}(t-3)^3\,N_{5,1}(t)$$

$$N_{3,4}(t)=\frac{1}{6}t^3\,N_{3,1}(t)+\left\{-(t-2)^3+\frac{2}{3}(t-3)^3-\frac{1}{6}(t-4)^3\right\}N_{4,1}(t)$$
$$+\left\{\frac{2}{3}(t-3)^3-\frac{1}{6}(t-4)^3\right\}N_{5,1}(t)-\frac{1}{6}(t-4)^3\,N_{6,1}(t)$$

Formulas (6.108) and (6.109) define B-spline functions for knot vectors that have multiple knots. In the case $M=4$, the effect of the multiple knots on the B-spline functions is discussed in Appendix C.

Next, let us give an example of how to compute B-spline functions when there are multiple knots.

Example 6.2. When the order is $M=4$ and the knot vector is:

$$T=[t_0\ t_1\ t_2\ t_3\ t_4\ t_5\ t_6\ t_7]=[0\ 0\ 0\ 0\ 1\ 1\ 1\ 1]$$

find the B-spline functions $N_{0,4}(t)$, $N_{1,4}(t)$, $N_{2,4}(t)$, $N_{3,4}(t)$.

Solution. Noting the condition on t in Eq. (6.80) and the convention that $0/0=0$, we have:

$$\left.\begin{array}{l} N_{0,1}(t)=N_{1,1}(t)=N_{2,1}(t)=N_{4,1}(t)=N_{5,1}(t)=N_{6,1}(t)=0 \\ N_{3,1}(t)=1 \quad (0\leqq t<1) \end{array}\right\}$$

Using these relations, we obtain:

$$N_{0,2}(t)=\frac{t-t_0}{t_1-t_0}N_{0,1}(t)+\frac{t_2-t}{t_2-t_1}N_{1,1}(t)=0$$

$$N_{1,2}(t)=\frac{t-t_1}{t_2-t_1}N_{1,1}(t)+\frac{t_3-t}{t_3-t_2}N_{2,1}(t)=0$$

$$N_{2,2}(t)=\frac{t-t_2}{t_3-t_2}N_{2,1}(t)+\frac{t_4-t}{t_4-t_3}N_{3,1}(t)=(1-t)N_{3,1}(t)$$

$$N_{3,2}(t)=\frac{t-t_3}{t_4-t_3}N_{3,1}(t)+\frac{t_5-t}{t_5-t_4}N_{4,1}(t)=tN_{3,1}(t)$$

$$N_{4,2}(t)=\frac{t-t_4}{t_5-t_4}N_{4,1}(t)+\frac{t_6-t}{t_6-t_5}N_{5,1}(t)=0$$

$$N_{5,2}(t)=\frac{t-t_5}{t_6-t_5}N_{5,1}(t)+\frac{t_7-t}{t_7-t_6}N_{6,1}(t)=0$$

Using the above results, we obtain for $N_{0,3}(t)$, $N_{1,3}(t)$, $N_{2,3}(t)$, $N_{3,3}(t)$, $N_{4,3}(t)$:

$$N_{0,3}(t)=\frac{t-t_0}{t_2-t_0}N_{0,2}(t)+\frac{t_3-t}{t_3-t_1}N_{1,2}(t)=0$$

$$N_{1,3}(t)=\frac{t-t_1}{t_3-t_1}N_{1,2}(t)+\frac{t_4-t}{t_4-t_2}N_{2,2}(t)=(1-t)^2N_{3,1}(t)$$

$$N_{2,3}(t)=\frac{t-t_2}{t_4-t_2}N_{2,2}(t)+\frac{t_5-t}{t_5-t_3}N_{3,2}(t)=2t(1-t)N_{3,1}(t)$$

$$N_{3,3}(t)=\frac{t-t_3}{t_5-t_3}N_{3,2}(t)+\frac{t_6-t}{t_6-t_4}N_{4,2}(t)=t^2N_{3,1}(t)$$

$$N_{4,3}(t)=\frac{t-t_4}{t_6-t_4}N_{4,2}(t)+\frac{t_7-t}{t_7-t_5}N_{5,2}(t)=0$$

Therefore:

$$N_{0,4}(t)=\frac{t-t_0}{t_3-t_0}N_{0,3}(t)+\frac{t_4-t}{t_4-t_1}N_{1,3}(t)=(1-t)^3N_{3,1}(t)$$

$$N_{1,4}(t)=\frac{t-t_1}{t_4-t_1}N_{1,3}(t)+\frac{t_5-t}{t_5-t_2}N_{2,3}(t)=3(1-t)^2tN_{3,1}(t)$$

$$N_{2,4}(t)=\frac{t-t_2}{t_5-t_2}N_{2,3}(t)+\frac{t_6-t}{t_6-t_3}N_{3,3}(t)=3(1-t)t^2N_{3,1}(t)$$

$$N_{3,4}(t)=\frac{t-t_3}{t_6-t_3}N_{3,3}(t)+\frac{t_7-t}{t_7-t_4}N_{4,3}(t)=t^3N_{3,1}(t)$$

From the above results, for $0\leqq t<1$ the *B*-spline functions become:

$$\begin{aligned}
N_{0,4}(t)&=(1-t)^3\equiv B_{0,3}(t)\\
N_{1,4}(t)&=3(1-t)^2t\equiv B_{1,3}(t)\\
N_{2,4}(t)&=3(1-t)t^2\equiv B_{2,3}(t)\\
N_{3,4}(t)&=t^3\equiv B_{3,3}(t).
\end{aligned}$$

It is known that these agree with Bernstein basis functions of degree 3 (refer to Sect. 6.9(8)).

6.9 *B*-Spline Functions and Their Properties (3)

When *B*-spline functions are calculated using formulas (6.108) and (6.109), multiple knots can be specified.

⑥ (Multiple knots and continuity of functions.)[20]
Suppose that k_i knots have the same value as t_i:

$$t_i=t_{i+1}=\ldots=t_{i+k_i-1}.$$

Then, at t_i, the continuity of the *B*-spline functions $N_{i,M}$ is reduced to C^{M-k_i-1}. A *B*-spline obtained with internal knots having a multiplicity k_i of 2 or more is called a subspline basis. In contrast, a basis which has $k_i=1$ for all i is called a full spline basis. The relation between multiple knots and continuity of functions is given both mathematically and graphically for the case $M=4$ in Appendix C.

A basis for which the multiplicity k_i is 0 for all i can be regarded as corresponding to a single polynomial. In this case, with the degree $M-1$ and continuity C^{M-1}, the knots are sometimes called pseudo-knots.

⑦ (Number of nonzero basis functions.)
For a t such that $t \neq t_i$, there are always M nonzero basis functions of degree $M-1$. For a t such that $t=t_i$, if k_i is the multiplicity of that t_i, the number of nonzero basis functions is $M-k_i$. Therefore, in the case of a full spline, there are $M-1$ nonzero basis functions at the knots and M between knots.

Because of this property, to calculate a point $t=t_s$ ($t_i \leqq t_s < t_{i+1}$) on a B-spline curve it is sufficient to calculate the M values of $N_{j,M}(t_s)$ for $j=(i-M+1), \ldots, i$. This is because the other $N_{j,M}(t_s)$ are zero (refer to Fig. 6.36). This implies that:

$$\boldsymbol{P}(t_s) = \sum_{j=i-M+1}^{i} N_{j,M}(t_s)\boldsymbol{Q}_j. \tag{6.110}$$

When $t_i \leqq t_s < t_{i+1}$ is specified, the B-spline functions for $M=1$ are zero, except for $N_{i,1}(t_s)=1$, so that, as shown in Fig. 6.37, the M values $N_{i-M+1,M}(t_s), \ldots, N_{i,M}(t_s)$ can be found all at once from formula (6.109), in sequence starting from $N_{i,1}(t_s)$.

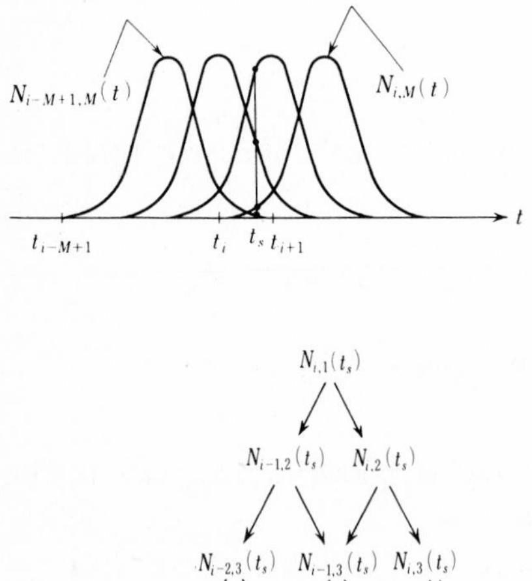

Fig. 6.36. Nonzero $N_{j,M}(t_s)$ when $t_i \leqq t_s < t_{i+1}$. Case of $M=4$

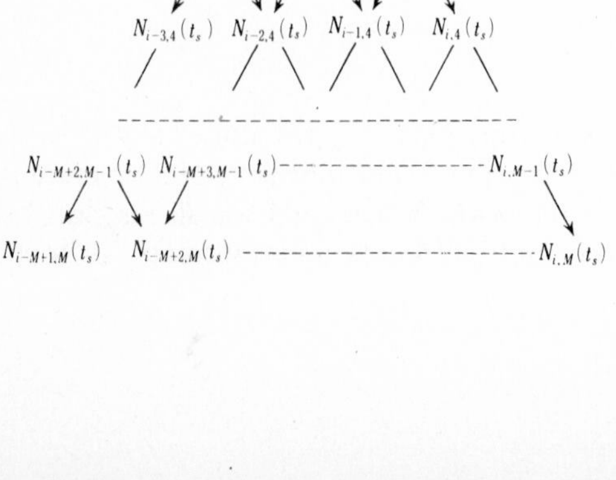

Fig. 6.37. Hierarchical relation among B-spline functions ($t_i \leqq t_s < t_{i+1}$)

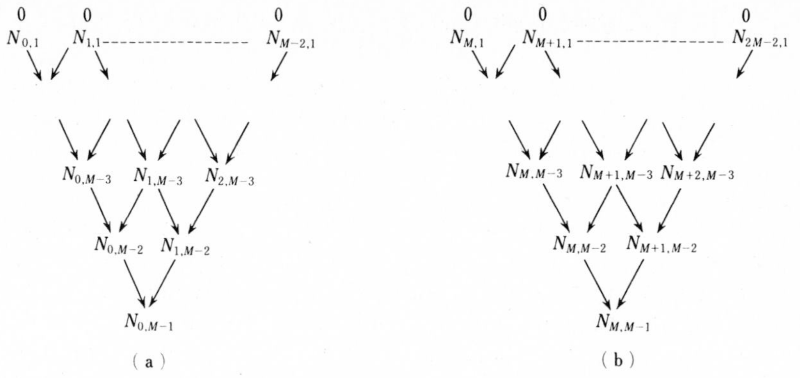

Fig. 6.38. Explanatory diagram for $N_{0,M-1}(t)=N_{M,M-1}(t)=0$

⑧ (Relation between *B*-splines and Bernstein basis.)
Suppose that an extended knot vector has internal knots [0 1] and the additional knots on both sides are [0 ... 0] and [1 ... 1] respectively:

$$\underbrace{\begin{matrix} t_0 & t_1 & \dots & t_{M-1} \\ 0 & 0 & \dots & 0 \end{matrix}}_{M \text{ knots}} \quad \underbrace{\begin{matrix} t_M & t_{M+1} & \dots & t_{2M-1} \\ 1 & 1 & \dots & 1 \end{matrix}}_{M \text{ knots}}$$

Let us look at the *B*-spline functions with such a knot vector. In formula (6.109) we have:

- for the case $j=0$.

As shown in Fig. 6.38(a), the value of $N_{0,M-1}(t)$ is related to $N_{0,1}(t)$, $N_{1,1}(t)$, ..., $N_{M-2,1}(t)$. From Eq. (6.79), $N_{0,1}(t)=\dots=N_{M-2,1}(t)=0$, so $N_{0,M-1}(t)=0$. We also have $t_{M-1}-t_0=0$, so, using the convention that $0/0=0$:

$$N_{j,M}(t)=(1-t)\,N_{j+1,M-1}(t) \quad (j=0) \tag{6.111}$$

- for the case $j=M-1$.

As shown in Fig. 6.38(b), the value of $N_{M,M-1}(t)$ is related $N_{M,1}(t)$, $N_{M+1,1}(t)$, ..., $N_{2M-2,1}(t)$. From Eq. (6.79), $N_{M,1}(t)=\dots=N_{2M-2,1}(t)=0$, so $N_{M,M-1}(t)=0$. We also have $t_{2M-1}-t_M=0$, so, using the convention that $0/0=0$:

$$N_{j,M}(t)=t\,N_{j,M-1}(t) \quad (j=M-1) \tag{6.112}$$

- for the case $j \neq 0$, $j \neq M-1$.

From formula (6.109):

$$N_{j,M}(t)=t\,N_{j,M-1}(t)+(1-t)\,N_{j+1,M-1}(t). \tag{6.113}$$

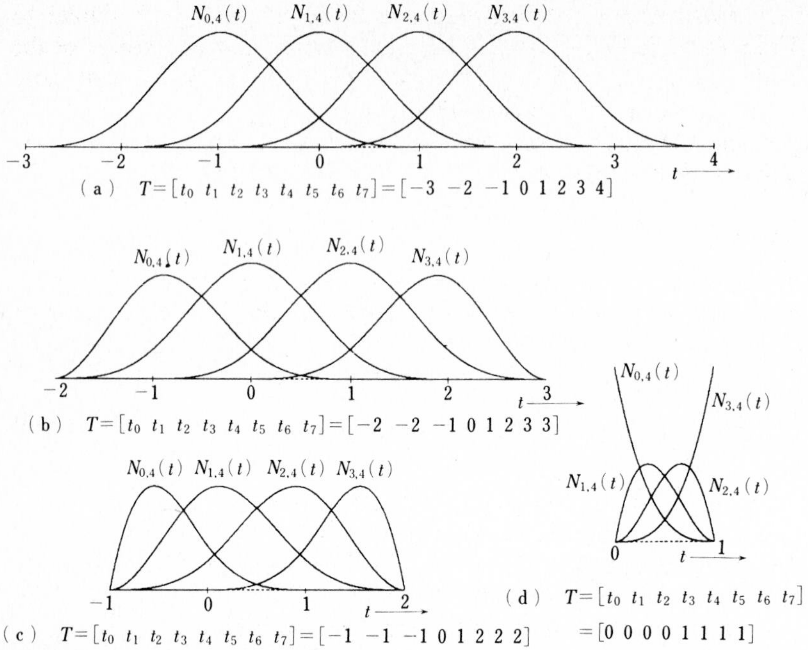

Fig. 6.39. Conversion from *B*-spline functions to Bernstein basis functions by increasing multiplicity of knots

It is known that Eqs. (6.111), (6.112) and (6.113) are the same as Bernstein basis equations (5.73) and (5.75). A Bernstein basis is a special case of *B*-splines (refer to Example 6.2). Figure 6.39 shows how a *B*-spline function approaches a Bernstein basis as the multiplicity is increased in the case $M=4$.

⑨ (Rigorous expression of a straight line.)
In Eq. (6.72), if the b_j are taken to be values on a straight line, $S(x)$ is that straight line.

⑩ The polynomial (6.72), which is a linear combination of *B*-spline functions, has the same variation diminishing property as a Bernstein polynomial:

$$v(S(x)) \leqq v(b_j).$$

The geometrical significance of this property is discussed in Sect. 5.1.4 ⑨.

6.10 *B*-Spline Curve Type (3)

In *B*-spline curve types (1) and (2), in general the curve does not pass through the points $\boldsymbol{Q}_j$, but, as discussed in Sect. 6.9, when there are M multiple knots at the start and end of a knot vector, then at the start and end of the curve the relation between the curve defining vectors and the curve shape is similar to that for Bézier curve segments. Specifically, the starting and end points of the curve coincide with $\boldsymbol{Q}_0$ and $\boldsymbol{Q}_n$, respectively and the slopes of the curve at those points are the directions of the vectors $\overrightarrow{\boldsymbol{Q}_0\boldsymbol{Q}_1}$ and $\overrightarrow{\boldsymbol{Q}_{n-1}\boldsymbol{Q}_n}$.

In curve types (1) and (2) the intervals between knots are all 1; but in the following curve type they are arbitrary.

(A) Open Curve

An open curve (of degree $M-1$) consisting of $n-M+2$ segments, with a curve defining polygon determined by the $n+1$ position vectors $\boldsymbol{Q}_0, \boldsymbol{Q}_1, \ldots, \boldsymbol{Q}_n$, is expressed by the following function:

$$\boldsymbol{P}(t)=\sum_{j=0}^{n} N_{j,M}(t)\boldsymbol{Q}_j \quad (a_0 \leqq t < a_{n-M+2}) \tag{6.114}$$

where the knot vector determining the *B*-spline functions is:

$$T=[t_0 \ t_1 \ \ldots \ t_{n+M}].$$

The knot values are determined as follows (refer to Fig. 6.35):

$$\left.\begin{array}{ll} t_i=a_0 & i=0, 1, \ldots, M-1 \\ t_{i+M}=a_{i+1} & i=0, 1, \ldots, n-M \\ t_{i+n+1}=a_{n-M+2} & i=0, 1, \ldots, M-1 \end{array}\right\} \tag{6.115}$$

The open curve expressed by function (6.114) has a starting point that coincides with $\boldsymbol{Q}_0$ and an end point that coincides with $\boldsymbol{Q}_n$. The relation between the knot values and curve segments is shown in Fig. 6.40.

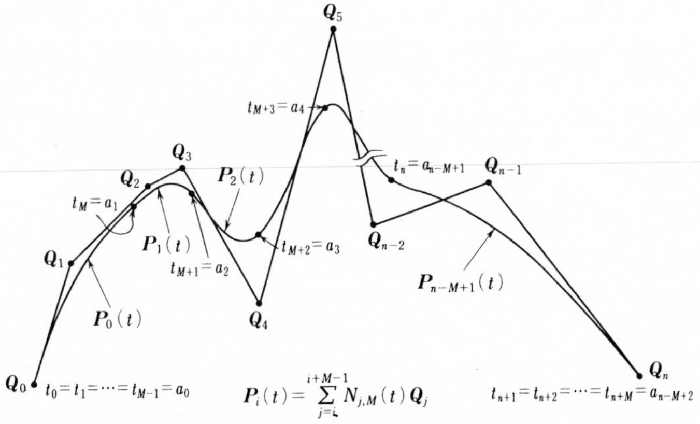

Fig. 6.40. Relation between curve knot values and curve segments (case of an open curve)

(B) Closed Curve

A class C^{M-2} closed curve (of degree $M-1$) consisting of $n+1$ segments, with a curve defining polygon determined by the $n+1$ position vectors $\boldsymbol{Q}_0, \boldsymbol{Q}_1, \ldots, \boldsymbol{Q}_n$ is expressed by the following function:

$$\boldsymbol{P}(t)=\sum_{j=0}^{n+M-1} N_{j,M}(t)\boldsymbol{Q}_{j\bmod(n+1)} \quad (a_0 \leqq t \leqq a_{n+1}) \tag{6.116}$$

The knot vector that determines the B-spline functions is:

$$T=[t_0\ t_1\ \ldots\ t_{n+2M-1}].$$

The knots are as follows (Fig. 6.35):

$$\begin{aligned} &t_i=a_0-(a_{n+1}-a_{i+n-M+2}) && i=0,1,\ldots,M-2 \\ &t_{i+M-1}=a_i && i=0,1,\ldots,n+1 \\ &t_{i+n+M+1}=a_{n+1}+(a_{i+1}-a_0) && i=0,1,\ldots,M-2 \end{aligned} \tag{6.117}$$

The relation between the knot values and curve segments is shown in Fig. 6.41.

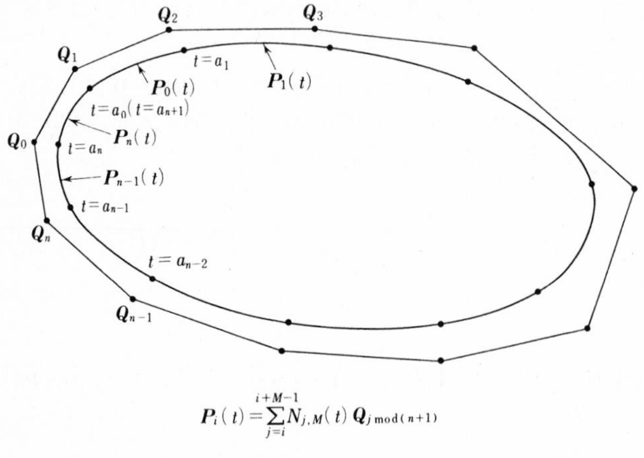

Fig. 6.41. Relation between curve knot values and curve segments (case of a closed curve)

In a B-spline curve, a curve generated by knot values corresponding to internal knots which are specified at equal intervals is called a uniform B-spline curve; a curve generated by knot values specified at unequal intervals is called a non-uniform B-spline curve.

Example 6.3. In the case $M=4$ $n=8$, find all of the uniform B-spline functions needed to generate an open curve of B-spline curve type (3).

Solution. With $a_i = i$ $(i = 0, 1, \ldots, n-M+2)$, the knot vector T is:

$$T = [t_0\ t_1\ t_2\ t_3\ t_4\ t_5\ t_6\ t_7\ t_8\ t_9\ t_{10}\ t_{11}\ t_{12}]$$
$$= [0\ 0\ 0\ 0\ 1\ 2\ 3\ 4\ 5\ 6\ 6\ 6\ 6]$$

There are multiple knots. The *B*-spline functions can be found from formula (6.109). Here we use the results from Appendix C.

$\boldsymbol{N_{0,4}(t)}$

$N_{0,4}(t)$ for the knot vector $[0\ 0\ 0\ 0\ 1]$ is, from Appendix C, Eq. (C.4):

$$N_{0,4}(t) = -(t-1)^3 \quad (0 \leqq t \leqq 1)$$

$\boldsymbol{N_{1,4}(t)}$

$N_{1,4}(t)$ is for the knot vector $[0\ 0\ 0\ 1\ 2]$. From Appendix C, Eq. (C.3) this is:

$$N_{1,4}(t) = \begin{cases} \dfrac{t}{4}(7t^2 - 18t + 12) & (0 \leqq t < 1) \\[2ex] -\dfrac{1}{4}(t-2)^3 & (1 \leqq t \leqq 2) \end{cases}$$

$\boldsymbol{N_{2,4}(t)}$

$N_{2,4}(t)$ is for the knot vector $[0\ 0\ 1\ 2\ 3]$. From Appendix C, Eq. (C.2) this is:

$$N_{2,4}(t) = \begin{cases} \dfrac{t^2}{12}(-11t + 18) & (0 \leqq t < 1) \\[2ex] \dfrac{7}{12}t^3 - 3t^2 + \dfrac{9}{2}t - \dfrac{3}{2} & (1 \leqq t < 2) \\[2ex] -\dfrac{1}{6}(t-3)^3 & (2 \leqq t \leqq 3) \end{cases}$$

$\boldsymbol{N_{3,4}(t)}$

$N_{3,4}(t)$ is for the knot vector $[0\ 1\ 2\ 3\ 4]$. From Appendix C, Eq. (C.1) this is:

$$N_{3,4}(t) = \begin{cases} \dfrac{1}{6}t^3 & (0 \leqq t < 1) \\[2ex] -(t-2)^3 + \dfrac{2}{3}(t-3)^3 - \dfrac{1}{6}(t-4)^3 & (1 \leqq t < 2) \\[2ex] \dfrac{2}{3}(t-3)^3 - \dfrac{1}{6}(t-4)^3 & (2 \leqq t < 3) \\[2ex] -\dfrac{1}{6}(t-4)^3 & (3 \leqq t \leqq 4) \end{cases}$$

$N_{4,4}(t)$

$N_{4,4}(t)$ is for the knot vector [1 2 3 4 5]. Appendix C, Eq. (C.1) gives the $N_{0,4}(t)$ for [0 1 2 3 4]. $N_{4,4}(t)$ is obtained by translating this function to the right 1 unit:

$$N_{4,4}(t)=\begin{cases} \frac{1}{6}(t-1)^3 & (1\leqq t<2) \\ -(t-3)^3+\frac{2}{3}(t-4)^3-\frac{1}{6}(t-5)^3 & (2\leqq t<3) \\ \frac{2}{3}(t-4)^3-\frac{1}{6}(t-5)^3 & (3\leqq t<4) \\ -\frac{1}{6}(t-5)^3 & (4\leqq t\leqq 5) \end{cases}$$

$N_{5,4}(t)$

$N_{5,4}(t)$ is for the knot vector [2 3 4 5 6]. This is obtained by translating $N_{4,4}(t)$ obtained above 1 unit to the right:

$$N_{5,4}(t)=\begin{cases} \frac{1}{6}(t-2)^3 & (2\leqq t<3) \\ -(t-4)^3+\frac{2}{3}(t-5)^3-\frac{1}{6}(t-6)^3 & (3\leqq t<4) \\ \frac{2}{3}(t-5)^3-\frac{1}{6}(t-6)^3 & (4\leqq t<5) \\ -\frac{1}{6}(t-6)^3 & (5\leqq t\leqq 6) \end{cases}$$

$N_{6,4}(t)$

$N_{6,4}(t)$ is the same as $N_{0,4}(t)$ for the knot vector [3 4 5 6 6]. Appendix C, Eq. (C.5) gives $N_{0,4}(t)$ for the knot vector [0 1 2 3 3]. $N_{6,4}(t)$ is found by translating this 3 units to the right:

$$N_{6,4}(t)=\begin{cases} \frac{1}{6}(t-3)^3 & (3\leqq t<4) \\ -\frac{7}{12}(t-3)^3+\frac{9}{4}(t-3)^2-\frac{9}{4}(t-3)+\frac{3}{4} & (4\leqq t<5) \\ \frac{1}{12}(6-t)^2(11t-48) & (5\leqq t\leqq 6) \end{cases}$$

$N_{7,4}(t)$

$N_{7,4}(t)$ is the same as $N_{0,4}(t)$ for the knot vector [4 5 6 6 6]. Appendix C,

Eq. (C.6) gives $N_{0,4}(t)$ for the knot vector [0 1 2 2 2]. $N_{7,4}(t)$ is found by translating this 4 units to the right:

$$N_{7,4}(t)=\begin{cases}\dfrac{1}{4}(t-4)^3 & (4\leqq t<5)\\[2mm] \dfrac{1}{4}(6-t)\{7(t-4)^2-10(t-4)+4\} & (5\leqq t\leqq 6)\end{cases}$$

$N_{8,4}(t)$

$N_{8,4}(t)$ is the same as $N_{0,4}(t)$ for the knot vector [5 6 6 6 6]. Appendix C, Eq. (C.7) gives $N_{0,4}(t)$ for the knot vector [0 1 1 1 1]. $N_{8,4}(t)$ is found by translating this 5 units to the right:

$$N_{8,4}(t)=(t-5)^3 \quad (5\leqq t<6).$$

Graphs of $N_{0,4}(t)$, $N_{1,4}(t)$, ..., $N_{8,4}(t)$ are shown in Fig. 6.42.

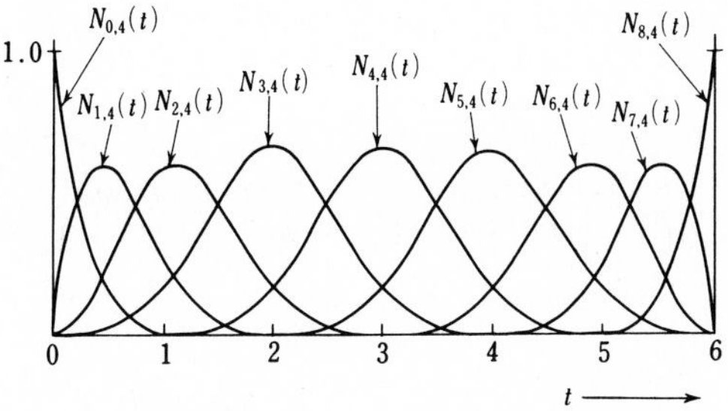

Fig. 6.42. *B*-spline functions for generating open curve for $M=4$, $n=8$ (curve type (3)) (Example 6.3)

Example 6.4. Using the results from Example 6.3, calculate the open curve for the case $\boldsymbol{Q}_0=[0\ 0]$, $\boldsymbol{Q}_1=[10\ 30]$, $\boldsymbol{Q}_2=[30\ 50]$, $\boldsymbol{Q}_3=[40\ 55]$, $\boldsymbol{Q}_4=[60\ 20]$, $\boldsymbol{Q}_5=[80\ 90]$, $\boldsymbol{Q}_6=[90\ 40]$, $\boldsymbol{Q}_7=[120\ 50]$, $\boldsymbol{Q}_8=[160\ 0]$.

Solution. From Eq. (6.114) the curve parameter range is from $a_0=0$ to a_{n-M+2} $=8-4+2=6$. Values calculated for the *B*-spline functions found in Example 6.3 at intervals of 0.25 in t are given in Table 6.1. Results of the calculation of:

$$\boldsymbol{P}(t)=\sum_{i=0}^{8} N_{i,4}(t)\boldsymbol{Q}_i$$

using the values in this table are given in Table 6.2. The *B*-spline curve is shown in Fig. 6.43.*)

*) In practice, it is desirable to use formula (6.109) and determine the point on the curve, as demonstrated in Example 6.7.

Table 6.1. Values of *B*-spline functions found in Example 6.3

t	$N_{0,4}$	$N_{1,4}$	$N_{2,4}$	$N_{3,4}$	$N_{4,4}$	$N_{5,4}$	$N_{6,4}$	$N_{7,4}$	$N_{8,4}$
0	1	0	0	0	0	0	0	0	0
0.25	0.422	0.496	0.079	0.003	0	0	0	0	0
0.5	0.125	0.594	0.260	0.021	0	0	0	0	0
0.75	0.0156	0.457	0.457	0.070	0	0	0	0	0
1	0	0.25	0.583	0.167	0	0	0	0	0
1.25	0	0.105	0.577	0.315	0.003	0	0	0	0
1.5	0	0.031	0.469	0.479	0.021	0	0	0	0
1.75	0	0.004	0.314	0.612	0.070	0	0	0	0
2	0	0	0.167	0.667	0.167	0	0	0	0
2.25	0	0	0.070	0.612	0.315	0.003	0	0	0
2.5	0	0	0.021	0.479	0.479	0.021	0	0	0
2.75	0	0	0.003	0.315	0.612	0.070	0	0	0
3	0	0	0	0.167	0.667	0.167	0	0	0
3.25	0	0	0	0.070	0.612	0.315	0.003	0	0
3.5	0	0	0	0.021	0.479	0.479	0.021	0	0
3.75	0	0	0	0.003	0.315	0.612	0.070	0	0
4	0	0	0	0	0.167	0.667	0.167	0	0
4.25	0	0	0	0	0.070	0.612	0.314	0.004	0
4.5	0	0	0	0	0.021	0.479	0.469	0.031	0
4.75	0	0	0	0	0.003	0.315	0.577	0.105	0
5	0	0	0	0	0	0.167	0.583	0.25	0
5.25	0	0	0	0	0	0.070	0.457	0.457	0.016
5.5	0	0	0	0	0	0.021	0.260	0.594	0.125
5.75	0	0	0	0	0	0.003	0.079	0.496	0.422
6	0	0	0	0	0	0	0	0	1

Table 6.2. Coordinates of curve of Example 6.4

t	$P(t)$		t	$P(t)$	
	x	y		x	y
0	0	0	3.25	65	44.6
0.25	7.49	19	3.5	69.8	54.7
0.5	14.6	32	3.75	74.3	64.3
0.75	21.1	40.4	4	78.4	70.1
1	26.7	45.8	4.25	81.9	69.2
1.25	31.1	49.4	4.5	85.5	63.8
1.5	34.8	51.1	4.75	90.0	56.8
1.75	38.1	50.9	5	95.8	50.9
2	41.7	48.4	5.25	104.1	47.4
2.25	45.7	44	5.5	116.4	42
2.5	50.2	38.9	5.75	134.4	28.2
2.75	55	36	6	160	0
3	60.1	37.6			

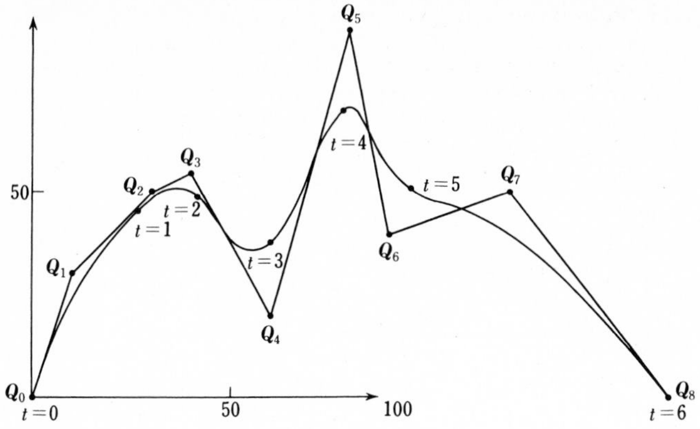

Fig. 6.43. Curve of Example 6.4

Example 6.5. For the case $M=4$, $n=3$, find all of the *B*-spline functions needed to generate a closed curve of *B*-spline curve type (3).

Solution. Setting $a_i=i$ $(i=0, 1, 2, \ldots, n+1)$, the knot vector T is, from Eq. (6.117):

$$\begin{aligned}T&=[t_0\ t_1\ t_2\ t_3\ t_4\ t_5\ t_6\ t_7\ t_8\ t_9\ t_{10}]\\&=[-3\ -2\ -1\ 0\ 1\ 2\ 3\ 4\ 5\ 6\ 7].\end{aligned}$$

$\boldsymbol{N_{0,4}(t)}$

$N_{0,4}(t)$ for the knot vector $[-3\ -2\ -1\ 0\ 1]$ is the same as $N_{0,4}(t)$ for the knot vector $[0\ 1\ 2\ 3\ 4]$ translated 3 units to the left. From Appendix C, Eq. (C.1) this is:

$$N_{0,4}(t)=-\frac{1}{6}(t-1)^3 \quad (0\leqq t\leqq 1)$$

$\boldsymbol{N_{1,4}(t)}$

$N_{1,4}(t)$ is the same as $N_{0,4}(t)$ for the knot vector $[-2\ -1\ 0\ 1\ 2]$. This can be obtained by translating $N_{0,4}(t)$ for the knot vector $[0\ 1\ 2\ 3\ 4]$ 2 units to the left. From Appendix C, Eq. (C.1) this is:

$$N_{1,4}(t)=\begin{cases}\dfrac{2}{3}(t-1)^3-\dfrac{1}{6}(t-2)^3 & (0\leqq t<1)\\[2ex] -\dfrac{1}{6}(t-2)^3 & (1\leqq t\leqq 2)\end{cases}$$

$N_{2,4}(t)$

$N_{2,4}(t)$ is the same as $N_{0,4}(t)$ for the knot vector [−1 0 1 2 3]. This is obtained by translating $N_{0,4}(t)$ for the knot vector [0 1 2 3 4] 1 unit to the left. From Appendix C, Eq. (C.1) this is:

$$N_{2,4}(t)=\begin{cases} -(t-1)^3+\dfrac{2}{3}(t-2)^3-\dfrac{1}{6}(t-3)^3 & (0\leqq t<1) \\ \dfrac{2}{3}(t-2)^3-\dfrac{1}{6}(t-3)^3 & (1\leqq t<2) \\ -\dfrac{1}{6}(t-3)^3 & (2\leqq t<3) \end{cases}$$

$N_{3,4}(t)$

$N_{3,4}(t)$ is the same as $N_{0,4}(t)$ for the knot vector [0 1 2 3 4]. From Appendix C, Eq. (C.1) this is:

$$N_{3,4}(t)=\begin{cases} \dfrac{1}{6}t^3 & (0\leqq t<1) \\ -(t-2)^3+\dfrac{2}{3}(t-3)^3-\dfrac{1}{6}(t-4)^3 & (1\leqq t<2) \\ \dfrac{2}{3}(t-3)^3-\dfrac{1}{6}(t-4)^3 & (2\leqq t<3) \\ -\dfrac{1}{6}(t-4)^3 & (3\leqq t\leqq 4) \end{cases}$$

$N_{4,4}(t)$

$N_{4,4}(t)$ is the same as $N_{0,4}(t)$ for the knot vector [1 2 3 4 5]. This is obtained by translating $N_{0,4}(t)$ for the knot vector [0 1 2 3 4] 1 unit to the right. From Appendix C, Eq. (C.1) this is:

$$N_{4,4}(t)=\begin{cases} \dfrac{1}{6}(t-1)^3 & (1\leqq t<2) \\ -(t-3)^3+\dfrac{2}{3}(t-4)^3-\dfrac{1}{6}(t-5)^3 & (2\leqq t<3) \\ \dfrac{2}{3}(t-4)^3-\dfrac{1}{6}(t-5)^3 & (3\leqq t\leqq 4) \end{cases}$$

$N_{5,4}(t)$

$N_{5,4}(t)$ is the same as $N_{0,4}(t)$ for the knot vector [2 3 4 5 6]. This is obtained by translating $N_{0,4}(t)$ for the knot vector [0 1 2 3 4] 2 units to the right. From Appendix C, Eq. (C.1) this is:

$$N_{5,4}(t)=\begin{cases}\frac{1}{6}(t-2)^3 & (2\leqq t<3)\\ -(t-4)^3+\frac{2}{3}(t-5)^3-\frac{1}{6}(t-6)^3 & (3\leqq t\leqq 4)\end{cases}$$

$N_{6,4}(t)$

$N_{6,4}(t)$ is the same as $N_{0,4}(t)$ for the knot vector [3 4 5 6 7]. This is obtained by translating $N_{0,4}(t)$ for the knot vector [0 1 2 3 4] 3 units to the right. From Appendix C, Eq. (C.1) this is:

$$N_{6,4}(t)=\frac{1}{6}(t-3)^3 \quad (3\leqq t\leqq 4).$$

Graphs of $N_{0,4}(t)$, $N_{1,4}(t)$, …, $N_{6,4}(t)$ are shown in Fig. 6.44.

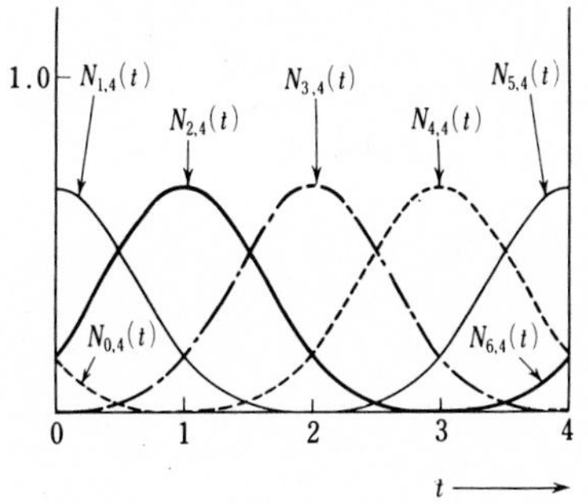

Fig. 6.44. *B*-spline functions for generating closed curve for $M=4$, $n=3$ (Example 6.5)

Table 6.3. Values of *B*-spline functions found in Example 6.5

t	$N_{0,4}(t)$	$N_{1,4}(t)$	$N_{2,4}(t)$	$N_{3,4}(t)$	$N_{4,4}(t)$	$N_{5,4}(t)$	$N_{6,4}(t)$
0	0.167	0.667	0.167	0	0	0	0
0.25	0.070	0.612	0.315	0.003	0	0	0
0.5	0.021	0.479	0.479	0.021	0	0	0
0.75	0.003	0.315	0.612	0.070	0	0	0
1	0	0.167	0.667	0.167	0	0	0
1.25	0	0.070	0.612	0.315	0.003	0	0
1.5	0	0.021	0.479	0.479	0.021	0	0
1.75	0	0.003	0.315	0.612	0.070	0	0
2	0	0	0.167	0.667	0.167	0	0
2.25	0	0	0.070	0.612	0.315	0.003	0
2.5	0	0	0.021	0.479	0.479	0.021	0
2.75	0	0	0.003	0.315	0.612	0.070	0
3	0	0	0	0.167	0.667	0.167	0
3.25	0	0	0	0.070	0.612	0.315	0.003
3.5	0	0	0	0.021	0.479	0.479	0.021
3.75	0	0	0	0.003	0.315	0.612	0.070
4	0	0	0	0	0.167	0.667	0.167

Example 6.6. Using the results from Example 6.5, calculate the closed curve for the case $Q_0=[-100\ -100]$, $Q_1=[-100\ 100]$, $Q_2=[100\ 100]$, $Q_3=[100\ -100]$.

Solution. From Eq. (6.116), the range of curve parameter is from $a_0=0$ to $a_{n+1}=a_4=4$. Values of the *B*-spline functions found in Example 6.5 calculated for intervals of 0.25 in t are given in Table 6.3. Values of

$$P(t)=\sum_{j=0}^{6} N_{j,4}(t)\,Q_{j\,\mathrm{mod}\,4}$$

calculated using the values in this table are given in Table 6.4. The *B*-spline curve is shown in Fig. 6.45*).

Table 6.4. Coordinates of curve of Example 6.6

t	$P(t)$		t	$P(t)$	
	x	y		x	y
0	−66.7	66.7	2	66.7	−66.7
0.25	−36.5	85.4	2.25	36.4	−85.4
0.5	0	91.7	2.5	0	−91.7
0.75	36.5	85.4	2.75	−36.4	−85.4
1	66.7	66.7	3	−66.7	−66.7
1.25	85.4	36.4	3.25	−85.5	−36.4
1.5	91.7	0	3.5	−91.6	0
1.75	85.4	−36.4	3.75	−85.5	36.4

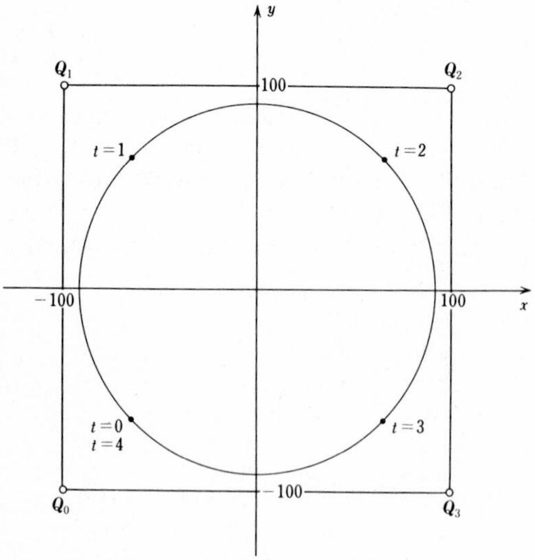

Fig. 6.45. Curve of Example 6.6

*) In practice, it is desirable to use formula (6.109) and determine the point on the curve, as demonstrated in Example 6.7.

Example 6.7. Find the values of the curve of Example 6.4 directly using the recurrence formula (6.109) rather than from the *B*-spline function formula.

Solution. When the curve parameter values t_s $(t_i \leqq t_s < t_{i+1})$ are given, the *B*-spline functions, which in general are nonzero, are:

$$N_{i-M+1,M}(t_s),\ N_{i-M+2,M}(t_s),\ \ldots,\ N_{i,M}(t_s).$$

The lower order *B*-spline functions that are used to compute these values are given in Fig. 6.37. Using these relations, we compute the *B*-spline function values below.

The knot vector for this example is:

$$\begin{aligned} T &= [t_0\ t_1\ t_2\ t_3\ t_4\ t_5\ t_6\ t_7\ t_8\ t_9\ t_{10}\ t_{11}\ t_{12}] \\ &= [0\ 0\ 0\ 0\ 1\ 2\ 3\ 4\ 5\ 6\ 6\ 6\ 6]. \end{aligned}$$

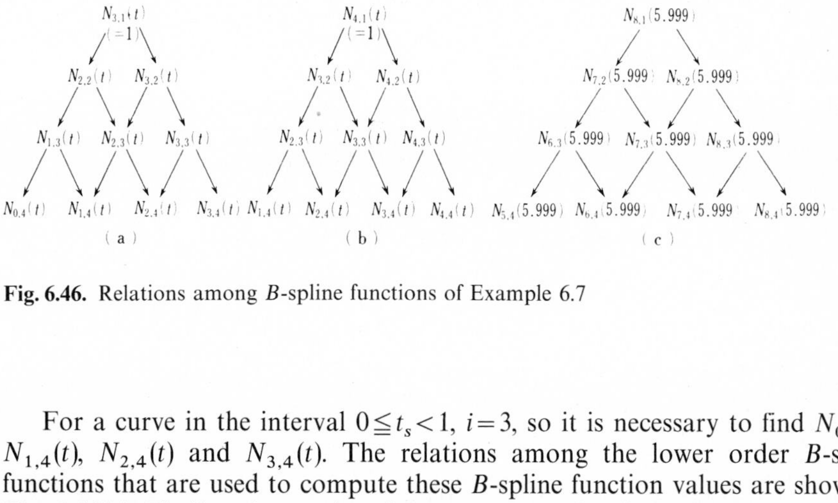

Fig. 6.46. Relations among *B*-spline functions of Example 6.7

For a curve in the interval $0 \leqq t_s < 1$, $i=3$, so it is necessary to find $N_{0,4}(t)$, $N_{1,4}(t)$, $N_{2,4}(t)$ and $N_{3,4}(t)$. The relations among the lower order *B*-spline functions that are used to compute these *B*-spline function values are shown in Fig. 6.46(a). Finding these values in sequence starting from the lower-order *B*-spline functions, since:

$$N_{2,2}(t) = \frac{t_4 - t}{t_4 - t_3} N_{3,1}(t) = (1-t) N_{3,1}(t)$$

we have:

$$N_{2,2}(0) = 1, \quad N_{2,2}(0.25) = 0.75, \quad N_{2,2}(0.5) = 0.5, \quad N_{2,2}(0.75) = 0.25$$

and:

$$N_{3,2}(t) = \frac{t - t_3}{t_4 - t_3} N_{3,1}(t) = t N_{3,1}(t).$$

Therefore:

$$N_{3,2}(0)=0,\ N_{3,2}(0.25)=0.25,\ N_{3,2}(0.5)=0.5,\ N_{3,2}(0.75)=0.75$$

$$N_{1,3}(t)=\frac{t_4-t}{t_4-t_2}\,N_{2,2}(t)=(1-t)\,N_{2,2}(t)$$

Therefore:

$$N_{1,3}(0)=(1-0)\cdot 1=1$$
$$N_{1,3}(0.25)=(1-0.25)\cdot 0.75=0.5625$$
$$N_{1,3}(0.5)=(1-0.5)\cdot 0.5=0.25$$
$$N_{1,3}(0.75)=(1-0.75)\cdot 0.25=0.0625$$

$$N_{2,3}(t)=\frac{t-t_2}{t_4-t_2}\,N_{2,2}(t)+\frac{t_5-t}{t_5-t_3}\,N_{3,2}(t)$$
$$=t\,N_{2,2}(t)+\frac{1}{2}\,(2-t)\,N_{3,2}(t)$$

Therefore:

$$N_{2,3}(0)=0\cdot 1+0.5\cdot(2-0)\cdot 0=0$$
$$N_{2,3}(0.25)=0.25\cdot 0.75+0.5\cdot(2-0.25)\cdot 0.25=0.40625$$
$$N_{2,3}(0.5)=0.5\cdot 0.5+0.5\cdot(2-0.5)\cdot 0.5=0.625$$
$$N_{2,3}(0.75)=0.75\cdot 0.25+0.5\cdot(2-0.75)\cdot 0.75=0.65625$$

$$N_{3,3}(t)=\frac{t-t_3}{t_5-t_3}\,N_{3,2}(t)=0.5t\,N_{3,2}(t)$$

Therefore:

$$N_{3,3}(0)=0.5\cdot 0\cdot 0=0$$
$$N_{3,3}(0.25)=0.5\cdot 0.25\cdot 0.25=0.03125$$
$$N_{3,3}(0.5)=0.5\cdot 0.5\cdot 0.5=0.125$$
$$N_{3,3}(0.75)=0.5\cdot 0.75\cdot 0.75=0.28125$$

$$N_{0,4}(t)=\frac{t_4-t}{t_4-t_1}\,N_{1,3}(t)=(1-t)\,N_{1,3}(t)$$

Therefore:

$$N_{0,4}(0)=(1-0)\cdot 1=1$$
$$N_{0,4}(0.25)=(1-0.25)\cdot 0.5625=0.422$$
$$N_{0,4}(0.5)=(1-0.5)\cdot 0.25=0.125$$

$$N_{0,4}(0.75)=(1-0.75)\cdot 0.0625=0.0156$$

$$N_{1,4}(t)=\frac{t-t_1}{t_4-t_1}N_{1,3}(t)+\frac{t_5-t}{t_5-t_2}N_{2,3}(t)$$

$$=t\,N_{1,3}(t)+\frac{1}{2}(2-t)\,N_{2,3}(t)$$

Therefore:

$$N_{1,4}(0)=0\cdot 1+0.5\cdot(2-0)\cdot 0=0$$
$$N_{1,4}(0.25)=0.25\cdot 0.5625+0.5\cdot(2-0.25)\cdot 0.40625=0.496$$
$$N_{1,4}(0.5)=0.5\cdot 0.25+0.5\cdot(2-0.5)\cdot 0.625=0.594$$
$$N_{1,4}(0.75)=0.75\cdot 0.0625+0.5\cdot(2-0.75)\cdot 0.65625=0.457$$

$$N_{2,4}(t)=\frac{t-t_2}{t_5-t_2}N_{2,3}(t)+\frac{t_6-t}{t_6-t_3}N_{3,3}(t)$$

$$=\frac{1}{2}t\,N_{2,3}(t)+\frac{1}{3}(3-t)\,N_{3,3}(t)$$

$$N_{2,4}(0)=0.5\cdot 0\cdot 0+0.333\cdot(3-0)\cdot 0=0$$
$$N_{2,4}(0.25)=0.5\cdot 0.25\cdot 0.40625+0.333\cdot(3-0.25)\cdot 0.03125=0.0794$$
$$N_{2,4}(0.5)=0.5\cdot 0.5\cdot 0.625+0.333\cdot(3-0.5)\cdot 0.125=0.260$$
$$N_{2,4}(0.75)=0.5\cdot 0.75\cdot 0.65625+0.333\cdot(3-0.75)\cdot 0.28125=0.457$$

$$N_{3,4}(t)=\frac{t-t_3}{t_6-t_3}N_{3,3}(t)=\frac{1}{3}t\,N_{3,3}(t)$$

Therefore:

$$N_{3,4}(0)=0.333\cdot 0\cdot 0=0$$
$$N_{3,4}(0.25)=0.333\cdot 0.25\cdot 0.03125=0.0026$$
$$N_{3,4}(0.5)=0.333\cdot 0.5\cdot 0.125=0.0208$$
$$N_{3,4}(0.75)=0.333\cdot 0.75\cdot 0.28125=0.0703$$

The curve can be generated in the range $t_3(=0)\leqq t<t_4(=1)$ by substituting the above values in:

$$\boldsymbol{P}(t)=\sum_{j=0}^{3}N_{j,4}(t)\boldsymbol{Q}_j.$$

The curve in the range $t_4(=1)\leqq t<t_5(=2)$ is found from the values of $N_{1,4}(t)$, $N_{2,4}(t)$, $N_{3,4}(t)$ and $N_{4,4}(t)$. The relations among the lower order *B*-spline functions that are relevant to these function values are shown in Fig. 6.46(b). By the same method as before we determine:

$$N_{1,4}(1),\ N_{1,4}(1.25),\ N_{1,4}(1.5),\ N_{1,4}(1.75)$$
$$N_{2,4}(1),\ N_{2,4}(1.25),\ N_{2,4}(1.5),\ N_{2,4}(1.75)$$
$$N_{3,4}(1),\ N_{3,4}(1.25),\ N_{3,4}(1.5),\ N_{3,4}(1.75)$$
$$N_{4,4}(1),\ N_{4,4}(1.25),\ N_{4,4}(1.5),\ N_{4,4}(1.75)$$

and thus generate the curve in $t_4(=1) \leqq t < t_5(=2)$.

Repeating a similar procedure, we determine the curve throughout $0 \leqq t < 6$. Since the B-spline function values at $t=6$ cannot be determined directly by using the recurrence formula (6.109), it is approximated by the value at a nearby point such as $t=5.999$. Since $t_i \leqq 5.999 < t_{i+1}$, $i=8$, so that it is necessary to find $N_{5,4}(5.999)$, $N_{6,4}(5.999)$, $N_{7,4}(5.999)$ and $N_{8,4}(5.999)$. The relations among the lower order B-spline functions that are relevant to these B-spline function values are shown in Fig. 6.46(c).

$$N_{7,2}(t) = \frac{t_9 - t}{t_9 - t_8} N_{8,1}(t)$$

$$\therefore \quad N_{7,2}(5.999) = \frac{6 - 5.999}{6 - 5} \cdot 1 = 0.001$$

$$N_{8,2}(t) = \frac{t - t_8}{t_9 - t_8} N_{8,1}(t)$$

$$\therefore \quad N_{8,2}(5.999) = \frac{5.999 - 5}{6 - 5} \cdot 1 = 0.999$$

$$N_{6,3}(t) = \frac{t_9 - t}{t_9 - t_7} N_{7,2}(t)$$

$$\therefore \quad N_{6,3}(5.999) = \frac{6 - 5.999}{6 - 4} \cdot 0.001 \fallingdotseq 0$$

$$N_{7,3}(t) = \frac{t - t_7}{t_9 - t_7} N_{7,2}(t) + \frac{t_{10} - t}{t_{10} - t_8} N_{8,2}(t)$$

$$\therefore \quad N_{7,3}(5.999) = \frac{5.999 - 4}{6 - 4} \cdot 0.001 + \frac{6 - 5.999}{6 - 5} \cdot 0.999 = 0.002$$

$$N_{8,3}(t) = \frac{t - t_8}{t_{10} - t_8} N_{8,2}(t)$$

$$\therefore \quad N_{8,3}(5.999) = \frac{5.999 - 5}{6 - 5} \cdot 0.999 = 0.998$$

$$N_{5,4}(t) = \frac{t_9 - t}{t_9 - t_6} N_{6,3}(t)$$

$$\therefore \quad N_{5,4}(5.999) = \frac{6 - 5.999}{6 - 3} \cdot 0 = 0$$

$$N_{6,4}(t)=\frac{t-t_6}{t_9-t_6}N_{6,3}(t)+\frac{t_{10}-t}{t_{10}-t_7}N_{7,3}(t)$$

$$\therefore \quad N_{6,4}(5.999)=\frac{5.999-3}{6-3}\cdot 0+\frac{6-5.999}{6-4}\cdot 0.002 \doteqdot 0$$

$$N_{7,4}(t)=\frac{t-t_7}{t_{10}-t_7}N_{7,3}(t)+\frac{t_{11}-t}{t_{11}-t_8}N_{8,3}(t)$$

$$\therefore \quad N_{7,4}(5.999)=\frac{5.999-4}{6-4}\cdot 0.002+\frac{6-5.999}{6-5}\cdot 0.998=0.003$$

$$N_{8,4}(t)=\frac{t-t_8}{t_{11}-t_8}N_{8,3}(t)$$

$$\therefore \quad N_{8,4}(5.999)=\frac{5.999-5}{6-5}\cdot 0.998=0.997$$

6.11 Differentiation of *B*-Spline Curves

Denote a *B*-spline curve by:

$$\boldsymbol{P}(t)=\sum_{j=i-M+1}^{i} N_{j,M}(t)\boldsymbol{Q}_j \tag{6.118}$$

where $t_i \leqq t < t_{i+1}$.

$N_{j,M}(t)$ can be expressed as (refer to Eq. (6.77)):

$$N_{j,M}(t)=M[t;\ t_{j+1},\ \ldots,\ t_{j+M}]-M[t;\ t_j,\ \ldots,\ t_{j+M-1}]$$

so that:

$$\begin{aligned} N_{j,M}^{(1)}(t) &= (d/dt)\{M[t;\ t_{j+1},\ \ldots,\ t_{j+M}]-M[t;\ t_j,\ \ldots,\ t_{j+M-1}]\} \\ &= -(M-1)[M_{j+1,M-1}(t)-M_{j,M-1}(t)]. \end{aligned}$$

Therefore:

$$\boldsymbol{P}^{(1)}(t)=(M-1)\sum_{j=i-M+1}^{i}[M_{j,M-1}(t)-M_{j+1,M-1}(t)]\boldsymbol{Q}_j$$

$$=(M-1)\sum_{j=i-M+1}^{i}\left[\frac{1}{t_{j+M-1}-t_j}N_{j,M-1}(t)\boldsymbol{Q}_j-\frac{1}{t_{j+M}-t_{j+1}}N_{j+1,M-1}(t)\boldsymbol{Q}_j\right]=$$

$$=(M-1)\sum_{j=i-M+2}^{i}\left[\frac{1}{t_{j+M-1}-t_j}N_{j,M-1}(t)\boldsymbol{Q}_j-\frac{1}{t_{j+M-1}-t_j}N_{j,M-1}(t)\boldsymbol{Q}_{j-1}\right]^{*)}$$

$$=(M-1)\sum_{j=i-M+2}^{i}N_{j,M-1}(t)\frac{\boldsymbol{Q}_j-\boldsymbol{Q}_{j-1}}{t_{j+M-1}-t_j}$$

$$\equiv(M-1)\sum_{j=i-M+2}^{i}N_{j,M-1}(t)\boldsymbol{Q}_j^{(1)}\quad(t_i\leqq t<t_{i+1}) \tag{6.119}$$

where:

$$\boldsymbol{Q}_j^{(1)}=\frac{\boldsymbol{Q}_j-\boldsymbol{Q}_{j-1}}{t_{j+M-1}-t_j}. \tag{6.120}$$

More generally, the r-th derivative of $\boldsymbol{P}(t)$ is:

$$\boldsymbol{P}^{(r)}(t)=(M-1)\dots(M-r)\sum_j N_{j,M-r}(t)\boldsymbol{Q}_j^{(r)} \tag{6.121}$$

where:

$$\boldsymbol{Q}_j^{(0)}=\boldsymbol{Q}_j$$

$$\boldsymbol{Q}_j^{(r)}=\frac{\boldsymbol{Q}_j^{(r-1)}-\boldsymbol{Q}_{j-1}^{(r-1)}}{t_{j+M-r}-t_j}\quad(r>0). \tag{6.122}$$

If the knots are uniformly spaced, that is, if $t_j=t_0+jh$ for all j, then, with ∇ as the backward finite difference operator:

$$\begin{aligned}\nabla\boldsymbol{Q}_j&=\boldsymbol{Q}_j^{(0)}-\boldsymbol{Q}_{j-1}^{(0)}\\&=(M-1)h\boldsymbol{Q}_j^{(1)}\\\nabla^2\boldsymbol{Q}_j&=\nabla\boldsymbol{Q}_j-\nabla\boldsymbol{Q}_{j-1}\\&=(M-1)h(\boldsymbol{Q}_j^{(1)}-\boldsymbol{Q}_{j-1}^{(1)})\\&=(M-1)(M-2)h^2\boldsymbol{Q}_j^{(2)}\\\nabla^3\boldsymbol{Q}_j&=\nabla^2\boldsymbol{Q}_j-\nabla^2\boldsymbol{Q}_{j-1}\\&=(M-1)(M-2)h^2(\boldsymbol{Q}_j^{(2)}-\boldsymbol{Q}_{j-1}^{(2)})\\&=(M-1)(M-2)(M-3)h^3\boldsymbol{Q}_j^{(3)}.\end{aligned}$$

Repeating a similar procedure, we find that, in general:

$$\nabla^r\boldsymbol{Q}_j=(M-1)(M-2)\dots(M-r)h^r\boldsymbol{Q}_j^{(r)}. \tag{6.123}$$

*) Note that in $t_i\leqq t<t_{i+1}$, both $N_{i+1,M-1}(t)$ and $N_{i-M+1,M-1}(t)$ are zero (refer to Fig. 6.37).

From this equation we can find $Q_j^{(r)}$ and substitute into Eq. (6.122) to obtain:

$$P^{(r)}(t)=h^{-r}\sum_j N_{j,M-r}(t)\nabla^r Q_j. \tag{6.124}$$

6.12 Geometrical Properties of *B*-Spline Curves

The general geometrical properties of *B*-spline curves are summarized below.

(1) Locality

A point on the curve is determined by only the M curve defining vectors Q_j in the immediate neighborhood of that point. Consequently, if a certain polygon vertex is varied, only the part of the curve in the immediate neighborhood of that vertex is affected (refer to Fig. 6.47).

(2) Continuity

In general C^{M-2} continuity between curve segments is maintained.

(3) Convex Hull Property

Equations (6.84), (6.85) and (6.86) hold for a *B*-spline function, so the curve segments that comprise the *B*-spline curve are convex combinations of the nearest M vectors Q_j. That is to say, each curve segment is contained inside a convex hull formed by the M points. Consequently, compared to a Bézier curve, a *B*-spline curve is more faithful to variations in the polygon shape.

As a special case of this property, if $Q_j = Q_{j+1} = \ldots = Q_{j+M-1}$, these convex hulls reduce to the points Q_j, the curve segments defined by this sequence of points degenerate to the points Q_j, and the curve passes through the Q_j.

Also, if $Q_j, Q_{j+1}, \ldots, Q_{j+M-1}$ all lie colinear straight line, from the convex hull property of a *B*-spline curve it follows that the *B*-spline curve segments are

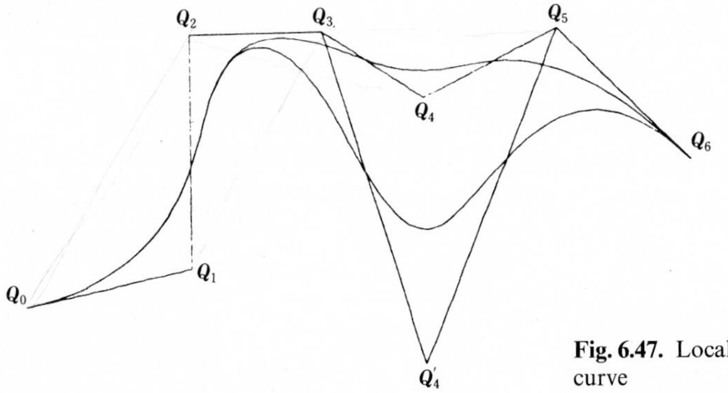

Fig. 6.47. Local control of a *B*-spline curve

straight line segments. Since a B-spline curve generally maintains C^{M-2} continuity at the connection points between curve segments, it can include straight line segments that are connected to curved segments with C^{M-2} continuity.

(4) Variation Diminishing Property

The number of intersections between an arbitrary straight line and a B-spline curve is not more than the number of intersections between that straight line and the curve defining polygon. Therefore, a B-spline curve assumes a shape that is a smoothed form of the curve defining polygon shape (refer to Sect. 6.9 ⑩).

6.13 Determination of a Point on a Curve by Linear Operations[29)]

De Boor proposed an algorithm for finding a point on a curve recursively by repeated application of linear operations, without calculating the B-spline function values (refer to Appendix B.2).

De Boor's algorithm

We are to find a point $\boldsymbol{P}(t_s)$ on a curve at the parameter value $t=t_s$.

Step 1: Find an i for which $t_i \leqq t_s < t_{i+1}$, then set:

$$r=i-M+1. \tag{6.125}$$

Step 2: Let:

$$\boldsymbol{Q}_j^{[0]}(t_s)=\boldsymbol{Q}_j \quad (j=r,\ r+1,\ \ldots,\ r+M-1). \tag{6.126}$$

Step 3: Repeatedly apply the formulas:

$$\boldsymbol{Q}_j^{[k]}(t_s)=(1-\lambda)\,\boldsymbol{Q}_{j-1}^{[k-1]}(t_s)+\lambda\,\boldsymbol{Q}_j^{[k-1]}(t_s) \tag{6.127}$$

$$\lambda=\frac{t_s-t_j}{t_{j+M-k}-t_j} \;{}^{*)} \tag{6.128}$$

to find:

$$\boldsymbol{P}(t_s)=\boldsymbol{Q}_i^{[M-1]}(t_s).$$

In this algorithm, in the case of a closed curve, the j in $\boldsymbol{Q}_j$ is replaced by $j \bmod (n+1)$.

---

*) If a uniform knot vector having a span of 1 is given, then:

$$\lambda=\frac{t_s-t_j}{M-k}.$$

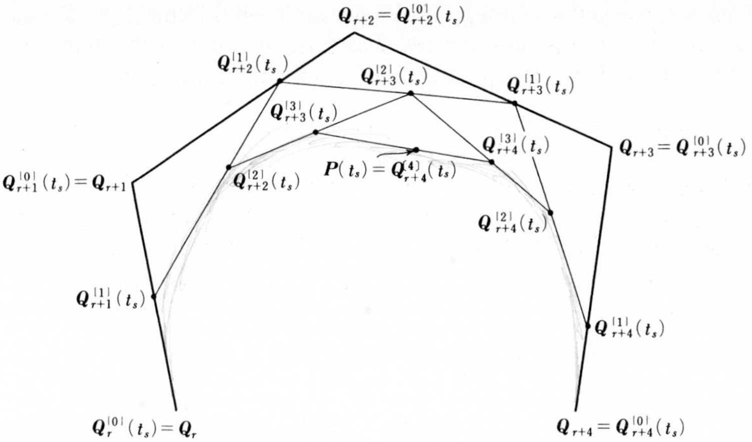

Fig. 6.48. Geometrical interpretation of De Boor's algorithm (case of $M=5$)

In the above algorithm, in step 1 we are finding the r in the expression Q_j ($j=r$, $r+1$, ..., $r+M-1$) for the curve defining vectors related to the curve segment in which t_s is located. Step 2 is the initial value setting. Steps 2 and 3 can be geometrically interpreted as follows (refer to Fig. 6.48). The case $M=5$ will be used as an example in this explanation. In Step 1, suppose that the curve defining vectors Q_r, Q_{r+1}, Q_{r+2}, Q_{r+3} and Q_{r+4} of the segment in which t_s is located are known. First, according to Step 2, these are the initial value vectors for the algorithm. That is,

$$Q_r^{[0]}(t_s) \leftarrow Q_r,\ Q_{r+1}^{[0]}(t_s) \leftarrow Q_{r+1},\ \dots,\ Q_{r+4}^{[0]}(t_s) \leftarrow Q_{r+4}.$$

By linear interpolation between these position vectors, new points

$$Q_{r+1}^{[1]}(t_s),\ Q_{r+2}^{[1]}(t_s),\ Q_{r+3}^{[1]}(t_s),\ Q_{r+4}^{[1]}(t_s)$$

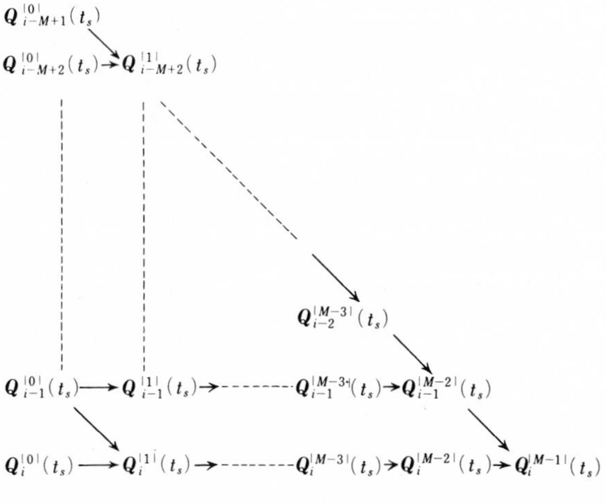

Fig. 6.49. Relations in De Boor's algorithm

are produced on the sides of the polygon. A new polygon can be formed from these 4 points. Again linearly interpolate between the points of this new polygon to obtain another new polygon consisting of the points

$$Q^{[2]}_{r+2}(t_s),\ Q^{[2]}_{r+3}(t_s),\ Q^{[2]}_{r+4}(t_s).$$

Again linearly interpolate between these points to obtain the 2 points $Q^{[3]}_{r+3}(t_s)$, $Q^{[3]}_{r+4}(t_s)$, which are the ends of a line segment. Finally, linearly interpolate between these 2 points to determine a single point $Q^{[4]}_{r+4}(t_s)$. This point is the point $P(t_s)$ on the curve that corresponds to $t=t_s$. In general, the linear interpolation parameter value λ is different in each interpolation.

The relationships in De Boor's algorithm are shown in Fig. 6.49.

Example 6.8. Find the point on the open *B*-spline curve found in Example 6.4 at $t=4.75$, using De Boor's algorithm.

Solution. From the knot vector of Example 6.3:

$$\begin{aligned} T &= [t_0\ t_1\ t_2\ t_3\ t_4\ t_5\ t_6\ t_7\ t_8\ t_9\ t_{10}\ t_{11}\ t_{12}] \\ &= [0\ 0\ 0\ 0\ 1\ 2\ 3\ 4\ 5\ 6\ 6\ 6\ 6] \end{aligned}$$

we see that, since $t_7 \leqq 4.75 < t_8$, $i=7$. Therefore $r=7-4+1=4$.

$$\begin{aligned} Q^{[1]}_5(4.75) &= (1-\lambda)\,Q^{[0]}_4(4.75)+\lambda\,Q^{[0]}_5(4.75) \qquad \left(\lambda=\frac{4.75-t_5}{t_8-t_5}=0.917\right) \\ &= (1-\lambda)\,Q_4+\lambda\,Q_5 \\ &= 0.083\,Q_4+0.917\,Q_5 \end{aligned}$$

$$\begin{aligned} Q^{[1]}_6(4.75) &= (1-\lambda)\,Q^{[0]}_5(4.75)+\lambda\,Q^{[0]}_5(4.75) \qquad \left(\lambda=\frac{4.75-t_6}{t_9-t_6}=0.583\right) \\ &= (1-\lambda)\,Q_5+\lambda\,Q_6 \\ &= 0.417\,Q_5+0.583\,Q_6 \end{aligned}$$

$$\begin{aligned} Q^{[1]}_7(4.75) &= (1-\lambda)\,Q^{[0]}_6(4.75)+\lambda\,Q^{[0]}_7(4.75) \qquad \left(\lambda=\frac{4.75-t_7}{t_{10}-t_7}=0.375\right) \\ &= (1-\lambda)\,Q_6+\lambda\,Q_7 \\ &= 0.625\,Q_6+0.375\,Q_7 \end{aligned}$$

$$\begin{aligned} Q^{[2]}_6(4.75) &= (1-\lambda)\,Q^{[1]}_5(4.75)+\lambda\,Q^{[1]}_6(4.75) \qquad \left(\lambda=\frac{4.75-t_6}{t_8-t_6}=0.875\right) \\ &= 0.125(0.083\,Q_4+0.917\,Q_5)+0.875(0.417\,Q_5+0.583\,Q_6) \\ &= 0.010\,Q_4+0.479\,Q_5+0.510\,Q_6 \end{aligned}$$

$$\begin{aligned} Q^{[2]}_7(4.75) &= (1-\lambda)\,Q^{[1]}_6(4.75)+\lambda\,Q^{[1]}_7(4.75) \qquad \left(\lambda=\frac{4.75-t_7}{t_9-t_7}=0.375\right) \\ &= 0.625(0.417\,Q_5+0.583\,Q_6)+0.375(0.625\,Q_6+0.375\,Q_7) \\ &= 0.261\,Q_5+0.598\,Q_6+0.141\,Q_7 \end{aligned}$$

$$Q_7^{[3]}(4.75) = (1-\lambda)\,Q_6^{[2]}(4.75) + \lambda Q_7^{[2]}(4.75) \qquad \left(\lambda = \frac{4.75 - t_7}{t_8 - t_7} = 0.75\right)$$

$$= 0.25(0.010\,Q_4 + 0.479\,Q_5 + 0.510\,Q_6)$$

$$+ 0.75(0.261\,Q_5 + 0.598\,Q_6 + 0.141\,Q_7)$$

$$= 0.0025\,Q_4 + 0.316\,Q_5 + 0.576\,Q_6 + 0.106\,Q_7$$

$$= 0.0025[60\ 20] + 0.316[80\ 90] + 0.576[90\ 40] + 0.106[120\ 50]$$

$$= [90.0\ 56.8]$$

This result agrees with the coordinates of the point at $t = 4.75$ in Table 6.2.

Example 6.9. Find the point on the closed B-spline curve found in Example 6.6 at $t = 3.5$, using DeBoor's algorithm.

Solution. From the knot vector of Example 6.5:

$$T = [t_0\ t_1\ t_2\ t_3\ t_4\ t_5\ t_6\ t_7\ t_8\ t_9\ t_{10}]$$
$$= [-3\ -2\ -1\ 0\ 1\ 2\ 3\ 4\ 5\ 6\ 7]$$

we see that, since $t_6 \leqq 3.5 < t_7$, $i = 6$. Therefore $r = 6 - 4 + 1 = 3$.

$$Q_4^{[1]}(3.5) = (1-\lambda)\,Q_3^{[0]}(3.5) + \lambda Q_4^{[0]}(3.5) \qquad \left(\lambda = \frac{3.5 - t_4}{4-1} = 0.833\right)$$

$$= 0.167\,Q_3 + 0.833\,Q_0$$

$$Q_5^{[1]}(3.5) = (1-\lambda)\,Q_4^{[0]}(3.5) + \lambda Q_5^{[0]}(3.5) \qquad \left(\lambda = \frac{3.5 - t_5}{4-1} = 0.5\right)$$

$$= 0.5\,Q_0 + 0.5\,Q_1$$

$$Q_6^{[1]}(3.5) = (1-\lambda)\,Q_5^{[0]}(3.5) + \lambda Q_6^{[0]}(3.5) \qquad \left(\lambda = \frac{3.5 - t_6}{4-1} = 0.167\right)$$

$$= 0.833\,Q_1 + 0.167\,Q_2$$

$$Q_5^{[2]}(3.5) = (1-\lambda)\,Q_4^{[1]}(3.5) + \lambda Q_5^{[1]}(3.5) \qquad \left(\lambda = \frac{3.5 - t_5}{4-2} = 0.75\right)$$

$$= 0.25(0.167\,Q_3 + 0.833\,Q_0) + 0.75(0.5\,Q_0 + 0.5\,Q_1)$$

$$= 0.583\,Q_0 + 0.375 Q_1 + 0.042\,Q_3$$

$$Q_6^{[2]}(3.5) = (1-\lambda)\,Q_5^{[1]}(3.5) + \lambda Q_6^{[1]}(3.5) \qquad \left(\lambda = \frac{3.5 - t_6}{4-2} = 0.25\right)$$

$$= 0.75(0.5\,Q_0 + 0.5\,Q_1) + 0.25(0.833 Q_1 + 0.167 Q_2)$$

$$= 0.375 Q_0 + 0.583 Q_1 + 0.042 Q_2$$

$$Q_6^{[3]}(3.5) = (1-\lambda)\,Q_5^{[2]}(3.5) + \lambda Q_6^{[2]}(3.5) = \qquad \left(\lambda = \frac{3.5 - t_6}{4-3} = 0.5\right)$$

$$
\begin{aligned}
&= 0.5(0.583\,\boldsymbol{Q}_0 + 0.375\,\boldsymbol{Q}_1 + 0.042\boldsymbol{Q}_3)\\
&\quad + 0.5(0.375\boldsymbol{Q}_0 + 0.583\,\boldsymbol{Q}_1 + 0.042\boldsymbol{Q}_2)\\
&= 0.479\,\boldsymbol{Q}_0 + 0.479\,\boldsymbol{Q}_1 + 0.021\,\boldsymbol{Q}_2 + 0.021\,\boldsymbol{Q}_3\\
&= 0.479\,[-100 \quad -100] + 0.479\,[-100 \quad 100]\\
&\quad + 0.021\,[100 \quad 100] + 0.021\,[100 \quad -100]\\
&= [-91.6 \quad 0]\,.
\end{aligned}
$$

The coordinate values we have found agree with those at $t = 3.5$ in Table 6.4.

6.14 Insertion of Knots[30)]

A Bézier curve segment can be divided into two curve segments at an arbitrary point while maintaining its shape. This property is very convenient in curve design when it is necessary to apply fine shape control. In the case of a non-uniform *B*-spline curve a new knot can be inserted to increase the number of curve defining vectors and the number of curve segments. The theory of this will now be given.

Suppose that we have a *B*-spline curve defined on a sequence of knots ..., t_i, t_{i+1}, ...:

$$\boldsymbol{P}(t) = \sum_j N_{j,M}(t)\,\boldsymbol{Q}_j\,. \tag{6.129}$$

Suppose now that we insert a new knot t_l between t_i and t_{i+1} $(t_i < t_l \leqq t_{i+1})$ to form a knot sequence $\hat{t}_j$. The relation between the old and new knots is:

$$\left.\begin{aligned}
&\hat{t}_j = t_j && (j \leqq i)\\
&\hat{t}_j = t_l && (j = i+1)\\
&\hat{t}_j = t_{i-1} && (j \geqq i+2)
\end{aligned}\right\} \tag{6.130}$$

The *B*-spline curve defined by the new knot sequence is:

$$\boldsymbol{P}(t) = \sum_j \hat{N}_{j,M}(t)\,\hat{\boldsymbol{Q}}_j\,. \tag{6.131}$$

$\hat{N}_{j,M}(t)$ are the *B*-spline functions defined by the new knot sequence $\hat{t}_j$. Let us now find the $\hat{\boldsymbol{Q}}_j$ that express the same curve in terms of the new knot sequence.

For the $M+2$ knots including the knots t_i, t_l and t_{i+1}:

$$t_j, \ldots, t_{j+M} \quad (i - M + 1 \leqq j \leqq i)$$

the following divided difference relationship holds.

$$(t_j - t_{j+M})\,M[t;t_j,\ldots,t_{j+M}] + (t_l - t_j)\,M[t;t_j,\ldots,t_{j+M-1},t_l] + (t_{j+M} - t_l)\,M[t;t_l,t_{j+1},\ldots,t_{j+M}] = 0 \tag{6.132}$$

This equation can be easily understood from the divided difference relations:

$$M[t;t_j,\ldots,t_{j+M}] = \frac{1}{t_{j+M}-t_j}\,(M[t;t_{j+1},\ldots,t_{j+M}] - M[t;t_j,\ldots,t_{j+M-1}])$$

$$M[t;t_j,\ldots,t_{j+M-1},t_l] = \frac{1}{t_l-t_j}\,(M[t;t_{j+1},\ldots,t_{j+M-1},t_l] - M[t;t_j,\ldots,t_{j+M-1}])$$

$$M[t;t_l,t_{j+1},\ldots,t_{j+M}] = \frac{1}{t_{j+M}-t_l}\,(M[t;t_{j+1},\ldots,t_{j+M}] - M[t;t_l,t_{j+1},\ldots,t_{j+M-1}]).$$

Setting $t_l = \hat{t}$ in Eq. (6.132) and rearranging it into a B-spline function relation gives:

$$(t_{j+M} - t_j)\,M_{j,M}(t) = (\hat{t} - t_j)\,\hat{M}_{j,M}(t) + (t_{j+M} - \hat{t})\,\hat{M}_{j+1,M}(t) \quad (i-M+1 \leqq j \leqq i).$$

This can be expressed as a relation among the normalized B-spline functions:

$$\begin{aligned} N_{j,M}(t) &= \frac{\hat{t}-t_j}{\hat{t}_{j+M}-\hat{t}_j}\,\hat{N}_{j,M}(t) + \frac{t_{j+M}-\hat{t}}{\hat{t}_{j+M+1}-\hat{t}_{j+1}}\,\hat{N}_{j+1,M}(t) \\ &= \frac{\hat{t}-\hat{t}_j}{\hat{t}_{j+M}-\hat{t}_j}\,\hat{N}_{j,M}(t) + \frac{\hat{t}_{j+M+1}-\hat{t}}{\hat{t}_{j+M+1}-\hat{t}_{j+1}}\,\hat{N}_{j+1,M}(t) \\ &\quad (i-M+1 \leqq j \leqq i). \end{aligned} \tag{6.133}$$

If j has a different value than the above we have:

$$\left.\begin{aligned} N_{j,M}(t) &= \hat{N}_{j,M}(t) \qquad (j \leqq i-M) \\ N_{j,M}(t) &= \hat{N}_{j+1,M}(t) \quad (j \geqq i+1) \end{aligned}\right\}. \tag{6.134}$$

Substituting $N_{j,M}(t)$ from Eq. (6.133) into Eq. (6.129) gives:

$$\begin{aligned} \boldsymbol{P}(t) &= \sum_j N_{j,M}(t)\,\boldsymbol{Q}_j \\ &= \sum_{j=i-M+1}^{i} \left(\frac{\hat{t}-\hat{t}_j}{\hat{t}_{j+M}-\hat{t}_j}\,\boldsymbol{Q}_j \hat{N}_{j,M}(t) + \frac{\hat{t}_{j+M+1}-\hat{t}}{\hat{t}_{j+M+1}-\hat{t}_{j+1}}\,\boldsymbol{Q}_j \hat{N}_{j+1,M}(t) \right) = \end{aligned}$$

$$= \sum_{j=i-M+1}^{i+1} \left(\frac{\hat{t}-\hat{t}_j}{\hat{t}_{j+M}-\hat{t}_j} \boldsymbol{Q}_j \hat{N}_{j,M}(t) + \frac{\hat{t}_{j+M}-\hat{t}}{\hat{t}_{j+M}-\hat{t}_j} \boldsymbol{Q}_{j-1} \hat{N}_{j,M}(t) \right)^{*)}$$

$$= \sum_{j=i-M+1}^{i+1} \hat{N}_{j,M}(t) \left[\frac{\hat{t}-\hat{t}_j}{\hat{t}_{j+M}-\hat{t}_j} \boldsymbol{Q}_j + \left(1-\frac{\hat{t}-\hat{t}_j}{\hat{t}_{j+M}-\hat{t}_j}\right) \boldsymbol{Q}_{j-1} \right] \qquad (t_i \leqq t < t_{i+1}) \tag{6.135}$$

Comparing this equation with Eq. (6.131) gives:

$$\hat{\boldsymbol{Q}}_j = (1-\alpha_j)\,\boldsymbol{Q}_{j-1} + \alpha_j \boldsymbol{Q}_j \tag{6.136}$$

where:

$$\alpha_j = \begin{cases} 1 & (j \leqq i-M+1) \\ \dfrac{\hat{t}-\hat{t}_j}{\hat{t}_{j+M}-\hat{t}_j} = \dfrac{\hat{t}-t_j}{t_{j+M-1}-t_j} & (i-M+2 \leqq j \leqq i) \\ 0 & (j \geqq i+1) \end{cases} \tag{6.137}$$

Equation (6.136) means that the new surface defining vectors $\hat{\boldsymbol{Q}}_j$ divide the line segments $\boldsymbol{Q}_{j-1}\boldsymbol{Q}_j$ in the ratio $\alpha_j : 1-\alpha_j$. This relation is shown in Fig. 6.50.

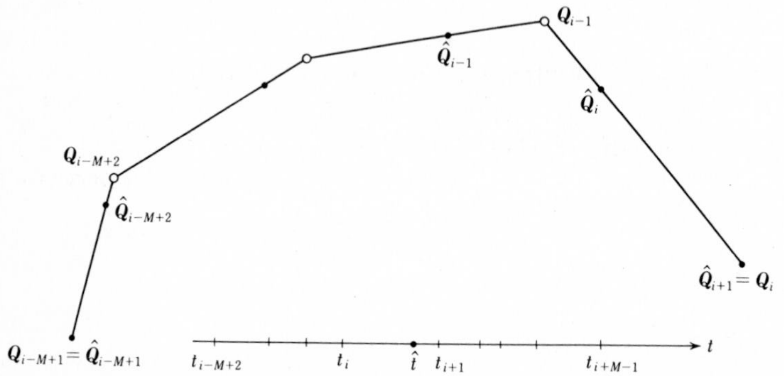

Fig. 6.50. Relation between old curve defining vectors and new curve defining vectors formed by adding another knot

Example 6.10. Add a knot to the *B*-spline curve in Example 6.4 at $t=2.8$.
Solution

$\hat{t}=2.8$ and $M=4$.

---

*) Note that $\hat{t}=\hat{t}_{i+1}$.

The B-spline curve knot vector in Example 6.4 is:

$$T=[t_0\ t_1\ t_2\ t_3\ t_4\ t_5\ t_6\ t_7\ t_8\ t_9\ t_{10}\ t_{11}\ t_{12}]$$
$$=[0\ 0\ 0\ 0\ 1\ 2\ 3\ 4\ 5\ 6\ 6\ 6\ 6]$$

Since $t_5<\hat{t}<t_6$, $i=5$.

Using these data, Eq. (6.137) becomes:

$$\alpha_j=\begin{cases}1 & (j\leqq 2)\\ \dfrac{2.8-t_j}{t_{j+3}-t_j} & (3\leqq j\leqq 5)\\ 0 & (j\geqq 6)\end{cases}$$

Let us now calculate α_j for $3\leqq j\leqq 5$.

$$\alpha_3=\frac{2.8-t_3}{t_6-t_3}=\frac{2.8-0}{3-0}=0.933$$

$$\alpha_4=\frac{2.8-t_4}{t_7-t_4}=\frac{2.8-1}{4-1}=0.6$$

$$\alpha_5=\frac{2.8-t_5}{t_8-t_5}=\frac{2.8-2}{5-2}=0.267$$

Using these values of α_j, we determine new curve defining vectors $\hat{\boldsymbol{Q}}_j$ from Eq. (6.136). Since $\alpha_j=1$ $(j\leqq 2)$, $\hat{\boldsymbol{Q}}_0=\boldsymbol{Q}_0$, $\hat{\boldsymbol{Q}}_1=\boldsymbol{Q}_1$, $\hat{\boldsymbol{Q}}_2=\boldsymbol{Q}_2$. $\alpha_j=0$ $(j\geqq 6)$, $\hat{\boldsymbol{Q}}_6=\boldsymbol{Q}_5$, $\hat{\boldsymbol{Q}}_7=\boldsymbol{Q}_6$, $\hat{\boldsymbol{Q}}_8=\boldsymbol{Q}_7$, $\hat{\boldsymbol{Q}}_9=\boldsymbol{Q}_8$. The remaining curve defining vectors are:

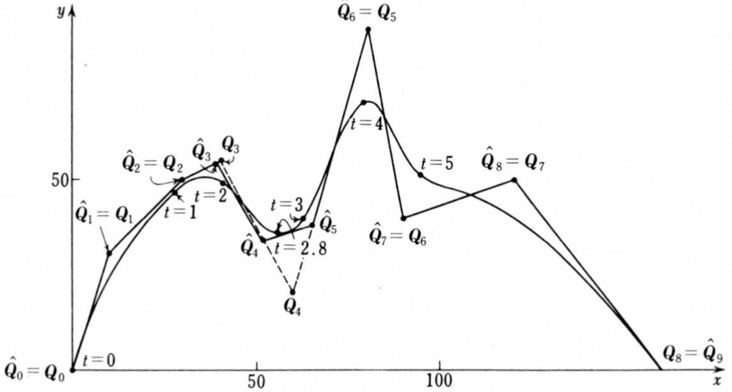

Fig. 6.51. Insertion of a knot at $t=2.8$ into the curve of Example 6.4 (Example 6.10)

$$\begin{aligned}\hat{Q}_3 &= (1-\alpha_3)\,Q_2+\alpha_3\,Q_3\\ &= (1-0.933)\,[30\ 50]+0.933\,[40\ 55]\\ &= [39.33\ 54.67]\\ \hat{Q}_4 &= (1-\alpha_4)\,Q_3+\alpha_4\,Q_4\\ &= (1-0.6)\,[40\ 55]+0.6\,[60\ 20]\\ &= [52\ 34]\\ \hat{Q}_5 &= (1-\alpha_5)\,Q_4+\alpha_5\,Q_5\\ &= (1-0.267)\,[60\ 20]+0.267\,[80\ 90]\\ &= [65.34\ 38.69]\end{aligned}$$

A graph of the curve after the knot is inserted is shown in Fig. 6.51.

6.15 Curve Generation by Geometrical Processing [31)]

G.M. Chaikin announced a method for generating curves procedurally, using a very simple geometrical algorithm, without depending on mathematical representations. This can be easily understood from DeBoor's algorithm in Sect. 6.13.

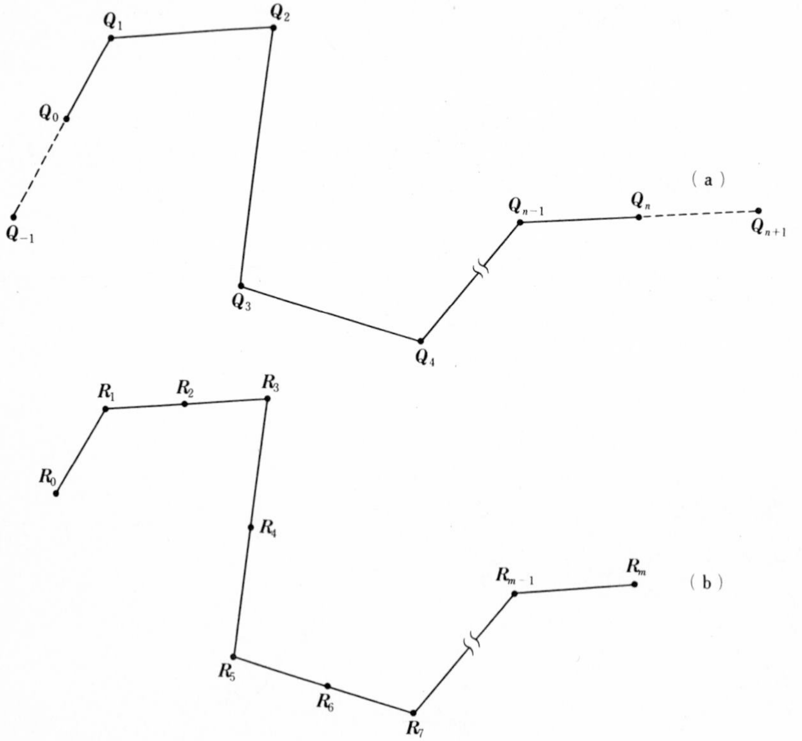

Fig. 6.52. Diagram explaining Chaikin's algorithm (1)

Suppose that a polygon $Q_0, Q_1, \ldots, Q_n$ is given, as shown in Fig. 6.52(a). Find the midpoints of all of the sides except for the two end sides, and label the polygon formed from these midpoints $R_0, R_1, \ldots, R_{2n-2}$ (Fig. 6.52(b)). Figure 6.53(a) shows just the vertex $R_i(B)$ of the initially given polygon and the midpoints $R_{i-1}(A)$ and $R_{i+1}(C)$ of the two sides attached to it.

Letting P be the midpoint of the line segment formed by connecting the midpoints of sides AB and BC, we have:

$$P = \frac{\dfrac{A+B}{2} + \dfrac{B+C}{2}}{2} = \frac{A+2B+C}{4}. \tag{6.138}$$

This point P becomes one point on the curve being generated. This division produces a set of 2 sides on either side of point P. Equation (6.138) can be applied to both of these sides. We push the set of 2 sides on the right side, for example, into the stack and then apply Eq. (6.138) again to the set of 2 sides on the left. This operation is repeated, pushing the set of 2 sides on the right into the stack each time, until the length of the set of 2 sides on the left reaches a certain limit. At this point the contents of the stack are poped up and similar

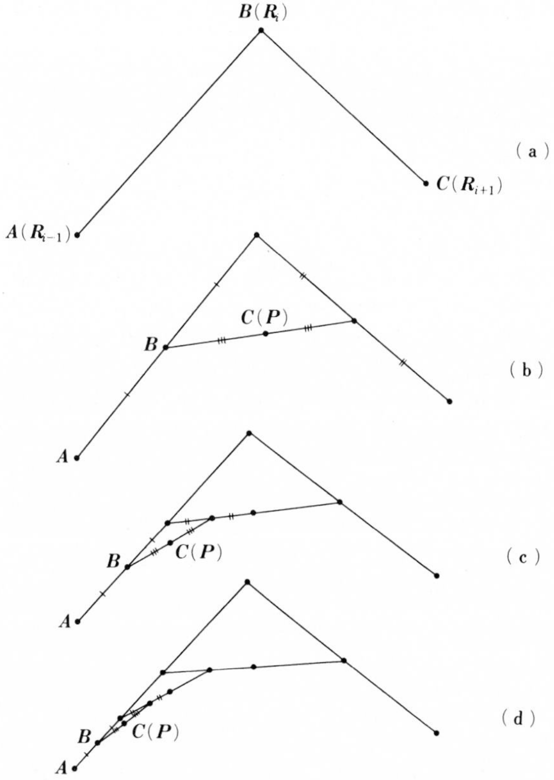

Fig. 6.53. Diagram explaining Chaikin's algorithm (2)

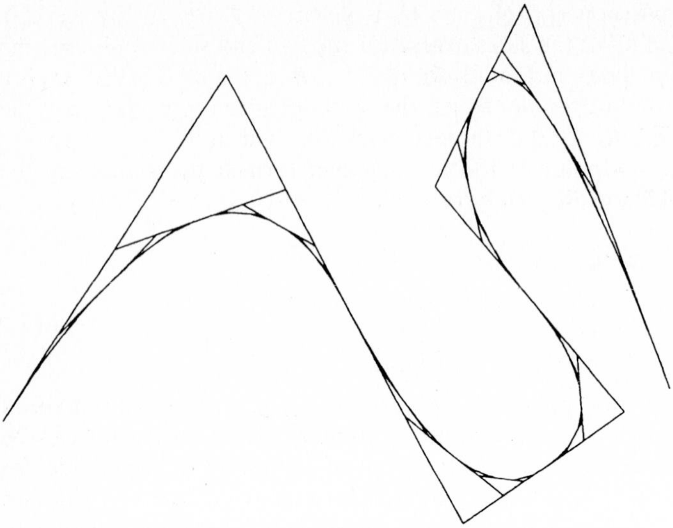

Fig. 6.54. Example of a curve generated by Chaikin's method

operations are performed on them until the stack is empty. This generates a sequence of lines on the curve corresponding to $\boldsymbol{R}_{i-1}$, $\boldsymbol{R}_i$, $\boldsymbol{R}_{i+1}$. If the above processing is performed for all groups of $\boldsymbol{R}_{i-1}$, $\boldsymbol{R}_i$, $\boldsymbol{R}_{i+1}$ ($i=1, 3, \ldots, 2n-3$), a curve corresponding to the given polygon $\boldsymbol{Q}_0$, $\boldsymbol{Q}_1$, ..., $\boldsymbol{Q}_n$ is generated. An example of such a curve is shown in Fig. 6.54.

As can be understood from the curve generation process, the curve is completely determined by only the 3 points $\boldsymbol{R}_{i-1}$, $\boldsymbol{R}_i$, $\boldsymbol{R}_{i+1}$. The curve segment has one end at $\boldsymbol{R}_{i-1}$ and the other end at $\boldsymbol{R}_{i+1}$. Its slope at $\boldsymbol{R}_{i-1}$ is in the direction of the line segment $\boldsymbol{R}_{i-1}\boldsymbol{R}_i$, and its slope at $\boldsymbol{R}_{i+1}$ is in the direction of the line segment $\boldsymbol{R}_i\boldsymbol{R}_{i+1}$. Since the curve passes through the point $\boldsymbol{P}$ calculated with Eq. (6.138), it is clear that this curve is a parabola (refer to Sect. 7.3). Let us find the equation of this parabola. Writing the curve function in the form

$$\boldsymbol{P}(t) = [t^2 \ \ t \ \ 1] \begin{bmatrix} \boldsymbol{A} \\ \boldsymbol{B} \\ \boldsymbol{C} \end{bmatrix} \tag{6.139}$$

we obtain:

$$\dot{\boldsymbol{P}}(t) = [2t \ \ 1 \ \ 0] \begin{bmatrix} \boldsymbol{A} \\ \boldsymbol{B} \\ \boldsymbol{C} \end{bmatrix}. \tag{6.140}$$

Since we have the relations $\boldsymbol{P}(0)=\boldsymbol{R}_{i-1}$, $\boldsymbol{P}(1)=\boldsymbol{R}_{i+1}$, $\dot{\boldsymbol{P}}(0)=p(\boldsymbol{R}_i-\boldsymbol{R}_{i-1})$, $\dot{\boldsymbol{P}}(1) = q(\boldsymbol{R}_{i+1}-\boldsymbol{R}_i)$, the following equation is obtained:

$$\begin{bmatrix} \boldsymbol{R}_{i-1} \\ \boldsymbol{R}_{i+1} \\ p(\boldsymbol{R}_i - \boldsymbol{R}_{i-1}) \\ q(\boldsymbol{R}_{i+1} - \boldsymbol{R}_i) \end{bmatrix} = \begin{bmatrix} 0 & 0 & 1 \\ 1 & 1 & 1 \\ 0 & 1 & 0 \\ 2 & 1 & 0 \end{bmatrix} \begin{bmatrix} \boldsymbol{A} \\ \boldsymbol{B} \\ \boldsymbol{C} \end{bmatrix}. \tag{6.141}$$

Converting the first matrix on the right-hand side into a square matrix gives:

$$\begin{bmatrix} \boldsymbol{R}_{i-1} \\ \boldsymbol{R}_{i+1} \\ p(\boldsymbol{R}_i - \boldsymbol{R}_{i-1}) \\ q(\boldsymbol{R}_{i+1} - \boldsymbol{R}_i) \end{bmatrix} = \begin{bmatrix} 0 & 0 & 1 & 1 \\ 1 & 1 & 1 & 1 \\ 0 & 1 & 0 & 1 \\ 2 & 1 & 0 & 1 \end{bmatrix} \begin{bmatrix} \boldsymbol{A} \\ \boldsymbol{B} \\ \boldsymbol{C} \\ \boldsymbol{O} \end{bmatrix}. \tag{6.142}$$

This implies:

$$\begin{bmatrix} \boldsymbol{A} \\ \boldsymbol{B} \\ \boldsymbol{C} \\ \boldsymbol{O} \end{bmatrix} = \begin{bmatrix} 0 & 0 & 1 & 1 \\ 1 & 1 & 1 & 1 \\ 0 & 1 & 0 & 1 \\ 2 & 1 & 0 & 1 \end{bmatrix}^{-1} \begin{bmatrix} \boldsymbol{R}_{i-1} \\ \boldsymbol{R}_{i+1} \\ p(\boldsymbol{R}_i - \boldsymbol{R}_{i-1}) \\ q(\boldsymbol{R}_{i+1} - \boldsymbol{R}_i) \end{bmatrix}$$

$$= \begin{bmatrix} 0 & 0 & -\frac{1}{2} & \frac{1}{2} \\ -1 & 1 & \frac{1}{2} & -\frac{1}{2} \\ 0 & 1 & -\frac{1}{2} & -\frac{1}{2} \\ 1 & -1 & \frac{1}{2} & \frac{1}{2} \end{bmatrix} \begin{bmatrix} \boldsymbol{R}_{i-1} \\ \boldsymbol{R}_{i+1} \\ p(\boldsymbol{R}_i - \boldsymbol{R}_{i-1}) \\ q(\boldsymbol{R}_{i+1} - \boldsymbol{R}_i) \end{bmatrix}$$

$$= \begin{bmatrix} \frac{p}{2}\boldsymbol{R}_{i-1} - \frac{1}{2}(p+q)\boldsymbol{R}_i + \frac{q}{2}\boldsymbol{R}_{i+1} \\ -\left(1+\frac{p}{2}\right)\boldsymbol{R}_{i-1} + \frac{1}{2}(p+q)\boldsymbol{R}_i + \left(1-\frac{q}{2}\right)\boldsymbol{R}_{i+1} \\ \frac{p}{2}\boldsymbol{R}_{i-1} + \frac{1}{2}(q-p)\boldsymbol{R}_i + \left(1-\frac{q}{2}\right)\boldsymbol{R}_{i+1} \\ \left(1-\frac{p}{2}\right)\boldsymbol{R}_{i-1} + \frac{1}{2}(p-q)\boldsymbol{R}_i - \left(1-\frac{q}{2}\right)\boldsymbol{R}_{i+1} \end{bmatrix} \tag{6.143}$$

Equating the 4-th rows of the matrices on both sides gives:

$$\left(1-\frac{p}{2}\right)\boldsymbol{R}_{i-1} + \frac{1}{2}(p-q)\boldsymbol{R}_i - \left(1-\frac{q}{2}\right)\boldsymbol{R}_{i+1} = \boldsymbol{O} \tag{6.144}$$

which implies:

$$p=q=2.$$

Therefore:

$$\begin{bmatrix} A \\ B \\ C \end{bmatrix} = \begin{bmatrix} R_{i-1}-2R_i+R_{i+1} \\ 2R_i-2R_{i-1} \\ R_{i-1} \end{bmatrix} \tag{6.145}$$

so the parabola function $P(t)$ becomes:

$$P(t)=[t^2 \ t \ 1]\begin{bmatrix} A \\ B \\ C \end{bmatrix}$$

$$=[t^2 \ t \ 1]\begin{bmatrix} R_{i-1}-2R_i+R_{i+1} \\ 2R_i-2R_{i-1} \\ R_{i-1} \end{bmatrix}$$

$$=[t^2 \ t \ 1]\begin{bmatrix} 1 & -2 & 1 \\ -2 & 2 & 0 \\ 1 & 0 & 0 \end{bmatrix}\begin{bmatrix} R_{i-1} \\ R_i \\ R_{i+1} \end{bmatrix}$$

$$=[(1-t)^2 \ 2t(1-t) \ t^2]\begin{bmatrix} R_{i-1} \\ R_i \\ R_{i+1} \end{bmatrix}$$

$$\equiv[B_{0,2}(t) \ B_{1,2}(t) \ B_{2,2}(t)]\begin{bmatrix} R_{i-1} \\ R_i \\ R_{i+1} \end{bmatrix} \quad (i=1,3,\ldots,2n-3) \tag{6.146}$$

It is clear that $P(t)$ is a quadratic Bézier curve corresponding to the sequence of vertices R_{i-1}, R_i, R_{i-1}. It is easy to confirm that the following relation holds in this case (Fig. 6.55):

$$P\left(\frac{1}{2}\right)=\frac{R_{i-1}+2R_i+R_{i+1}}{4}, \quad \dot{P}\left(\frac{1}{2}\right)=R_{i+1}-R_{i-1}.$$

Let us now convert this into an expression in terms of the initially given sequence of vertices R_{i-2}, R_i, R_{i+2} rather than R_{i-1}, R_i, R_{i+1}. Since we have the relations:

$$R_{i-1}=\frac{R_{i-2}+R_i}{2},$$

$$R_{i+1}=\frac{R_i+R_{i+2}}{2}$$

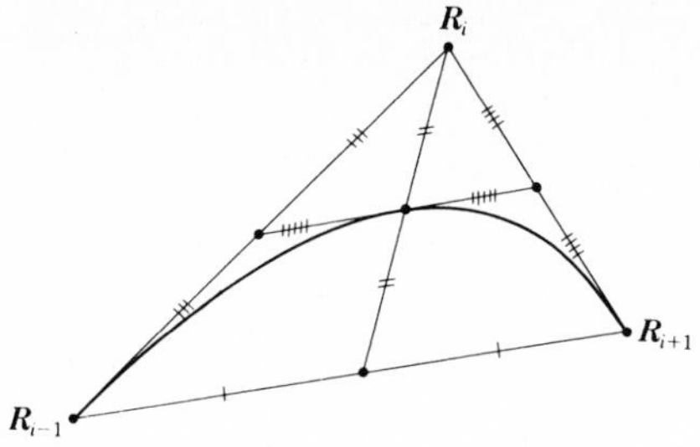

Fig. 6.55.
Relation between Chaikin's curve and polygon

the curve becomes:

$$P(t) = [(1-t)^2 \;\; 2t(1-t) \;\; t^2] \begin{bmatrix} R_{i-1} \\ R_i \\ R_{i+1} \end{bmatrix}$$

$$= [(1-t)^2 \;\; 2t(1-t) \;\; t^2] \begin{bmatrix} \frac{1}{2} & \frac{1}{2} & 0 \\ 0 & 1 & 0 \\ 0 & \frac{1}{2} & \frac{1}{2} \end{bmatrix} \begin{bmatrix} R_{i-2} \\ R_i \\ R_{i+2} \end{bmatrix}$$

$$= \left[\frac{1}{2}(1-t)^2 \;\; -t^2+t+\frac{1}{2} \;\; \frac{1}{2}t^2 \right] \begin{bmatrix} R_{i-2} \\ R_i \\ R_{i+2} \end{bmatrix}$$

$$\equiv [N_{0,3}(t) \;\; N_{1,3}(t) \;\; N_{2,3}(t)] \begin{bmatrix} R_{i-2} \\ R_i \\ R_{i+2} \end{bmatrix} \qquad (6.147)$$

(refer to Eqs. (6.17) and (6.18)). This means that the function expresses a uniform B-spline curve of type (1) and order 3 with respect to $Q_{-1}, Q_1, Q_2, \ldots, Q_{n-1}, Q_{n+1}$ $(Q_{-1} = 2Q_0 - Q_1,\ Q_{n+1} = 2Q_n - Q_{n-1})$.

6.16 Interpolation of a Sequence of Points with a *B*-Spline Curve

Consider the generation of a B-spline curve (curve type (3)) by interpolation of a sequence of points $P_0, P_1, \ldots, P_n$. The following discussion is for the case of $M=4$ (degree 3).

The relation among the points to be interpolated, interior knots and extended knots (with * symbol) is shown in Fig. 6.56. From property ⑦ of a B-spline function, there are 3 nonzero B-spline functions at each knot. We call the

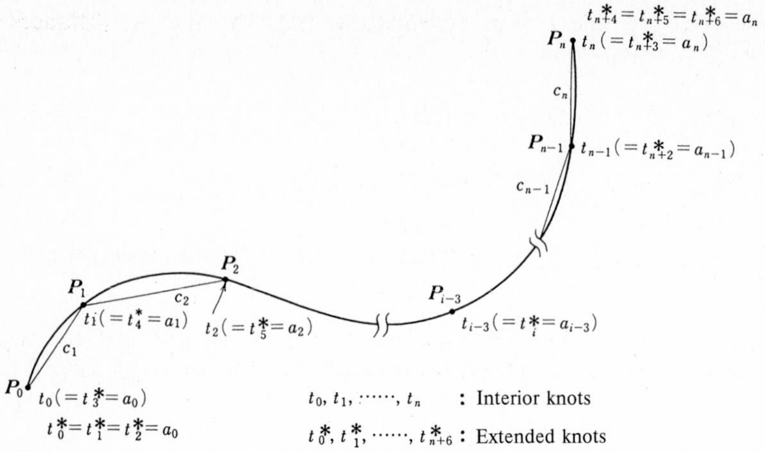

Fig. 6.56. Interpolation of a sequence of points with a B-spline curve (curve type (3)) (case of $M=4$)

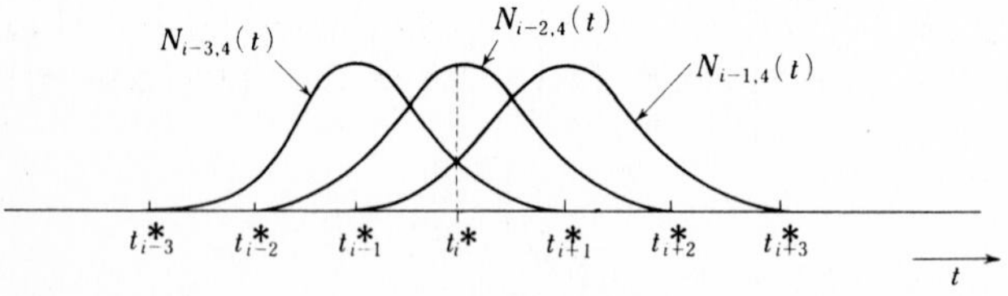

Fig. 6.57. Nonzero B-spline functions at $t=t^*_i$

3 nonzero functions at knot t^*_i, $N_{i-3,4}(t)$, $N_{i-2,4}(t)$ and $N_{i-1,4}(t)$ (refer to Fig. 6.57). Therefore, in order for the curve to interpolate point $\boldsymbol{P}_{i-3}$ at knot t^*_i, the following equation must hold:

$$N_{i-3,4}(a_{i-3})\boldsymbol{Q}_{i-3}+N_{i-2,4}(a_{i-3})\boldsymbol{Q}_{i-2}+N_{i-1,4}(a_{i-3})\boldsymbol{Q}_{i-1}=\boldsymbol{P}_{i-3}$$
$$(3\leqq i\leqq n+3). \tag{6.148}$$

This equation represents $n+1$ conditions, but there are $n+3$ unknown vectors $\boldsymbol{Q}_0$, $\boldsymbol{Q}_1$, ..., $\boldsymbol{Q}_{n+2}$, so we are 2 conditions short. So it is necessary to add 2 conditions. We will set the 2-nd derivative vectors of the generated curve $\boldsymbol{P}(t)$ equal to $\boldsymbol{O}$ at both ends:
at knot t^*_3:

$$\ddot{N}_{0,4}(a_0)\boldsymbol{Q}_0+\ddot{N}_{1,4}(a_0)\boldsymbol{Q}_1+\ddot{N}_{2,4}(a_0)\boldsymbol{Q}_2=\boldsymbol{O} \tag{6.149}$$

at knot t^*_{n+3}:

$$\ddot{N}_{n,4}(a_0)\boldsymbol{Q}_n+\ddot{N}_{n+1,4}(a_n)\boldsymbol{Q}_{n+1}+\ddot{N}_{n+2,4}(a_n)\boldsymbol{Q}_{n+2}=\boldsymbol{O}. \tag{6.150}$$

For values of extended knots, it is sufficient to take the sum of distances between points (Fig. 6.56):

$$
\begin{aligned}
t_i^* &= a_0 = 0 && (i=0, 1, 2, 3) \\
t_{i+4}^* &= a_{i+1} = \sum_{j=1}^{i+1} c_j && (i=0, 1, \ldots, n-2) \\
t_{i+n+3}^* &= a_n = \sum_{j=1}^{n} c_j && (i=0, 1, 2, 3)
\end{aligned}
\tag{6.151}
$$

$\boldsymbol{Q}_0, \boldsymbol{Q}_1, \ldots, \boldsymbol{Q}_{n+2}$ can be found by solving Eqs. (6.148), (6.149) and (6.150), thus determining the *B*-spline curve that passes through the specified points $\boldsymbol{P}_0, \boldsymbol{P}_1, \ldots, \boldsymbol{P}_n$.

6.17 Matrix Expression of *B*-Spline Curves

Let us espress the *B*-spline curve segment given by function (6.102) by a matrix, as we expressed a Bézier curve segment in Sect. 5.1.5 ③[23]. If we set $M=n+1$ in order to establish a correspondence with a Bézier curve, the curve segment (6.102) becomes:

$$
\begin{aligned}
\boldsymbol{P}_i(t) &= [N_{0,n+1}(t)\ N_{1,n+1}(t)\ \ldots\ N_{n,n+1}(t)] \begin{bmatrix} \boldsymbol{Q}_{i-1} \\ \boldsymbol{Q}_i \\ \vdots \\ \boldsymbol{Q}_{i+n-1} \end{bmatrix} \\
&= [t^n\ t^{n-1}\ \ldots\ t\ 1]\,\gamma \begin{bmatrix} \boldsymbol{Q}_{i-1} \\ \boldsymbol{Q}_i \\ \vdots \\ \boldsymbol{Q}_{i+n-1} \end{bmatrix}
\end{aligned}
\tag{6.152}
$$

where γ is an $(n+1)\times(n+1)$ matrix.

Setting $M=n+1$ in formula (6.101) gives:

$$
N_{j,n+1}(t) = \frac{1}{n!} \sum_{i=0}^{n+1} (-1)^i \binom{n+1}{i} (t+n-i-j)_+^n \quad (0 \leqq t \leqq 1). \tag{6.153}
$$

Therefore, we have:

$$
\begin{aligned}
&[N_{0,n+1}(t)\ N_{1,n+1}(t)\ N_{2,n+1}(t)\ \ldots\ N_{n,n+1}(t)] \\
&= \left[\frac{1}{n!} \sum_{i=0}^{n+1} (-1)^i \binom{n+1}{i} (t+n-i)_+^n \quad \frac{1}{n!} \sum_{i=0}^{n+1} (-1)^i \binom{n+1}{i} (t+n-i-1)_+^n \right. \\
&\qquad \left. \ldots\ \frac{1}{n!} \sum_{i=0}^{n+1} (-1)^i \binom{n+1}{i} (t-i)_+^n \right] =
\end{aligned}
$$

$$= [(t+n)^n \ (t+n-1)^n \ \dots \ t^n] \frac{1}{n!}$$

$$\times \begin{bmatrix} (-1)^0\binom{n+1}{0} & 0 & \dots & 0 & 0 \\ (-1)^1\binom{n+1}{i} & (-1)^0\binom{n+1}{0} & \dots & 0 & 0 \\ \vdots & \vdots & & \vdots & \vdots \\ \vdots & \vdots & \dots & (-1)^0\binom{n+1}{0} & 0 \\ (-1)^n\binom{n+1}{n} & (-1)^{n-1}\binom{n+1}{n-1} & \dots & (-1)^1\binom{n+1}{1} & (-1)^0\binom{n+1}{0} \end{bmatrix}$$

$$= \left[\sum_{i=0}^{n} \binom{n}{i} t^{n-i} n^i \ \sum_{i=0}^{n} \binom{n}{i} t^{n-i}(n-1)^i \ \dots \ t^n \right] \frac{1}{n!}$$

$$\times \begin{bmatrix} (-1)^0\binom{n+1}{0} & 0 & \dots & 0 & 0 \\ (-1)^1\binom{n+1}{i} & (-1)^0\binom{n+1}{0} & \dots & 0 & 0 \\ \vdots & \vdots & & \vdots & \vdots \\ \vdots & \vdots & \dots & (-1)^0\binom{n+1}{0} & 0 \\ (-1)^n\binom{n+1}{n} & (-1)^{n-1}\binom{n+1}{n-1} & \dots & (-1)^1\binom{n+1}{1} & (-1)^0\binom{n+1}{0} \end{bmatrix}$$

$$= [t^n \ t^{n-1} \ \dots \ t \ 1] \frac{1}{n!} \begin{bmatrix} \binom{n}{0} n^0 & \binom{n}{0}(n-1)^0 & \dots & \binom{n}{0} 1^0 & \binom{n}{0} \\ \binom{n}{1} n^1 & \binom{n}{1}(n-1)^1 & \dots & \binom{n}{1} 1^1 & 0 \\ \vdots & \vdots & \vdots & \vdots & \vdots \\ \binom{n}{n} n^n & \binom{n}{n}(n-1)^n & \dots & \binom{n}{n} 1^n & 0 \end{bmatrix} =$$

$$\times\begin{bmatrix} (-1)^0\binom{n+1}{0} & 0 & \dots & 0 & 0 \\ (-1)^1\binom{n+1}{1} & (-1)^0\binom{n+1}{0} & \dots & 0 & 0 \\ \vdots & \vdots & & \vdots & \vdots \\ \vdots & \vdots & \dots & (-1)^0\binom{n+1}{0} & 0 \\ (-1)^n\binom{n+1}{n} & (-1)^{n-1}\binom{n+1}{n-1} & \dots & (-1)^1\binom{n+1}{1} & (-1)^0\binom{n+1}{0} \end{bmatrix}$$

$$\equiv [t^n \; t^{n-1} \; \dots \; t \; 1]\,\gamma$$

$$\gamma = \{\gamma_{ij}\} \quad (i, j = 0, 1, 2, \dots, n)$$

$$\gamma_{ij} = \frac{1}{n!}\binom{n}{i}\sum_{k=j}^{n}(n-k)^i(-1)^{k-j}\binom{n+1}{k-j} \tag{6.154}$$

6.18 Expression of the Functions $C_{0,0}(t)$, $C_{0,1}(t)$, $C_{1,0}(t)$ and $C_{1,1}(t)$ by *B*-Spline Functions[28)]

The blending functions that we used to make the higher-order derivative vectors of a Coons surface continuous (refer to Sects. 3.3.2 and 3.3.3) can be expressed using *B*-spline curves.

In the following discussion we consider functions which satisfy the blending function conditions that will make the 2nd derivative vectors continuous:

$$\left.\begin{aligned} &[C_{0,0}(0)\; C_{0,0}(1)\; \dot{C}_{0,0}(0)\; \dot{C}_{0,0}(1)\; \ddot{C}_{0,0}(0)\; \ddot{C}_{0,0}(1)] \\ &\quad = [1\;0\;0\;0\;0\;0] \\ &[C_{0,1}(0)\; C_{0,1}(1)\; \dot{C}_{0,1}(0)\; \dot{C}_{0,1}(1)\; \ddot{C}_{0,1}(0)\; \ddot{C}_{0,1}(1)] \\ &\quad = [0\;1\;0\;0\;0\;0] \\ &[C_{1,0}(0)\; C_{1,0}(1)\; \dot{C}_{1,0}(0)\; \dot{C}_{1,0}(1)\; \ddot{C}_{1,0}(0)\; \ddot{C}_{1,0}(1)] \\ &\quad = [0\;0\;1\;0\;0\;0] \\ &[C_{1,1}(0)\; C_{1,1}(1)\; \dot{C}_{1,1}(0)\; \dot{C}_{1,1}(1)\; \ddot{C}_{1,1}(0)\; \ddot{C}_{1,1}(1)] \\ &\quad = [0\;0\;0\;1\;0\;0] \end{aligned}\right\} \tag{6.155}$$

If we wish to derive ordinary polynomials which will satisfy these conditions, they will be 5-th degree polynomials such as those derived in Sect. 3.2.3.

Let us express $C_{0,0}(t)$, $C_{0,1}(t)$, $C_{1,0}(t)$ and $C_{1,1}(t)$ as uniform cubic *B*-spline curves of type (1) (refer to Sect. 6.1.1). Take the coordinates of $\boldsymbol{Q}_0, \boldsymbol{Q}_1, \dots, \boldsymbol{Q}_5$ in Fig. 6.58 to be:

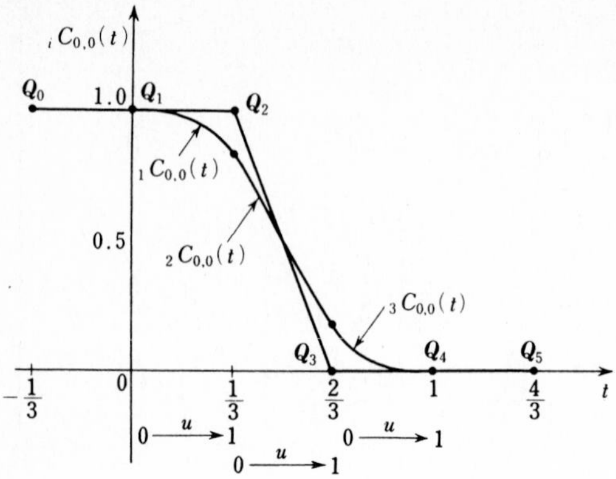

Fig. 6.58. Expression of the function $C_{0,0}(t)$ in terms of *B*-spline functions

$$\begin{bmatrix} Q_0 \\ Q_1 \\ Q_2 \\ Q_3 \\ Q_4 \\ Q_5 \end{bmatrix} = \begin{bmatrix} -\frac{1}{3} & 1 \\ 0 & 1 \\ \frac{1}{3} & 1 \\ \frac{2}{3} & 0 \\ 1 & 0 \\ \frac{4}{3} & 0 \end{bmatrix}$$

In this case, 3 curve segments are produced by the sequence of vertices Q_0, Q_1, Q_2, Q_3, Q_4, Q_5. From the properties of a *B*-spline curve, it is clear that the curve which is made up of these segments satisfies the condition on function $C_{0,0}(t)$ in Eq. (6.155). Let us write these 3 functions as $_1C_{0,0}(t)$, $_2C_{0,0}(t)$ and $_3C_{0,0}(t)$.

Now we introduce the local parameter $u\,(0 \leqq u \leqq 1)$ shown in Fig. 6.58. The curve segment $\boldsymbol{P}_i(u)$ $(i=1, 2, 3)$ formed by the vertices Q_{i-1}, Q_i, Q_{i+1} and Q_{i+2} is:

$$\boldsymbol{P}_i(u) = [t \;\; _iC_{0,0}(t)]$$

$$= \frac{1}{6}[u^3 \; u^2 \; u \; 1] \begin{bmatrix} -1 & 3 & -3 & 1 \\ 3 & -6 & 3 & 0 \\ -3 & 0 & 3 & 0 \\ 1 & 4 & 1 & 0 \end{bmatrix} \begin{bmatrix} \frac{1}{3}(i-2) & Q_{i-1,y} \\ \frac{1}{3}(i-1) & Q_{i,y} \\ \frac{1}{3}i & Q_{i+1,y} \\ \frac{1}{3}(i+1) & Q_{i+2,y} \end{bmatrix} \tag{6.156}$$

$$t = \frac{1}{3}(u-1+i). \tag{6.157}$$

From Eq. (6.156) we have:

$$[{}_1C_{0,0}(t)\ {}_2C_{0,0}(t)\ {}_3C_{0,0}(t)]$$

$$=\frac{1}{6}[u^3\ u^2\ u\ 1]\begin{bmatrix}-1 & 3 & -3 & 1\\ 3 & -6 & 3 & 0\\ -3 & 0 & 3 & 0\\ 1 & 4 & 1 & 0\end{bmatrix}\begin{bmatrix}Q_{0,y} & Q_{1,y} & Q_{2,y}\\ Q_{1,y} & Q_{2,y} & Q_{3,y}\\ Q_{2,y} & Q_{3,y} & Q_{4,y}\\ Q_{3,y} & Q_{4,y} & Q_{5,y}\end{bmatrix}$$

$$=\frac{1}{6}[u^3\ u^2\ u\ 1]\begin{bmatrix}-1 & 3 & -3 & 1\\ 3 & -6 & 3 & 0\\ -3 & 0 & 3 & 0\\ 1 & 4 & 1 & 0\end{bmatrix}\begin{bmatrix}1 & 1 & 1\\ 1 & 1 & 0\\ 1 & 0 & 0\\ 0 & 0 & 0\end{bmatrix}$$

$$=\frac{1}{6}[u^3\ u^2\ u\ 1]\begin{bmatrix}-1 & 2 & -1\\ 0 & -3 & 3\\ 0 & -3 & -3\\ 6 & 5 & 1\end{bmatrix}$$

$$\left(0\leqq u\leqq 1;\ t=\frac{1}{3}(u-1+i)\right). \tag{6.158}$$

Similarly, by expressing the position vectors in Fig. 6.59 as:

$$\begin{bmatrix}\boldsymbol{Q}_0\\ \boldsymbol{Q}_1\\ \boldsymbol{Q}_2\\ \boldsymbol{Q}_3\\ \boldsymbol{Q}_4\\ \boldsymbol{Q}_5\end{bmatrix}=\begin{bmatrix}-\frac{1}{3} & 0\\ 0 & 0\\ \frac{1}{3} & 0\\ \frac{2}{3} & 1\\ 1 & 1\\ \frac{4}{3} & 1\end{bmatrix}$$

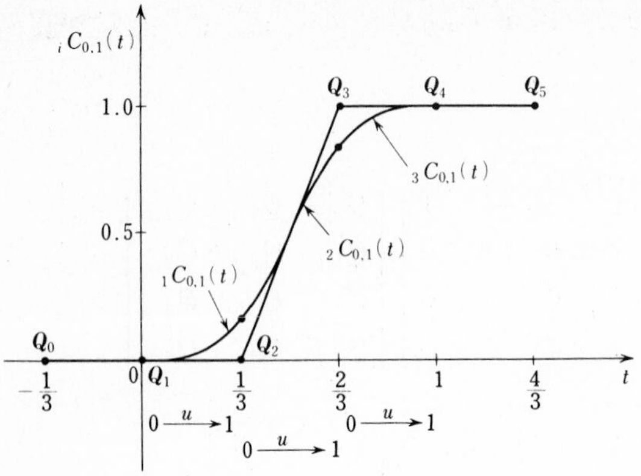

Fig. 6.59. Expression of the function $C_{0,1}(t)$ in terms of B-spline functions

$C_{0,1}(t)$ is expressed as a function consisting of the 3 functions $_1C_{0,1}(t)$, $_2C_{0,1}(t)$ and $_3C_{0,1}(t)$ connected together. The forms of these functions are as follows:

$$[{}_1C_{0,1}(t)\ \ {}_2C_{0,1}(t)\ \ {}_3C_{0,1}(t)]$$

$$=\frac{1}{6}[u^3\ u^2\ u\ 1]\begin{bmatrix}-1 & 3 & -3 & 1\\ 3 & -6 & 3 & 0\\ -3 & 0 & 3 & 0\\ 1 & 4 & 1 & 0\end{bmatrix}\begin{bmatrix}0 & 0 & 0\\ 0 & 0 & 1\\ 0 & 1 & 1\\ 1 & 1 & 1\end{bmatrix}$$

$$=\frac{1}{6}[u^3\ u^2\ u\ 1]\begin{bmatrix}1 & -2 & 1\\ 0 & 3 & -3\\ 0 & 3 & 3\\ 0 & 1 & 5\end{bmatrix}$$

$$\left(0\leqq u\leqq 1;\ t=\frac{1}{3}(u-1+i)\right). \tag{6.159}$$

Similarly, by setting:

$$\begin{bmatrix}Q_0\\ Q_1\\ Q_2\\ Q_3\\ Q_4\\ Q_5\end{bmatrix}=\begin{bmatrix}-\frac{1}{3} & -\frac{1}{3}\\ 0 & 0\\ \frac{1}{3} & \frac{1}{3}\\ \frac{2}{3} & 0\\ 1 & 0\\ \frac{4}{3} & 0\end{bmatrix} \qquad \begin{bmatrix}Q_0\\ Q_1\\ Q_2\\ Q_3\\ Q_4\\ Q_5\end{bmatrix}=\begin{bmatrix}-\frac{1}{3} & 0\\ 0 & 0\\ \frac{1}{3} & 0\\ \frac{2}{3} & -\frac{1}{3}\\ 1 & 0\\ \frac{4}{3} & \frac{1}{3}\end{bmatrix}$$

respectively (Figs. 6.60, 6.61), the functions $C_{1,0}(t)$ and $C_{1,1}(t)$ are expressed in terms of the following:

$$[{}_1C_{1,0}(t)\ {}_2C_{1,0}(t)\ {}_3C_{1,0}(t)] = \frac{1}{6}[u^3\ u^2\ u\ 1]\begin{bmatrix} -1 & 3 & -3 & 1 \\ 3 & -6 & 3 & 0 \\ -3 & 0 & 3 & 0 \\ 1 & 4 & 1 & 0 \end{bmatrix}$$

$$\times \begin{bmatrix} -\frac{1}{3} & 0 & \frac{1}{3} \\ 0 & \frac{1}{3} & 0 \\ \frac{1}{3} & 0 & 0 \\ 0 & 0 & 0 \end{bmatrix}$$

$$= \frac{1}{18}[u^3\ u^2\ u\ 1]\begin{bmatrix} -2 & 3 & -1 \\ 0 & -6 & 3 \\ 6 & 0 & -3 \\ 0 & 4 & 1 \end{bmatrix}$$

$$\left(0 \leqq u \leqq 1;\ t = \frac{1}{3}(u-1+i)\right) \tag{6.160}$$

and:

$$[{}_1C_{1,1}(t)\ {}_2C_{1,1}(t)\ {}_3C_{1,1}(t)] = \frac{1}{6}[u^3\ u^2\ u\ 1]\begin{bmatrix} -1 & 3 & -3 & 1 \\ 3 & -6 & 3 & 0 \\ -3 & 0 & 3 & 0 \\ 1 & 4 & 1 & 0 \end{bmatrix} \times$$

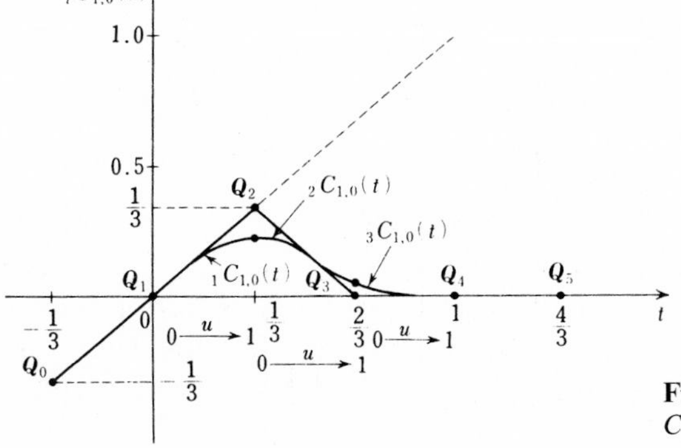

Fig. 6.60. Expression of the function $C_{1,0}(t)$ in terms of B-spline functions

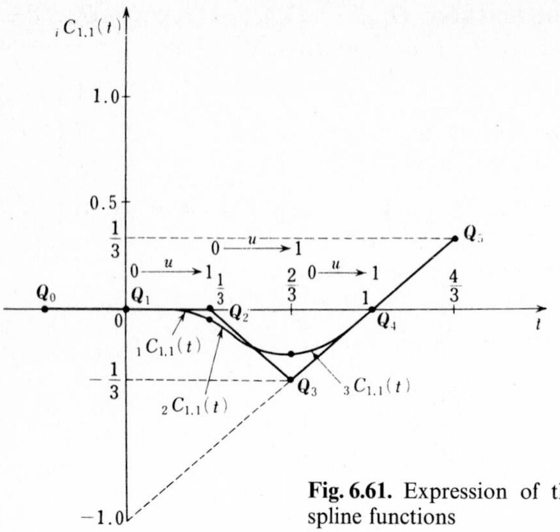

Fig. 6.61. Expression of the function $C_{1,1}(t)$ in terms of *B*-spline functions

$$\times\begin{bmatrix} 0 & 0 & 0 \\ 0 & 0 & -\dfrac{1}{3} \\ 0 & -\dfrac{1}{3} & 0 \\ -\dfrac{1}{3} & 0 & \dfrac{1}{3} \end{bmatrix}$$

$$=\frac{1}{18}[u^3 \; u^2 \; u \; 1]\begin{bmatrix} -1 & 3 & -2 \\ 0 & -3 & 6 \\ 0 & -3 & 0 \\ 0 & -1 & -4 \end{bmatrix}$$

$$\left(0 \leqq u \leqq 1;\; t=\frac{1}{3}(u-1+i)\right). \tag{6.161}$$

6.19 General *B*-Spline Surfaces

B-spline surfaces can be defined using *B*-spline basis functions, just as Bézier surfaces can be defined using Bernstein basis functions. The following discussion introduces the most frequently used type of expression, of *B*-spline surfaces in cartesian product form.

Given a lattice of position vectors $\boldsymbol{Q}_{00}, \boldsymbol{Q}_{01}, \ldots, \boldsymbol{Q}_{0n}, \ldots, \boldsymbol{Q}_{m0}, \ldots, \boldsymbol{Q}_{mn}$, the corresponding cartesian product type B-spline surface is given by:

$$\boldsymbol{P}(u, w) = \sum_{i=0}^{m} \sum_{j=0}^{n} N_{i,K}(u)\, N_{j,L}(w)\, \boldsymbol{Q}_{ij}. \tag{6.162}$$

Let $N_{i,K}(u)$ and $N_{j,L}(w)$ be the B-spline functions of order K and L, respectively, used to define a B-spline surface in type (3).

Formula (6.162) can be expressed in matrix form as:

$$\boldsymbol{P}(u, w) = [N_{0,K}(u)\ N_{1,K}(u)\ \ldots\ N_{m,K}(u)]$$

$$\times \begin{bmatrix} \boldsymbol{Q}_{00} & \boldsymbol{Q}_{01} & \cdots & \boldsymbol{Q}_{0n} \\ \boldsymbol{Q}_{10} & \boldsymbol{Q}_{11} & \cdots & \boldsymbol{Q}_{1n} \\ & \cdots & & \\ & \cdots & & \\ \boldsymbol{Q}_{m0} & & \cdots & \boldsymbol{Q}_{mn} \end{bmatrix} \begin{bmatrix} N_{0,L}(w) \\ N_{1,L}(w) \\ \vdots \\ N_{n,L}(w) \end{bmatrix}. \tag{6.163}$$

Examples of B-spline surfaces described by function (6.162) or (6.163) are shown in Fig. 6.62. Some properties are analogous to those of a Bézier surface patch, such as the fact that $\boldsymbol{Q}_{00}, \boldsymbol{Q}_{0n}, \boldsymbol{Q}_{m0}$ and $\boldsymbol{Q}_{mn}$ are the 4 corners of the surface, and the shape of the surface in the neighborhood of the 4 corners is similar to that of a Bézier surface patch. However, whereas a Bézier surface patch is a single surface patch, a B-spline surface, consists of a number of surface patches which are connected together smoothly; moreover, a B-spline surface patch is defined locally by nearby surface defining vectors. Consequently, on a B-spline surface, unlike a Bézier surface patch, the effect of varying one vector $\boldsymbol{Q}_{ij}$ is locally limited.

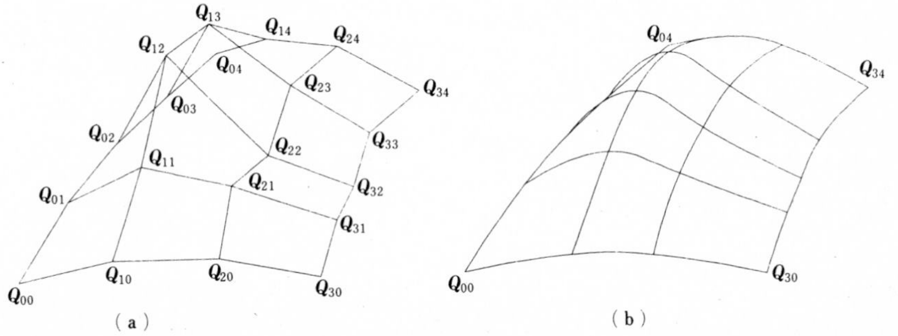

Fig. 6.62. Example of a B-spline surface. (a) surface defining net; (b) B-spline surface formed by the surface defining net in (a)

References (Chap. 6)

26) Fujio Yamaguchi and Kazumasa Konishi: A method of designing free form surfaces (in Japanese), *Reports of the Man-Machine System Study Group, Information Processing Society,* 75–19, 1975.
27) Fujio Yamaguchi: A method of designing free form surfaces by computer display (1st Report) (in Japanese) *Precision Machinery 43* (2), 1977.
28) Coons, S.A.: "Surface Patches and *B*-spline Curves", Computer Aided Geometric Design, Academic Press, 1974.
29) DeBoor, C.: "On Calculating with *B*-splines", J. Approx. Th. 6, 50–62.
30) Böhm, W.: "Inserting new knots into *B*-spline curves", Computer-Aided Design, 12–4, 1980.
31) Chaikin, G.M.: "An Algorithm for High-Speed Curve Generation", Computer Graphics and Image processing, 3, 1974.

7. The Rational Polynomial Curves

7.1 Derivation of Parametric Conic Section Curves

Conic section curves are in a mutual central projection relationship*⁾. Consequently, arbitrary conic section curves can be derived by performing a suitable affine transformation and then a central projection on one conic section curve. For the initial conic section curve, let us use the simplest one to express, the parabola shown in Fig. 7.1:

$$[x \;\; y \;\; 1] = [t^2 \;\; t \;\; 1].$$

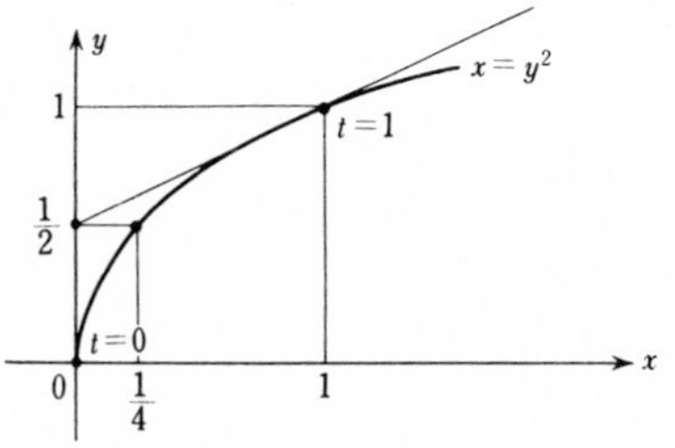

Fig. 7.1. The parabola $x = y^2$

If the conic section curve segment to be found is denoted by $\mathbf{P}(t) = [x^* \;\; y^* \;\; 1]$, the transformation relation become, with $\boldsymbol{M}$ as the matrix which performs the projection:

$$[t^2 \;\; t \;\; 1]\boldsymbol{M} = w^* \mathbf{P}(t) = [X^* \;\; Y^* \;\; w^*]. \tag{7.1}$$

As shown in Fig. 7.2, the conic section curve $\mathbf{P}(t)$ to be found passes through point $\mathbf{Q}_0$ and has a slope $\overrightarrow{\mathbf{Q}_0\mathbf{Q}_T}$ at $t=0$, and passes through point $\mathbf{Q}_1$ and has a slope $\overrightarrow{\mathbf{Q}_T\mathbf{Q}_1}$ at $t=1$. These conditions are mathematically expressed as:

$$[0 \;\; 0 \;\; 1]\; \boldsymbol{M} = w_0^* \mathbf{Q}_0$$

$$\left[0 \;\; \frac{1}{2} \;\; 1\right] \boldsymbol{M} = w_T^* \mathbf{Q}_T$$

$$[1 \;\; 1 \;\; 1]\; \boldsymbol{M} = w_1^* \mathbf{Q}_1$$

---

*⁾ Refer to the author's book *Graphics Processing Engineering* (in Japanese) (Nikkan Kogyo Shimbun Sha), see Sect. 3.3.3.

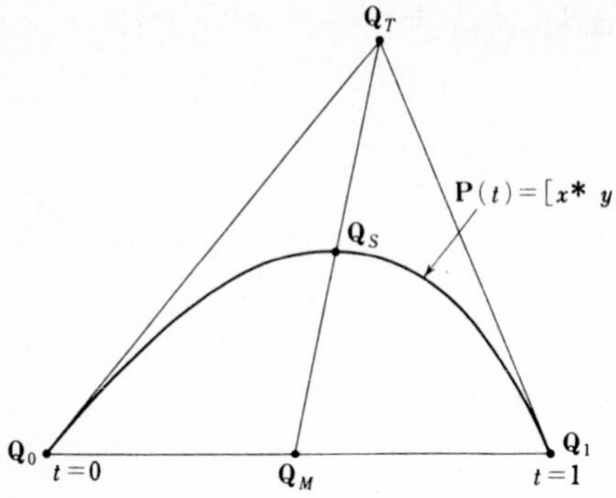

Fig. 7.2. The conic section curve $\mathbf{P}(t)$ found by a projective transformation

which becomes:

$$\begin{bmatrix} 0 & 0 & 1 \\ 0 & \frac{1}{2} & 1 \\ 1 & 1 & 1 \end{bmatrix} \boldsymbol{M} = \begin{bmatrix} w_0^* & 0 & 0 \\ 0 & w_T^* & 0 \\ 0 & 0 & w_1^* \end{bmatrix} \begin{bmatrix} \mathbf{Q}_0 \\ \mathbf{Q}_T \\ \mathbf{Q}_1 \end{bmatrix}.$$

Solving for $\boldsymbol{M}$ gives:

$$\boldsymbol{M} = \begin{bmatrix} 0 & 0 & 1 \\ 0 & \frac{1}{2} & 1 \\ 1 & 1 & 1 \end{bmatrix}^{-1} \begin{bmatrix} w_0^* & 0 & 0 \\ 0 & w_T^* & 0 \\ 0 & 0 & w_1^* \end{bmatrix} \begin{bmatrix} \mathbf{Q}_0 \\ \mathbf{Q}_T \\ \mathbf{Q}_1 \end{bmatrix}$$

$$= \begin{bmatrix} 1 & -2 & 1 \\ -2 & 2 & 0 \\ 1 & 0 & 0 \end{bmatrix} \begin{bmatrix} w_0^* & 0 & 0 \\ 0 & w_T^* & 0 \\ 0 & 0 & w_1^* \end{bmatrix} \begin{bmatrix} \mathbf{Q}_0 \\ \mathbf{Q}_T \\ \mathbf{Q}_1 \end{bmatrix}. \tag{7.2}$$

In Fig. 7.2, $\mathbf{Q}_M$ is the midpoint of the line segment connecting $\mathbf{Q}_0$ and $\mathbf{Q}_1$, and $\mathbf{Q}_S$ is the intersection of the line segment connecting $\mathbf{Q}_M$ and $\mathbf{Q}_T$ with the conic section curve.

If we assume that we have the additional condition that the conic section curve segment passes through point $\mathbf{Q}_S$ at $t=1/2$, then substituting (7.2) into (7.1) gives the following equation.

$$w_S^* \mathbf{Q}_S = \begin{bmatrix} \frac{1}{4} & \frac{1}{2} & 1 \end{bmatrix} \begin{bmatrix} 1 & -2 & 1 \\ -2 & 2 & 0 \\ 1 & 0 & 0 \end{bmatrix} \begin{bmatrix} w_0^* & 0 & 0 \\ 0 & w_T^* & 0 \\ 0 & 0 & w_1^* \end{bmatrix} \begin{bmatrix} \mathbf{Q}_0 \\ \mathbf{Q}_T \\ \mathbf{Q}_1 \end{bmatrix}$$

$$= \begin{bmatrix} \frac{1}{4} w_0^* & \frac{1}{2} w_T^* & \frac{1}{4} w_1^* \end{bmatrix} \begin{bmatrix} \mathbf{Q}_0 \\ \mathbf{Q}_T \\ \mathbf{Q}_1 \end{bmatrix}. \tag{7.3}$$

Let p be the ratio between the length of $\mathbf{Q}_M\mathbf{Q}_T$ and the length of $\mathbf{Q}_M\mathbf{Q}_S$:

$$\begin{aligned}\mathbf{Q}_S &= \mathbf{Q}_M + p(\mathbf{Q}_T - \mathbf{Q}_M)\\ &= \frac{1-p}{2}\mathbf{Q}_0 + p\mathbf{Q}_T + \frac{1-p}{2}\mathbf{Q}_1 .\end{aligned} \tag{7.4}$$

Substituting this equation into (7.3) gives:

$$w_S^*\begin{bmatrix}\frac{1-p}{2} & p & \frac{1-p}{2}\end{bmatrix}\begin{bmatrix}\mathbf{Q}_0\\ \mathbf{Q}_T\\ \mathbf{Q}_1\end{bmatrix} = \begin{bmatrix}\frac{1}{4}w_0^* & \frac{1}{2}w_T^* & \frac{1}{4}w_1^*\end{bmatrix}\begin{bmatrix}\mathbf{Q}_0\\ \mathbf{Q}_T\\ \mathbf{Q}_1\end{bmatrix}.$$

Since $\mathbf{Q}_0$, $\mathbf{Q}_T$ and $\mathbf{Q}_1$ are homogeneous coordinate vectors, $\begin{bmatrix}\mathbf{Q}_0\\ \mathbf{Q}_T\\ \mathbf{Q}_1\end{bmatrix}$ is a 3×3 matrix. Multiplying both sides of the above equation by $\begin{bmatrix}\mathbf{Q}_0\\ \mathbf{Q}_T\\ \mathbf{Q}_1\end{bmatrix}^{-1}$ from the right and setting $w_S^* = 1/2$ (the value of w_S is arbitrary), we obtain:

$$w_0^* = 1-p, \quad w_T^* = p, \quad w_1^* = 1-p .$$

Substituting these relations into Eq. (7.2), $\boldsymbol{M}$ becomes:

$$\boldsymbol{M} = \begin{bmatrix}1 & -2 & 1\\ -2 & 2 & 0\\ 1 & 0 & 0\end{bmatrix}\begin{bmatrix}1-p & 0 & 0\\ 0 & p & 0\\ 0 & 0 & 1-p\end{bmatrix}\begin{bmatrix}\mathbf{Q}_0\\ \mathbf{Q}_T\\ \mathbf{Q}_1\end{bmatrix}$$

the conic section curve segment to be found becomes:

$$\begin{aligned}w\mathbf{P}(t) &= [X \;\; Y \;\; w]\\ &= [t^2 \;\; t \;\; 1]\begin{bmatrix}1 & -2 & 1\\ -2 & 2 & 0\\ 1 & 0 & 0\end{bmatrix}\begin{bmatrix}1-p & 0 & 0\\ 0 & p & 0\\ 0 & 0 & 1-p\end{bmatrix}\begin{bmatrix}\mathbf{Q}_0\\ \mathbf{Q}_T\\ \mathbf{Q}_1\end{bmatrix}\end{aligned} \tag{7.5}$$

(the asterisks have been omitted from X^*, Y^* and w^*).

7.2 Classification of Conic Section Curves[32)]

Expressing Eq. (7.5) in ordinary coordinates gives:

$$x(t)=\frac{X(t)}{w(t)}=\frac{\{(1-p)Q_{0,x}-2pQ_{T,x}+(1-p)Q_{1,x}\}t^2-2\{(1-p)Q_{0,x}-pQ_{T,x}\}t+(1-p)Q_{0,x}}{2(1-2p)t^2-2(1-2p)t+1-p}$$

$$y(t)=\frac{Y(t)}{w(t)}=\frac{\{(1-p)Q_{0,y}-2pQ_{T,y}+(1-p)Q_{1,y}\}t^2-2\{(1-p)Q_{0,y}-pQ_{T,y}\}t+(1-p)Q_{0,y}}{2(1-2p)t^2-2(1-2p)t+1-p} \tag{7.6}$$

The discriminant D for finding the roots of the denominator $w(t)$ is:

$$D=(1-2p)^2-2(1-2p)(1-p)$$
$$=2p-1$$

so we have:

(i) When $p>\frac{1}{2}$, the denominator has different real roots t_1 and t_2:

$$t_1, t_2=\frac{1}{2}\pm\frac{1}{2\sqrt{2p-1}}.$$

Having real roots means that the curve has asymptotes. If $p>1$, the two real roots satisfy $0<t_1, t_2<1$. If $1/2<p<1$, the roots are not in the range 0 to 1. If $p=1$, $t_1=0$ and $t_2=1$.

(ii) When $p<\frac{1}{2}$, the denominator does not have any real roots, so the curve does not have an asymptote.

(iii) When $p=\frac{1}{2}$, the denominator becomes a constant independent of t. For $t=\pm\infty$, $x(t)$ and $y(t)$ become infinite.

From the above considerations, the derived conic section curve segment formula (7.5) can be classified as follows depending on the value of p.

① $p=0$: degenerates to a straight line (concides with the line segment $\mathbf{Q}_0\mathbf{Q}_1$).

② $0<p<\frac{1}{2}$: ellipse

③ $p=\frac{1}{2}$: parabola

④ $\frac{1}{2}<p<1$: hyperbola

⑤ $p=1$: degenerates to 2 straight lines ($\mathbf{Q}_0\mathbf{Q}_T$ and $\mathbf{Q}_T\mathbf{Q}_1$).

If $p < 0$, the curve becomes an ellipse on the opposite side of $\mathbf{Q}_0\mathbf{Q}_1$ from the triangle shown in the figure. All of these cases are shown in Fig. 7.3.

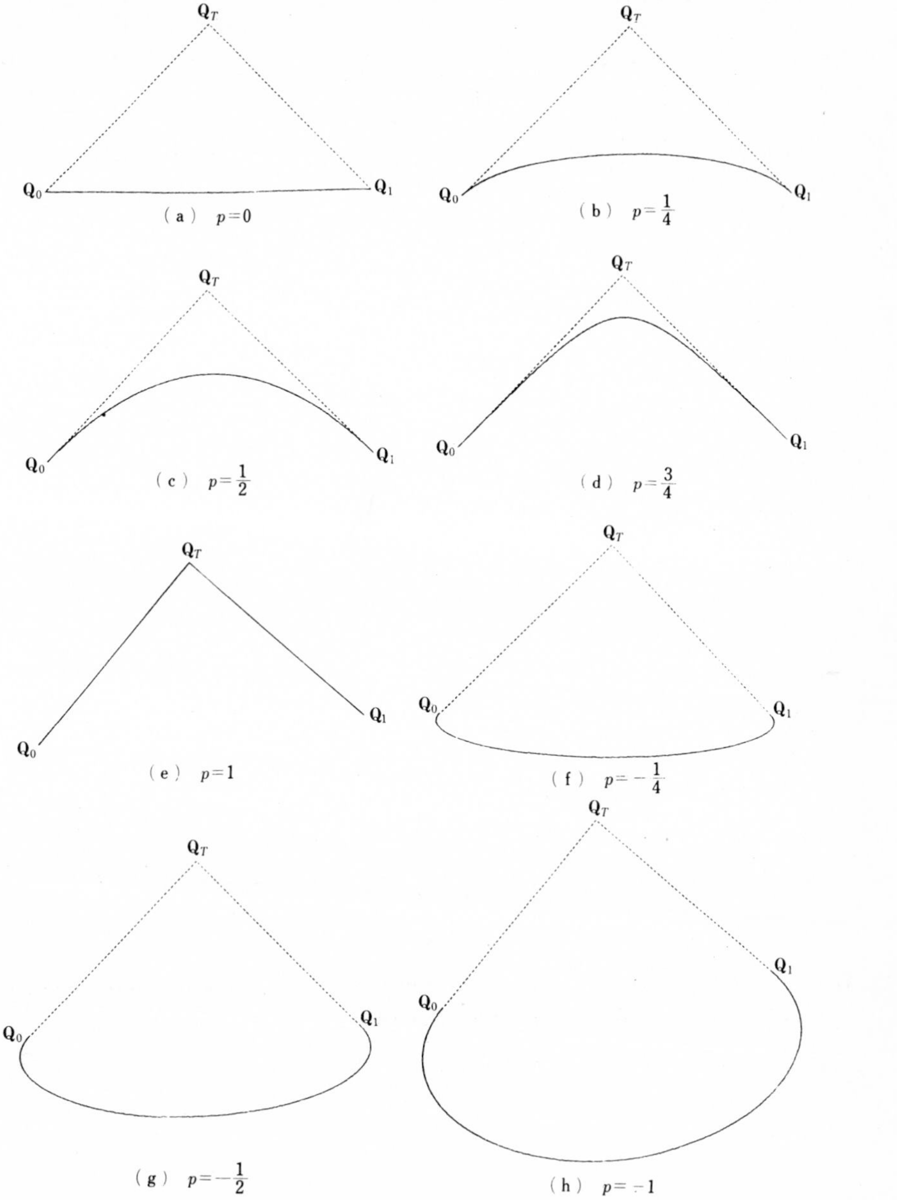

Fig. 7.3. Conic section curve segments. (a) degenerate straight line; (b) ellipse; (c) parabola; (d) hyperbola; (e) 2 degenerate straight lines; (f) ellipse; (g) ellipse; (h) ellipse

7.3 Parabolas

A parabola is obtained by setting $p=1/2$ in Eq. (7.5). In this case, the homogeneous coordinate w becomes a constant independent of t, so the parabola can be expressed in terms of ordinary coordinate vectors $\mathbf{Q}_0$, $\mathbf{Q}_T$, $\mathbf{Q}_1$ as follows:

$$\mathbf{P}(t)=(1-t)^2\mathbf{Q}_0+2(1-t)t\mathbf{Q}_T+t^2\mathbf{Q}_1. \tag{7.7}$$

Evaluating this formula for the parameter value $t=1/2$ gives:

$$\mathbf{P}\left(\frac{1}{2}\right)=\frac{1}{4}\mathbf{Q}_0+\frac{1}{2}\mathbf{Q}_T+\frac{1}{4}\mathbf{Q}_1 = \frac{\dfrac{\mathbf{Q}_0+\mathbf{Q}_T}{2}+\dfrac{\mathbf{Q}_T+\mathbf{Q}_1}{2}}{2}. \tag{7.8}$$

The tangent vectors at $t=0$, 1/2 and 1 are:

$$\left.\begin{aligned}\dot{\mathbf{P}}(0)&=2(\mathbf{Q}_T-\mathbf{Q}_0)\\ \dot{\mathbf{P}}\left(\frac{1}{2}\right)&=\mathbf{Q}_1-\mathbf{Q}_0\\ \dot{\mathbf{P}}(1)&=2(\mathbf{Q}_1-\mathbf{Q}_T)\end{aligned}\right\} \tag{7.9}$$

The relations expressed by formulas (7.8) and (7.9) are shown graphically in Fig. 7.4.

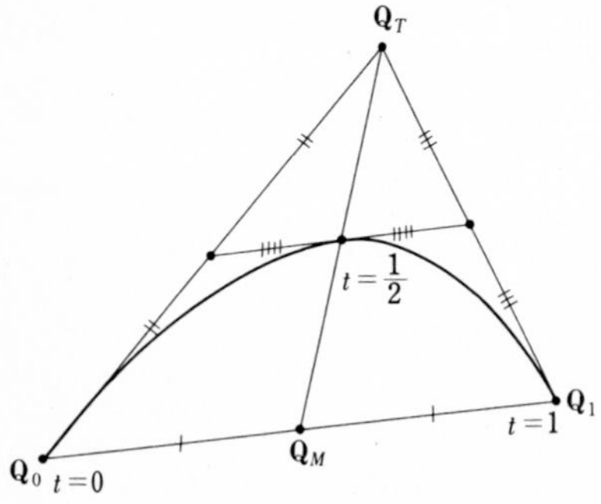

Fig. 7.4. Relation between parabola defining vectors and the curve

7.4 Circular Arc Formulas

A circular arc having center $\mathbf{Q}_c$, starting point $\mathbf{Q}_0$ and end point $\mathbf{Q}_1$ is shown in Fig. 7.5. Letting R be the radius, θ_0 the angle between the radius vector at the starting point and the horizontal, and θ the angle subtended by the arc at the center, we have:

$$p=\frac{|\mathbf{Q}_S-\mathbf{Q}_M|}{|\mathbf{Q}_T-\mathbf{Q}_M|}=\frac{\cos\frac{\theta}{2}}{1+\cos\frac{\theta}{2}}. \tag{7.10}$$

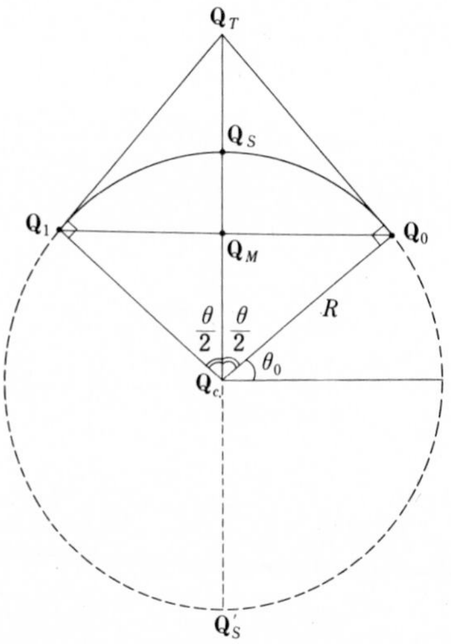

Fig. 7.5. Generation of a circular arc

From Fig. 7.5 we have:

$$\left.\begin{aligned}
\mathbf{Q}_0&=[Q_{cx}+R\cos\theta_0 \quad Q_{cy}+R\sin\theta_0 \quad 1]\\
\mathbf{Q}_T&=\left[Q_{cx}+R\,\frac{\cos\left(\theta_0+\frac{\theta}{2}\right)}{\cos\frac{\theta}{2}} \quad Q_{cy}+R\,\frac{\sin\left(\theta_0+\frac{\theta}{2}\right)}{\cos\frac{\theta}{2}} \quad 1\right]\\
\mathbf{Q}_1&=[Q_{cx}+R\cos(\theta_0+\theta) \quad Q_{cy}+R\sin(\theta_0+\theta) \quad 1]
\end{aligned}\right\} \tag{7.11}$$

Substituting Eqs. (7.10) and (7.11) into (7.5) gives:

$$w(t)\mathbf{P}(t)=\frac{1}{1+\cos\dfrac{\theta}{2}}[t^2\ t\ 1]\begin{bmatrix}1&-2&1\\-2&2&0\\1&0&0\end{bmatrix}\begin{bmatrix}\mathbf{Q}_0\\\left(\cos\dfrac{\theta}{2}\right)\mathbf{Q}_T\\\mathbf{Q}_1\end{bmatrix}$$

$$\doteq^{*)}[t^2\ t\ 1]\begin{bmatrix}1&-2&1\\-2&2&0\\1&0&0\end{bmatrix}\begin{bmatrix}\mathbf{Q}_0\\\left(\cos\dfrac{\theta}{2}\right)\mathbf{Q}_T\\\mathbf{Q}_1\end{bmatrix}$$

$$=[t^2\ t\ 1]\begin{bmatrix}1&-2&1\\-2&2&0\\1&0&0\end{bmatrix}$$

$$\times\begin{bmatrix}Q_{cx}+R\cos\theta_0 & Q_{cy}+R\sin\theta_0 & 1\\ Q_{cx}\cos\dfrac{\theta}{2}+R\cos\left(\theta_0+\dfrac{\theta}{2}\right) & Q_{cy}\cos\dfrac{\theta}{2}+R\sin\left(\theta_0+\dfrac{\theta}{2}\right) & \cos\dfrac{\theta}{2}\\ Q_{cx}+R\cos(\theta_0+\theta) & Q_{cy}+R\cos(\theta_0+\theta) & 1\end{bmatrix} \tag{7.12}$$

which is a parametric expression for the circular arc. As a specific example, consider the expression for a 1-st quadrant quarter circle with $\theta_0=0°$, $\theta=90°$, $Q_{cx}=Q_{cy}=0$ and $R=1$**):

$$w(t)\mathbf{P}(t)=[t^2\ t\ 1]\begin{bmatrix}1&-2&1\\-2&2&0\\1&0&0\end{bmatrix}\begin{bmatrix}1&0&1\\\dfrac{1}{\sqrt{2}}&\dfrac{1}{\sqrt{2}}&\dfrac{1}{\sqrt{2}}\\0&1&1\end{bmatrix}$$

$$=[t^2\ t\ 1]\begin{bmatrix}1-\sqrt{2}&1-\sqrt{2}&2-\sqrt{2}\\\sqrt{2}-2&\sqrt{2}&\sqrt{2}-2\\1&0&1\end{bmatrix}$$

*) The symbol $\doteq$ indicates that the corresponding ordinary coordinates are equal.

**) Refer to the author's book *Graphics Processing Engineering* (in Japanese) (Nikkan Kogyo Shimbun Sha), p. 102 Eq. (3.92).

Next, let us find the circular arcs shown by dotted lines in Fig. 7.5. Setting:

$$p = -\frac{|\mathbf{Q}_S' - \mathbf{Q}_M|}{|\mathbf{Q}_T - \mathbf{Q}_M|} = -\frac{\cos\frac{\theta}{2}}{1-\cos\frac{\theta}{2}} \tag{7.13}$$

and substituting Eqs. (7.11) and (7.13) into (7.5) gives:

$$w(t)\mathbf{P}(t) = [t^2 \;\; t \;\; 1] \begin{bmatrix} 1 & -2 & 1 \\ -2 & 2 & 0 \\ 1 & 0 & 0 \end{bmatrix}$$

$$\times \begin{bmatrix} Q_{cx} + R\cos\theta_0 & Q_{cy} + R\sin\theta_0 & 1 \\ -Q_{cx}\cos\frac{\theta}{2} - R\cos\left(\theta_0 + \frac{\theta}{2}\right) & -Q_{cy}\cos\frac{\theta}{2} - R\sin\left(\theta_0 + \frac{\theta}{2}\right) & -\cos\frac{\theta}{2} \\ Q_{cx} + R\cos(\theta_0 + \theta) & Q_{cy} + R\sin(\theta_0 + \theta) & 1 \end{bmatrix} \tag{7.14}$$

This arc starts from point $\mathbf{Q}_0$ $(t=0)$, and turns clockwise until it reaches point $\mathbf{Q}_1$ $(t=1)$.

In order to obtain the arc expression proceeding counterclockwise from $\mathbf{Q}_1\,(t=0)$ to $\mathbf{Q}_0\,(t=1)$ we substitute

$$t = -t' + 1 \tag{7.15}$$

in Eq. (7.14); this new parameter t' is 0 at $\mathbf{Q}_1$ and 1 at $\mathbf{Q}_0$. Then replacing t' by t gives:

$$w(t)\mathbf{P}(t) = [t^2 \;\; t \;\; 1] \begin{bmatrix} 1 & -2 & 1 \\ -2 & 2 & 0 \\ 1 & 0 & 0 \end{bmatrix}$$

$$\times \begin{bmatrix} Q_{cx} + R\cos(\theta_0 + \theta) & Q_{cy} + R\sin(\theta_0 + \theta) & 1 \\ -Q_{cx}\cos\frac{\theta}{2} - R\cos\left(\theta_0 + \frac{\theta}{2}\right) & -Q_{cy}\cos\frac{\theta}{2} - R\sin\left(\theta_0 + \frac{\theta}{2}\right) & -\cos\frac{\theta}{2} \\ Q_{cx} + R\cos\theta_0 & Q_{cy} + R\sin\theta_0 & 1 \end{bmatrix} \tag{7.16}$$

Examples of arcs generated by Eqs. (7.12) and (7.16) are shown in Fig. 7.6.

If a circular arc is expressed as a rational polynomial, as the center angle becomes large (especially if it exceeds 180°), in the central part of the arc the intervals between points on the arc corresponding to equidistant parameter values become widely spaced. Figure 7.7 shows the case with parameter intervals of 0.1.

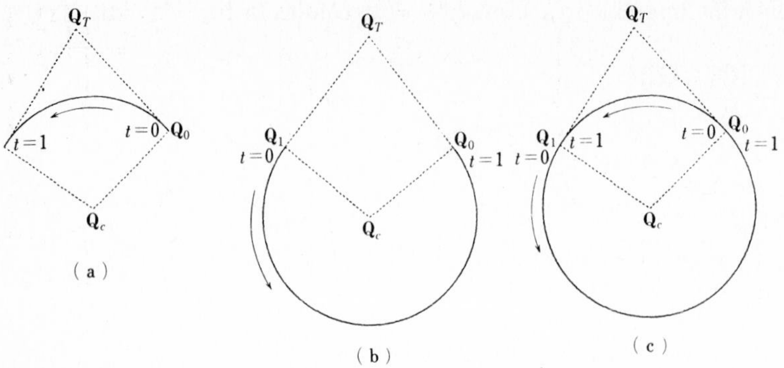

Fig. 7.6. Generation of a circle

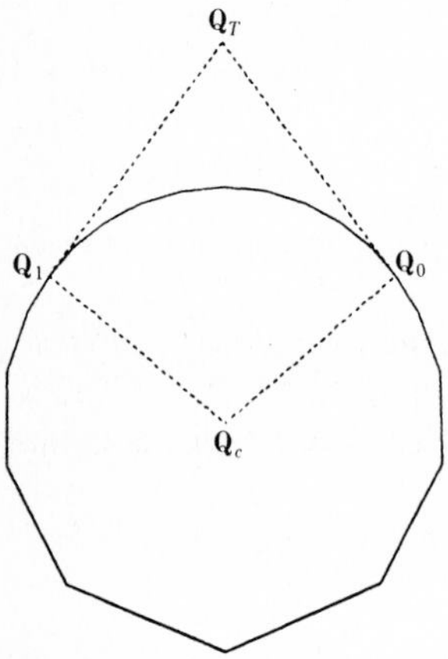

Fig. 7.7. Generation of a circle

7.5 Cubic/Cubic Rational Polynomial Curves

Quadratic/quadratic rational polynomial curves expressing conic sections could be derived by performing an affine transformation and projective transformation on a curve of the form $[x\ y\ 1]=[t^2\ t\ 1]$. By a similar method, an affine transformation and projective transformation can be performed on a curve of the form $[x\ y\ z\ 1]=[t^3\ t^2\ t\ 1]$ to derive cubic/cubic rational polynomial curves. Here we only show the results*).

Let the position vectors and tangent vectors of the curve at $t=0$ and $t=1$ be $\mathbf{Q}_0$, $\dot{\mathbf{Q}}_0$ and $\mathbf{Q}_1$, $\dot{\mathbf{Q}}_1$, respectively. The curve is supposed to pass through point $\mathbf{Q}_S$ at $t=t_s$. Then the cubic/cubic rational polynomial curve is expressed by the following equation:

---

*) Refer to the present author's book *Graphics Processing Engineering* (in Japanese) (Nikkan Kogyo Shimbun Sha), Sect. 3.5.3.

$$w(t)\mathbf{P}(t)=[t^3\ t^2\ t\ 1]\begin{bmatrix}2&-2&1&1\\-3&3&-2&-1\\0&0&1&0\\1&0&0&0\end{bmatrix}\begin{bmatrix}w_0&0&0&0\\0&w_1&0&0\\\dot{w}_0&0&w_0&0\\0&\dot{w}_1&0&w_1\end{bmatrix}\begin{bmatrix}\mathbf{Q}_0\\\mathbf{Q}_1\\\dot{\mathbf{Q}}_0\\\dot{\mathbf{Q}}_1\end{bmatrix}$$

$$=[H_{0,0}(t)\ H_{0,1}(t)\ H_{1,0}(t)\ H_{1,1}(t)]\begin{bmatrix}w_0&0&0&0\\0&w_1&0&0\\\dot{w}_0&0&w_0&0\\0&\dot{w}_1&0&w_1\end{bmatrix}\begin{bmatrix}\mathbf{Q}_0\\\mathbf{Q}_1\\\dot{\mathbf{Q}}_0\\\dot{\mathbf{Q}}_1\end{bmatrix}\tag{7.17}$$

where w_0, w_1, $\dot{w}_0$ and $\dot{w}_1$ are given by the following formula:

$$[w_0\ w_1\ \dot{w}_0\ \dot{w}_1]=\mathbf{Q}_S\begin{bmatrix}\mathbf{Q}_0\\\mathbf{Q}_1\\\dot{\mathbf{Q}}_0\\\dot{\mathbf{Q}}_1\end{bmatrix}^{-1}\begin{bmatrix}H_{0,0}(t_s)&0&H_{1,0}(t_s)&0\\0&H_{0,1}(t_s)&0&H_{1,1}(t_s)\\H_{1,0}(t_s)&0&0&0\\0&H_{1,1}(t_s)&0&0\end{bmatrix}$$

ok

If we set $w_0=w_1$ and $\dot{w}_0=\dot{w}_1=0$ in the cubic/cubic rational polynomial curve equation (7.17), since (7.17) is expressed in terms of homogeneous coordinates, when converted into ordinary coordinates it agrees with a Hermite interpolation curve (Ferguson curve segment).

7.6 *T*-Conic Curves [10)]

Let us impose the condition that a cubic/cubic rational polynomial curve be able to express a conic section curve.

Consider, for example, the x component of Eq. (7.17) in the following form.

$$x(t)=\frac{w_0Q_{0,x}H_{0,0}(t)+w_1Q_{1,x}H_{0,1}(t)+(\dot{w}_0Q_{0,x}+w_0\dot{Q}_{0,x})H_{1,0}(t)+(\dot{w}_1Q_{1,x}+w_1\dot{Q}_{1,x})H_{1,1}(t)}{w_0H_{0,0}(t)+w_1H_{0,1}(t)+\dot{w}_0H_{1,0}(t)+\dot{w}_1H_{1,1}(t)}$$

The denominator can be expressed as a polynomial in t as follows:

$$(2w_0-2w_1+\dot{w}_0+\dot{w}_1)t^3-(3w_0-3w_1+2\dot{w}_0+\dot{w}_1)t^2+\dot{w}_0t+w_0.$$

In order to express a conic section curve, the denominator must agree with the denominator in Eq. (7.5):

$$(2w_0-2w_1+\dot{w}_0+\dot{w}_1)t^3-(3w_0-3w_1+2\dot{w}_0+\dot{w}_1)t^2+\dot{w}_0t+w_0$$
$$\equiv 2(1-2p)t^2-2(1-2p)t+1-p.$$

Equating coefficients of like powers of t gives:

$$\begin{aligned}
&2w_0 - 2w_1 + \dot{w}_0 + \dot{w}_1 = 0\\
&-(3w_0 - 3w_1 + 2\dot{w}_0 + \dot{w}_1) = 2(1-2p)\\
&\dot{w}_0 = -2(1-2p)\\
&w_0 = 1-p
\end{aligned} \tag{7.18}$$

Solving Eqs. (7.18) gives:

$$w_0 = 1-p, \quad w_1 = 1-p, \quad \dot{w}_0 = -2(1-2p), \quad \dot{w}_1 = 2(1-2p).$$

Substituting these values into Eq. (7.17) gives:

$$w(t)\,\mathbf{P}(t) = [t^3 \;\; t^2 \;\; t \;\; 1] \begin{bmatrix} 2 & -2 & 1 & 1\\ -3 & 3 & -2 & -1\\ 0 & 0 & 1 & 0\\ 1 & 0 & 0 & 0 \end{bmatrix}$$

$$\times \begin{bmatrix} 1-p & 0 & 0 & 0\\ 0 & 1-p & 0 & 0\\ -2(1-2p) & 0 & 1-p & 0\\ 0 & 2(1-2p) & 0 & 1-p \end{bmatrix} \begin{bmatrix} \mathbf{Q}_0\\ \mathbf{Q}_1\\ \dot{\mathbf{Q}}_0\\ \dot{\mathbf{Q}}_1 \end{bmatrix}. \tag{7.19}$$

If we set $p = 1/2$ in Eq. (7.19), it reduces to a cubic Hermite interpolation curve (Ferguson curve). In other words, the curve expressed by Eq. (7.19), can express both a cubic Hermite interpolation curve and a conic section curve. This makes it a very interesting curve. This curve, which was developed at Boeing Aircraft Company, is called a twisted conic curve or *T*-conic curve.

In order for Eq. (7.19) to express a conic section curve, certain relations must hold among the specified data $\mathbf{Q}_0$, $\mathbf{Q}_1$, $\dot{\mathbf{Q}}_0$ and $\dot{\mathbf{Q}}_1$. Equating the coefficient of t^3 in (7.19) to 0 gives:

$$\text{coefficient of } t^3 = 2p(\mathbf{Q}_0 - \mathbf{Q}_1) + (1-p)\,(\dot{\mathbf{Q}}_0 + \dot{\mathbf{Q}}_1) = 0.$$

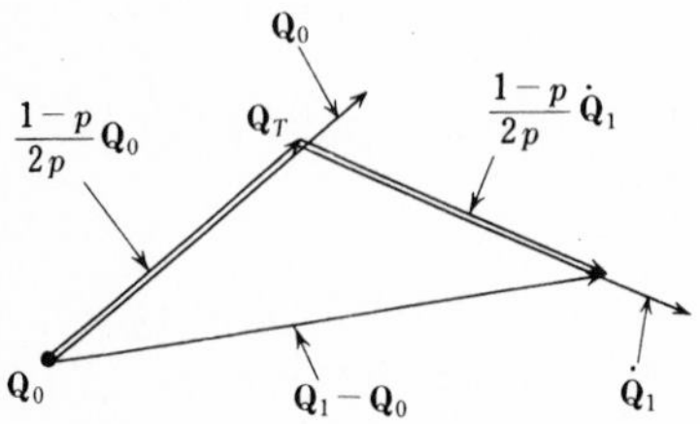

Fig. 7.8. Geometrical relation among $\mathbf{Q}_0$, $\mathbf{Q}_1$, $\dot{\mathbf{Q}}_0$ and $\dot{\mathbf{Q}}_1$ to express a conic section curve

From this we obtain:

$$\mathbf{Q}_1 - \mathbf{Q}_0 = \frac{1-p}{2p}(\dot{\mathbf{Q}}_0 + \dot{\mathbf{Q}}_1). \tag{7.20}$$

This relation is shown in Fig. 7.8. Letting $\mathbf{Q}_T$ be the intersection of the tangents at points $\mathbf{Q}_0$ and $\mathbf{Q}_1$, we have:

$$\begin{aligned} \mathbf{Q}_T - \mathbf{Q}_0 &= \frac{1-p}{2p}\dot{\mathbf{Q}}_0 \\ \mathbf{Q}_1 - \mathbf{Q}_T &= \frac{1-p}{2p}\dot{\mathbf{Q}}_1 \end{aligned} \tag{7.21}$$

Solving these equations for $\dot{\mathbf{Q}}_0$ and $\dot{\mathbf{Q}}_1$ and substituting into (7.19) gives:

$$w(t)\mathbf{P}(t) = [t^3 \; t^2 \; t \; 1]\begin{bmatrix} 2 & -2 & 1 & 1 \\ -3 & 3 & -2 & -1 \\ 0 & 0 & 1 & 0 \\ 1 & 0 & 0 & 0 \end{bmatrix}$$

$$\times \begin{bmatrix} 1-p & 0 & 0 & 0 \\ 0 & 1-p & 0 & 0 \\ -2(1-2p) & 0 & 1-p & 0 \\ 0 & 2(1-2p) & 0 & 1-p \end{bmatrix} \begin{bmatrix} \mathbf{Q}_0 \\ \mathbf{Q}_1 \\ \dfrac{2p}{1-p}(\mathbf{Q}_T - \mathbf{Q}_0) \\ \dfrac{2p}{1-p}(\mathbf{Q}_1 - \mathbf{Q}_T) \end{bmatrix}$$

$$= [t^3 \; t^2 \; t \; 1]\begin{bmatrix} 2 & -2 & 1 & 1 \\ -3 & 3 & -2 & -1 \\ 0 & 0 & 1 & 0 \\ 1 & 0 & 0 & 0 \end{bmatrix}$$

$$\times \begin{bmatrix} 1-p & 0 & 0 & 0 \\ 0 & 1-p & 0 & 0 \\ -2(1-2p) & 0 & 1-p & 0 \\ 0 & 2(1-2p) & 0 & 1-p \end{bmatrix}$$

$$\times \begin{bmatrix} 1 & 0 & 0 \\ 0 & 0 & 1 \\ -\dfrac{2p}{1-p} & \dfrac{2p}{1-p} & 0 \\ 0 & -\dfrac{2p}{1-p} & \dfrac{2p}{1-p} \end{bmatrix} \begin{bmatrix} \mathbf{Q}_0 \\ \mathbf{Q}_T \\ \mathbf{Q}_1 \end{bmatrix} =$$

$$= [t^3 \; t^2 \; t \; 1]$$

$$\times \begin{bmatrix} 0 & 0 & 0 \\ 1-p & -2p & 1-p \\ -2(1-p) & 2p & 0 \\ 1-p & 0 & 0 \end{bmatrix} \begin{bmatrix} \mathbf{Q}_0 \\ \mathbf{Q}_T \\ \mathbf{Q}_1 \end{bmatrix}$$

$$= [t^2 \; t \; 1] \begin{bmatrix} 1 & -2 & 1 \\ -2 & 2 & 0 \\ 1 & 0 & 0 \end{bmatrix} \begin{bmatrix} 1-p & 0 & 0 \\ 0 & p & 0 \\ 0 & 0 & 1-p \end{bmatrix} \begin{bmatrix} \mathbf{Q}_0 \\ \mathbf{Q}_T \\ \mathbf{Q}_1 \end{bmatrix}$$

showing that this agrees with Eq. (7.5) for a conic section curve.

References (Chap. 7)

32) Forrest, A.R.: "Conic Sections", Draft of Computer-Aided Geometric Design, Dec. 1978.

Appendix A: Vector Expression of Simple Geometrical Relations

Straight Line Vector Formulas

A straight line which passes through the point $\boldsymbol{P}_0 = [x_0\ y_0\ z_0]$ in the direction of the vector $\boldsymbol{L}$ can be expressed in terms of the parameter t as:

$$\boldsymbol{P}(t) = \boldsymbol{P}_0 + \boldsymbol{L}t. \tag{A.1}$$

When $\boldsymbol{L}$ is a unit vector, the distance between points $\boldsymbol{P}(t)$ and $\boldsymbol{P}_0$ along the straight line becomes $|\boldsymbol{P}(t) - \boldsymbol{P}_0| = |\boldsymbol{L}t| = |t|$.

The vector formula for the straight line passing through the two points $\boldsymbol{P}_0 = [x_0\ y_0\ z_0]$ and $\boldsymbol{P}_1 = [x_1\ y_1\ z_1]$ is, with $\boldsymbol{L} = \boldsymbol{P}_1 - \boldsymbol{P}_0$:

$$\boldsymbol{P}(t) = (1-t)\boldsymbol{P}_0 + t\boldsymbol{P}_1. \tag{A.2}$$

Points on the line segment $\boldsymbol{P}_0\boldsymbol{P}_1$ correspond to values of the parameter t in the range $0 \leqq t \leqq 1$.

Perpendicular Bisector of a Line Segment

Let $\boldsymbol{N}$ be a vector normal to a plane on which lie the two points $\boldsymbol{P}_0$ and $\boldsymbol{P}_1$. Since the vector $(\boldsymbol{P}_1 - \boldsymbol{P}_0) \times \boldsymbol{N}$ is in the direction of the perpendicular bisector,

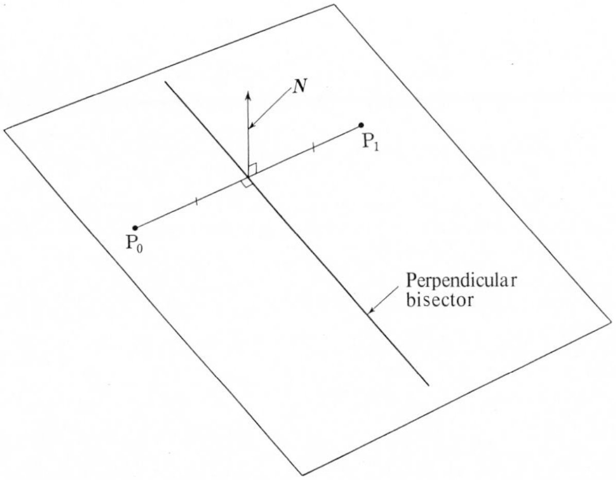

Fig. A.1

the formula for the perpendicular bisector is found by setting $\boldsymbol{L}=(\boldsymbol{P}_1-\boldsymbol{P}_0)\times\boldsymbol{N}$ and by replacing $\boldsymbol{P}_0$ by $(\boldsymbol{P}_1-\boldsymbol{P}_0)/2$ in Eq. (A.1), (Fig. A.1):

$$\boldsymbol{P}(t)=\frac{\boldsymbol{P}_1-\boldsymbol{P}_0}{2}+(\boldsymbol{P}_1-\boldsymbol{P}_0)\times\boldsymbol{N}t\,. \tag{A.3}$$

Length of a Common Perpendicular to Two Non-Parallel Lines

Let $\boldsymbol{P}_0\boldsymbol{P}_1$ and $\boldsymbol{P}_0'\boldsymbol{P}_1'$ be two straight lines. Let the intersections of the common perpendicular with these lines be $\boldsymbol{H}$ and $\boldsymbol{H}'$ (Fig. A.2). From Eq. (A.1):

$$\boldsymbol{H}=\boldsymbol{P}_0+(\boldsymbol{P}_1-\boldsymbol{P}_0)t$$
$$\boldsymbol{H}'=\boldsymbol{P}_0'+(\boldsymbol{P}_1'-\boldsymbol{P}_0')t'\,.$$

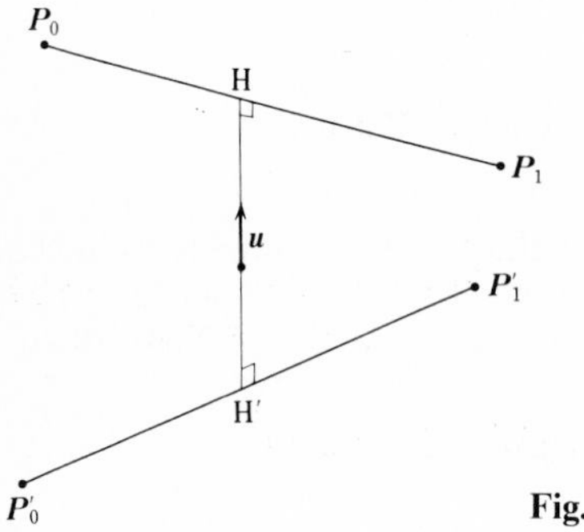

Fig. A.2

Therefore:

$$\boldsymbol{H}-\boldsymbol{H}'=\boldsymbol{P}_0-\boldsymbol{P}_0'+(\boldsymbol{P}_1-\boldsymbol{P}_0)t-(\boldsymbol{P}_1'-\boldsymbol{P}_0')t'\,.$$

Let l be the length of the common perpendicular and $\boldsymbol{u}$ be the unit vector in the direction $\boldsymbol{H}-\boldsymbol{H}'$. Then we have:

$$\begin{aligned} l&=(\boldsymbol{H}-\boldsymbol{H}')\cdot\boldsymbol{u}\\ &=(\boldsymbol{P}_0-\boldsymbol{P}_0')\cdot\boldsymbol{u}+(\boldsymbol{P}_1-\boldsymbol{P}_0)t\cdot\boldsymbol{u}-(\boldsymbol{P}_1'-\boldsymbol{P}_0')t'\cdot\boldsymbol{u}\\ &=(\boldsymbol{P}_0-\boldsymbol{P}_0')\cdot\boldsymbol{u}\end{aligned}$$

$\boldsymbol{u}$ is a unit vector perpendicular to $\boldsymbol{P}_1-\boldsymbol{P}_0$ and $\boldsymbol{P}_1'-\boldsymbol{P}_0'$:

$$\boldsymbol{u}=\frac{(\boldsymbol{P}_1-\boldsymbol{P}_0)\times(\boldsymbol{P}_1'-\boldsymbol{P}_0')}{|(\boldsymbol{P}_1-\boldsymbol{P}_0)\times(\boldsymbol{P}_1'-\boldsymbol{P}_0')|}\,.$$

Therefore:

$$l=\frac{(P_0-P_0')\cdot[(P_1-P_0)\times(P_1'-P_0')]}{|(P_1-P_0)\times(P_1'-P_0')|}$$

$$=\frac{[P_0-P_0', P_1-P_0, P_1'-P_0']^{*)}}{|(P_1-P_0)\times(P_1'-P_0')|}. \tag{A.4}$$

Plane Vector Formulas

A plane is expressed by $ax+by+cz+d=0$. Since $[a\ b\ c]^T$ is a normal vector N, if $P=[x\ y\ z]$ is a point on the plane, we have:

$$P\cdot N+d=0. \tag{A.5}$$

Formula for a Plane That Passes Through 3 Points not on a Straight Line

Consider the 3 points Q_0, Q_1, Q_2 that lie on a plane but not colinear. $(Q_1-Q_0)\times(Q_2-Q_0)$ is a vector normal to the plane. Therefore, from Eq. (A.5) we have:

$$P\cdot[(Q_1-Q_0)\times(Q_2-Q_0)]=Q_0\cdot[(Q_1-Q_0)\times(Q_2-Q_0)].$$

Expanding the right-hand side gives:

$$\begin{aligned}\text{right side}&=Q_0\cdot[(Q_1-Q_0)\times(Q_2-Q_0)]\\&=Q_0\cdot(Q_1\times Q_2-Q_1\times Q_0-Q_0\times Q_2+Q_0\times Q_0)\\&=Q_0\cdot(Q_1\times Q_2)\\&=[Q_0, Q_1, Q_2].\end{aligned}$$

Therefore, the formula for the plane that passes through the three points Q_0, Q_1, Q_2 is:

$$[P, Q_1-Q_0, Q_2-Q_0]=[Q_0, Q_1, Q_2]. \tag{A.6}$$

*) The brackets in the numerator indicate the triple scalar product.

Intersection Point of 3 Planes

Consider 3 planes having the formulas:

$$\boldsymbol{P}\cdot \boldsymbol{N}_1 + d_1 = 0$$

$$\boldsymbol{P}\cdot \boldsymbol{N}_2 + d_2 = 0$$

$$\boldsymbol{P}\cdot \boldsymbol{N}_3 + d_3 = 0\,.$$

The intersection point is:

$$\boldsymbol{P} = -\frac{d_1(\boldsymbol{N}_2\times \boldsymbol{N}_3) + d_2(\boldsymbol{N}_3\times \boldsymbol{N}_1) + d_3(\boldsymbol{N}_1\times \boldsymbol{N}_2)}{\boldsymbol{N}_1\cdot(\boldsymbol{N}_2\times \boldsymbol{N}_3)}\,. \tag{A.7}$$

It is necessary to have $\boldsymbol{N}_1\cdot(\boldsymbol{N}_2\times \boldsymbol{N}_3)\neq 0$.

A Circle That Passes Through 3 Points

Let us find the circle that passes through the 3 points $\boldsymbol{Q}_0$, $\boldsymbol{Q}_1$, $\boldsymbol{Q}_2$ (Fig. A.3). The center of the circle is at the intersection of the perpendicular bisector L_1 of the line segment $\boldsymbol{Q}_0\boldsymbol{Q}_1$ and the perpendicular bisector L_2 of the line segment $\boldsymbol{Q}_0\boldsymbol{Q}_2$. Assuming that the points $\boldsymbol{Q}_0$, $\boldsymbol{Q}_1$, $\boldsymbol{Q}_2$ are on a plane in 3-dimensional space, the center $\boldsymbol{Q}_c$ of the circle to be found is at the intersection of 3 planes, namely, the plane in which the circle lies, the plane which includes L_1 and is perpendicular to the circle plane, and the plane which includes L_2 and is perpendicular to the circle plane.

The plane in which the circle lies is expressed by the equation:

$$(\boldsymbol{P}-\boldsymbol{Q}_0)\cdot\{(\boldsymbol{Q}_1-\boldsymbol{Q}_0)\times(\boldsymbol{Q}_2-\boldsymbol{Q}_0)\} = 0\,. \tag{A.8}$$

The equations of the planes perpendicular to the circle plane and including L_1 and L_2, respectively:

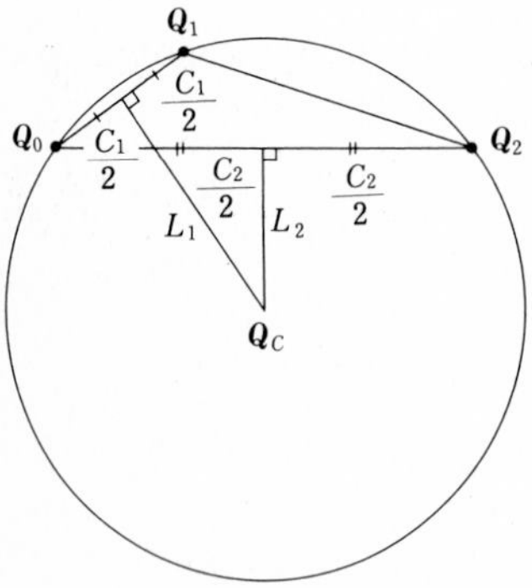

Fig. A.3

$$(\boldsymbol{P}-\boldsymbol{Q}_0)\cdot(\boldsymbol{Q}_1-\boldsymbol{Q}_0)-\frac{c_1^2}{2}=0 \tag{A.9}$$

$$(\boldsymbol{P}-\boldsymbol{Q}_0)\cdot(\boldsymbol{Q}_2-\boldsymbol{Q}_0)-\frac{c_2^2}{2}=0 \tag{A.10}$$

so that substituting $\boldsymbol{P}-\boldsymbol{Q}_0$ for $\boldsymbol{P}$ in Eq. (A.7), $(\boldsymbol{Q}_1-\boldsymbol{Q}_0)\times(\boldsymbol{Q}_2-\boldsymbol{Q}_0)$ for N_1, $\boldsymbol{Q}_1-\boldsymbol{Q}_0$ for N_2, $\boldsymbol{Q}_2-\boldsymbol{Q}_0$ for N_3, 0 for d_1, $-c_1^2/2$ for d_2 and $-c_2^2/2$ for d_3, we obtain:

$$\boldsymbol{P}-\boldsymbol{Q}_0=\frac{c_1^2(\boldsymbol{Q}_2-\boldsymbol{Q}_0)\times[(\boldsymbol{Q}_1-\boldsymbol{Q}_0)\times(\boldsymbol{Q}_2-\boldsymbol{Q}_0)]+c_2^2[(\boldsymbol{Q}_1-\boldsymbol{Q}_0)\times(\boldsymbol{Q}_2-\boldsymbol{Q}_0)]\times(\boldsymbol{Q}_1-\boldsymbol{Q}_0)}{2\,|(\boldsymbol{Q}_1-\boldsymbol{Q}_0)\times(\boldsymbol{Q}_2-\boldsymbol{Q}_0)|^2}$$

$$=\frac{c_2^2\,[c_1^2-(\boldsymbol{Q}_1-\boldsymbol{Q}_0)\cdot(\boldsymbol{Q}_2-\boldsymbol{Q}_0)]\,(\boldsymbol{Q}_1-\boldsymbol{Q}_0)+c_1^2\,[c_2^2-(\boldsymbol{Q}_1-\boldsymbol{Q}_0)\cdot(\boldsymbol{Q}_2-\boldsymbol{Q}_0)]\,(\boldsymbol{Q}_2-\boldsymbol{Q}_0)}{2\,|(\boldsymbol{Q}_1-\boldsymbol{Q}_0)\times(\boldsymbol{Q}_2-\boldsymbol{Q}_0)|^2} \tag{A.11}$$

Letting R be the radius of the circle, we have:

$$R^2=(\boldsymbol{P}-\boldsymbol{Q}_0)^2=\frac{\begin{array}{l}c_2^2[c_1^2-(\boldsymbol{Q}_1-\boldsymbol{Q}_0)\cdot(\boldsymbol{Q}_2-\boldsymbol{Q}_0)](\boldsymbol{Q}_1-\boldsymbol{Q}_0)\cdot(\boldsymbol{P}-\boldsymbol{Q}_0)\\ +c_1^2[c_2^2-(\boldsymbol{Q}_1-\boldsymbol{Q}_0)\cdot(\boldsymbol{Q}_2-\boldsymbol{Q}_0)](\boldsymbol{Q}_2-\boldsymbol{Q}_0)\cdot(\boldsymbol{P}-\boldsymbol{Q}_0)\end{array}}{2\,|(\boldsymbol{Q}_1-\boldsymbol{Q}_0)\times(\boldsymbol{Q}_2-\boldsymbol{Q}_0)|^2}$$

Using the relations in Eqs. (A.9) and (A.10) gives:

$$R^2=\frac{c_1^2c_2^2\,\{c_1^2-2(\boldsymbol{Q}_1-\boldsymbol{Q}_0)\cdot(\boldsymbol{Q}_2-\boldsymbol{Q}_0)+c_2^2\}}{4\,|(\boldsymbol{Q}_1-\boldsymbol{Q}_0)\times(\boldsymbol{Q}_2-\boldsymbol{Q}_0)|^2}$$

$$=\frac{c_1^2c_2^2\,\{(\boldsymbol{Q}_1-\boldsymbol{Q}_0)-(\boldsymbol{Q}_2-\boldsymbol{Q}_0)\}^2}{4\,|(\boldsymbol{Q}_1-\boldsymbol{Q}_0)\times(\boldsymbol{Q}_2-\boldsymbol{Q}_0)|^2}$$

$$\therefore\quad R=\frac{|\boldsymbol{Q}_1-\boldsymbol{Q}_0|\;|\boldsymbol{Q}_2-\boldsymbol{Q}_0|\;|\boldsymbol{Q}_1-\boldsymbol{Q}_2|}{2\,|(\boldsymbol{Q}_1-\boldsymbol{Q}_0)\times(\boldsymbol{Q}_2-\boldsymbol{Q}_0)|}. \tag{A.12}$$

In this case, the circular arc $\boldsymbol{Q}_0\boldsymbol{Q}_1=a_1$ has the length:

$$a_1=2R\sin^{-1}\left(\frac{c_1}{2R}\right). \tag{A.13}$$

Appendix B: Proofs of Formulas Relating to *B*-Spline Functions

B.1 Proof of Eq. (6.83)

The Leibnitz formula holds for the divided difference of a product of functions. For a function $h(u)$ expressed in the form:

$$h(u) = f(u)\, g(u)$$

the M-th order divided difference of $h(u)$ is given by:

$$h[u_0, \ldots, u_M] = \sum_{r=0}^{M} f[u_0, \ldots, u_r]\, g[u_r, \ldots, u_M]. \tag{B.1}$$

Applying Leibnitz' formula to the function:

$$M_M(t; u) = (u-t)_+^{M-1} = (u-t)_+^{M-2}(u-t)$$

the 2-nd order and higher divided difference of $(u-t)$ vanish, so we have:

$$\begin{aligned} &M_M[t;\ t_j, \ldots, t_{j+M}]^{*)} \\ &= M_{M-1}[t;\ t_j, \ldots, t_{j+M-1}] \cdot 1 + M_{M-1}[t;\ t_j, \ldots, t_{j+M}]\,(t_{j+M}-t) \\ &= M_{M-1}[t;\ t_j, \ldots, t_{j+M-1}] \\ &\quad + \frac{M_{M-1}[t;\ t_{j+1}, \ldots, t_{j+M}] - M_{M-1}[t;\ t_j, \ldots, t_{j+M-1}]}{t_{j+M}-t_j}\,(t_{j+M}-t). \end{aligned}$$

Expressing this formula in the notation of Eq. (6.75) gives:

$$\begin{aligned} M_{j,M}(t) &= M_{j,M-1}(t) + \frac{t_{j+M}-t}{t_{j+M}-t_j}\,(M_{j+1,M-1}(t) - M_{j,M-1}(t)) \\ &= \frac{t-t_j}{t_{j+M}-t_j}\, M_{j,M-1}(t) + \frac{t_{j+M}-t}{t_{j+M}-t_j}\, M_{j+1,M-1}(t). \end{aligned} \tag{B.2}$$

Putting this into the form of normalized *B*-spline functions gives, from Eq. (6.76):

---

*) The subscript M of M_M is the M of $(u-t)_+^{M-1}$.

$$M_{j,M}(t) = \frac{1}{t_{j+M}-t_j} N_{j,M}(t)$$

$$M_{j,M-1}(t) = \frac{1}{t_{j+M-1}-t_j} N_{j,M-1}(t)$$

$$M_{j+1,M-1}(t) = \frac{1}{t_{j+M}-t_{j+1}} N_{j+1,M-1}(t)$$

leading to the following formula:

$$N_{j,M}(t) = \frac{t-t_j}{t_{j+M-1}-t_j} N_{j,M-1}(t) + \frac{t_{j+M}-t}{t_{j+M}-t_{j+1}} N_{j+1,M-1}(t). \tag{B.3}$$

B.2 Proof of Formula (6.127)

The *B*-spline curve formula:

$$\boldsymbol{P}(t) = \sum_{j=i-M+1}^{i} N_{j,M}(t)\boldsymbol{Q}_j \quad (t_i \leqq t < t_{i+1})$$

can be transformed as follows using formula (B.2):

$$\begin{aligned}
\boldsymbol{P}(t) &= \sum_{j=i-M+1}^{i} [(t-t_j)M_{j,M-1}(t) + (t_{j+M}-t)M_{j+1,M-1}(t)]\boldsymbol{Q}_j \\
&= \sum_{j=i-M+1}^{i+1} [(t-t_j)\boldsymbol{Q}_j M_{j,M-1}(t) + (t_{j+M-1}-t)\boldsymbol{Q}_{j-1}M_{j,M-1}(t)]^{*)} \\
&= \sum_{j=i-M+1}^{i+1} M_{j,M-1}(t)\,[(t-t_j)\boldsymbol{Q}_j + (t_{j+M-1}-t)\boldsymbol{Q}_{j-1}] \\
&= \sum_j N_{j,M-1}(t)\frac{1}{t_{j+M-1}-t_j}[(t-t_j)\boldsymbol{Q}_j + (t_{j+M-1}-t)\boldsymbol{Q}_{j-1}] \\
&= \sum_j N_{j,M-1}(t)\boldsymbol{Q}_j^{[1]}(t)
\end{aligned} \tag{B.4}$$

where:

$$\boldsymbol{Q}_j^{[1]}(t) = \left(1 - \frac{t-t_j}{t_{j+M-1}-t_j}\right)\boldsymbol{Q}_{j-1} + \frac{t-t_j}{t_{j+M-1}-t_j}\boldsymbol{Q}_j.$$

---

*) Note that $M_{i+1,M-1}(t)$ and $M_{i-M+1,M-1}(t)$ are both zero in $t_i \leq t < t_{i+1}$.

In general, setting:

$$\boldsymbol{Q}_j^{[k]}(t)=\begin{cases}\boldsymbol{Q}_j & (k=0)\\ \left(1-\dfrac{t-t_j}{t_{j+M-k}-t_j}\right)\boldsymbol{Q}_{j-1}^{[k-1]}(t)+\dfrac{t-t_j}{t_{j+M-k}-t_j}\boldsymbol{Q}_j^{[k-1]}(t) & (k>0)\end{cases} \tag{B.5}$$

we have:

$$\boldsymbol{P}(t)=\sum_j N_{j,M-k}(t)\,\boldsymbol{Q}_j^{[k]}(t). \tag{B.6}$$

For $t_i \leqq t < t_{i+1}$, $N_{i,1}(t)=1$ and the other B-spline functions are zero, so:

$$\boldsymbol{P}(t)=\boldsymbol{Q}_i^{[M-1]}(t) \qquad (t_i \leqq t < t_{i+1}) \tag{B.7}$$

B.3 Proof of Formula (6.86)

In the proof of formula (6.86) we use the Marsden identity:

$$(u-t)^{M-1}=\sum_j N_{j,M}(t)\,W_{j,M}(u) \tag{B.8}$$

$$W_{j,M}(u)=\prod_{r=1}^{M-1}(u-t_{j+r}), \quad W_{j,1}(u)\equiv 1$$

so let us prove it first.

In the case of $M=1$, the right-hand side of Eq. (B.8) becomes:

$$\sum_j N_{j,1}(t)\,W_{j,1}(u)=\sum_j N_{j,1}(t)\equiv 1$$

so Eq. (B.8) holds.

Of course, Eq. (B.5) also holds for scalar functions. Letting:

$$Q_j^{[k]}(t)=\begin{cases}Q_j & (k=0)\\ \left(1-\dfrac{t-t_j}{t_{j+M-k}-t_j}\right)Q_{j-1}^{[k-1]}(t)+\dfrac{t-t_j}{t_{j+M-k}-t_j}Q_j^{[k-1]}(t) & (k>0)\end{cases} \tag{B.9}$$

we have, from Eq. (B.6):

$$P(t)=\sum_j N_{j,M}(t)\,Q_j=\sum_j N_{j,M-k}(t)\,Q_j^{[k]}(t)$$

where:

$$Q_j^{[0]}(t)=Q_j=W_{j,M}(u) \quad \text{for all } j.$$

Then, from Eq. (B.9), we have:

$$\begin{aligned}
Q_j^{[1]}(t) &= [(t_{j+M-1}-t)\,Q_{j-1}+(t-t_j)\,Q_j]/(t_{j+M-1}-t_j)\\
&= [(t_{j+M-1}-t)\,W_{j-1,M}(u)+(t-t_j)\,W_{j,M}(u)]/(t_{j+M-1}-t_j)\\
&= [(t_{j+M-1}-t)\,(u-t_j)\,W_{j,M-1}(u)\\
&\quad +(t-t_j)\,(u-t_{j+M-1})\,W_{j,M-1}(u)]/(t_{j+M-1}-t_j)\\
&= W_{j,M-1}(u)\,(u-t)
\end{aligned}$$

Therefore:

$$\begin{aligned}
\sum_j N_{j,M}(t)\,W_{j,M}(u) &= \sum_j N_{j,M-1}(t)\,Q_j^{[1]}(t)\\
&= (u-t)\sum_j N_{j,M-1}(t)\,W_{j,M-1}(u). \qquad \text{(B.10)}
\end{aligned}$$

Accordingly, if we assume that Eq. (B.8) holds when M is replaced by $M-1$, that is, if we let:

$$(u-t)^{M-2}=\sum_j N_{j,M-1}(t)\,W_{j,M-1}(u)$$

then, from Eq. (B.10):

$$\begin{aligned}
\sum_j N_{j,M}(t)\,W_{j,M}(u) &= (u-t)\sum_j N_{j,M-1}(t)\,W_{j,M-1}(u)\\
&= (u-t)^{M-1}.
\end{aligned}$$

From this the Marsden identity can be proved by mathematical induction.

Then comparing the coefficients of u^{M-1} on the left and right-hand sides of the Marsden identity, we immediately obtain:

$$\sum_j N_{j,M}(t)\equiv 1. \qquad \text{(B.11)}$$

Appendix C:
Effect of Multiple Knots on *B*-Spline Functions

Here we show the effect of multiple knots on *B*-spline functions for the case $M=4$.

(1) Knot vector $T=[t_0\ t_1\ t_2\ t_3\ t_4]=[0\ 1\ 2\ 3\ 4]$
Case without multiple knots

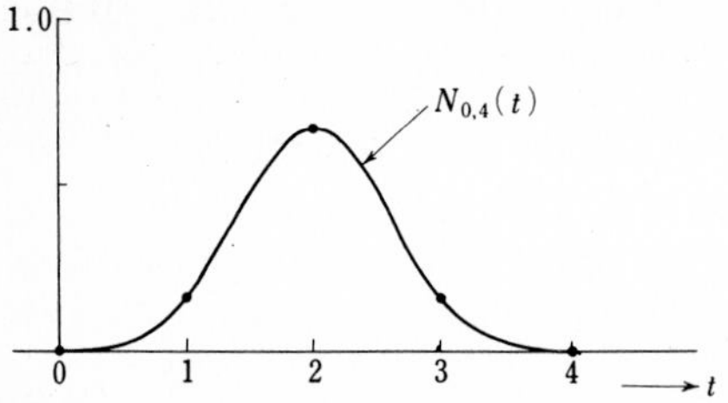

$$N_{0,4}(t)=\begin{cases} \frac{1}{6}t^3 & (0\leqq t<1) \\ -(t-2)^3+\frac{2}{3}(t-3)^3-\frac{1}{6}(t-4)^3 & (1\leqq t<2) \\ \frac{2}{3}(t-3)^3-\frac{1}{6}(t-4)^3 & (2\leqq t<3) \\ -\frac{1}{6}(t-4)^3 & (3\leqq t\leqq 4) \end{cases} \tag{C.1}$$

(2) Knot vector $T=[t_0\ t_1\ t_2\ t_3\ t_4]=[0\ 0\ 1\ 2\ 3]$

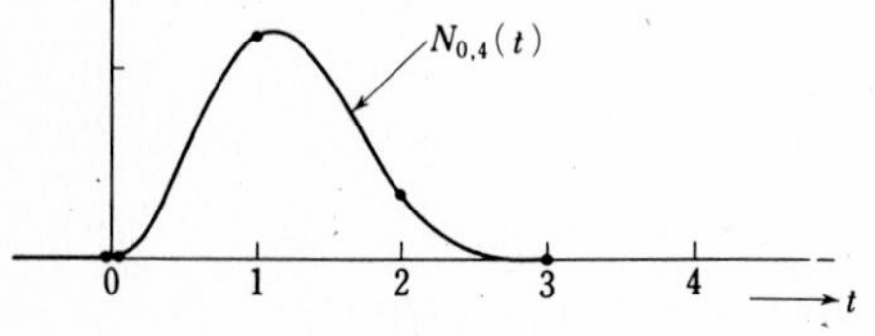

$$N_{0,4}(t)=\begin{cases} \frac{t^2}{12}(-11t+18) & (0\leqq t<1) \\ \frac{7}{12}t^3-3t^2+\frac{9}{2}t-\frac{3}{2} & (1\leqq t<2) \\ -\frac{1}{6}(t-3)^3 & (2\leqq t\leqq 3) \end{cases} \tag{C.2}$$

(3) Knot vector $T=[t_0\ t_1\ t_2\ t_3\ t_4]=[0\ 0\ 0\ 1\ 2]$

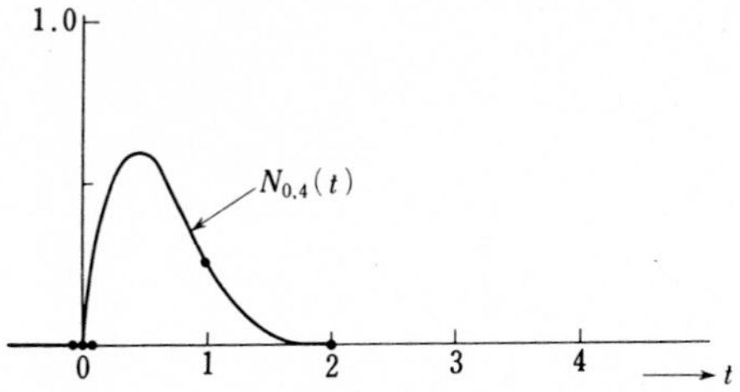

$$N_{0,4}(t)=\begin{cases} \frac{t}{4}(7t^2-18t+12) & (0\leqq t<1) \\ -\frac{1}{4}(t-2)^3 & (1\leqq t\leqq 2) \end{cases} \tag{C.3}$$

(4) Knot vector $T=[t_0\ t_1\ t_2\ t_3\ t_4]=[0\ 0\ 0\ 0\ 1]$

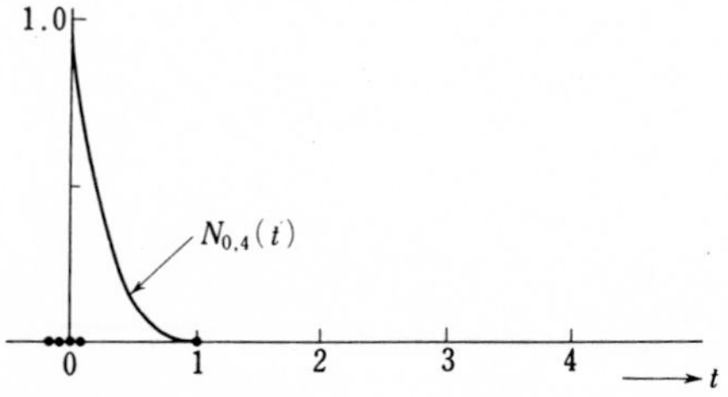

$$N_{0,4}(t)=-(t-1)^3 \qquad (0\leqq t\leqq 1) \tag{C.4}$$

(5) Knot vector $T=[t_0\ t_1\ t_2\ t_3\ t_4]=[0\ 1\ 2\ 3\ 3]$

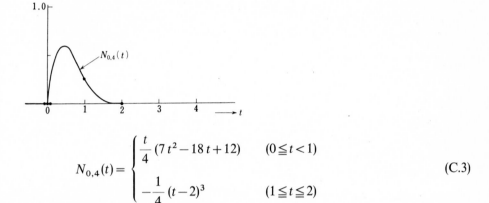

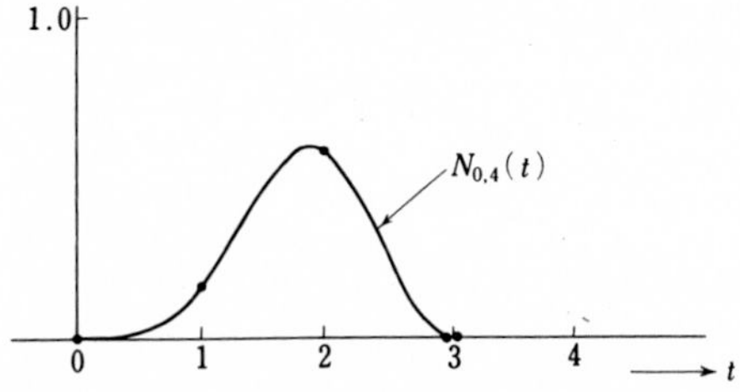

$$N_{0,4}(t)=\begin{cases}\frac{1}{6}t^3 & (0\leqq t<1)\\ \frac{1}{12}(-7t^3+27t^2-27t+9) & (1\leqq t<2)\\ \frac{1}{12}(3-t)^2(11t-15) & (2\leqq t\leqq 3)\end{cases} \tag{C.5}$$

(6) Knot vector $T=[t_0\ t_1\ t_2\ t_3\ t_4]=[0\ 1\ 2\ 2\ 2]$

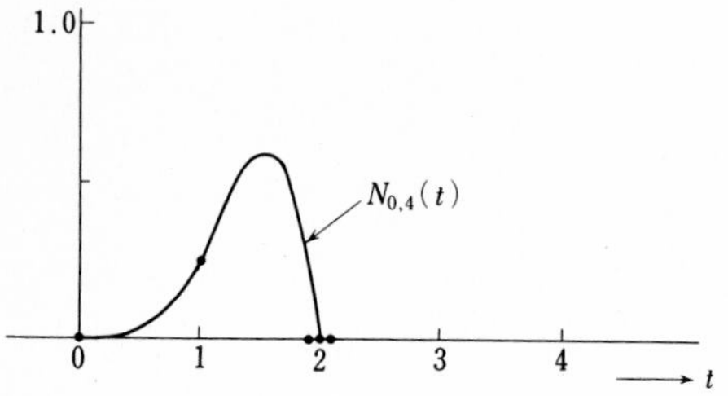

$$N_{0,4}(t)=\begin{cases}\frac{1}{4}t^3 & (0\leqq t<1)\\ \frac{1}{4}(2-t)(7t^2-10t+4) & (1\leqq t\leqq 2)\end{cases} \tag{C.6}$$

(7) Knot vector $T=[t_0\ t_1\ t_2\ t_3\ t_4]=[0\ 1\ 1\ 1\ 1]$

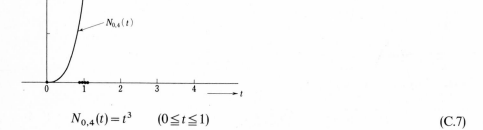

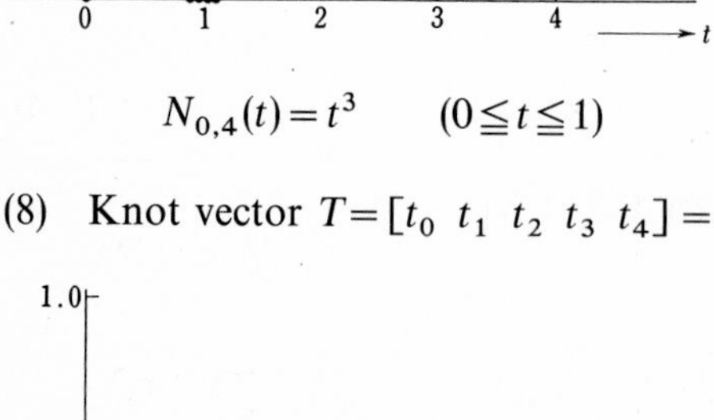

$$N_{0,4}(t)=t^3 \qquad (0\leqq t\leqq 1) \tag{C.7}$$

(8) Knot vector $T=[t_0\ t_1\ t_2\ t_3\ t_4]=[0\ 0\ 1\ 2\ 2]$

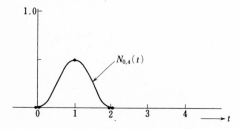

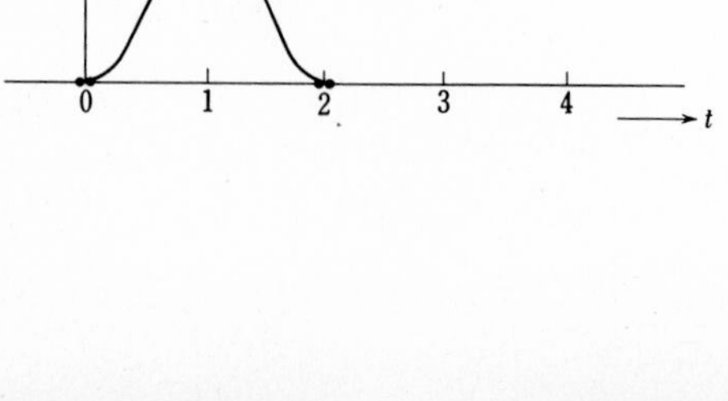

$$N_{0,4}(t)=\begin{cases}\frac{1}{2}t^2(3-2t) & (0\leqq t<1)\\ \frac{1}{2}(2-t)(-2t^2+5t-2) & (1\leqq t\leqq 2)\end{cases} \tag{C.8}$$

(9) Knot vector $T=[t_0\ t_1\ t_2\ t_3\ t_4]=[0\ 1\ 2\ 2\ 3]$

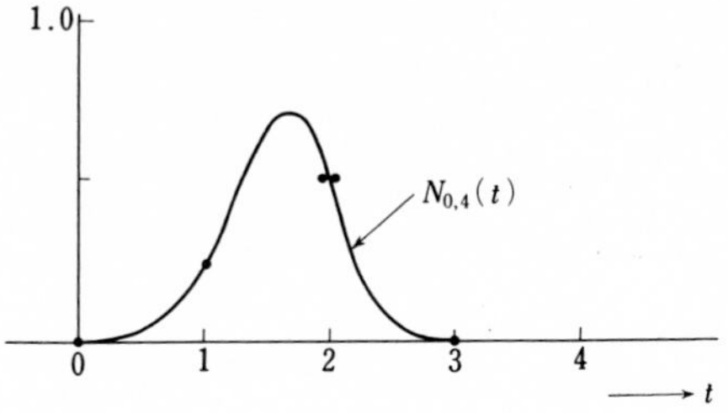

$$N_{0,4}(t)=\begin{cases}\frac{1}{4}t^3 & (0\leqq t<1)\\ \frac{1}{4}t^2(2-t)+\frac{1}{2}t(t-1)(2-t)+\frac{1}{2}(3-t)(t-1)^2 & (1\leqq t<2)\\ -\frac{1}{2}(t-3)^3 & (2\leqq t\leqq 3)\end{cases} \tag{C.9}$$

(10) Knot vector $T=[t_0\ t_1\ t_2\ t_3\ t_4]=[0\ 1\ 1\ 2\ 3]$

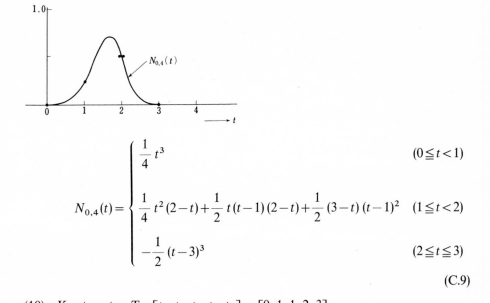

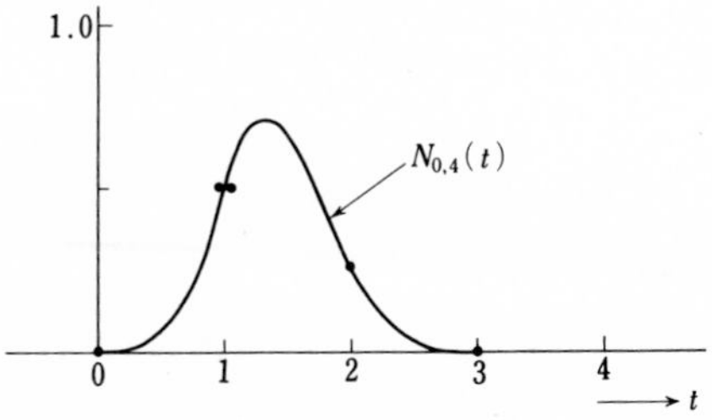

$$N_{0,4}(t)=\begin{cases}\frac{1}{2}t^3 & (0\leqq t<1)\\ \frac{1}{2}t(2-t)^2+\frac{1}{2}(t-1)(2-t)(3-t)+\frac{1}{4}(t-1)(3-t)^2 & (1\leqq t<2)\\ -\frac{1}{4}(t-3)^3 & (2\leqq t\leqq 3)\end{cases} \tag{C.10}$$

(11) Knot vector $T=[t_0\ t_1\ t_2\ t_3\ t_4]=[0\ 1\ 1\ 2\ 2]$

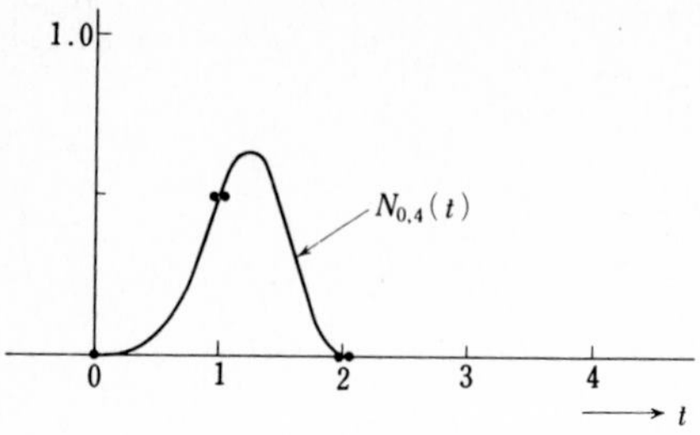

$$N_{0,4}(t)=\begin{cases}\dfrac{1}{2}t^3 & (0\leqq t<1)\\[2ex] \dfrac{1}{2}t(2-t)^2+(t-1)(2-t)^2+(t-1)(2-t)^2 & (1\leqq t\leqq 2)\end{cases} \tag{C.11}$$

(12) Knot vector $T=[t_0\ t_1\ t_2\ t_3\ t_4]=[0\ 0\ 1\ 1\ 2]$

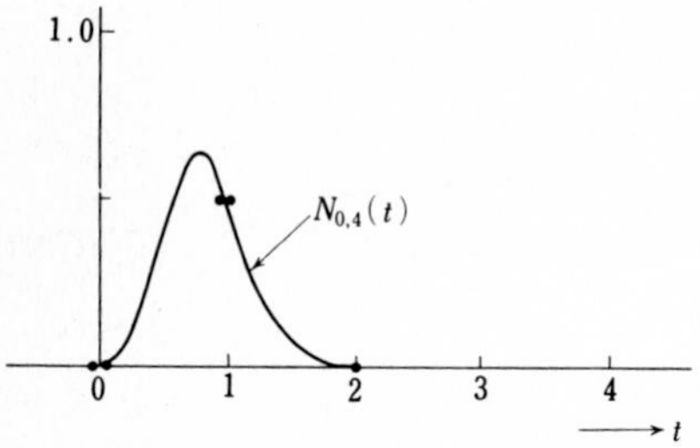

$$N_{0,4}(t)=\begin{cases}\dfrac{1}{2}t^2(6-5t) & (0\leqq t<1)\\[2ex] -\dfrac{1}{2}(t-2)^3 & (1\leqq t\leqq 2)\end{cases} \tag{C.12}$$

(13) Knot vector $T=[t_0\ t_1\ t_2\ t_3\ t_4]=[0\ 1\ 1\ 1\ 2]$

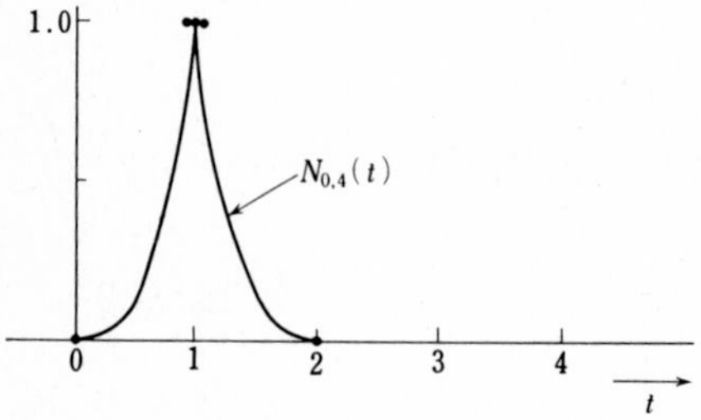

$$N_{0,4}(t)=\begin{cases} t^3 & (0\leqq t<1) \\ -(t-2)^3 & (0\leqq t\leqq 2) \end{cases} \tag{C.13}$$

(14) Knot vector $T=[t_0\ t_1\ t_2\ t_3\ t_4]=[0\ 0\ 1\ 1\ 1]$

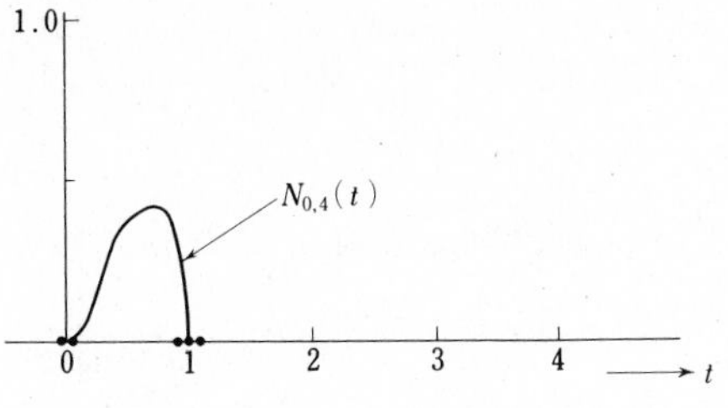

$$N_{0,4}(t)=3t^2(1-t) \qquad (0\leqq t\leqq 1) \tag{C.14}$$

(15) Knot vector $T=[t_0\ t_1\ t_2\ t_3\ t_4]=[0\ 0\ 0\ 1\ 1]$

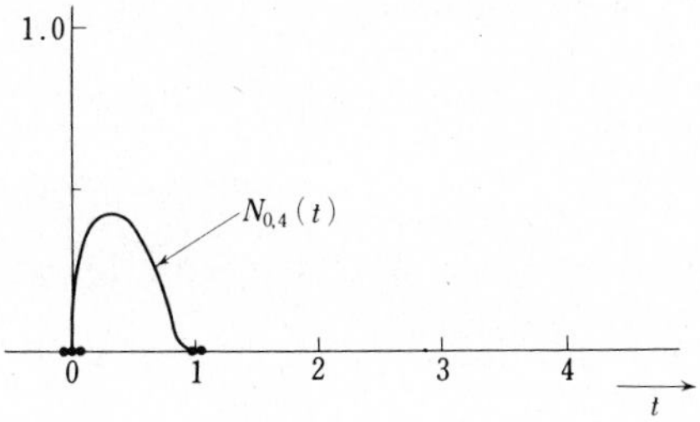

$$N_{0,4}(t)=3t(1-t)^2 \qquad (0\leqq t\leqq 1) \tag{C.15}$$

(16) Knot vector $T=[t_0\ t_1\ t_2\ t_3\ t_4]=[0\ 0\ 0\ 0\ 0]$

$$N_{0,4}(t)=0 \tag{C.16}$$

Bibliography

Additional References

33) Armit, A. P.: "Computer Systems for Interactive Design of Three-dimensional Shapes", Ph. D. thesis, Univ. Cambridge, November 1970.
34) Armit, A.P.: "Multipatch and Multiobject Design Systems", *Proc. R. Soc. Lond.,* **A 321** (1545): 235, 1971.
35) Armit, A.P.: "The Interactive Languages of Multipatch and Multiobject Design Systems", *Comput. Aided Design,* 4(1): 10, Autumn 1971.
36) Armit, A.P., and A.R. Forrest: "Interactive Surface Design", *Comput. Graphics 1970,* Brunel University, April 1970.
37) Armit, A.P.: "Example of an Existing System in University Research, Multipatch and Multiobject Design Systems", *Proc. Roy. Soc. Lond.,* **A 321,** 235–242, 1971.
38) Armit, A.P.: "Interactive 3D Shape Design–MULTIPATCH and MULTIOBJECT", in *Curved Surfaces in Engineering* (proc. Conference at Churchill College, Cambridge, 1972), IPC Science and Technology Press Ltd. 1972.
39) Armit, A.P.: "A Language for Interactive Design", in *Graphic Languages,* F. Nake and A. Rosenfeld (eds.), North-Holland, pp. 369–380, 1972.
40) Armit, A.P.: "Multipatch design program–users guide", Cambridge University Computer-Aided Design Group, September 1968.
41) Armit, A.P.: "A Multipatch design system for Coons' patches", IEEE Conference Publication No. 51, April 1969.
42) Armit, A.P.: "Curve and surface design using Multipatch and Multiobject design systems", Computer-Aided Design 3 No. 3, 3–12, 1971.
43) Armit, A.P., and A.R. Forrest: "Interactive surface design", in *Advanced Computer Graphics–Economics Techniques and Applications* (Parslow and Elliot Green, Eds.), Plenum Press, London 1971.
44) Akima, H.: "A Method of Bivariate Interpolation and Smooth Surface Fitting Based on Local Procedures", *CACM,* **17**(1): 18, January 1974.
45) Akima, H.: "A New Method of Interpolation and Smooth Curve Fitting Based on Local Procedures", *JACM,* **17**(4): 589, October 1970.
46) Akima, H.: "A Method of Bivariate Interpolation and Smooth Surface Fitting for Irregularly Distributed Data Points", *ACM Trans. Math. Software,* **4**, 2, 148–159, 1978.
47) Adams, J.A.: "The Intrinsic Method for Curve Definition", *Computer Aided Design,* **7**, 4, 243–249, 1975.
48) Agin, G.J.: "Representation and Description of Curved Objects", *Stanford Univ., Comput. Sci. Dept.,* STAN-CS-72-305, October 1972.
49) Ahlberg, J.H., E.N. Nilson, and J.I. Walsh: *"Theory of Splines and their Applications",* Academic, New York, 1967.
50) Ahuja, D.V.: "An Algorithm for Generating Spline-like Curves", *IBM Syst. J.,* **7**(3/4): 206, 1968.
51) Ahuja, D.V., and S.A. Coons: "Geometry for Construction and Display", *IBM Syst. J.,* **7**(3/4): 188, 1968.
52) Bézier, P.E.: *"Emploi des Machines a Commande Numérique",* Masson et Cie., Paris, 1970.
53) Bézier, P.E.: "Mathematical and Practical Possibilities of UNISURF", in R.E. Barnhill and R.F. Riesenfeld (eds.), *Computer Aided Geometric Design,* Academic, New York, 1974.
54) Bézier, P.E.: *"Numerical Control–Mathematics and Applications",* A.R. Forrest (trans.), Wiley, London, 1972.

55) Bézier, P.E.: "How Renault uses Numerical Control for Car Body Design and Tooling", SAE Paper 680010, 1968.
56) Bézier, P.E.: "UNISURF System: Principles, Program, Language", *Proc. 1973 PROLAMAT Conference, Budapest* (J. Hatvany, ed.), North-Holland Publ. Co., Amsterdam, 1974b.
57) Bézier, P.E.: "Numerical control in automobile design and manufacture of curved surfaces", in *Curved Surfaces in Engineering,* IPC Science and Technology Press, London, 1972.
58) Birkhoff, G., and H.L. Garabedian: "Smooth surface interpolation", *J. Maths. Phys.,* **39,** 258–268, 1960.
59) Bloor, M.S., A. de Pennington, and J.R. Woodwark: "Bridging the Gap between Conventional Surface Elements", *Proc. CAD 78 Conference, Brighton, 1978,* IPC Science and Technology Press Ltd., 1978.
60) Boere, H.: "The design of surfaces with the Coons' method", Thesis Technical University Delft, November 1970.
61) Burton, W.: "A Fixed Grid Curve Representation for Efficient Processing", *Computer Graphics,* Vol. 12, No. 3, Aug. pp. 64–49, 1978.
62) Butterfield, K.R.: "The Computation of all the Derivatives of a *B*-spline Basis", *J. Inst. Maths. Applics.* **17,** 15–25, 1976.
63) Butterfield, K.R.: Ph.D. Thesis, Brunel University, Uxbridge, Middlesex, 1978.
64) Barnhill, R.E.: "Smooth Interpolation over Triangles", in *Computer-Aided Geometric Design,* (R.E. Barnhill and R.F. Riesenfeld, eds.), Academic Press, 1974.
65) Barnhill, R.E., Birkhoff, G. and Gordon, W.J.: "Smooth Interpolation in Triangles", *J. Approx. Th.* **8,** 114–128, 1973.
66) Bolton, K.M.: "Biarc Curves", *Computer-Aided Design* **7,** 89–92, 1975.
67) Bates, K.J.: "The AUTOKON AUTOMOTIVE and AEROSPACE Packages", In *Curved Surfaces in Engineering* (proc. Conference at Churchill College, Cambridge, 1972), IPC Science and Technology Press Ltd., 1972.
68) Botting, R.J.: "A Theory of Parametric Curve Plotting", *Comp. Graphics and Image Proc.,* Vol. 7, No. 1, Feb. 139–145, 1978.
69) Ball, A.A.: "CONSURF. Part 1: Introduction of the Conic Lofting Tile", *Computer Aided Design,* **6,** 4, 243–249, 1974.
70) Ball, A.A.: "CONSURF. Part 2: Description of the Algorithms", *Computer Aided Design,* **7,** 4, 237–242, 1975.
71) Ball, A.A.: "CONSURF. Part 3: How the Program is used", *Computer Aided Design,* **9,** 1, 9–12, 1977.
72) Ball, A.A.: "A Simple Specification of the Parametric Cubic Segment", *Computer Aided Design,* **10,** 3, 181–182, 1978.
73) Bär, G.: "Parametrische Interpolation empirischer Raumkurven", *ZAMM,* **57,** 305–314, 1977.
74) Böhn, W.: "Über die Konstruktion von *B*-Spline-Kurven", *Computing,* **18,** 161–166, 1977.
75) Böhm, W.: "Generating the Bézier points of *B*-spline curves and surfaces", *Computer Aided Design,* Vol. 13, No. 6, pp. 365–366, Nov. 1981.
76) Belser, K.: "Comment on an Improved Algorithm for the Generation of Nonparametric Curves", *IEEE Trans. Computers,* Vol. C-25, No. 1, p. 103, Jan. 1976.
77) Cox, M.G.: "An Algorithm for approximating Convex Functions by means of First-Degree Splines", *Comput. J.* **14,** 3, 272–275, 1971.
78) Cox, M.G.: "The Numerical Evaluation of *B*-Splines", *J. Inst. Maths. Applics.* **10,** 134–149, 1972.
79) Cox, M.G.: "An Algorithm for Spline Interpolation", *J. Inst. Maths. Applics.* **15,** 95–108, 1975.
80) Cox, M.G.: "The Numerical Evaluation of a Spline from its *B*-Spline Representation", NPL Report NAC 68, National Physical Laboratory, Teddington, Middlesex, 1976.
81) Cox, M.G., and J.G. Hayes: "Curve Fitting: A Guide and Suite of Programs for the Non-specialist User", NPL Report NAC 26, National Physical Laboratory, Teddington, Middlesex, 1973.
82) Curry, H.B., and I.J. Schoenberg: "On Polya Frequency Functions IV: The Fundamental Spline Functions and their Limits", *J. Analyse Math.* **17,** 71–107, 1966.

83) Crocker, J.W., and B. Binnard: "Automated lofting using computer graphics", McDonnell-Douglas Report G230, March 1968.
84) Carlson, R.E., and C.A. Hall: "On piecewise polynomial interpolation in rectangular polygons", Westinghouse Bettis Nuclear Power Station, WAPD-T-2160, November 1969.
85) Cerny, H.F.: "*F*-curve and *F*-surface layout", Boeing Co. Doc. D2-23924-1, March, 1965.
86) C.A.D. Centre: *"An Introduction to Numerical Master Geometry"*, Computer Aided Design Centre, Madingley Road, Cambridge, 1972.
87) Collins, P.S., and S.S. Gould: "Computer-Aided Design and Manufacture of Surfaces for Bottle Moulds", *Proc. CAM 74 Conference on Computer Aided Manufacture and Numerical Control,* Strathclyde University, 1974.
88) Cline, A.K.: "Scalar and Planar valued Curve Fitting using Splines under Tension", *Comm. ACM* **17,** 4, 218–220, 1974.
89) CAM-I: "User Documentation for Sculptured Surfaces Releases SSX5 and SSX5A", Publication No. PS-76-SS-02, Computer Aided Manufacturing International, Inc., Arlington, Texas, 1976.
90) CAM-I: "Sculptured Surfaces Users Course", Publication No. TM-77-SS-01, Computer Aided Manufacturing International, Inc., Arlington, Texas, 1977.
91) Cohen, E.H.: "Discrete *B*-Splines and subdivision techniques in computer-aided geometric design and computer graphics", *Computer Graphics and Image Processing,* Vol. 14, pp. 87–111, 1980.
92) Cohen, D.: "On Linear Difference Curves", *Dept. Eng. Appl. Math. Harvard Univ.,* 1969.
93) Cohen, D., and T.M.P. Lee: "Fast Drawing of Curves for Computer Display", *SJCC 1969,* AFIPS Press, Montvale, N.J., p. 297.
94) Cohen, D.: "Incremental Methods for Computer Graphics", *Dept. Eng. Appl. Math. Harvard Univ.,* ESD-TR-69-193, April 1969.
95) Catmull, E.E.: "A Subdivision Algorithm for Computer Display of Curved Surfaces", *Univ. Utah Comput. Sci. Dept.,* UTEC-CSc-74-133, December 1974, NTIS A-004968/Ad/A-004973.
96) Catmull, E., and J. Clark: "Recursively Generated *B*-Spline Surfaces on Arbitrary Topological Meshes", *Comput. Aided Design,* **10**(6), November 1978.
97) Coons, S.A.: "Surfaces", University of Michigan Engineering Summer School Notes, June 1969.
98) Coons, S.A.: "Rational bicubic surface patches", MIT, Project MAC, Internal Note, November 1968.
99) Coons, S.A., and B. Herzog: "Surfaces for computer-aided aircraft design", Paper 67–895, American Institute of Aeronautics and Astronautics, 1967.
100) Coons, S.A.: "Modification of the Shape of Piecewise Curves", *Computer-Aided Design,* **9,** 178–180, 1977.
101) Coons, S.A.: "Multivariable Relations, in *Investigations in Computer-Aided Design*", M.I.T. Interim Engineering Report 8436-IR-1, 132–137, 141–145, January 1961.
102) Ciaffi, F., and G. Valle: "Interactive surface molding for a car body design", Computer Graphics Symposium, Berlin 255–274, 1971.
103) Clark, J.H.: "Parametric Curves, Surfaces, and Volumes in Computer Graphics and Computer-aided Geometric Design", NASA Ames Research Center, 1978.
104) Clark, J.H.: "3-D Design of Free-Form *B*-Spline Surfaces", *Univ. Utah Comput. Sci Dept.,* UTEC-CSc-74-120, September 1975, NTIS A002736 / AD/A002736.
105) Clark, J.H.: "Designing Surfaces in 3-D", *CACM,* **19**(8): 454, August 1976.
106) Dimsdale, B.: "Bicubic Patch Bounds", *Comp. & Maths. with Appls.* **3,** 2, 95–104, 1977.
107) Dimsdale, B.: "Convex Cubic Splines", *IBM J. Res. Develop.* **22,** 168–178, 1978.
108) Dimsdale, B., and R.M. Burkly: "Bicubic Patch Surfaces for High-speed Numerical Control Processing", *IBM J. Res. Develop.* 358–367, 1976.
109) Dimsdale, B., and K. Johnson: "Multiconic Surfaces", *IBM J. Res. Develop.* **19,** 6, 523–529, 1975.
110) Dollries, J.F.: "Three-dimensional Surface Fit and Numerically Controlled Machining from a Mesh of Points", Technical Information Series Report No. R63FPD319, General Electric Co., 1963.

111) De Boor, C.: "Splines as Linear Combinations of *B*-Splines. A Survey", In *Approximation Theory II* (G.G. Lorentz, C.K. Chui and L.L. Schumaker, eds.), Academic Press, 1976.
112) De Boor, C.: "Package for Calculating with *B*-Splines", *SIAM J. Numer. Anal.* **14,** 3, 441–472, 1977.
113) De Boor, C.: "Bicubic spline interpolation", *J. Maths. Phys.* **41,** 1962.
114) Dube, R., G.J. Herron, F.F. Little, and R.F. Riesenfeld: "SURFED: An Interactive Editor for Free-Form Surfaces", *Comput. Aided Design,* **10**(2): 111–115, March 1978.
115) Dube, R.P.: "Local Schemes for Computer Aided Geometric Design", PhD thesis, Univ. Utah, Dept. of Mathematics, June 1975.
116) Doo, D.: "A method of smoothing highly irregular polyhedrons", Interactive Techniques in Computer Aided Design Conference, Bologna, September 1978.
117) Doo, D., Ph.D. dissertation, Brunel University, to appear.
118) Ellis, T.M.R., and D.H. McLain: "A New Method of Cubic Curve Fitting using Local Data", *ACM Trans. Math. Software* **3,** 2, 175–178, 1977.
119) Einar, H., and E. Skappel: "FORMELA: A general Design and Production Data System for Sculptured Products", *Computer-Aided Design* **5,** 2, 68–76, 1973.
120) Epstein, M.P.: "On the Influence of Parametrisation in Parametric Interpolation", *SIAM J. Numer. Anal.* **13,** 2, 261–268, 1976.
121) Forrest, A.R.: "A Unified Approach to Geometric Modeling", *Comput. Graphics.* **12**(3): 264, August 1978.
122) Forrest, A.R.: "On Coons and Other Methods for the Representation of Curved Surfaces", *Comput. Graphics Image Processing,* **1**(4): 341, 1972.
123) Forrest, A.R.: "The Twisted Cubic Curve", *Univ. Cambridge, Comput. Lab. CAD Doc.* 50, November 1970.
124) Forrest, A.R.: "Computational Geometry", *Proc. Roy. Soc. Lond.* **A321,** 187–195, 1971.
125) Forrest, A.R.: "Computational Geometry–Achievements and Problems", In *Computer-Aided Geometric Design* (R.E. Barnhill and R.F. Riesenfeld, eds.), Academic Press, 1974.
126) Forrest, A.R.: "Curves for Computer Graphics", in *Pertinent Concepts in Computer Graphics,* M. Faiman and J. Nievergelt (eds.), U. Illinois Press, pp. 31–47, 1969.
127) Forrest, A.R.: "Coons Surfaces and Multivariable Functional Interpolation", Cambridge University CAD Group Document 38, (Rev.), December 1971.
128) Forrest, A.R.: "The computation of bicubic twist terms", Cambridge University Computer Aided Design Group, CAD Doc. 20, January 1969.
129) Forrest, A.R.: "Computer-aided design of three-dimensional objects: a survey", ACM/AICA International Computing Symposium, Venice, April 1972.
130) Forrest, A.R.: "Notes on Chaikin's algorithm", University of East Anglia Computational Geometry Project CGP 74/1, Dec. 1974.
131) Forrest, A.R.: "Multivariate approximation problems in computational geometry", In *Multivariate Approximation,* D.C. Handscomb (Ed.), Academic Press, London, 1978.
132) Faux, I.D.: "Simple Cross-Sectional Designs", *Proc. CAM 78 Conference, National Engineering Laboratory, East Kilbride,* 1978.
133) Freemantle, A.C., and P.L. Freeman: "The Evolution and Application of Lofting Techniques at Hawker Siddeley Aviation", In *Curved Surfaces in Engineering* (proc. Conference at Churchill College, Cambridge, 1972), IPC Science and Technology Press Ltd., 1972.
134) Flutter, A.G.: "The POLYSURF System", *Proc. 1973 PROLAMAT Conference, Budapest* (J. Hatvany, ed.), North Holland Publ. Co., Amsterdam, 1976.
135) Flutter, A.G., and R.N. Rolph: "POLYSURF: An Interactive System for Computer-Aided Design and Manufacture of Components", *CAD 76 Proceedings,* 150–158; CAD Centre, Madingley Road, Cambridge, 1976.
136) Flanagan, D.L., and O.V. Hefner: "Surface Moulding–New Tool for the Engineer", *Aeronautics,* 58–62, April 1967.
137) Ferguson, J.C.: "Multivariable Curve Interpolation", Report No. D2-22504, The Boeing Co., Seattle, Washington, 1963.
138) Ferguson, J.C.: "Multivariate Curve Interpolation", *Journal ACM,* **11,** 2, 221–228, 1964.
139) Greville, T.N.E.: "On the Normalisation of the *B*-splines and the Location of the Nodes for the case of Unequally Spaced Knots", Supplement to the paper 'On Spline Functions' by I.J. Schoenberg in *Inequalities* (O. Schisha, ed.), Academic Press, 1967.

140) Greville, T. N. E. (Ed.): "*Theory and Applications of Spline Functions*", Academic Press, 1969.
141) Gordon, W. J., and R. F. Riesenfeld: "Bernstein-Bézier Methods for the Computer-aided Design of Free-Form Curves and Surfaces", *JACM*, **21** (2): 293–310, April 1974.
142) Gordon, W. J., and R. F. Riesenfeld: "*B*-Spline Curves and Surfaces", in R. E. Barnhill and R. F. Riesenfeld (eds.), *Computer Aided Geometric Design*, Academic, New York, 1974.
143) Gordon, W. J.: "Spline-blended Surface interpolation through Curve Networks", *Journal of Mathematics and Mechanics*, **18**, 10, 931–952, 1969.
144) Gordon, W. J., and J. A. Wixom: "Shephard's Method of "Metric Interpolation" to Bivariate and Multivariate Interpolation", *Mathematics of Computation*, **32**, 141, 253–264, 1978.
145) Gordon, W. J.: "Blending-functions methods of bivariate and multivariate interpolation and approximation", *SIAM J. Numer. Anul.* **8**, 158–177, 1971.
146) Gordon, W. J.: "Distributive lattices and the approximation of multivariate functions", in Proc. Symposium on *Approximation with Special Emphasis on Spline Functions* (I. J. Schoenberg, Ed.), Academic Press, New York, 1969.
147) Gregory, J. A.: "Smooth Interpolation without Twist Constraints", In *Computer-Aided Geometric Design* (R. E. Barnhill and R. F. Riesenfeld, eds.), Academic Press, 1974.
148) Gaffney, P. W.: "To Compute the Optimum Interpolation Formula?" Report CSS 52, Atomic Energy Research Establishment, Harwell, 1977.
149) Gylys, V. B.: "Evaluation of sculptured surface techniques", IIT Research Inst. APT Project Report 1967.
150) Ghezzi, C., and F. Tisato: "Interactive Computer-Aided Design for Sculptured Surfaces", *Proc. 1973 PROLAMAT Conference*, Budapest (J. Hatvany, ed.), North Holland Publishing Co., Amsterdam, 1973.
151) Gossling, T. H.: "The 'Duct' System of Design for Practical Objects", *Proc. World Congress on the Theory of Machines and Mechanisms, Milan*, 1976.
152) Hosaka, M.: "Synthesis and Smoothing theories of curves and surfaces" (in Japanese), Johoshori, Vol. 10, No. 3, 1969.
153) Hosaka, M., and F. Kimura: "Synthesis Methods of Curves and Surfaces in Interactive CAD", Proc. Conf. Interactive Techniques in CAD, IEEE Computer Society 78CH 1289-8C, 1978.
154) Hosaka, M., and F. Kimura: "Interactive Input Methods for Free-Form Shape Design", IFIP W.G.5.2-W.G.5.3 Working Conference, Tokyo, 1980.
155) Herzog, B. W., and G. Valle: "Interactive control of surface patches (from a remote graphics terminal)", presented at CG 70, Brunel University, U.K., April 1970.
156) Hayes, J. G.: "Available Algorithms for Curve and Surface Fitting", NPL Report NAC 39, National Physical Laboratory, Teddington, Middlesex, 1973.
157) Hayes, J. G.: "New Shapes from Bicubic Splines", NPL Report NAC 58, National Physical Laboratory, Teddington, Middlesex, 1974.
158) Hayes, J. G., and J. Halliday: "The Least-squares Fitting of Cubic Spline Surfaces to General Data Sets", *J. Inst. Maths. Applics.* **14**, 89–103, 1974.
159) Hart, W. B: "Glider Fuselage Design with the Aid of Computer Graphics", *Computer-Aided Design*, **3**, 2, 3–8, 1971.
160) Hart, W. B.: "Current and Potential Applications to Industrial Design and Manufacture", In *Curved Surfaces in Engineering* (proc. Conference at Churchill College, Cambridge, 1972), IPC Science and Technology Press Ltd., 1972.
161) Hart, W. B.: "The Application of Computer-Aided Design Techniques to Glassware and Mould Design", *Computer-Aided Design*, **4**, 2, 57–66, 1972.
162) Hulme, B. L.: "Piecewise bicubic methods for plate blending problems", Harvard University, Ph. D. Thesis, April 1969.
163) Hartley, P. J., and C. J. Judd: "Parametrisation of Bézier-type *B*-spline Curves and Surfaces", *Computer-Aided Design*, **10**, 2, 130–134, 1978.
164) Inselberg, A.: "Cubic Splines with Infinite Derivatives at Some Knots", *IBM J. Res. Develop.*, 430–436, September 1976.
165) Jordan, B. W., W. J. Lennon, and B. C. Holm: "An Improved Algorithm for the Generation of Non-parametric Curves", *IEEE Trans.*, **C-22** (12): 1052–1060, December 1973.
166) Kuo, C.: "*Computer Methods for Ship Surface Design*", Longman, 1971.

167) Knott, G.D., and D.K. Reece: "Modelab: A Civilized Curve-fitting System", *Proc. ONLINE 72,* Uxbridge, England, September 1972.
168) Liming, R.A.: "*Practical Analytical Geometry with Applications to Aircraft*", Macmillan, New York, 1944.
169) Lane, J.M., and R.F. Riesenfeld: "The Application of Total Positivity to Computer-aided Curve and Surface Design", to appear in *JACM.*
170) Lane, J.M., and R.F. Riesenfeld: "A Theoretical Development for the Computer Generation and Display of Piecewise Polynomial Surfaces", *IEEE Trans. Pattern Anal. Machine Intell.* to appear.
171) Levin, J.: "A Parametric Algorithm for drawing Pictures of Solid Objects Composed of Quadric Surfaces", *Comm. ACM,* **19,** 10, 555–563, 1976.
172) Lee, T.M.P.: "A Class of Surfaces for Computer Display", *SJCC 1969,* AFIPS Press, Montvale, N.J., p. 309.
173) Lee, T.M.P.: "Three Dimensional Curves and Surfaces for Rapid Computer Display", *Harvard Univ., Dept. Eng. Appl. Phys.* ESD-TR-69–189, April 1969.
174) Lee, T.M.P.: "Analysis of an Efficient Homogeneous Tensor Representation of Surfaces for Computer Display", In *Advanced Computer Graphics* (R.D. Parslow and R. Elliott Green, eds.), Plenum Press, London and New York, 1971.
175) Meinardus, G.: "Algebräische Formulierung von Spline-Interpolationen", International Series in Numerical Mathematics, Vol. 32, *Moderne Methoden der numerischen Mathematik,* 125–138, Birkhäuser Verlag, Basel & Stuttgart, 1976.
176) Marks, R.E.: "Surface fitting for Coons patches", Cambridge University Computer-Aided Design Group, CAD Doc. 46, October 1970.
177) Moore, C.L.: "Method of Fitting a Smooth Surface to a Mesh of Points", Technical Information Series Report No. R59FPD927, General Electric Co., 1959.
178) MacCallum, K.J.: "Surfaces for Interactive Graphical Design", *Comput. J.,* **13**(4): 352, November 1970.
179) MacCallum, K.J.: "Mathematical Design of Hull Surfaces", *The Naval Architect,* 359–373, July 1972.
180) MacCallum, K.J.: "The application of computer graphics to the preliminary design of ship hulls", Ph. d. Thesis, Imperial College, London, 1970.
181) Mehlum, E.: "A Curve-Fitting Method based on a Variational Criterion", *BIT,* **4,** 213–223, 1964.
182) Mehlum, E: "*Curve and Surface Fitting based on a Variational Criterion for Smoothness*", Central Institute of Industrial Research, Oslo, 1969.
183) Mehlum, E: "Non-linear Splines", In *Computer-Aided Geometric Design* (R.E. Barnhill and R.F. Riesenfeld, eds.), Academic Press, 1974.
184) Mehlum, E., and P.F. Sørenson: "Example of an Existing System in the Shipbuilding Industry: the AUTOKON system", *Proc. Roy. Soc. Lond.* **A 321,** 219–233, 1971.
185) Malcolm, M.A.: "On the Computation of Nonlinear Spline Functions", *SIAM J. Numer. Anal.,* **14,** 254–282, 1977.
186) Manning, J.R.: "Continuity Conditions for Spline Curves", *Comput. J.* **17,** 181–186, 1974.
187) Miyamoto, E., and T.O. Binford: "Display Generated by a Generalized Cone Representation", *Proc. IEEE Conf. on Computer Graphics, Pattern Recognition, and Data Structure,* May 1975, pp. 385–387.
188) Nielson, G.M.: "Multivariate Smoothing and Interpolating Splines", *SIAM J. Numer. Anal.,* **11** (2): 435, April 1974.
189) Nielson, G.M.: "Some Piecewise Polynomial Alternatives to Splines in Tension", In *Computer-Aided Geometric Design* (R.E. Barnhill and R.F. Riesenfeld, eds.), Academic Press, 1974.
190) Nielson, G.M.: "Surface approximation and data smoothing using generalized spline functions", Ph. D. Thesis, University of Utah, June 1970.
191) Nicolo, V.: "A preliminary study for the choice of the algorithm to be used in a 3-D shape description language", Cambridge University Computer-Aided Design Group, CAD Doc. 67, August 1972.
192) Nutbourne, A.W.: "A Cubic Spline Package. Part 2–The Mathematics", *Computer-Aided Design,* **5,** 1, 7–13, 1973.

193) Nutbourne, A.W., P.M. McLellan, and R.M.L. Kensit: "Curvature Profiles for Plane Curves", *Computer-Aided Design,* **4,** 4, 176–184, 1972.
194) Overhauser, A.W.: "Analytic definition of curves and surfaces by parabolic blending", Ford Motor Co., Dearborn, Science Lab. Tech. Report, SL-68-40, March 1968.
195) Payne, P.J.: "A contouring program for joined surface patches", Cambridge University Computer-Aided Design Group CAD Doc. 58, June 1971.
196) Payne, P.J.: "A program for the design of ducts", Cambridge University Computer-Aided Design Group Doc. 61, October 1971.
197) Pal, T.K.: "Intrinsic Curve with Local Control", *Computer-Aided Design,* **10,** 1, 19–29, 1978.
198) Pal, T.K.: "Mean Tangent Rotational Angles and Curvature Integration", *Computer-Aided Design,* **10,** 1, 30–34, 1978.
199) Pal, T.K., and A.W. Nutbourne: "Two-dimensional Curve Synthesis using Linear Curvature Elements", *Computer-Aided Design,* **9,** 2, 121–134, 1977.
200) Peters, G.J.: "Interactive Computer Graphics Application of the Parametric Bicubic Surface to Engineering Design Problems", In *Computer-Aided Geometric Design* (R.E. Barnhill and R.F. Riesenfeld, eds.), Academic Press, 1974.
201) Pruess, S.: "Properties of Splines in Tension", *J. Approx. Th.,* **17,** 86–96, 1976.
202) Porter Goff, R.F.D.: "The representation of surfaces by Coons' patch technique", University of Leicester, England, Rpt. 69–3, February 1969.
203) Riesenfeld, R.F.: "Non-uniform *B*-Spline Curves", *Proc. 2nd USA-Japan Computer Conf.,* p. 551, 1975.
204) Riesenfeld, R.F.: "On Chaikin's Algorithm", *Comput. Graphics Image Processing,* **4** (3): 304–310, 1975.
205) South, N.E., and J.P. Kelly: "Analytic surface methods", Ford Motor Co., Dearborn, N/C Development Unit, December 1965.
206) Shu, H., S. Hori, W.R. Mann, and R.N. Little: "The synthesis of sculptured surfaces", in *Numerical Control Programming Languages* (W.H.P. Leslie, Ed.), North-Holland, Amsterdam, 358–375, 1970.
207) Sabin, M.A.: "Parametric Surface Equations for Non-rectangular Regions", Report No. VTO/MS/147, Dynamics and Mathematical Services Dept., British Aircraft Corporation, Weybridge, 1968.
208) Sabin, M.A.: "An existing System in the Aircraft Industry. The British Aircraft Corporation Numerical Master Geometry System", *Proc. Roy. Soc. Lond.* **A 321,** 197–205, 1971.
209) Sabin, M.A.: "Comments on some Algorithms for the Representation of Curves by Straight Line Segments", letter in *Comput J.* **15,** 2, 104, 1972.
210) Sabin, M.A.: "A Method for displaying the Intersection Curve of two Quadric Surfaces", *Comput. J.* **19,** 336–338, 1976.
211) Sabin, M.A.: "Numerical Master Geometry", in *Curved Surfaces in Engineering,* IPC Science and Technology Press, London 1972.
212) Sabin, M.A.: "Numerical Master Geometry", Design Office Manual Vol. 7 Section 73, British Aircraft Corporation, Preston Division, June 1969.
213) Sabin, M.A.: "Two basic interrogations of parametric surfaces", BAC Weybridge VTO/MS/148, October 1968.
214) Sabin, M.A.: "Offset parametric surfaces", BAC Weybridge VTO/MS/149, September 1968.
215) Sabin, M.A.: "General interrogations of parametric surfaces", BAC Weybridge VTO/MS/150, October 1968.
216) Sabin, M.A.: "Conditions for continuity of surface normal between adjacent parametric surfaces", BAC Weybridge VTO/MS/151, October 1968.
217) Sabin, M.A.: "The use of vectors and parameters to describle geometrical concepts", BAC Weybridge VTO/MS/152, December 1965.
218) Sabin, M.A.: "A 16-Point bicubic formulation suitable for multipatch surfaces", BAC Weybridge VTO/MS/155, March 1969.
219) Sabin, M.A.: "Spline surfaces", BAC Weybridge VTO/MS/156, June 1969.
220) Sabin, M.A.: "Interrogation techniques for parametric surfaces", in *Advanced Computer Graphics–Economics, Techniques and Applications* (Parslow and Elliot Green, Eds.), Plenum Press, London, 1971.

221) Sabin, M.A.: "Trinomial basis functions for interpolation in triangular regions (Bézier triangles)", BAC VTO/MS/188, July 1971.
222) Sabin, M.A.: "*B*-spline interpolation over regular triangular lattices", BAC VTO/MS/195, Oct. 1972.
223) Sabin, M.A.: "A triangular element giving slope continuity over all boundaries using piecewise cubic interior", BAC VTO/MS/198, July 1973.
224) Sabin, M.A.: "Two slope-continuous triangular elements constructed from low order polynomial pieces", BAC VTO/MS/199, July 1973.
225) Sabin, M.A.: "A Bézier-like surface definition controlled by points joined in an arbitrary network", Kongsberg U.K. Ltd., Sept. 1976.
226) Sabin, M.A.: "The use of piecewise forms for the numerical representation of shape", Ph. D. dissertation, Computer and Automation Institute, Hungarian Academy of Sciences, 1977.
227) Shiotani, et al.: A method of generating a free form curve for CAD, Conference of Seikigakkai, 1981.
228) Thomas, D.H.: "Pseudospline Interpolation for Space Curves", *Mathematics of Computation,* **30,** 133, 58–67, 1976.
229) Talbot, J.: "Experiments towards interactive graphical design of motor bodies", Cambridge University Computer-Aided Design Group CAD Doc. 56, December 1970.
230) Sablonnière, P.: "Spline and Bézier polygons associated with a polynomial spline curve", *Computer-Aided Design,* Vol. 10, No. 4, pp. 257–261, July 1978.
231) Smith, D.J.L., and H. Merryweather: "The Use of Analytic Surfaces for the Design of Centrifugal Impellers by Computer Graphics", *Int. J. Num. Meth. Eng.* **17,** 137–154, 1973.
232) Smith, L.B.: "Drawing Ellipses, Hyperbolas or Parabolas with a Fixed Number of Points and Maximum Inscribed Area", *Comput. J.* **14,** 81–86, 1971.
233) Schechter, A.: "Synthesis of 2D Curves by Blending Piecewise Linear Curvature Profiles", *Computer-Aided Design,* **10,** 1, 8–18, 1978.
234) Schechter, A.: "Linear Blending of Curvature Profiles", *Computer-Aided Design,* **10,** 2, 101–109, 1978.
235) Schoenberg, I.J.: "*Cardinal Spline Interpolation*", SIAM, Philadelphia, 1973.
236) Schoenberg, I.J., and A. Whitney: "On Polya Frequency Functions III: The Positivity of Translation Determinants with an Application to the Interpolation Problem by Spline Curves", *Trans. Amer. Math. Soc.* **74,** 246–259, 1953.
237) Schumaker, L.L.: "Some Algorithms for the Computation of Interpolating and Approximating Spline Functions", In *Theory and Applications of Spline Functions,* (T.N.E. Greville, ed.), Academic Press, 1969.
238) Schumaker, L.L.: "Fitting Surfaces to Scattered Data", In *Approximation Theory II* (G.G. Lorentz, C.K. Chui and L.L. Schumaker, eds.), Academic Press, 1976.
239) Späth, H.: "*Spline Algorithms for Curves and Surfaces*", Utilitas Mathematica Publishing Inc., Winnipeg, Canada, 1974.
240) Shephard, D.: "A Two-dimensional Interpolation Function for Irregularly Spaced Data", *Proc. ACM National Conference,* 517–524, 1968.
241) Varah, J.M.: "On the Condition Number of Local Bases for Piecewise Cubic Polynomials", *Mathematics of Computation,* **31,** 137, 37–44, 1977.
242) Veron, M., G. Ris, and J.P. Musse: "Continuity of Biparametric Surface Patches", *Computer-aided Design,* **8,** 4, 267–273, 1976.
243) Woodsford, P.A.: "Computer programs for curve and surface manipulation", Bell Telephone Labs. Holmdel, New Jersey, Summer 1970.
244) Woodsford, P.A.: "Mathematical methods in computer graphics–a survey", Computer Graphics Symposium, 234–253, Berlin 1971.
245) Wilson, H.B., and D.S. Farrior: "Computation of Geometrical and Inertial Properties for General Areas and Volumes of Revolution", *Computer-Aided Design,* **8,** 4, 257–263, 1976.
246) Walker, L.F.: "Curved Surfaces in Shipbuilding Design and Production", In *Curved Surfaces in Engineering* (proc. Conference at Churchill College, Cambridge, 1972), IPC Science and Technology Press, 1972.
247) Walter, H.: "Computer-Aided Design in the Aircraft Industry", In *Computer-Aided Design* (J. Vlietstra and R.F. Wielinga, eds.), North-Holland Publishing Co., 1973.
248) Wielinga, R.F.: "Constrained Interpolation using Bézier Curves as a New Tool in

Computer-Aided Geometric Design", In *Computer-Aided Geometric Design* (R. E. Barnhill and R. F. Riesenfeld, eds.), Academic Press, 1974.
249) Yamaguchi, F.: "A new Curve Fitting Method Using a CRT Computer Display", *Comput. Graphics Image Processing,* **7**(3): 425–437, June 1978.
250) Yamaguchi, F.: "A method of designing free form shapes by computer display", (3rd Report) (in Japanese), *Precision Machinery* **43**(10), 1977.
251) Yuille, I. M.: "Ship Design", In *Curved Surfaces in Engineering* (proc. Conference at Churchill College, Cambridge, 1972), IPC Science and Technology Press Ltd., 1972.

Books

252) Kazuto, Togino: *Automatic Design Methods* (in Japanese) Corona Sha, 1969.
253) Hosaka, M.: *Computer Graphics* (in Japanese) Sangyo Tosho, 1974.
254) Yamaguchi, Fujio: *Computer Graphics* (Japanese translation of reference 261), Nikkan Kogyo Shimbun Sha.
255) Yamaguchi, Fujio: *Graphics Processing Engineering* (in Japanese) Nikkan Kogyo Shimbun Sha, 1981.
256) Yamaguchi, Fujio: *Introduction to CAD/CAM* (in Japanese) Kogyo Chosa Kai, 1982.
257) Sakurai, Akira: *Introduction to Spline Functions* (in Japanese) Tokyo Electrotechnical University, 1981.
258) Bézier, P.: "Numerical Control–Mathematics and Applications", John Wiley & Sons, 1972.
259) Barnhill, R., and R. Riesenfeld: "Computer Aided Geometric Design", Academic Press, 1974.
260) Prenter, P.: "Splines and Variational Methods", John Wiley & Sons, 1975.
261) Rogers, D., and J. Adams: "Mathematical Elements for Computer Graphics", McGraw-Hill, 1976.
262) Davis, P.: "Interpolation & Approximation", Dover Publications, 1963.
263) Chasen, S.: "Geometric Principles and Procedures for Computer Graphic Applications", Prentice-Hall, 1978.
264) Giloi, W.: "Interactive Computer Graphics", Prentice Hall, 1978.
265) Newman, W., and R. Sproull: "Principles of Interactive Computer Graphics" (2nd edition), McGraw-Hill, 1979.
266) Freeman, H.: "Interactive Computer Graphics", IEEE Computer Society, 1980.
267) Liming, R.: "Mathematics for Computer Graphics", Aero Publishers, 1979.
268) Brodlie, K.: "Mathematical Methods in Computer Graphics and Design", Academic Press, 1980.
269) Faux, I., and M. Pratt: "Computational Geometry for Design and Manufacture", Ellis Horwood, 1979.
270) DeBoor, C.: "A Practical Guide to Splines", *Applied Mathematical Sciences,* Vol. 27, Springer-Verlag, 1978.
271) Goos, G., and J. Hartmanis: "Computer Aided Design, Modelling, Systems Engineering, CAD-Systems", Lecture Notes in *Computer Science,* No. 89, Springer-Verlag, 1980.
272) Schumaker, L.: "Spline Functions: Basic Theory", John Wiley & Sons, 1981.
273) Foley, J., and A. van Dam: "Fundamentals of Interactive Computer Graphics", Addison-Wesley, 1982.

Subject Index